Social Welfare Policy and Advocacy

Second Edition

This book is dedicated to social workers who engage in micro, mezzo, and macro policy advocacy.

Sara Miller McCune founded SAGE Publishing in 1965 to support the dissemination of usable knowledge and educate a global community. SAGE publishes more than 1000 journals and over 800 new books each year, spanning a wide range of subject areas. Our growing selection of library products includes archives, data, case studies and video. SAGE remains majority owned by our founder and after her lifetime will become owned by a charitable trust that secures the company's continued independence.

Los Angeles | London | New Delhi | Singapore | Washington DC | Melbourne

Social Welfare Policy and Advocacy

Advancing Social Justice Through Eight Policy Sectors

Second Edition

Bruce S. Jansson, Ph.D.
University of Southern California

Los Angeles | London | New Delhi
Singapore | Washington DC | Melbourne

FOR INFORMATION:

SAGE Publications, Inc.
2455 Teller Road
Thousand Oaks, California 91320
E-mail: order@sagepub.com

SAGE Publications Ltd.
1 Oliver's Yard
55 City Road
London, EC1Y 1SP
United Kingdom

SAGE Publications India Pvt. Ltd.
B 1/I 1 Mohan Cooperative Industrial Area
Mathura Road, New Delhi 110 044
India

SAGE Publications Asia-Pacific Pte. Ltd.
18 Cross Street #10-10/11/12
China Square Central
Singapore 048423

Printed in Canada

Library of Congress Cataloging-in-Publication Data

ISBN 978-1-5063-8406-1

This book is printed on acid-free paper.

MIX
Paper from
responsible sources
FSC® C004071

Acquisitions Editor: Joshua Perigo
Editorial Assistant: Noelle Cumberbatch
Production Editor: Rebecca Lee
Copy Editor: Pam Schroeder
Typesetter: Hurix Digital
Proofreader: Scott Oney
Indexer: Wendy Allex
Cover Designer: Dally Verghese
Marketing Manager: Jenna Retana

19 20 21 22 23 10 9 8 7 6 5 4 3 2

BRIEF CONTENTS

DETAILED CONTENTS

PREFACE

I published the first edition of this book in April 2015. It was the first text to present a multilevel policy advocacy framework that links micro policy advocacy, mezzo policy advocacy, and macro policy advocacy—and demonstrates how these three kinds of advocacy can and should be used by social workers in the health care, gerontology, safety net, child and family, mental health, education, immigration, and criminal justice sectors. It was the first text that identified "Red Flag Alerts" that highlighted opportunities for policy advocacy at three levels in each of the eight policy sectors.

The imperative to engage in policy advocacy at three levels stems not only from the Code of Ethics of the National Association of Social Workers (NASW) but from policy shortcomings in each of the eight sectors where most social workers are employed. Underfunded programs in the *safety-net sector* are partly responsible for extreme economic inequality that is approaching levels of the Gilded Age of the 1890s. Millions of children lack sufficient or adequate childcare, and thousands of children graduate from foster care only to become homeless in the *child and youth sector*. Persons with chronic mental conditions lack sufficient community-based services to help them in the *mental health and substance abuse sector* where epidemics of drug abuse, alcoholism, anxiety, and depression are inadequately addressed. Roughly one-half of minority students fail to graduate from many high schools in the *education sector*, and relatively few of them graduate from junior colleges and colleges. The nation has not prepared sufficient numbers of social workers, nurses, and gerontologists to help roughly 30 million baby boomers as they become older in the *gerontology sector*. Warehousing of inmates and lack of preventive programs have contributed to high rates of recidivism in the *criminal justice sector*. The United States has readily used immigrants for labor in agricultural, tourism, construction, and caregiving for seniors but has often failed to provide them with adequate human services and violated their human rights in the *immigration sector*. Moreover, the Code of Ethics of NASW *requires* social workers to engage in policy advocacy with respect to vulnerable and marginalized populations.

I've made a number of changes in this edition. I place the multilevel advocacy framework in Chapter 1 rather than in Chapter 3 to highlight it. This framework was first introduced in the first edition of this book. This framework informs the discussion in every succeeding chapter unlike many other texts on policy sectors that restrict policy advocacy to a single chapter or section of a chapter. I discuss how policy advocates can move between micro, mezzo, and macro levels with case examples in Chapter 1. I discuss Red Flag Alerts that are specific manifestations of the core issues such as "a clinic or hospital fails to link its patients to community-based preventive services" under Core Problem 7 in the health sector.

I discuss in Chapter 1, as well, how I developed the multilevel policy advocacy framework in three phases. By examining 800 articles and books, I identified seven core issues discussed in health literature and research. These seven core issues include (1) ethical rights, human rights, and economic justice; (2) the quality of services and programs; (3) cultural responsiveness of services and programs; (4) preventive strategies and programs; (5) the affordability and accessibility of social programs; (6) the scope and effectiveness of programs that address consumers'/clients' mental distress; and (7) linkages between social programs and services with clients' households and communities. Qualitative interviews with hospital social workers and other frontline professionals confirmed that they frequently address the seven core issues in their advocacy interventions. My work in Phase 1 led to the publication of *Improving Healthcare Through Advocacy* (John Wiley & Sons, 2011) that Dr. Gary Rosenberg, the director of the Division of Social Work and Behavioral Science at Mount Sinai Hospital in New York City, called "by far the best advocacy book I have seen."

In Phase 2, I engaged in quantitative research that was funded by the Patient-Centered Outcomes Research Institute (PCORI). This grant funded the gathering of data from 300 frontline health professionals in eight hospitals in Los Angeles County that include social workers, nurses, and medical residents. The survey confirmed that frontline professionals frequently engage in policy advocacy on micro, mezzo, and macro levels with respect to the seven core issues. The data showed that these professionals engaged in micro policy advocacy for each of the manifestations of the seven core issues. Data was gathered, as well, about the extent to which the 300 frontline health professions engaged in macro policy advocacy with respect to each of the seven core problems. As I suspected, respondents reported far lower levels of macro policy advocacy as well as with respect to mezzo policy advocacy, but significant numbers *did* engage in mezzo and micro policy advocacy as I discuss in Chapter 1. I cite four peer-reviewed articles in Chapter 1 that discuss validated scales that measure frontline professionals' engagement in micro and macro policy advocacy as well as validated scales that predict the extent specific frontline health professionals engage in micro and macro policy advocacy.

In Phase 3, I approached experts in mental health, child welfare, gerontology, education, and criminal justice sectors to see if the multilevel advocacy framework describes the advocacy interventions of frontline professionals in their areas. They responded in the affirmative as can be seen in chapters on these sectors in this text that contain many Red Flag Alerts that describe social workers' micro, mezzo, and macro policy advocacy with respect to the seven core issues in these various sectors.

I also discuss in Chapter 1 how the Code of Ethics of NASW requires social workers to engage in policy advocacy at micro, mezzo, and macro levels and requires social workers to engage in advocacy with vulnerable populations, such as the 18 ones I identified in *The Reluctant Welfare State* (Cengage, 2018), including women; African Americans; Asian Americans; seniors; Native Americans; Latinos; children and adolescents; persons with physical and mental disabilities; persons with substance abuse and mental health challenges; lesbian, gay, bisexual, and transgender persons; persons accused of violating laws and residing in or released from correctional institutions; immigrants; low-income persons; homeless people; and white low-income and blue-collar people in rural areas.

I devote Chapter 2 to the seven core issues that provide the basis for policy advocacy. I discuss how ethicists, the United Nations, religious leaders, researchers, courts, the Code of Ethics of NASW, historical events, and cultural anthropologists analyze them.

I've inserted a chapter on the evolution of the American welfare state in Chapter 3 of the second edition for several reasons. I discuss why policy advocacy is particularly needed in a nation that currently lags behind the social policies of 20 other industrialized nations. I discuss the pendulum swings between relatively liberal and conservative periods in the United States, such as the relatively liberal New Deal of the 1930s and the Great Society of the 1960s as well as the relatively conservative Gilded Age of the late 19th century and the presidencies of Ronald Reagan in the 1980s and Donald Trump in the contemporary period. I introduce them to social workers who engaged in policy advocacy during both periods, including Jane Addams, Harry Hopkins, Whitney Young, Representative Ron Dellums, Senator Barbara Mikulski, and Senator Debbie Stabenow. I also identify many advocates who worked closely with social workers and their causes, including Eleanor Roosevelt, Frances Perkins, Martin Luther King, and Cesar Chavez. I want contemporary social workers to realize that social reforms emanate from the efforts of thousands of policy advocates in both liberal and conservative periods. I discuss how the Social Work Code of Ethics requires social workers to engage in micro, mezzo, and macro policy advocacy no matter the political milieu while also suggesting that social workers can seek to move the nation toward humane policies.

I've retained Chapters 4, 5, and 6 of the first edition that respectively discuss micro policy advocacy, mezzo policy advocacy, and macro policy advocacy. They discuss policy practice skills at each of these levels. They provide vignettes that illustrate policy advocacy strategies and skills, including deciding whether to proceed, assessing the context, placing issues on agendas, engaging in policy analysis, implementing policies, and evaluating policies.

Each of the eight policy sector chapters is organized in the following way. They begin with "empowerment" sections that discuss how social workers can assume critical advocacy roles. I discuss some key policy defects in each sector as well as promising policy reforms including evidence-based initiatives. Each sector chapter contains a historical timeline that describes how policies evolved and how the nation's extreme income inequality impacts clients in each sector. I discuss the political economy of each sector by identifying key players, key interest groups, and important advocacy groups. Each of the sector chapters identifies and discusses Red Flag Alerts under each of the core problems. They all end with a challenge to students to think big by proposing major policy initiatives.

This is a user-friendly book. It begins with the multilevel policy advocacy framework that is illustrated by scores of vignettes that show how specific social workers engaged in micro, mezzo, and macro policy advocacy with respect to many Red Flag Alerts in each of the policy sector chapters. The text identifies numerous controversies in each of the sectors while pointing the way to possible solutions. I updated the policy sector chapters to include recent policy enactments during the presidency of Donald Trump.

I describe this book as an "advocacy passport" to different policy sectors. It allows social workers to obtain an overview of different sectors as they move among them during their careers and as they make referrals to clients who need services and resources from different policy sectors. Although this text has extensive materials about policy advocacy practice, it also contains information about scores of policies that impact the lives of clients.

ACKNOWLEDGMENTS

The first edition of this book, published in April 2015, presented the first multilevel policy advocacy framework. This second edition presents empirical findings from 300 frontline health professionals that validate this framework in a federal research project and four published articles in refereed journals.

Many people helped me develop innovations in this book that include identification of seven core problems that exist in each policy sector, Red Flag Alerts, and the division of policy advocacy into three elements: micro, mezzo, and macro policy advocacy. They also helped me embed these innovations in specific policy sectors and gather empirical data that support the innovations in this second edition.

I am indebted to Dr. Sarah-Jane Dodd, Ph.D., associate professor at the Silverman School of Social Work at Hunter and the CUNY Graduate Center. Her dissertation on advocacy by social workers and nurses at the Suzanne Dworak-Peck School of Social Work was a precursor to the funded research and research articles that I led between 2013 and 2016 as well as the first edition of this text.

I linked case advocacy (or patient advocacy) with macro policy advocacy in my book, *Improving Healthcare Through Advocacy* (John Wiley & Sons, 2011). I thank Dr. Gary Rosenberg, former Edith J. Baerwald Professor of Preventive Medicine at the Icahn School of Medicine at Mount Sinai Hospital, for his review of this book.

I gathered empirical data from roughly 300 frontline health professionals in a research project funded by the Patient-Centered Outcomes Research Institute (PCORI) to test concepts in the aforementioned book. The members of the PCORI research team and coauthors of four published articles in referred journals included:

- Gretchen Heidemann, MSW, Ph.D., adjunct instructor, USC School of Social Work, and special projects manager, USC Department of Psychiatry
- Lei Duan, Ph.D., biostatistician at the University of Southern California Hamovitch Research Center
- Adeline Nyamathi, Ph.D., professor in the School of Nursing, University of California at Los Angeles
- Charles Kaplan, Ph.D., associate dean of Research Hamovitch Center, USC Suzanne Dworak-Peck School of Social Work

I trial tested the concept of Red Flag Alerts with doctoral students as well as the division of policy advocacy into micro, mezzo, and macro policy advocacy. I also trial tested concepts in the first edition of this book with MSW students.

I would like to thank the following persons for demonstrating in the first edition of this book that micro, mezzo, and macro policy advocacy are deeply embedded in social workers' practice in (at least) the eight policy sectors discussed in Chapters 7 through 14 of this book:

Dawn Joosten, Ph.D., clinical professor, Suzanne Dworak-Peck School of Social Work, University of Southern California

Eri Nakagami, Ph.D., clinical social worker, Santa Monica, California

James David Simon, Ph.D., assistant professor, California State University San Bernardino

Vivian Villaverde, MSW, clinical associate professor, Suzanne Dworak-Peck School of Social Work, University of Southern California

Judy DeBonis, Ph.D., assistant professor, Department of Social Work, California State University at Northridge

Gretchen Heidemann, Ph.D., special projects manager at the Department of Psychiatry, School of Medicine, University of Southern California

Anamika Barman-Adhikari, Ph.D., assistant professor, School of Social Work, University of Denver

Elaine Sanchez, MPP, independent researcher and journalist

Thanks to Melissa Bird, MSW, Ph.D., for her outstanding discussion of her macro policy advocacy as chief lobbyist for Planned Parenthood of Utah.

Rafael Angulo, MSW, clinical professor at the Suzanne Dworak-Peck School of Social Work, University of Southern California, developed many video clips for the first edition of this book.

SAGE Publications and myself thank the following reviewers for the second edition of this book:

Leslie D. Cook, LMSW, PsyD (ABD), assistant professor/director of field education, Western New Mexico University School of Social Work

David Estringel, LCSW, BCD, C-ASWCM, CART, University of Texas–Rio Grande Valley

Ebony L. Hall, Tarleton State University

John Q. Hodges, University of North Alabama

Pilar Horner, Michigan State University

The Rev. Susan Hrostowski, Ph.D., LMSW, associate professor, the University of Southern Mississippi

Joyce Litten, Lourdes University

Teresa M. Reinders, University of Wisconsin Parkside

Dr. Anita Sharma, LCSW, University of Louisiana at Monroe

Rebecca Wade-Rancourt, LCSW, Western Connecticut State University

Thanks to my wife, Betty Ann, for her constant support.

A BRIEF BIO-SKETCH OF BRUCE S. JANSSON

Bruce S. Jansson is currently the Margaret W. Driscoll/Louise M. Clevenger Professor of Social Policy and Social Administration, Suzanne Dworak-Peck School of Social Work, University of Southern California. He has made these contributions to the theory and practice of policy advocacy during his career:

- Invented the term "policy practice" in *The Theory and Practice of Social Welfare Policy* (Wadsworth, 1984, pp. 24–28)—and then continued to refine his reconceptualization of policy practice in two subsequent editions of *Social Welfare Policy: From Theory to Practice* (1990, 1994 Brooks/Cole) and in six editions of *Becoming an Effective Policy Advocate: From Policy Practice to Social Justice* (Brooks/Cole, 1999, 2003, 2008, 2011, 2012, 2014, and 2018).
- Analyzed the evolution of the American welfare state in nine editions of *The Reluctant Welfare State* from 1988 to 2019 so that social workers can critically analyze and seek to reform it with use of a multidisciplinary and strategic framework (Brooks/Cole/Cengage).
- Wrote the first critical analysis of budget priorities in the United States during seven decades from FDR through Clinton in *The Sixteen-Trillion-Dollar Mistake: How the U.S. Bungled Its National Priorities From the New Deal to the Present* (Columbia University Press, 2001)—and linked these priorities to inadequate funding of American social policies and programs. This book received the Red Star Review from Publishers Weekly and many positive reviews from historians and political scientists.
- Wrote the first detailed analysis of patient advocacy and policy advocacy in the American health care system in *Improving Healthcare Through Advocacy: A Guide for the Health and Helping Professions* (John Wiley & Sons, 2011) that was called the "best advocacy book by far that I have seen" by Dr. Gary Rosenberg, director of social work and behavioral science at Mount Sinai School of Medicine.
- Obtained funding for a research project titled "Improving Healthcare Through Advocacy" that was funded by the Patient-Centered Outcomes Research Institute (PCORI).

- Wrote *Reducing Inequality: Addressing the Wicked Problems Across Professions and Disciplines* (Cognella Academic Press, 2019) that analyzes why the United States has greater income inequality than 20 industrialized nations and how policy advocates can develop and fund policy proposals to reduce it.
- Wrote *Social Welfare Policy and Advocacy* (SAGE, 2016 and 2020) that developed the first multilevel policy advocacy framework that analyzes how social workers engage in micro, mezzo, and macro policy advocacy in eight policy sectors where most social workers are employed.
- Wrote many book chapters and articles throughout his career including, most recently, four peer-reviewed articles that created many scales that measured levels of frontline health professionals' micro and macro policy advocacy and that predicted levels of health professionals' micro and macro policy advocacy.

BECOMING A POLICY ADVOCATE IN EIGHT POLICY SECTORS

LEARNING OBJECTIVES

In this chapter, you will learn to:

1. Engage social welfare policy whether you are micro or macro
2. Conceptualize policy practice
3. Understand how a multilevel policy advocacy framework was developed
4. Link the multilevel policy advocacy framework to eight policy sectors
5. Use this book as a road map for your student and professional career
6. Contrast micro, mezzo, and macro policy advocacy with clinical practice
7. Understand how the social workers' code of ethics requires micro, mezzo, and macro policy advocacy
8. Use policy advocacy to help marginalized and vulnerable populations
9. Analyze a multilevel policy advocacy framework
10. Provide policy advocacy at three levels
11. Link three levels of advocacy for pregnant teens and teen mothers
12. Develop policy advocacy Red Flag Alerts at three levels

Social workers engage in humanitarian work in many kinds of social agencies. They work with people from all social classes, racial and ethnic groups, genders, ages, and nationalities. They work with active and retired military personnel. They work with residents of urban, suburban, and rural areas. They work with people with myriad social problems.

They often encounter obstacles as they engage in their work such as adverse social policies and difficult work environments that stem from insufficient funding, punitive policies, and heavy workloads. Their clients, too, are often impacted by hardships, such as poverty, mental illness, disability, excessive incarceration, deportation, and discrimination—and large numbers of them live in the lower 50% of the economic distribution.

ENGAGE SOCIAL WELFARE POLICY WHETHER YOU ARE MICRO OR MACRO

Because their clients are profoundly impacted by social policies that emanate from social agencies, communities, states, the federal government, and courts, social workers often engage in three kinds of policy advocacy:

- helping their clients navigate social policies in eight sectors that personally impact them (*micro policy advocacy*)
- reforming dysfunctional agency and community policies in eight sectors (*mezzo policy advocacy*)
- changing policies that emanate from local, state, and federal governments as well as courts (*macro policy advocacy*)

After providing orienting materials about social policy in its first three chapters, this book provides in-depth discussion of micro policy advocacy in Chapter 4, mezzo policy advocacy in Chapter 5, and macro policy advocacy in Chapter 6. It applies the multilevel policy advocacy framework to health, gerontology, safety net, mental health, child and family, education, immigration, and criminal justice sectors in Chapters 7 through 14. Because you will probably work in one or more of these sectors and will often refer clients to programs in different sectors, this book provides a road map to your career.

Social policy was widely viewed *not* as a practice discipline but as a descriptive and analytic discipline prior to the 1980s. Social work scholars described myriad policies at local, state, and federal levels. They evaluated many of these policies by engaging in policy analysis. They focused on government policies with little attention to agency policies or policies impacting communities. These activities have merit, but they failed to make social policy sufficiently relevant to many social work students. This book aims to open up social policy to *all* social workers including to ones in direct service, community organization, and administration.

Three changes took place in social policy that expanded its relevance to all social workers: conceptualizing policy practice, developing a multilevel policy advocacy framework, and linking this framework to eight policy sectors.

CONCEPTUALIZING POLICY PRACTICE

The term *policy practice* first emerged in social work in 1984 to describe policy as a *practice* discipline (Jansson, 1984). Discussion of ways that social workers could participate in *making* social policies hardly existed in the profession's scholarly literature prior to 1984. Rather, existing policy literature was mostly confined to defining social policy, studying the history of policy, analyzing the philosophical underpinning of policy choices, and policy analysis. These topics are important, but do not sufficiently discuss how social workers work to change policies in different venues, such as agencies, communities, and government entities.

Policy practice describes roles, tasks, skills, and strategies that policy practitioners need to read contexts as well as develop, propose, enact, implement, and evaluate policies in specific settings. It describes different styles of policy practice, such as ones that involve social action, rational deliberations, implementation of polices, or combinations of these and other styles. Emerging policy practice literature discusses how social workers read the context to identify constraints that can be surmounted or opportunities that they can seize. It discusses how to place issues on policy agendas, develop policy proposals, engage in policy analysis, enact policies, implement policies, and evaluate policies. It describes skills needed by policy practitioners, including analytic, ethical, political, and interactional ones. It describes different models of policy practice, such as ones that emphasize analytic skills (such as think tanks), political skills (such as campaigns to pressure public officials to enact specific policies), interactional skills (such as developing coalitions to develop and pressure public officials to enact a policy), and ethical skills (such as developing policies that advance social justice).

This redefinition of policy as a practice discipline raised its stature in a profession oriented to *practice*, whether direct service or clinical practice, administrative practice, or practice of community organizers. It facilitated social workers' engagement in policy practice in community-based agencies; community boards; government agencies at local, state, and federal levels; legislatures; and political campaigns. The Council of Social Work Education mandated that schools of social work include policy practice in their curriculums in the 1980s—a requirement that currently exists in its accreditation standards for schools of social work. A national organization of social work policy faculty, known as Influencing State Policy, was established in the 1980s to encourage the teaching, research, and practice of social policy. To clarify that important social policies are developed not just at the level of states but also at levels of local and federal governments, this organization changed its name to Influencing Social Policy (http://www.influencingsocialpolicy.org). It maintains a website and convenes an annual national conference where it awards prizes to the best policy advocacy projects of BSW, MSW, and doctoral students.

Policy practice also includes involvement in political campaigns whether working on campaigns, running for office, or voting. Elected officials develop policies that shape and fund American social policies. Social workers need to work to improve these policies by placing pressure on elected officials, helping elect promising ones, or running for office themselves. The Nancy A. Humphreys Institute for Political Social Work at the School of Social Work at the University of Connecticut, for example, trains hundreds of social workers to work in campaigns, to run for political office, and to hold leadership positions in local, state, and federal governments.

UNDERSTANDING HOW A MULTILEVEL POLICY ADVOCACY FRAMEWORK WAS DEVELOPED

A multilevel policy advocacy framework was developed in a book that was published in April 2015 but copyrighted in 2016 that describes the policy advocacy of social workers no matter in which sector they are employed (Jansson, 2016). It includes *micro policy advocacy* at the level of individuals and families, *mezzo policy advocacy* at the level of organizations and communities, and *macro policy advocacy* at the level of government agencies, legislative and executive branches of government, and political campaigns.

This framework was developed in three stages. First, a review of 800 citations in health care literature identified seven core issues that frontline health professionals address in their professional work, including social workers, nurses, and medical residents:

- protecting patients' ethical rights
- improving patients' quality of care
- helping patients receive culturally competent health care
- helping patients receive preventive health care
- helping patients finance their health care
- helping patients obtain mental health services
- helping patients link their health care to their households and communities (Jansson, 2011)

Second, empirical research was initiated to measure the extent frontline health professionals engage in micro, mezzo, and macro policy advocacy with respect to these seven core issues with a grant obtained from the federally funded Patient-Centered Outcomes Research Institute (PCORI). A research team surveyed 300 frontline health professionals in eight major hospitals to measure the extent they engaged in micro policy advocacy with respect to the seven core problems. These health professionals included 100 social workers, 100 nurses, and 100 medical residents. It also measured their involvement in mezzo policy advocacy because health literature frontline professionals and patients often navigate and contend with hospital policies as well as policies of community agencies. It also measured their involvement in macro policy advocacy to change policies of local, state, and federal agencies, courts, and accreditation bodies.

The data obtained from the PCORI survey demonstrated that frontline professionals help patients at the micro policy advocacy level frequently as can be seen in Table 1.1 (Jansson, Nyamathi, Heidemann, Duan, & Kaplan, 2015a). They frequently help patients get their ethical rights honored; find evidence-based treatments; receive culturally responsive care; receive preventive treatments; finance their medical bills; obtain mental health services; and receive medical care linked to their households and neighborhoods.

Third, with assistance from an expert panel, the research team identified four to seven manifestations of each of the core problems as can be seen in Table 1.1, in which they are numbered from 1 to 33 (Jansson et al., 2015a). The expert panel identified five manifestations of Core Problem 1 (patients' or clients' rights), for example, such as whether patients need assistance in obtaining "informed consent to a medical intervention," "accurate medical information," protection of "confidentiality of (their) medical information," "advance directives," and "care from professionals with competence to make medical decisions" (see Items 1 through 5 in Table 1.1 where asterisks signify half or more of the 300 respondents selected "sometimes," "frequently," or "always").

TABLE 1.1 ■ Frontline Health Professionals' Patient Advocacy Engagements Regarding 33 Types of Patients' Unresolved Problems in Seven Categories

Item	Mean (SD)	Never	Seldom	Sometimes	Frequently	Always
Core Problem 1: Patients' Rights	**2.97 (0.99)**					
1. Informed consent to a medical intervention	2.81 (1.3)	61	66	68	67	33
*2. Accurate medical information	3.26 (1.18)	24	54	88	78	51
3. Confidential medical information	2.81 (1.31)	52	85	66	50	42
*4. Advance directives	2.95 (1.38)	62	52	71	60	50
*5. Competence to make medical decisions	3.0 (1.28)	45	60	83	63	44
Core Problem 2: Quality Care	**2.49 (0.90)**					
6. Lack of evidence-based health care	2.3 (1.12)	88	87	71	41	8
7. Medical errors	2.22 (1.1)	88	104	64	27	12
8. Whether to take specific diagnostic tests	2.62 (1.17)	66	64	99	48	18
*9. Fragmented care	2.95 (1.21)	46	56	90	74	29
10. Non-beneficial treatment	2.37 (1.15)	81	86	81	31	16
Core Problem 3: Culturally Competent Care	**2.87 (0.90)**					
*11. Information in patients' preferred language	3.3 (1.2)	24	58	68	95	50
*12. Communication with persons with limited literacy or health knowledge	3.38 (1.1)	16	46	90	95	48
13. Religious, spiritual, and cultural practices	2.68 (1.15)	49	82	103	35	26
14. Use of complementary and alternative medicine	2.12 (1.07)	105	91	68	22	9
Core Problem 4: Preventive Care	**2.98 (1.02)**					
15. Wellness exams	2.28 (1.35)	120	63	44	44	24
*16. Extent factors known to cause poor health not addressed	3.51 (1.26)	32	31	58	103	71
*17. Chronic disease care	3.58 (1.17)	23	28	68	107	69
18. Immunizations	2.55 (1.44)	104	52	50	51	38

(Continued)

TABLE 1.1 ■ Frontline Health Professionals' Patient Advocacy Engagements Regarding 33 Types of Patients' Unresolved Problems in Seven Categories (Continued)

Item	Mean (SD)	Never	Seldom	Sometimes	Frequently	Always
Core Problem 5: Affordable Care	**3.04 (1.15)**					
*19. Financing medications and health care needs	3.36 (1.28)	30	49	67	82	67
*20. Use of publicly funded programs	3.18 (1.34)	44	51	69	70	61
21. Coverage from private insurance companies	2.57 (1.32)	77	82	60	42	34
Core Problem 6: Mental Health Care	**2.67 (1.10)**					
22. Screening for specific mental health conditions	2.83 (1.33)	63	59	81	50	42
23. Treatment of mental health conditions while hospitalized	2.72 (1.31)	65	71	79	41	39
24. Follow-up treatment for mental health conditions after discharge	2.52 (1.33)	85	75	65	36	34
25. Medications for mental health conditions	2.47 (1.24)	82	76	78	35	24
*26. Mental distress stemming from health conditions	3.08 (1.38)	49	62	62	61	61
27. Availability of individual counseling and or group therapy	2.6 (1.36)	80	76	60	41	38
28. Availability of support groups	2.48 (1.29)	80	89	62	33	31
Core Problem 7: Community-Based Care	**3.12 (1.12)**					
*29. Discharge planning	3.49 (1.32)	36	31	60	89	79
*30. Transitions between community-based levels of care	3.04 (1.33)	51	53	72	71	48
*31. Referrals to services in communities	3.25 (1.34)	41	49	66	74	65
*32. Reaching out to referral sources on behalf of the patient	3.13 (1.37)	48	54	66	66	61
33. Assessment of home, community, and work environments	2.71 (1.39)	76	68	62	45	44

As can be seen in Table 1.1, the research team with a panel of experts also identified multiple manifestations for the other six core problems, such as five manifestations under Core Problem 2. Advocacy interventions with respect to each of the 33 manifestations of the seven core problems are *micro* policy advocacy because they take place at the level of individuals and families. All of these manifestations are widely discussed in evidence-based literature, hospital accreditation standards, and public statutes. With respect to Core Problem 1, for example, federal policies and statutes, as well as ethical experts in health care, require health professionals to help patients obtain their rights regarding each of the five manifestations of patients' rights, such as giving their informed consent to a medical intervention (Manifestation 1).

The PCORI data demonstrates that micro policy advocacy lies at the heart of frontline professionals' work in hospitals as illustrated by data in Table 1.1. For example, more than half of the respondents selected sometimes (3), frequently (4), or always (5) with respect to the each of the five manifestations of patients' rights over a two-month period as can be seen, for example, with respect to Items 1 through 5 under patients' rights in Table 1.1. We discovered similar findings for each of the manifestations of the remaining six core problems.

Fourth, the researchers anticipated that frontline professionals in the PCORI project measured the extent frontline professionals engage in mezzo and macro policy advocacy. They hypothesized they would engage in lower levels of mezzo and macro policy advocacy than micro policy advocacy because frontline health professionals see many patients each month that need micro policy advocacy assistance with the 33 manifestations of the seven core problems in Table 1.1. Many of these micro policy engagements are relatively brief, such as ones that require only short discussions. By contrast, mezzo and macro policy engagements involve talking with many people, attending meetings, gathering information, and developing strategy. They do not usually occur multiple times in a given day. The researchers therefore asked frontline professionals to indicate the frequency of their mezzo and macro policy interventions over the prior six months rather than the two-month interval that was used to measure the frequency of micro policy advocacy engagements.

The data confirmed that frontline health professionals engaged in mezzo and macro policy advocacy with less frequency than micro policy advocacy. Yet it also confirmed that considerable numbers of them engaged in mezzo and macro policy advocacy during the prior six months. It also confirmed that many frontline health professionals want to engage in mezzo and macro policy engagements, mostly believe they have the requisite skills, and mostly believe they are effective.

Recall that social workers engage in mezzo policy advocacy when they seek to change agency policies. Whereas large majorities selected "never" or "seldom" with respect to discussing a hospital policy with an administrator (67%), developing a protocol to improve patient services (70%), or developing a multi-professional training program (84%), the remaining respondents said they had engaged in these kinds of mezzo policy advocacy. Although not measured in this survey, health policy literature identifies community-based projects where frontline professionals have major roles, such as organizing health fairs where community residents receive free medical tests; community services to educate patients about diet, exercise, and other strategies for improving their health; and visits to homes of patients, such as seniors, to gain information about their ability to engage in daily activities.

When asked whether they had engaged "sometimes, frequently, or always" in macro policy advocacy with respect to the seven core problems in the prior six months, 37% said they had engaged in macro policy advocacy with respect to patients' rights, 60% with respect to quality care, 46% with respect to culturally competent care, 42% with respect to affordable or accessible care, 40% with respect to mental health, and 42% with respect to linking hospital care to the patients' households and communities. The PCORI project developed variables that predict the extent to which frontline professionals engage in micro policy advocacy (Jansson et al., 2016) as well as a scale that measures their engagement in micro policy advocacy (Jansson et al., 2015). A majority of the frontline health professionals ranked the extent they possessed 13 skills often linked to mezzo and macro policy advocacy in existing literature at relatively high levels (Jansson, 2011). For example, more than 75% selected "somewhat," "quite a bit," and "a great deal" when asked if they could "influence other people to work with me to change specific policies," "mediate conflicts," and "discuss specific kinds of unresolved patient issues with hospital administrators" whereas more than 60% of them gave these rankings to "initiate policy changing interventions," "negotiate or bargain to achieve my policy goals," "help patients become policy advocates," and "develop better coordination between units or departments of my hospital." More than 50% gave these high rankings to their ability to "communicate with public officials," "change policies in my hospital," and "establish multidisciplinary training sessions in my hospital." They gave far lower rankings to "make budget suggestions in my hospital" (29%) and "changing protocols or operating procedures in my hospital" (41%).

Respondents generally expressed ethical commitment to micro, mezzo, and macro policy advocacy. For example, almost all respondents selected "quite a bit" or "a great deal" to characterize their belief that members of their profession "have an ethical duty to engage in micro policy advocacy (patient advocacy)," "are mandated by (their) profession's code of ethics to engage in mezzo policy advocacy," "should change organizational policies, including their budgets and procedures, to make patient advocacy less likely," "should develop multidisciplinary training programs to enhance policy advocacy skills," and "should work to correct flaws in current public policies."

When asked to indicate to what extent they believe mezzo and macro policy advocacy are effective at organizational, community, and government levels, more than 90% of respondents selected "somewhat," "quite a bit," and "a great deal"—and 85% selected these responses with respect to government policies.

Macro policy advocates work on many fronts: They try to influence public policies in local, state, and federal governments; initiate court proceedings to protect patients' rights; work to change specific government regulations; and seek additional funding for specific services from external sources. Staff in public and not-for-profit agencies cannot work on elections and campaigns due to federal laws that prohibit this activity during their work hours but can engage in political activities on their own time. They can inform residents about initiatives on ballots if they do not recommend that they vote for or against them. The PCORI project developed a scale to measure macro policy advocacy engagement by frontline health professionals (Jansson, Nyamathi, Heidemann, Duan, & Kaplan, 2015b) as well as methods of predicting levels of macro policy advocacy engagement among frontline health professionals (Jansson et al., 2016).

As discussed subsequently, I call "manifestations" of the seven core problems in Table 1.1 "Red Flag Alerts" to draw social workers' attention to them. Experts in the remaining seven sectors in this book identified Red Flag Alerts in those sectors based upon their professional work in them and professional and research literature as I now discuss.

LINKING THE MULTILEVEL ADVOCACY FRAMEWORK TO EIGHT POLICY SECTORS

The PCORI data demonstrated that micro, mezzo, and macro policy are integral to the work of frontline health professionals in hospitals. It did not tell us, however, to what extent social workers engage in micro, mezzo, and macro policy advocacy in sectors other than the health sector, including gerontology, mental health, child and family, education, safety net, immigration, and criminal justice ones.

Because it was not possible to replicate the PCORI survey in these sectors, I turned to experts in each of the remaining seven sectors to gauge to what extent social workers engage in micro, mezzo, and macro policy advocacy in them. I gave experts in these seven sectors the same list of seven core issues that we used in the PCORI project. (I list these experts at the bottom of the first page of each of the chapters in each of these sectors.) I asked them to list specific manifestations of the seven core issues in each of these sectors based not only on their own professional experience and expertise in them but from relevant research and practice literature as well. I asked them to select ones that frontline social workers frequently encounter. Their findings are presented in Chapters 8 through 14 in this book, except for immigration and safety net sectors, where I drew upon existing research and professional literature.

USING THIS BOOK AS A ROAD MAP FOR YOUR STUDENT AND PROFESSIONAL CAREER

Consider this book, then, to be a road map for your student and professional career because social workers often work in several sectors during their careers and often work across sectors when they make referrals. You can use it as a guide to your fieldwork placement, no matter in which sector it is positioned. You can use it when you make referrals that cross sectors, such as when you might refer clients from the mental health sector to the health sector—or when you might refer a child in the child welfare sector to the mental health or health sectors.

CONTRASTING MICRO POLICY ADVOCACY WITH CLINICAL PRACTICE

Micro policy practice and clinical practice are different. Clinicians do not usually view themselves as advocates because they focus on helping clients improve their mental condition within the counseling relationship including their personal emotions, beliefs, and actions. Clinicians often seek internal changes in their clients, such as helping them resolve conflicts, develop personal strategies, and surmount fears. Or they help them address conflicts within families. By contrast, policy advocates help consumers of service obtain widely accepted services or rights, such as the manifestations of seven core problems in Table 1.1. Or they try to change organizational policies that prevent clients' receipt of manifestations of the seven core problems. Clinicians do not usually view themselves as representing clients or populations as they deal with service providers or governments unlike micro policy and macro policy advocates. I devote Chapter 4 to micro policy advocacy because it is rarely discussed in direct-service social work literature, as content analysis of widely used textbooks reveals (Gambrill, 2006; Perlman, 1971; Hepworth & Larsen, 2006; Woods & Hollis, 1990). The fullest discussion of micro policy advocacy exists in Jansson (2011, pp. 23–57), Schneider and Lester (2000), Sunley (1983), and Ezell (1991). Some authors used the term "cause advocacy" to describe "macro policy advocacy" (Schneider & Lester, 2000; Sunley, 1983).

UNDERSTANDING HOW THE NASW CODE OF ETHICS REQUIRES USE OF MICRO, MEZZO, AND MACRO POLICY ADVOCACY

The social workers' code of ethics, promulgated by the National Association of Social Workers (NASW), asks social workers to engage in micro, mezzo, and macro policy advocacy. I place some text from the code under the headings of micro, mezzo, and macro policy advocacy. (You can access the full code at https://www.socialworkers.org/About/Ethics.)

Micro Policy Advocacy. "Social workers promote social justice and social change with and on behalf of clients. . . . Social workers' primary goal is to help people and to address social problems. . . . Social workers treat each person in a caring and respectful fashion mindful of individual differences and culture and ethnic diversity. . . . Social workers respect and promote the right of clients to self-determination (and) when appropriate (to) valid informed consent. . . . Social workers should ensure that (their) colleagues understand social workers' obligation to respect confidentiality. . . . Social workers should have a knowledge base of their clients' cultures and be able to demonstrate competence in the provision of services that are sensitive to client's culture."

Mezzo Policy Advocacy. "'Clients' is used inclusively to refer to individuals, families, groups, organizations, and communities. . . . Social workers' activities include community organizing (and) administration. . . . Social workers should advocate for resource allocation procedures that are open and fair. . . . Social workers should work to improve employing agencies' policies and procedures (and) should not allow an employing organization's policies, procedures, regulations, and administrative orders to interfere with their ethical practice of social work. . . . Social workers should act to prevent and eliminate discrimination in the employing organization's work assignments."

Macro Policy Advocacy. "Social workers' activities include social and political action, policy development and implementation. . . . Social workers challenge social injustice. . . . Social workers' social change efforts are focused primarily on issues of poverty, unemployment, discrimination, and other forms of social injustice. . . . Social workers should promote the general welfare of society (and) should advocate for living conditions conducive to the fulfillment of basic human needs. . . . Social workers should engage in social and political action that seeks to ensure that all people have equal access to the resources, employment, services, and opportunities they require to meet their basic human needs. . . . Social workers should act to prevent and eliminate domination of, exploitation of, and discrimination against any person, group, or class on the basis of race, ethnicity, national origin, color, sex, sexual orientation, gender identity, age, marital status, political belief, religion, immigration status, or mental or physical ability."

USING POLICY ADVOCACY TO HELP VULNERABLE POPULATIONS

Social workers need to prioritize micro, mezzo, and macro policy advocacy to those marginalized populations that encounter many kinds of discrimination. At least 16 of these populations disproportionately reside in the lower economic strata of the United States. They are subject to policy discrimination that fails to fund programs and services they need to improve their condition. They are subject to discrimination in educational institutions in relation to police and courts, employment, voting rights, and segregated housing (Jansson, 2019). Here are some examples:

- It will take 400 years for African Americans to approach economic levels of white Americans because they were denied mortgages from the end of the Civil War until (at least) the 1960s—and it will take 84 years for the average Latino family to make these gains (Holland, 2016).
- The median net worth for Hispanic households was $42,500 in 2014 as compared with $53,700 for all households as compared to $141,900 for white households (Krogstad, 2016).
- Native Americans have the highest rate of poverty at 14.7% of any large racial group in the nation (Wilson, 2014).
- Single mothers experience discrimination in labor markets where they must often work two or three jobs to survive (Law Office of Cohen and Jaffe, 2017).
- Hmong, Bangladesh, and Cambodian citizens, respectively, have poverty rates of 25.6%, 24.6%, and 19.9% (Wilson, 2014).
- Failure to hire and retain ex-felons may cost the U.S. economy as much as $65 billion because they must use safety net programs to survive, do not pay taxes, and are more likely to return to jail (Prison Legal News, 2011).
- Many seniors rely on Social Security benefits that averaged $15,528 in 2014 for single persons and $25,332 for couples when poverty levels were $11,173 for single persons and $14,095 for couples (Jansson, 2014).

- One-third of immigrants' children live in poverty compared with 10% of adults born in the United States (Nisbet, 2013).
- White males who have lost jobs in rural and semi-rural areas have shorter life spans, high rates of addiction to opioids, and high rates of suicide (Chen, 2016; Eberstadt, 2017).
- Disabled people earn on average 37% less than persons without disabilities (American Institutes for Research, 2014).
- High school dropouts come disproportionately from families of color. They are far more likely to be incarcerated, homeless, engage in violence, and have substance abuse problems (GradNation, 2016).
- People become homeless for a variety of reasons, including financial crises, medical emergencies, mental illness, substance abuse, trauma incurred from military service, unemployment, and eviction from rental units (Desmond, 2016).
- LGBTQ persons are subject to many kinds of discrimination including lack of employment protections, inability of transgender persons to use bathrooms consistent with their gender, discrimination in health programs, and bullying in schools and elsewhere (Williams Institute, 2015).
- Fifteen million children live in poverty—or 21% of all children. And 43% of children live in families that lack resources to meet their basic needs (Child Poverty, 2017).
- Millennial persons, born between 1980 and 2000, have higher levels of debt, poverty, and unemployment than the two predecessor generations (Pew Research Center, 2014).

ANALYZING A MULTILEVEL POLICY ADVOCACY FRAMEWORK

A multilevel policy advocacy framework is presented in Figure 1.1 that portrays micro, mezzo, and macro policy advocacy. Advocates at each of these levels do the following:

- Engage in eight tasks, which are portrayed around the outer edge of the circle
- Contend with a policy context that sometimes assists them (assets) and sometimes provides roadblocks (constraints), which are portrayed outside the circle
- Use political, interactional, value clarifying, and analytic skills as they implement each of these eight tasks
- Help persons individually (micro policy advocacy) or collectively (mezzo and macro policy advocacy) surmount the seven core challenges discussed in Chapter 1, including advocating for ethical rights, human rights, and economic justice; improving the quality of social programs; making social programs more culturally responsive; increasing preventive strategies to decrease social problems; improving access to social programs; increasing the scope and effectiveness of mental health programs; and making social programs more relevant to households

Advocates at micro, mezzo, and macro levels undertake eight tasks, including whether to proceed (Challenge 1), where to focus (Challenge 2), obtaining recognition that a client has an unresolved problem from other staff in an agency (micro policy advocacy in Challenge 3) or securing a decision maker's attention for a policy issue or problem (mezzo or macro policy advocacy in Challenge 3), analyzing or diagnosing why a client has an unresolved problem (micro policy advocacy in Challenge 4) or why a dysfunctional policy has developed (mezzo or macro policy advocacy in Challenge 4), developing a strategy to address a client's unresolved problem (micro policy advocacy in Challenge 5) or a proposal to address a policy-related problem (mezzo or macro policy advocacy in Challenge 5), developing support for a strategy to resolve a client's unresolved problem (micro policy advocacy in Challenge 6) or to enact a policy proposal (mezzo or macro policy advocacy in Challenge 6), implementing a strategy (micro policy advocacy in Challenge 7) or an enacted proposal (mezzo or macro policy advocacy in Challenge 7), and assessing whether their implemented strategy (micro policy advocacy in Challenge 8) or policy has been effective (mezzo or macro policy advocacy in Challenge 8).

FIGURE 1.1 ■ A Multilevel Policy Advocacy Framework

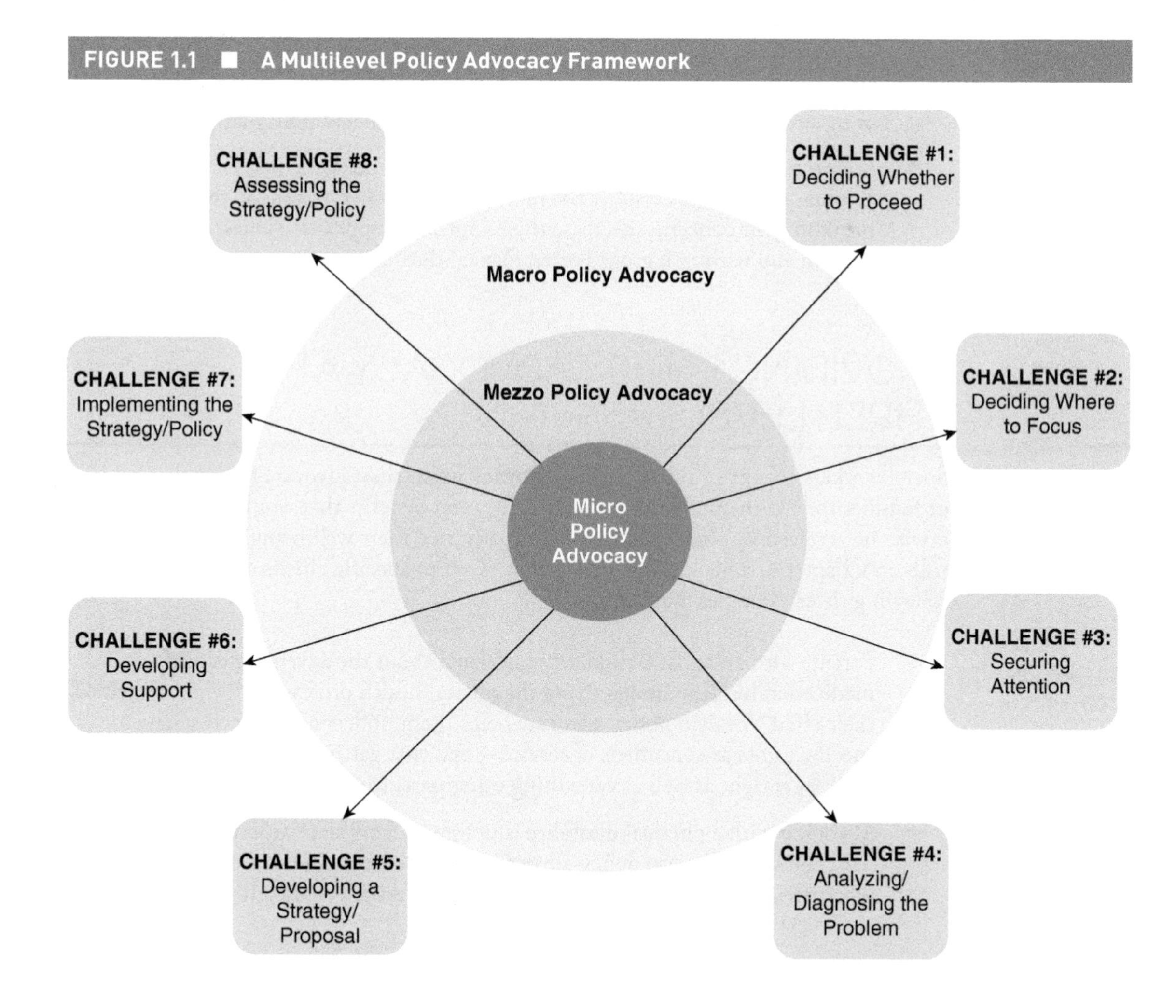

Advocates use four kinds of skills as they undertake these—skills that we discuss in more detail in Chapters 4, 5, and 6:

- *Value-clarifying skills* to determine whether to initiate an advocacy intervention in the first place and to conduct their advocacy ethically, such as by not using deceptive or dishonest tactics whenever possible—they should empower persons to be their own advocates whenever possible while realizing that some persons need some or considerable assistance.
- *Political or influence-using skills* to surmount the disinclination of specific persons to agree with specific policy advocacy initiatives at micro, mezzo, or macro levels—resistance to changes sought by advocates can be mild or sometimes intense as illustrated by opposition of many conservatives to President Obama's Affordable Care Act (ACA) and subsequent attempts by many attorneys general to overturn portions of the legislation in 2010 and 2011 as well as involvement by the U.S. Supreme Court. Obama, in turn, countered with use of his own political skills.
- *Analytic skills* to analyze situations and issues to decide what remedies will improve the well-being of specific persons, such as appealing the denial of eligibility to a program by a government official during micro policy advocacy—or to develop policy proposals during mezzo and macro policy advocacy.
- *Interactional skills* to communicate effectively to persuade others to take specific actions—they must decipher the motivations of other persons so that they can speak to their concerns, decrease their anger, and appeal to values, and they must work in and with task groups often formed during advocacy projects.

PROVIDING POLICY ADVOCACY AT THREE LEVELS

Social workers engage in micro policy advocacy when they advocate for specific persons or families to help them obtain services, rights, and benefits that would (likely) not otherwise be received by them and that would advance their well-being. I discuss it more fully in Chapter 4. This kind of advocacy is given to specific clients and families as the following three examples illustrate:

- Parents with an autistic child are concerned about the adverse effects of medication but fear antagonizing the mental health professional who has helped their child. A micro policy advocate helps them understand that they have specific rights as consumers of service—and that getting a second opinion is their legal right as well as consulting other parents with similar concerns.
- A woman with a physical disability is not given workplace accommodations for her condition. A micro policy advocate refers her to a public interest attorney who specializes in cases related to the rights of disabled persons.

- A woman mistreats her elderly husband with dementia by giving him inadequate nutrition and medical care. (The elderly man and his wife have no living relatives.) A micro policy advocate informs his wife about her husband's legal rights and, when no improvement occurs, refers the case to an agency that investigates cases of elder abuse.

Social workers engage in mezzo policy advocacy at the organizational level when they seek to change dysfunctional policies in agencies and communities that may create the need for micro policy advocacy in the first place and that impede the provision of needed services, benefits, and opportunities as well as the protection of clients' rights. These dysfunctional policies can include standard operating procedures, budgets, mission and organizational culture, eligibility requirements, selection of staff, allocation and training of staff, evaluation procedures, planning mechanisms, official organizational policies, and informal policies. We illustrate mezzo policy advocacy with the following two examples:

- A woman does not receive translation services to allow her to understand her transactions with a service provider. A mezzo policy advocate coaches her to request translation services that are mandated by state and federal law.
- Because a health clinic does not use a team approach when helping persons with diabetes, patients fail to receive integrated services needed for their well-being, including physical therapy, occupational therapy, counseling, preventive services, and medical assistance. A mezzo policy advocate works with clinic administrators to develop protocols for integrated services for persons with diabetes.

Social workers who engage in mezzo policy advocacy at the community level seek to change dysfunctional policies in specific communities—policies that include funding, zoning, land-use planning decisions, policies of community-based public agencies, allocation and training of first responders in police and other agencies, community social and other services, housing inspections, and repairing of infrastructure. We illustrate mezzo policy advocacy at the community level with the following two examples.

- A city has no regulations that limit the number of fast-food outlets in specific neighborhoods—leading to disproportionate location of them in low-income areas. Alarmed about high rates of obesity in these low-income areas, a mezzo policy advocate works to allow the city to establish limits on placement of fast-food outlets in low-income areas.
- The well-being of many low-income persons is jeopardized by the failure of a specific city to monitor and enforce housing regulations for their apartments. A mezzo policy advocate establishes a community coalition to pressure the city council and mayor to replace the current director of the city's housing agency, who, they believe, receives kickbacks from some landlords.

We discuss mezzo policy advocacy in more detail in Chapter 5.

Social workers engage in macro policy advocacy when they seek to change dysfunctional policies in government that may create the need for micro and mezzo policy advocacy in the first place and that impede the provision of needed services, benefits, and opportunities as well as the protection of clients' rights. These dysfunctional policies can include unwise budget priorities and allocations, statutes, regulations, administrative decisions, court rulings, and planning decisions. Macro policy advocates work to change policies and decisions in local, state, and federal governments. We discuss macro policy advocacy in more detail in Chapter 6. We illustrate macro policy advocacy with the following examples:

- A social worker who is the chief lobbyist for Planned Parenthood of Utah lobbies the Utah legislature to enact legislative measures to protect women's reproductive rights, including a law that protects their right to end pregnancies under certain conditions.
- A social worker develops a coalition to raise the rates paid to foster parents with infants to a sufficient level to reimburse the full costs of this care from the current level that only reimburses them for half of this care.
- A coalition of mental health advocates secures the enactment of a proposition on the statewide ballot that sets aside a large and guaranteed sum of money each year for the treatment of persons with mental health problems in that state.
- A state chapter of the NASW endorses candidates who endorse the chapter's policy and budget priorities—and gives them resources to help fund their campaigns.

LINKING THREE LEVELS OF ADVOCACY FOR PREGNANT TEENS AND TEEN MOTHERS

Social workers sometimes move among micro, mezzo, and macro advocacy as illustrated by strategies that social workers have used or could use to improve educational and other services for teenage women who become pregnant.

POLICY ADVOCACY LEARNING CHALLENGE 1.1

PROVIDING MICRO POLICY ADVOCACY FOR A PREGNANT TEENAGER

Many social workers are placed in schools and hospitals where they meet teenage mothers and develop strategies for addressing their needs. When they assess their needs, they discuss their support systems, psychological status, substance abuse history, education, current resources, and any other relevant factors.

Many pregnant teens present themselves at hospitals. In some urban areas, they disproportionately are Latinas from low-income families, have low education, and hold multiple jobs. Many of them did not know about the pregnancy until the fourth or fifth month. Some of them have sexually transmitted diseases. Many of them report they had little sex education that taught them about risk factors associated with unprotected sex, including from failure to use contraceptives.

Many of them report that they did not seek or obtain prenatal care for fear of deportation and their wish to keep their pregnancy confidential. Many inner-city schools have few nurses.

Many teen moms go to "continuation schools" for students who fall behind in the credits they need to graduate—and because they do not receive support from their regular high school. They often have to travel long distances to reach these schools, requiring them to wake up early in the morning and take two or more buses. Sometimes teen moms are "prodded" to leave their normal school because administrators don't want pregnant moms to walk across the stage before a large audience at graduation.

Teen moms need advocates to help them decide whether to stay in their normal school rather than moving to continuation schools. They mix with students with behavior problems in continuation schools, who often have been dismissed from their normal schools. Continuation schools often do not have college counselors, advanced courses, or work-ahead students. Data is often lacking about the performance of continuation schools as compared with regular schools. Some research suggests many of them have inferior teaching and curriculum (Butrymowicz, 2015). Many teen moms do not know if their state prohibits school administrators from requiring them to attend continuation schools, such as whether a state's education code prohibits discrimination against pregnant teens.

Teen moms need support from social workers. They are subjected to stigma even though they are like other students but just happened to become pregnant. They often need to work part time to support their parents. Immigrants often fear getting medical help because they believe medical staff may convey their names to Immigration Control Enforcement (ICE).

Once teen moms have their children, they run into other challenges. How will they find and fund childcare? Will they receive birth control? Will they receive ongoing advocacy from a social worker as they obtain assistance from Child Protection Services, whose staff make certain that the teen mother and her baby do not experience neglect or abuse. They need to obtain medical help from Medicaid, the Supplemental Nutrition Program for Women, Infants, and Children (WIC) to receive nutritional assistance, diapers, and an infant car seat. They need help from public health nurses, who regularly meet with the teen mom regarding baby care and parenting after discharge from the hospital. Other resources include the Teen Mothers Resource Center and legal aid with respect to the teen's right to remain in or return to her normal school.

Social workers' advocacy with teen mothers may allow some of them to beat the odds. Only 40% of teen moms finish high school. Less than 2% finish college by age 30. Women who give birth while attending community colleges are 65% less likely to complete their degrees than other women. Children of teen moms are 50% more likely to repeat a grade and more likely to drop out of high school than children of older mothers (National Conference of State Legislators, 2013).

Sources: Butrymowicz, S. (2015, July 6). Do California's continuation schools really work? *Tribune News Service*. Retrieved from http://www.governing.com/topics/.../do-californias-continuation-schools really-work.html, National Conference of State Legislators. (2013, July 17). Postcard: Teen pregnancy affects graduation rates. Retrieved from http://www.ncsl.org/research/health/teen-pregnancy-affects-graduation-rates-postcard.aspx

POLICY ADVOCACY LEARNING CHALLENGE 1.2

MOVING TOWARD MEZZO POLICY ADVOCACY TO HELP PREGNANT HIGH SCHOOL STUDENTS

Social workers engage in mezzo policy practice to help pregnant high school students when they seek to change policies and procedures in specific high schools or school districts. Despite the importance of teen education and equal education requirements of Title IX, many guidance counselors still informally counsel pregnant students to leave their high school for alternative schools without providing them assistance or resources and telling them they have the option to stay put. Official school policies could be established that prohibit encouraging pregnant students to leave high schools for alternative schools.

Sex education can be improved in specific schools or school districts by making it more effective with teenagers by developing or using models that have been proven to be effective in preventing or delaying teen pregnancy. Do specific policy and program deficiencies impede preventive strategies, such as sex education programs that discuss not only abstinence but also birth control strategies? Do schools have nurses on the premises who distribute condoms? Do schools inform teenagers to consult medical staff if they have unprotected sexual encounters to see if they wish to use medications to avert pregnancy? Are schools linked to Planned Parenthood to obtain information about their options?

Learning Exercise

Address the following questions with respect to mezzo policy advocacy with teenagers in schools:

- Do social workers frequently engage in micro policy advocacy to help pregnant adolescents gain their rights by identifying and addressing systemic defects in organizational policies, such as prejudice by school staff against this population or lack of quality education programs geared to the needs of this population?
- Do pregnant adolescents drop out of a specific school due to hostile treatment by a specific teacher or guidance counselor or due to defective policies in a specific school (organizational factors), in the school district (community factors), or at the State Department of Education (government factors)—or some combination of these factors?
- Did deficiencies in the policy and regulatory context contribute to the problem, such as lack of guidelines to protect the rights to education from the school district, the State Department of Education, or the federal Department of Education?
- Do budgets of specific schools or school districts prioritize services for pregnant teenagers—or sex education or nurses in schools?
- Are pregnant students of color treated differently in specific schools or school districts than white pregnant students? Are low-income pregnant students treated differently than more affluent pregnant students?
- Do schools keep data on the educational paths of pregnant teens?
- Do specific schools give pregnant adolescents special accommodations, allowing them to be tardy or absent when obtaining medical care?
- What policies have specific schools or school districts developed to help young women remain in school after they have given birth, such as assistance with childcare, supportive counseling, and special accommodations?

POLICY ADVOCACY LEARNING CHALLENGE 1.3

MOVING TOWARD MACRO POLICY ADVOCACY TO HELP PREGNANT HIGH SCHOOL STUDENTS

The United States has the highest rates of teen pregnancies of industrialized nations even though the pregnancy rate has markedly declined for teens from age 15 to 19. Only one-third of teen mothers finish high school and only 1.5% have a college degree by age 30.

Public schools differ markedly in their policies regarding pregnant teen mothers partly because of the absence or lack of clear state laws or federal policies. Some send them to continuation schools during their pregnancy where they are separated from their friends, not even inviting them back to their regular school when they have given birth. Continuation schools are of uncertain quality partly because their standards are not well defined by state law. State laws are often unclear about whether adolescents can remain in continuation schools even after giving birth. Laws forbid schools from expelling teen mothers, but they receive little policy guidance otherwise. Little case law enforces or guides the provision of educational services for teen mothers in many localities and states. Some evidence suggests, as well, that African American and low-income pregnant adolescents are treated more harshly than white and affluent adolescents. The laws and policies of some states do not require schools to teach sexual education. Many schools ignore the importance of preventive health education and comprehensive sex education. Many states do not require schools to keep data on the educational trajectories of teen mothers prior to or after giving birth.

Nor is it clear to what extent some states fund special programs for teen pregnant mothers and teen mothers with children. Although some teens can count on support from their parents and relatives, others lack such support—and may particularly need financial assistance from schools for medical care, childcare, counseling, and other assistance.

Nor is it clear what budgetary and policy roles exist for school districts as compared with state educational agencies and policies. Some state officials may wish to cede responsibility to school districts that lack resources and staff to help pregnant teens and teen mothers.

Advocates need to consider, as well, whether and under what circumstances pregnant teens can seek termination of their pregnancies. What laws in their states impact these decisions—and do these laws need to be reformed? What positions do Planned Parenthood and other advocacy groups take in specific states on this issue? These kinds of systemic policy factors can only be addressed through macro policy advocacy.

Identify some dysfunctional policies in your locality, region, or state that might be addressed through macro policy advocacy by social workers working with teenagers in schools or other settings.

Our discussion suggests that advocacy at micro, mezzo, and macro levels can be linked. Discuss how a social worker might move among micro, mezzo, and macro policy advocacy.

DEVELOPING MICRO POLICY ADVOCACY RED FLAG ALERTS

We now return to the seven core problems that we discussed earlier in this chapter. We need to move beyond general descriptions of these seven core problems to identify specific manifestations or examples so that we anticipate them. In the research reported earlier in this chapter, we identified an array of problems and issues that frontline workers

POLICY ADVOCACY LEARNING CHALLENGE 1.4

IDENTIFYING SOME RED FLAG ALERTS IN SPECIFIC AGENCIES

Take any setting that provides human services with which you are familiar, whether your field agency or where you have volunteered or worked. Develop a list of a specific manifestations for each of the seven core problems that are listed in Table 1.1. Then ask a professional who works in the setting to discuss with you the extent your list of manifestations of the seven core problems are relatively common. Also ask whether clients would be adversely impacted if social workers failed to address them. Ask this professional to augment your list with additional problems even as this professional might delete one or more of the problems that you identified from your list.

Learning Exercise

Discuss the following questions:

- Is it possible to develop specific Red Flag Alerts in the setting that you have chosen?
- Would these Red Flag Alerts facilitate the use of micro policy advocacy by social workers in this setting? Or might they facilitate mezzo and macro policy advocacy?
- To what extent did you use research findings, ethical principles, or pragmatic factors to develop these Red Flag Alerts?

are likely to encounter in hospitals with help from a panel of experts, that is, 33 of them as described in Table 1.1. Specific manifestations of the seven core problems become Red Flag Alerts when they occur relatively frequently in a population of clients and when they have negative effects if they are not addressed or resolved.

You can use the same methodology to identify a set of problems and issues in your own practice in any agency—and then change the list as you discover which ones reappear frequently in your practice. Or you can discuss with your supervisor or executives polling social workers and other frontline staff not only to identify a list of manifestations of the seven core problems but also to measure their relative incidence in your agency. You might also convene some consumers of service to obtain their designation of specific issues as ones they frequently encounter in a specific agency or organization. You can designate some as Red Flag Alerts based upon their frequency and their negative impact upon clients when they are not addressed. Frontline professionals can develop workshops to discuss how to help clients/patients address Red Flag Alerts. They can ask whether agency policies might be changed to help staff prioritize them.

DEVELOPING RED FLAG ALERTS AT THREE LEVELS

It is also possible to identify specific manifestations of one of the seven core problems that could require advocacy at micro, mezzo, and macro levels. Let's use an example drawn from schools. Assume that you have read extensive research literature that documents that many children are overmedicated for specific mental problems like autism, attention deficit hyperactivity disorder (ADHD), or behavioral problems. You could make this a micro advocacy policy Red Flag Alert falling under the sixth core problem (failure to address mental problems of clients) by identifying the following problem: "Micro Advocacy Red Flag Alert: Children who are overmedicated for behavioral and mental health problems including autism, ADHD, or behavioral problems." Were you to work in a mental health or education setting, you would be alert to this issue and might launch a micro policy advocacy intervention to help the child's parents obtain second opinions to ascertain if their child is overmedicated. You transform this micro advocacy Red Flag Alert into a mezzo Red Flag Alert and a macro advocacy Red Flag Alert by placing it in an organizational, community, and government context. Now assume that many children may be overmedicated in a large school system. The micro Red Flag Alert can be changed to read: "Mezzo Advocacy Red Flag Alert: Many schoolchildren who have been diagnosed with autism, ADHD, or behavioral problems have been overmedicated with respect to type of medication, number of medications, and dosage." A social worker could consider launching a mezzo policy advocacy intervention such as one of the following:

- Developing training programs for teachers, social workers, school counselors, school speech therapists, and other staff in a school district to recognize signs of overmedication
- Developing guidelines in the state's board of education about overmedication of children

A social worker could engage in macro policy advocacy by contacting a children's advocacy group in Washington, D.C., to see if specific regulations by the Food and Drug Administration (FDA) could be developed to regulate the use of medications with children under a specific age—much as the federal government is now attempting to limit prescriptions of pain medications that led to the preventable deaths of roughly 70,000 persons in 2017 (Centers for Disease Control and Prevention, 2018). This could be a "Macro Advocacy Red Flag Alert: Many schoolchildren are overmedicated for autism, ADHD, or behavioral problems due to lack of federal regulations about the use of medications with children under a specific age for these problems."

POLICY ADVOCACY LEARNING CHALLENGE 1.5

LOCATING INFORMATION ABOUT SPECIFIC SOCIAL PROBLEMS

Other manifestations of the seven core problems can be obtained by consulting the Internet, such as by typing problems into online search sites such as Google Search, Google Advanced Search, Google Scholar, or Microsoft's Bing. Assume, for example, that you wonder if you are likely to encounter malnutrition among schoolchildren in a particular low-income neighborhood. You can find information about malnutrition among low-income

(Continued)

(Continued)

children generally or in specific regions. This information may not accurately predict or measure malnutrition in a specific geographic area or a specific school district, so you would need to interview or contact researchers with geographic-specific knowledge about childhood malnutrition.

Take a stab at obtaining information about one of the following problems—and deciding if they should be designated as Red Flag Alerts. Identify where social workers might encounter persons with these problems by sector and geographic area as well as by the type of agency or hospital where they work.

- Extent to which veterans receive services for brain trauma or for mental problems
- Extent to which homeless persons receive affordable housing
- Extent to which sufficient services are given to truants in schools
- Extent to which released prisoners receive assistance with finding employment

We use the term *connected policy interventions* to describe linked policy interventions. Social workers can begin with micro, mezzo, or macro policy intervention and then consider progressing to other levels. Assume, for example, that a local child welfare department receives many reports from neighbors, school officials, and others that they suspect that specific children have been abused or neglected by their parent or parents. Also assume that a social worker discovers that many of these children, as well as their families, possess serious mental health and substance abuse problems even though child abuse and neglect are not discovered—and finds that his or her perceptions are supported by evidence-based research. Yet the social worker finds that these children and their families receive no assistance from child welfare workers once no evidence of abuse or neglect is found. The social worker may begin with a single child or family and work to get them supportive services from community agencies (a micro policy advocacy intervention) and *then* decide that the child welfare department should develop a policy to facilitate these referrals for many or all of these children (a mezzo policy advocacy intervention). Or perhaps the social worker could inquire whether child welfare regulations at the state level sufficiently mandate the provision of preventive services for children who are referred to child welfare departments. She might launch a macro policy advocacy intervention to modify these regulations.

We will refer in this book to:

- Micro policy advocacy interventions
- Mezzo policy advocacy interventions
- Macro policy advocacy interventions
- Connected policy advocacy interventions
- Red Flag Alerts

You have now learned that policy practice at micro, mezzo, and macro policy advocacy levels will be integral to your professional work, no matter whether you engage in direct service, administration, community organizing, or advocacy with legislators or other public officials—and no matter whether you work in health, gerontology, child and family, mental health, safety net, education, immigration, or criminal justice sectors. You will need to develop skills in each of these kinds of policy advocacy. You will join legions of social workers and other frontline professionals who enrich the human condition by developing these competencies.

Learning Outcomes

You are now equipped to:

- Engage social welfare policy more fully in agency, community, legislative, and other settings
- Conceptualize policy practice
- Understand how a multilevel policy advocacy framework was developed
- Link the multilevel policy advocacy framework to eight sectors
- Understand how this book provides a road map for your student and professional career
- Contrast micro policy advocacy with clinical practice
- Understand how the social workers' code of ethics requires micro, mezzo, and macro policy advocacy
- Use policy advocacy to help members of marginalized and vulnerable populations
- Analyze a multilevel policy advocacy framework
- Provide policy advocacy at three levels
- Link three levels of advocacy for pregnant teens and teen mothers
- Develop Red Flag Alerts at micro, mezzo, and macro levels

References

American Institutes for Research. (2014, December 14). Those with disabilities earn 37% less on average [Newsletter]. Retrieved from https://www.air.org/news/press-release/those-disabilities-earn-37-less-average-gap-even-wider-some-states

Centers for Disease Control and Prevention (CDC). (2018). Retrieved from https://www.cdc.gov. See Drug Overdose deaths in 2017.

Chen, V. T. (2016, January 16). All hollowed out: The lonely poverty of America's white working class. *The Atlantic.* Retrieved from https://theatlantic.com/business/archive/2016/11/white-working-claa.../424341/

Child Poverty. (2017). National Center for Children in Poverty. Retrieved from http://www.nccp.org/topics/childpoverty.html

Desmond, M. (2016). *Evicted: Poverty and profit in the American city* (New York; Penguin Random House.

Eberstadt, N. (2017, February 15). Our miserable 21st century. *Commentary.* Retrieved from https://aei.org/publication/our-miserable-21st-century

Ezell, M. (1991). *Administrators as advocates.* Philadelphia: Taylor and Francis Group.

Gambrill, E. (2006). *Social work practice: A critical thinker's guide.* New York: Oxford University Press.

GradNation. (2016). Retrieved from gradnation.americaspromise.org

Hepworth, D. & Larsen, J. (2006). *Direct social work practice: Theory and skills.* Belmont, CA: Wadsworth.

Holland, J. (2016, August 6). The average black family would need 228 years to build the wealth of a white family today. *Nation.* Retrieved from https://www.thenation.com/article/the-average-black-family-would-need-228-years-to-build-the-wealth-of-a-white-family-today/

Jansson, B. (1984). *The theory and practice of social welfare policy: Analysis, processes, and current issues.* Belmont, CA: Wadsworth Publishing Company.

Jansson, B. (2011). *Improving healthcare through advocacy: A guide for the health and helping professions.* Hoboken, NJ: John Wiley & Sons.

Jansson, B. (2014). *The reluctant welfare state.* Boston, MA: Cengage.

Jansson, B. (2016). *Social welfare policy and advocacy: Advancing social justice through 8 policy sectors.* Thousand Oaks, CA: Sage Publications.

Jansson, B. (2019). *Reducing inequality: Addressing the wicked problems across professions and disciplines.* San Diego: Cognella Academic Press.

Jansson, B., Nyamathi, A., Duan, L., Kaplan, C., Heidemann, G., & Ananias, D. (2015). Validation of the patient advocacy engagement scale. *Research in Nursing & Health, 38*(2), 162–172.

Jansson, B., Nyamathi A., Heidemann, G., Bird, M., Ward, C. R., Brown-Saltzman, K., Duan, L., & Kaplan, C. (2016). Predicting levels of policy advocacy engagement among acute-care health professionals. *Policy, Politics, and Nursing Practice, 17*(1), 43–55.

Jansson, B., Nyamathi, A., Heidemann, G., Duan, L., & Kaplan, C. (2015a). Predicting patient advocacy engagement: A multiple regression using data from health professionals in acute-care hospitals. *Social Work in Health Care, 54*(7), 559–581.

Jansson, B., Nyamathi, A., Heidemann, G., Duan, L., & Kaplan, C. (2015b). Validation of the policy advocacy engagement scale for frontline healthcare professionals. *Nursing Ethics.* Advance online publication. doi:10.1177/0969733015603443

Krogstad, J. (2016, July 15). The economy is a top issue for Latinos, and they're more upbeat about it. FACTTANK, Pew Research Center. Retrieved from http://www.pewresearch.org/fact-tank/2016/07/15/the-economy-is-a-top-issue-for-latinos-and-theyre-more-upbeat-about-it/

Law Office of Cohen & Jaffe. (2017). Marital status employment discrimination against single moms. Retrieved from cohenjaffe.com → Employment Discrimination

Nisbet, E. (2013). Earnings gap between undocumented and documented U.S. farm workers: 1990 to 2008. National Poverty Center Working Paper Series. Retrieved from http://npc.umich.edu/publications/u/2013-07-npc-working-paper.pdf

Perlman, H. (1971). *Perspectives on social casework.* Philadelphia: Temple University Press.

Pew Research Center. (2014, March 7). Millennials in adulthood. Retrieved from http://www.pewsocialtrends.org/2014/03/07/millennials-in-adulthood/

Prison Legal News. (2011, December 15). Study shows ex-offenders have greatly reduced employment rates. Retrieved from https://www.prisonlegalnews.org/news/2011/dec/15/study-shows-ex-offenders-have-greatly-reduced-employment-rates/

Schneider, R. & Lester, L. (2001). *Social work advocacy: A new framework for action.* Belmont, CA: Brooks/Cole.

Sunley, R. (1983). *Advocating today: A human service practitioner's handbook.* New York: Family Service of America.

Williams Institute. (2015). The LGBT divide in California. Retrieved from https://williamsinstitute.law.ucla.edu/wp-content/uploads/California-LGBT-Divide-Jan-2016.pdf

Wilson, V. (2014). 2013 ACS shows depth of Native American poverty and different degrees of economic well-being for Asian ethnic groups. *Economic Policy Institute.* Retrieved from https://www.epi.org/blog/2013-acs-shows-depth-native-american-poverty/

Woods, M. & Hollis, F. (1990). *Casework: A psychosocial therapy.* New York: McGraw Hill.

2

ADVANCING SOCIAL JUSTICE WITH SEVEN CORE PROBLEMS

LEARNING OBJECTIVES

In this chapter, you will learn to:

1. Define social policies and identify their origins
2. Define social justice and why we seek to advance it
3. Advance social justice with respect to seven core problems
4. Advance ethical rights, human rights, and economic justice
5. Improve the quality of social programs
6. Make social programs and policies more culturally responsive
7. Develop preventive strategies to decrease social problems
8. Improve affordability and access to social programs
9. Increase the scope and effectiveness of mental health programs
10. Make social programs more relevant to households and communities
11. Recognize the seven core problems in eight policy sectors
12. Join the reform tradition of social work

DEFINING SOCIAL POLICIES

We define social policies as "collective strategies to prevent and address social problems." They are collective because they are binding on those persons, populations, communities, companies, and jurisdictions to which they apply. When Congress enacts and the president signs a statute, for example, such as the Affordable Care Act (ACA), persons, health providers, states, and others must adhere to its provisions under penalty of law because statutes are binding laws. They can challenge provisions of a federal law, such as the ACA, through legislatures and the courts. When the federal government enacts a regulation, such as requiring health and mental health providers to protect the confidentiality of clients' data, social workers must adhere to it. They cannot argue that "they didn't know about it" in court if they are charged with violating it. When a state declares the growing and distribution of marijuana to be legal "under certain circumstances," growers and distributors have to engage in this practice in conformity with the legislation.

Social policies take specific and tangible form:

- *Constitutions* define the social policy powers of government at the federal and state levels. States, too, possess constitutions that establish important duties of state governments as well as how they govern themselves.
- *Public policies* are enacted in local, state, or federal legislatures. The Chinese Exclusion Act of 1882, the Social Security Act of 1935, the Adoption Assistance and Child Welfare Act of 1980, the Americans with Disabilities Act of 1991, and the ACA of 2010 are examples of public laws, as are the state and local laws that established poorhouses and mental institutions in the 19th century.
- *Court decisions* play important roles in American social policy. By overruling, upholding, and interpreting the federal and state constitutions, statutes of legislatures, ordinances of local government, and practices of public agencies such as mental health, police, and welfare departments, courts establish policies that significantly influence the American response to social needs. For example, in the 1980s, the courts required the Reagan administration to award disability benefits to many disabled persons even though many administration officials opposed this policy. Decisions of the U.S. Supreme Court upheld key provisions of the ACA while diluting other provisions such as the requirement that all states had to raise their Medicaid eligibility levels to guarantee medical services for a larger number of low-income persons.
- *Budget, spending, and tax programs* are also an expression of policy, as are the budget priorities established by the nature of budget allocations and tax policies. For example, Americans chose not to expend a major share of the gross national product on social programs prior to the 1930s but greatly increased levels of spending during the Great Depression and succeeding decades. Despite the large increases in spending on social programs in the 1960s and the 1970s, for example, the nation chose to devote a significant portion of its federal budget to military spending during the Cold War and also to make successive tax cuts—policies that greatly reduced the resources available for social programs.

- *International treaties, as well as policies of the United Nations,* govern an array of economic, social, migration, environmental, and national security issues in an era of globalization.
- *Regulations* are established by government departments and heads of government with input from the public, such as those that require hospitals to report medical errors. Government agencies issue administrative regulations to guide the implementation of policies that have the force of law. Persons who fail to follow government regulations can suffer serious penalties including fines and imprisonment.
- *Stated or implied objectives* also constitute a form of policy. For example, the preambles and titles of social legislation suggest broad purposes or goals. Thus, as its title suggests, the Personal Responsibility and Work Opportunity and Reconciliation Act that Bill Clinton signed in August 1996 emphasized punitive rules and procedures for getting welfare recipients off welfare rolls rather than greatly expanding training, education, or services for recipients.
- *Rules and procedures* define the way in which policies are to be implemented. Legislation often prescribes, for example, the rules or procedures to be used by agency staff in determining applicants' eligibility for specific programs. Courts often prescribe procedures that the staff of social agencies must employ to safeguard the rights of clients, patients, and consumers, such as requiring that persons be given legal counsel before involuntarily committing them to mental institutions.
- *Informal policies* as compared with *written or official policies* are subjective views of persons and groups that influence whether and how they implement specific policies. If we want to know how the poorhouses of the 19th century worked—or how social agencies implemented the ACA—we have to examine how their staff implemented formal policies that were given to them by legislatures and public officials. Informal and formal policies sometimes work in tandem, such as when the line staff of agencies fully understand and agree with official policies. They sometimes clash, however, when staff do not fully implement official policies because they disagree with them such as when local police fail to wear body cameras even when told that this policy is mandatory.

Social policy surrounds and envelops social workers, as well as the persons and communities that they help, at virtually every point in their professional work. It describes the benefits their clients can receive from many social programs—whether material benefits from programs like food stamps or services from mental health, vocational, or education programs. It describes clients' rights through regulations, legislation, and court rulings. It gives opportunities to clients, such as education, preschool, and vocational programs. It provides Americans with tax benefits that help them purchase homes and accumulate savings. It gives civil rights to persons of color, women, and disabled persons. It provides preventive services to persons, such as primary care medical services and nutritional benefits to pregnant women. It helps persons survive disasters like floods, hurricanes, and tornados. It funds social programs and determines the purpose or mission of specific social programs and agencies.

Social policy shapes the nature of society itself. It helps determine, for example, the extent of income inequality in a community, state, or nation. If the United States fails to fund education, social programs, and safety net programs sufficiently, it decreases the chances that low-income persons can improve their lot. If the nation gives affluent persons excessive tax breaks, it increases the odds that they will retain their dominant economic position. If a state possesses inferior vocational and job training programs, low-income persons will find it difficult to improve their economic standing.

Policies are vertically distributed at federal, state, and local government levels and community and agency or organizational levels. The federal government funds myriad entitlements, such as Social Security, Medicare, and Medicaid. It funds many means-tested programs like the Supplemental Nutrition Assistance Program (SNAP or food stamps) and the Supplemental Security Income Program (SSI). It funds the Earned Income Tax Credit (EITC) that gives tax rebates to many families. It funds hundreds of smaller programs through its annual budget, such as health prevention programs, Head Start, and block grant programs that give funds to states for mental health, childcare, and many additional programs. The U.S. Justice Department and the federal Equal Employment and Occupational Commission (EEOC) monitor violations of civil rights and take corrective action. The federal government funds the bulk of the American welfare state because its resources from the federal tax system and other kinds of taxation far exceed resources of the states.

State governments often share most of the costs of public education with the federal government covering only 10% of their costs. They also share the costs of the nation's huge correctional system, including costs of prisons, parole departments, and local police—save for federal prisons and federal law enforcement through the Federal Bureau of Investigation (FBI) and other federal police functions. States often fund many public health programs. They often inspect health facilities, such as hospitals, clinics, and nursing homes. States have been given many additional roles in overseeing the health system under the ACA. (Considerable variation exists among states regarding which policy functions reside at state vs. local levels and the extent states and local governments share policy responsibilities and costs.) States provide direction for child welfare, mental health, public health, and public systems of care. They often are conduits for federal resources that they supplement with their own funds. They set standards for many public services provided by counties and municipalities. State governments contribute significant resources as well to social programs and policies. They co-fund programs such as Medicaid and other federal-state programs. They establish and fund many of their own programs, such as public health programs that provide health clinics to low-income areas and inspect food facilities for safety.

Local governments determine how land can be used in their jurisdictions through zoning and tax policies. They determine if specific social agencies can locate in specific areas, such as halfway homes for persons released from state mental health facilities. They provide police and fire services. Cities and counties raise taxes and decide how to use tax revenues, such as for police and fire services, schools, recreation programs, and social service programs. They orchestrate many housing programs. Many county or municipal agencies distribute welfare benefits to persons under Temporary Assistance for Needy Families (TANF) and general assistance programs. Many counties and cities administer public systems of health care and assume important roles in child welfare, mental health, and many other social programs.

Considerable variation exists among states regarding which services are administered by state agencies or by local agencies. New York State runs the state's child welfare services, for example, as compared with their administration by counties in California.

Public and nonpublic organizations are largely funded by local, state, and federal governments to implement programs defined by public statutes. They have considerable leeway, however, in making many implementation decisions. Not-for-profit agencies, which are exempted from paying taxes by local, state, and federal governments, raise their own resources from public and private resources as well as from fees paid by consumers of service. They select a mission that shapes what programs they will fund by grants from public agencies, consumer fees, or resources from private donors. They hire staff to implement these resources. For-profit organizations and agencies have owners or shareholders who seek profits in the marketplace. For-profit agencies provide a wide array of social services, such as childcare, nursing homes, job training, education, and other services—and sometimes receive contracts from public authorities to implement welfare, correctional, educational, and other services.

DEFINING SOCIAL JUSTICE

Social injustice is widespread in the United States. Recall the sheer number of vulnerable populations that exist in the United States that we identified in Chapter 1. Many members of these vulnerable populations thrive despite adversity they encounter, but many others have disproportionate levels of lower income, shorter lives, higher levels of disability, greater unemployment, higher levels of incarceration, and other social problems because of social injustice embedded in the broader society. They often receive services, benefits, housing, opportunities, preventive services, ethical rights, health care, and other amenities that are inferior to persons not members of these marginalized populations. Social injustice is illustrated by the following inequities:

- The rights of women not to be sexually harassed as illustrated by scores of examples that surfaced in 2017 and 2018, including dignitaries such as Harvey Weinstein, Charlie Rose, and Bill Cosby
- The rights of persons of color not to be killed by police when they have not engaged in violence against police officers or others as occurred in scores of examples often recorded by cell phones from 2014 to the present
- The rights of immigrant children brought to the United States to not be separated from their parents—a right upheld by a judge in 2018 but not honored by federal agencies leading to the separation of thousands of children from their parents
- The rights of African Americans and Latinos not to be incarcerated at high levels for nonviolent crimes—and their rights to receive prison sentences not greater than white or affluent people who commit the same offenses
- The rights of high school students not to be subjected to massacres from military weapons

- The rights of single mothers not to have to work three jobs to survive
- The rights of women born into the lower tenth of the economic distribution not to live (on average) lives 12 years shorter than women born into the top tenth of the economic distribution

POLICY ADVOCACY LEARNING CHALLENGE 2.1

IDENTIFYING POLICIES TO HELP VULNERABLE POPULATIONS

As an exercise, take any of the following statements and ask what kinds of policies would be helpful to the specific vulnerable populations in them.

- The American poverty rate was 12.67 percent in 2006—or almost one percentage point higher than in 2001. It rose to 15.1 percent by 2010—the highest level since 1992. It fell to 12.7% in 2016 based on data of the U.S. Census Bureau—or 43.1 million Americans.
- About 533,742 people were homeless on any given night in the United States in 2017 according to the National Alliance to End Homelessness.
- In 2016, 28.6 million (9%) of persons of all ages were uninsured according to the Centers for Disease Control and Prevention (CDC).
- About 20,000 of the nation's 500,000 foster children "graduate" from foster care each year at age 18—only to encounter a difficult transition into life in the community. In the first two years, about 65% become homeless, 65% are arrested, and 53% suffer from mental illness (see data in Chapter 11 of this book).
- Seventy percent of the 650,000 people released annually from state and federal prisons will commit new crimes within three years. The vast majority of them receive no reentry help before they leave prison (see data in Chapter 12 of this book).
- The poverty rate for all women 18 years and older was 13% in 2016 compared with 10% for men according to the Henry J. Kaiser Family Foundation—accessed on October 22, 2018, from https://www.kff.org/other/state-indicator/adult-poverty-rate-by-gender/
- More than 55 million immigrants have settled in the United States since its founding, yet every wave of immigrants has encountered hostility. The roughly 12 million undocumented immigrants in the United States are no exception. American corporations, food growers and processors, contractors, restaurants, and hotels depend on their labor, yet Americans grant them few rights. Congress has repeatedly failed to enact legislation to clarify their rights, most recently in 2007. Many immigrants work for the minimum wage—and, in some cases, even less when employers' reimbursement of them is not monitored by the Department of Labor. Many immigrants do not receive some or all of their paychecks from fraudulent employers who believe they will not dare go to authorities for fear of being deported. Despite polls that show that most Americans favor giving a path to citizenship to roughly 800,000 immigrants brought to the United States as children, the United States had not yet enacted legislation to allow this policy.
- Americans have prided themselves on the ability of low-income persons to be upwardly mobile. Generations of Americans have contended that persons had only to work hard to enter the middle or even upper reaches of their society. Had not this been the script for tens of millions of Americans from the colonial period onward? Had not many of

the great corporate leaders—including most recently Steve Jobs—become highly successful entrepreneurs? This American dream has recently been tarnished by the findings of five large studies in recent years that find that the United States has lower rates of mobility than some European nations and Canada (DeParle, 2012). One research project discovered that 42% of males born in the bottom fifth of incomes stayed there as adults compared with only 25% in Denmark and 30% in Britain. Research by the Pew Charitable Trusts found that 62% of males and females raised in the top fifth of incomes stay in the top two-fifths—and 65% born in the bottom fifth stay in the bottom two-fifths. Another study found that 22% of Americans born in the bottom tenth of the income distribution stay there as adults compared with only 16% of Canadians (DeParle, 2012). On a more positive note, researchers discovered more mobility by persons who are raised in the middle fifth of the income distribution as compared with the highly "sticky" top and bottom rungs, where persons tend to stay where they were raised.

- Criminalizing poor people on the street? During and after the Great Recession of 2007 to 2009 and beyond, many jurisdictions enacted ordinances that criminalized poor persons according to the National Law Center on Homelessness and Poverty. These included bans on begging and crackdowns on indigent persons lying on a sidewalk; sitting, sleeping, lying on, or loitering on streets by poor people; and raiding shelters at night to find men with outstanding warrants. Some cities made it a crime to share food with persons in public places even as a federal judge declared it to be unconstitutional in Orlando, Florida (Ehrenreich, 2009).

- Using different police standards for persons of color? In an incident that reminded Congressman John Lewis, an African American leader in the civil rights movement, of racial violence of the 1950s and 1960s, Trayvon Martin, a high school student in Sanford, Florida, was shot to death on February 26, 2012, by George Zimmerman, a white person who served as a neighborhood watch captain. Zimmerman had called 911 to report a suspicious person—and had been told by police not to follow Martin but to leave the matter to local police, who would soon arrive. Zimmerman followed Martin. Zimmerman claimed Martin, who bloodied Zimmerman's nose, threw Zimmerman to the ground and beat him. Zimmerman says he opened fire on the unarmed Martin for self-defense, as allowed under a Florida statute called "Stand Your Ground." Local police did not question him at the scene, take evidence, take photographs, or even ask him, "What happened?" They did not take him into custody as a suspect in a homicide, so he remained free a month later. They took no steps to establish the identity of the slain teenager—not even calling numbers on his cell phone—so his parents did not learn he had died for three days. The case became a national one as the press came to learn these details. African Americans protested across the nation that Zimmerman was not arrested pending trial. They doubted the local police would investigate the case in a fair way because they had not investigated cases of homicides against other black youth in the area in recent years. Many persons wondered if the police would have taken different actions and launched more thorough investigations if the alleged murderer had been an African American and the victim a white person. The decision by a jury to acquit Zimmerman during the summer of 2013 appeared to justify the fears of many persons and brought renewed energy to the movement to repeal "Stand Your Ground" statutes in Florida and elsewhere.

- Since the death of Trayvon Martin, scores of African Americans have been slain by police

(Continued)

(Continued)

including the murder of unarmed Stephon Clark in his grandmother's backyard in Sacramento, California, on March 28, 2018. Data posted by the Washington Post lists more than 984 people fatally shot by police officers in 2015, 963 in 2016, and 987 in 2017 (database accessed on June 15, 2018, from the *Washington Post* at https://www.washingtonpost.com/graphics/national/police-shootings-2017/).

- Restricting black voting? Conservatives launched an ambitious effort to cut the voting rates of low-income persons in many jurisdictions in 2011 and 2012—persons likely to vote for Democratic candidates. They enacted state ordinances to require voters to show identification photographs at polls before voting. This seemingly innocuous policy would have cut voting rates in low-income areas because considerable numbers of low-income persons do not have documents with photographs. Many of them do not have passports or licenses because they do not own cars. Conservatives claimed the measure would cut voter fraud, but they lacked evidence that fraud was a significant problem in the United States. Enactment of laws to restrict voting continued into 2013 and 2014 in the wake of a ruling by the U.S. Supreme Court that they did not violate the federal Voting Rights Act. The U.S. Supreme Court ruled in June 2018 that states can remove voters from voter rolls who have failed to vote in a single federal election if they do not reply to a notice in the mail asking if they want to stay registered. Critics noted that at least 144,000 persons had been removed from voting rolls in recent years in Cleveland, Cincinnati, and Columbus, Ohio. The Supreme Court made this ruling even though the National Voter Registration Act says states may not remove nonvoting persons from voter rolls.

Extreme inequality violates ethical standards established by many religions, professions, and ethicists. It is even more ethically problematic in industrialized nations, such as the United States, that possess sufficient resources to create more egalitarian economic and social systems.

We should not imply that members of vulnerable populations are passive victims of circumstances. Members of these groups often contend with adversity in many ways, such as through churches, small businesses, neighborhood associations, and advocacy groups. Many persons better their conditions by perseverance and resilience. Members of vulnerable populations often join forces to achieve legislative reforms, such as when African Americans and Latinos have opposed efforts to curtail their voting rights.

Let's discuss the seven core problems identified in Chapter 1 from a social justice perspective.

CORE PROBLEM 1: ADVANCING ETHICAL RIGHTS, HUMAN RIGHTS, AND ECONOMIC JUSTICE

Ethicists mostly agree that service providers, as well as nations, should meet persons' basic survival needs, honor their self-determination, provide them with accurate and honest information, preserve the confidentiality of personal information, treat persons equitably, honor persons' human rights, and advance social justice (Holland, 2012).

Meeting Basic Survival Needs

The United States has made important strides in protecting basic survival needs during the past 60 years. It protected millions of Americans from starvation with cash relief and work relief in the Great Depression as well as formed federal welfare programs and unemployment insurance that supplemented welfare programs. It established the Food Stamps Program in 1964 and expanded it in subsequent decades, recently changing its name to SNAP. It established the Medicaid and Medicare programs in 1965 to expand health coverage for low-income persons and elderly persons—and added the Children's Health Insurance Program (CHIP) for low-income children in 1997. It developed federal housing programs in the 1930s and added other rent-subsidy programs in the 1970s. Many private organizations give cash, food, and shelter relief to Americans. It established a federal minimum wage in 1938 that has risen to $7.25 in 2014 even as many states mandated higher wages such as California with a minimum wage of $9 per hour. It planned to provide about 32 million additional Americans with health insurance had the ACA been fully implemented. The United States has developed federal emergency programs to help residents in the wake of natural disasters.

Much remains to be done to meet Americans' survival needs. Considerable numbers of persons possess income just above, at, or below federal poverty levels (FPLs) that are set at very low levels, such as $19,790 for families of three and $23,850 for families of four in 2014. Disparities in income between the bottom fifth of the population and the top fifth have widened since 1978—and the gap between the income of the bottom fifth and the top 2% has widened even more since that date. Many Americans are malnourished. Large numbers of Americans do not receive sufficient medical services. The mental health problems of many persons are not addressed. Many Americans do not use social programs that could markedly improve their well-being. Only about one-half of eligible persons use, for example, SNAP, Medicaid, or the EITC, as well as the CHIP program—depriving low-income persons of billions of dollars of income, food, and health resources.

Providing Persons With Opportunities

If persons do not receive education, job training, and employment, they can meet their survival needs only through assistance from government or private charities. Many people rely on welfare and assistance from the private and public sectors during economic downturns, in the wake of disabling health and physical problems, and during bouts of unemployment, but most people should have education and skills to support themselves and their dependents through gainful employment. Their employment should provide sufficient wages to meet their survival needs and to provide them with other amenities, such as savings, recreation, and decent housing. Their employment, too, should give them the ability to steadily improve their wages through time—and should meet other personal needs such as creativity and fulfillment.

Millions of Americans lack skills to find employment in the global economy. Persons who possess only a high school education or less are seriously compromised in finding work or in finding employment that meets their basic needs. The wages of many workers have been relatively stagnant during the past three decades. The quality of education varies widely in the United States. Inner-city schools often have more crowded classrooms and poorer teachers than suburban schools. Many students do not achieve national norms in math and reading, particularly low-income, African American, Latino, and Native American students—as well as many white people in rural areas. More than 50% of students drop out of many inner-city high schools and roughly 25% nationally. About 50% of students who enter community colleges drop out before the second year. Less than 60% of students who enter colleges have graduated in six years. Persons of color disproportionately do not enter educational programs after high school and fail to graduate from them. The learning of many students is compromised by their lack of nutrition, homelessness, and mental conditions.

Many laws protect vulnerable populations from discrimination. The Americans with Disabilities Act (ADA) prohibits discrimination by employers against persons with disabilities. The Civil Rights Acts of 1964 and 1965 prohibit discrimination by schools and employers against persons of color and women. Some states prohibit discrimination against members of the lesbian, gay, bisexual, transgender, and queer (LGBTQ) population in places of employment. Scores of additional laws protect the rights of women, disabled persons, children, and seniors.

Honoring Self-Determination

Many court rulings, as well as writings of ethical philosophers and religious leaders, support the right of individuals to decide whether, when, and how to seek services or use social programs in the United States. The Supreme Court has ruled, for example, that persons who are cognitively competent can decline medical treatments even when their physicians believe that this decision could cause them physical harm or even death. Specific laws, court rulings, and accreditation standards require agencies and programs to honor self-determination through *informed* consent that includes informing clients about the following:

- Likely outcomes and side effects of specific treatments
- Likely monetary costs of specific treatments
- Likely side effects or threats to their well-being from specific treatments or courses of action, including death and injury
- Possible alternative treatments or services
- Evidence-based findings relevant to their treatment

Providers who do not give individuals full and accurate information about treatment options so that they can make informed decisions are liable to penalties that can include loss of their licenses. Providers do not have to obtain informed consent if a court has ruled that specific clients are mentally incompetent and where the court itself, or guardians appointed by it, acts in the best interests of persons with specific problems like some neurological disorders, limited intelligence, and head trauma injuries.

States usually allow parents or guardians of minors to make decisions for them as long as they advance their well-being. Under the doctrine of *parens patriae*, however, public authorities can investigate whether parents or guardians neglect or abuse children and can take custody of them or order specific ameliorating actions or treatment when they discover that they have harmed them or the children might suffer harm.

Most states allow public authorities to commit persons involuntarily to hospitals, correctional facilities, or other settings when they are suicidal or have already attempted to harm themselves. They also allow public authorities to restrain or incarcerate persons who have threatened or harmed other persons. These actions of public authorities are circumscribed, however, by mandated legal representation for persons who are involuntarily admitted to institutions, required reviews of their cases by courts, and other procedural safeguards.

Many obstacles exist to self-determination. Some providers fail to involve their clients sufficiently in the helping process. Providers may be less likely to provide informed consent to persons from vulnerable populations, such as women and persons of color. Persons with mental conditions, such as schizophrenia, may be viewed as unable to make specific choices even when they are capable of making them. Relatively submissive persons may not assert their rights.

Honoring Consumers' Right to Accurate and Honest Information

Persons have an ethical right to know the nature of their mental health and other health problems in clear and understandable terms. They need to know their prognosis. They need accurate and honest information throughout their treatments or services, such as how their condition has changed, outcomes of their current treatments, and whether different treatments should be selected as they progress forward. Persons also need accurate and honest information from personnel in financial institutions, car dealers, landlords, and employers.

Many obstacles prevent the provision of accurate and honest information. Providers sometimes want to steer persons to treatments and facilities that will financially benefit them by giving them higher reimbursements. They may give them false statistics about the success of their services or treatments in the hope that they will be more likely to use them than otherwise. They may overstate the seriousness of specific conditions or prognoses to frighten persons into using services or treatments. They may deceive persons with low levels of education, such as when car dealers, landlords, and employers victimize them. Providers sometimes do not divulge conflicts of interest. Many policies, court rulings, regulations, and laws protect a person's right to accurate and honest information including prohibiting false advertising, requiring disclosures of conflicts of interest, and requiring disclosure of treatment options. Professionals may sometimes give inaccurate information because they are misinformed. Perhaps they are not aware of new research that identifies specific evidence-based treatments.

Protecting the Ethical Right to Confidentiality

Consumers of service are entitled to confidentiality—and would likely not seek assistance for many of their problems if they did not trust their providers to maintain it. The federal Health Insurance Portability and Accountability Act (HIPAA) of 1996 protects the confidentiality of clients' information in health systems. It requires health organizations and health personnel to adhere to numerous procedures to protect patients' confidentiality. Its standards extend to other service delivery systems, such as mental health. Other protections for confidentiality include state laws, accreditation standards, court rulings, and professional licensing standards and National Association of Social Workers (NASW) Code of Ethics. Other standards protect the confidentiality of students.

Exceptions to confidentiality exist. If someone threatens to injure or kill another person or persons, for example, health and mental health providers are required to disclose this information to the specific external authorities, such as the police. Some health information must be disclosed to public health authorities, such as HIV/AIDS infections and tuberculosis. Social workers must sometimes verify to courts that specific persons actually receive specific health and mental health services that courts mandated them to use.

Many factors jeopardize the confidentiality of clients' information. Some providers are insufficiently aware of the hazards that specific consumers may experience if their private information is divulged to specific persons or organizations—such as possible loss of employment and disruption of relationships and marriages. Providers are sometimes careless in the ways they transmit information about specific clients to other persons, such as by telephone in ways that third parties can hear. Hackers can sometimes gain illicit access to clients' information that is stored electronically.

Providing Equitable Treatment

All persons have a right to receive the quality, kind, and duration of services, benefits, or treatments from providers as other persons with the same problems or conditions regardless of their race, gender, sexual orientation, place of national origin, age, social class, intelligence, religion, disability, lack of literacy, inability to speak English, genetic characteristics, mental illness, poverty, or other personal characteristics. Many state and federal civil rights policies prohibit inequitable treatment including the following:

- The federal Civil Rights Acts of 1964 and 1965 that prohibit discrimination against persons of color, women, and persons born in other nations
- The ADA of 1990 that bans discrimination against persons with physical and mental disabilities
- State laws banning discrimination in housing and employment against persons on the basis of sexual orientation
- State and federal laws banning discrimination in housing for many vulnerable populations
- State and federal laws banning discrimination against persons with HIV/AIDS
- Policies that require affirmative action and quotas for women, persons of color, and veterans
- State and federal laws that protect members of vulnerable populations from discrimination at their places of work

Professionals sometimes discriminate on the basis of prejudice against specific kinds of persons. Persons with schizophrenia often receive inferior health care, even when they possess heart disease. African Americans and women are less likely than European Americans and men to obtain access to some advanced health technology. Disabled persons are often treated paternalistically with emphasis on their physical limitations rather than on their strengths. Persons with criminal offenses are often subject to discrimination in places of work. Homeless youth often find it difficult to enroll in secondary education. Persons of color are more likely to be sentenced to death than whites—and incarcerated at higher rates and for longer periods for specific offenses. African Americans who use illicit drugs are more likely than European Americans to receive prison sentences. Persons of color are subject to more police brutality than white persons. Students of color are more likely to be expelled from secondary schools than white students. Employers often use race as a shortcut measure to assess applicants' capabilities for employment rather than objectively viewing their qualifications. Educators often convey their lower expectations to persons of color and low-income whites.

Honoring Human Rights

Human rights are broadly defined as establishing basic rights and freedoms including the right to free speech, the ability to travel without restriction, protections of persons from human trafficking, freedom from violence, and the right to free association. They include the right to humane treatment by employers and fellow employees in places of work. They prohibit genocide. They include protections of children and others from bullying in schools and places of employment. They include protections against sexual harassment at places of work, verbal or physical abuse within families, and sex trafficking.

It is difficult to curtail violations of human rights for many reasons. Some behaviors are difficult to change, such as bullying of children in schools, sexual harassment of women in places of work or gang violence. Many persons profit monetarily from violating the rights of others, such as pimps and international traffickers of girls and women for prostitution. Violations of human rights often occur in developing nations such as genocide in Rwanda, the Sudan, and Syria. They often occur in nations with dictatorships where freedom of speech, freedom of movement, and freedom of assembly are often curbed or prohibited.

The fundamental rights of human beings are embodied in international law; constitutions, statutes, and regulations; and court rulings (Barria & Roper, 2010; Tomuschat, 2008):

- Civil rights laws in local, state, and federal jurisdictions
- Article 1 of the United Nations Universal Declaration of Rights that defines "basic rights and freedoms" in response to atrocities during World War II
- The four treaties and three protocols of the Geneva Convention (1949), which defined rights of combatants and civilians during wars
- The UN Convention on the Elimination of All Forms of Discrimination Against Women (1981)
- The UN Convention Against Torture (1984)
- The International Convention on Protection of the Rights of All Migrant Workers and Members of Their Families (1990)

- The Convention on the Elimination of All Forms of Discrimination Against Women (1975)
- The UN Protocol to Prevent, Suppress, and Punish Trafficking in Persons (2000)
- The UN Convention on the Rights of the Child
- The Victims of Trafficking and Violence Protection Act of the United States (2000) (Gelb & Palley, 2009)

President Donald Trump chose to remove the United States from membership in the United Nations' Human Rights Council in early July 2018. He argued that several of the 47 members of the council had unfairly issued resolutions against Israel for occupying portions of Palestinian territory and its construction of housing in Palestinian territory. Trump's critics contend that Israel wrongly killed Palestinians protesting Israel's actions. They also argued that the UN's Human Rights Council has representatives who monitor actions of 12 nations charged with violations of human rights. They note, as well, that the United States has a long history of involvement in the Human Rights Council (Quigley, 2018).

Provisions of the American Constitution include rights such as "due process," "equal protection under the law," freedom of speech, free press, and free association in the Bill of Rights. Many state and federal laws protect human rights, such as anti-trafficking laws, laws prohibiting discrimination against persons in their places of work, and laws protecting children and students from bullying in schools.

Promiting Social Justice

Social injustice occurs when specific populations, such as women, members of racial minorities, and low-income persons are subject to violation of civil rights and human rights, violations of life conditions, and violations of access to opportunities. We have already discussed disparities in civil and human rights between persons of color, women, LGBTQ persons, disabled persons, and the mainstream population.

Social injustice also occurs when specific populations suffer poorer life conditions than other persons, such as when low-income persons suffer from greater physical and mental illness, shorter life expectancy, more disabling conditions, higher rates of unemployment, and lower levels of school achievement than relatively affluent persons. Disabled persons have considerably less income, higher rates of unemployment, higher rates of mental illness, and shorter life expectancies than persons who are not disabled. Persons with schizophrenia have an average life expectancy of 63 years as compared with 77.9 years for other Americans. Persons of color are more likely to reside in crime-infested areas than white persons, making life hazardous on a daily basis. Female single heads of households are among the poorest members of our society—to the point that an academic coined the term the *feminization of poverty*. They often work multiple jobs in a desperate effort to meet their children's basic needs, wreaking a toll on their mental and physical health, not to mention their ability to obtain further schooling that might allow them to obtain a higher-wage job.

Social injustice imposes unnecessary costs on society. When specific vulnerable populations are denied equal access to opportunities, they are more likely than other persons to need financial and medical assistance from the broader society. They are likely to contribute lower taxes to the revenues of local, state, and federal governments.

Extreme inequality breeds social problems because it marginalizes large numbers of persons who are acutely aware that they are poorer and sicker than mainstream populations. It breeds desperation in the case of persons who cannot meet the survival needs of themselves and their children. It decreases participation in elections among low-income persons—making it harder for them to elect public officials who would work to secure enhancements of their rights, life conditions, and opportunities.

Extreme inequality violates ethical standards established by many religions, professions, and ethicists. It is even more ethically problematic in industrialized nations, such as the United States, that possess sufficient resources to create more egalitarian economic and social systems.

CORE PROBLEM 2: IMPROVING THE QUALITY OF SOCIAL PROGRAMS

Consumers have an ethical right to services and programs that ameliorate or solve social problems that they experience, but they often do not receive effective services. Perhaps providers give them interventions that have not been empirically evaluated. Perhaps service providers are not sufficiently skilled in providing specific services. Perhaps persons who help them are not qualified to provide specific services. Perhaps providers give them services that yield substantial revenues but are not effective.

Consumers of service, as well as taxpayers, have a right, as well, to cost-effective services. Some medications may prolong life, for example, but only for several months at a cost exceeding $100,000. Some interventions are far more expensive than ones that are as or more effective than other interventions, as was recently discovered when researchers found that acupuncture was more effective and less costly than surgery for some kinds of lower-back pain.

We can sometimes determine whether specific programs, treatments, interventions, or policies are effective or cost-effective by using research methodology. In so-called gold standard research, for example, researchers use random controlled trials (RCTs) in which they compare outcomes of groups of patients who receive specific assistance with outcomes of groups who receive different kinds of assistance, or no assistance. They might, for example, give one group a specific intervention for depression (the experimental group) and compare its outcomes with a group not receiving this intervention (the control group). Researchers sometimes cannot use RCTs but compare two or more groups that do and do not receive a specific intervention in quasi-experimental research. Researchers sometimes use natural experiments to gauge the effectiveness of specific policies such as determining if motorcyclists' death rate from accidents declines after the enactment of a state law that requires motorcyclists to wear helmets.

Surveys of consumer satisfaction with services provide another measure of their quality. Medicare officials routinely ask patients to evaluate the Medicare services they receive from physicians and hospitals, for example, and place the results on the Internet so that patients can use this information to decide where to seek services.

The quality of services is often impeded by their fragmentation when consumers receive care from specific providers that is not coordinated. Assume, for example, that a child receives help with convulsions that sometimes accompany autism but not for his or her emotional outbursts. Or an elderly person may receive medication to control depression that adversely interacts with other medications from other physicians.

Case management and navigation models of service have evolved to circumvent fragmentation. If case managers try to orchestrate a package of services for specific consumers, patient navigators help persons keep appointments, adhere to medications, and follow treatments. Only a small fraction of consumers receive these services, however, due to lack of staff and funding to provide them even for persons with serious problems. Some state and federal laws require providers to develop comprehensive plans for clients, such as when they mainstream disabled or developmentally challenged children into the educational system. Comprehensive plans are ineffective, however, if schools and other agencies lack sufficient staff to implement them. Inter-agency collaborations, such as ones developed between schools and mental health agencies, also curtail fragmentation.

Social injustice occurs when relatively affluent people receive services of higher quality than less affluent people—or if persons of specific ethnic or racial backgrounds receive services of higher quality than other people. Inner-city schools that primarily serve persons of color often receive fewer resources, for example, than suburban ones primarily attended by white students.

CORE PROBLEM 3: MAKING SOCIAL PROGRAMS AND POLICIES MORE CULTURALLY RESPONSIVE

The United States has been an immigrant nation since its inception with a mix of forced immigration of slaves, subjugation of native peoples, and voluntary immigration from Europe, Russia, Mexico and Central America, Asia, the Middle East, Africa, and elsewhere over several centuries. The United States had 199 million non-Hispanic whites, 45 million Latinos, 37 million African Americans, 13.1 million Asians, 2 million American Indian and Alaska Natives, and 402,000 Native Hawaiian and other Pacific Islander people in 2007. Many persons have limited English proficiency (LEP).

It is challenging to define culturally responsive care partly because it is difficult to define culture. Brislin (2000) defines culture as the "shared values and concepts among people who most often speak the same language in proximity to each other (and) . . . are transmitted for generations, and they provide guidance for everyday behaviors" (p. 4). It contains a language dimension. It requires that providers must respect cultural views about health and other social problems held by specific persons, such as how they view certain health conditions, how they wish to receive health care, terminology that they prefer, how they wish to communicate with health care providers, and what treatments they wish to have. Culturally competent care is needed in every policy sector.

Culturally competent services must be provided not only to LEPs but to persons from many groups that differ from mainstream ones, including illiterate or semi-literate persons, LGBTQ persons, elderly persons, disabled persons, persons from different religions, and women. Many persons want complementary and alternative medicine (CAM) in health care, such as use of herbs, acupuncture, meditation and yoga, and other nontraditional methods of preventing or treating health problems. They sometimes want CAM instead of traditional medicine or often use CAM and traditional medicine in tandem.

Persons who do not receive culturally competent services may suffer adverse consequences. They may not seek or return for needed services. They may not adhere to treatments that can improve their conditions. They may sue providers, such as when they are not given translation services as required by federal and state laws. They may believe that they have not given informed consent when they are unable to communicate with providers.

Culturally competent care is enhanced when providers learn to be self-aware, respect other cultures, have cultural awareness, possess cultural knowledge, and develop cultural skills with which they "negotiate or facilitate relationships between consumers and providers" (Kao & Jansson, 2011, p. 184). Providers need skills that are described by Galanti (2008) as the four Cs of culture, including the following:

- Call: What do you call the problem? What do you think is wrong?
- Cause: What do you think caused your problem?
- Cope: How do you cope with your condition? What have you done to make it better? Who else have you been to for treatment?
- Concerns: What concerns do you have regarding the condition? How serious do you think this is? What potential complications do you fear? How does it interfere with your life or your ability to function? What are your concerns regarding the recommended treatment?

Title VI of the federal Civil Rights Act of 1964 declares that no one in the United States can be excluded from participating in, or denied benefits of, any program or activity receiving federal financial assistance on the grounds of race, color, or national origin, which has been interpreted by courts to include an individual's primary language (Perkins, Youdelman, & Wong, 2003; Perkins & Youdelman, 2008). Subsequent presidential executive orders have required federal agencies and federally funded programs to provide "meaningful access" to LEP persons (Kao & Jansson, 2011). Many states have enacted statutes and regulations that also require provision of translation services to LEP persons. Accreditation standards of hospitals and clinics require use of translation services—and the ACA mandates greater collection of data on race, ethnicity, gender, primary language, disability status, and underserved rural populations in health settings in 2012.

Service organizations should use census materials, as well as analyses of their clients, to determine their ethnicity and other demographic characteristics. They should hire persons from these backgrounds. They should provide in-service training about cultural competence.

CORE PROBLEM 4: DEVELOPING PREVENTIVE STRATEGIES TO DECREASE SOCIAL PROBLEMS

Theorists have distinguished among primary, secondary, and tertiary prevention. Primary prevention seeks to prevent the emergence of specific social problems such as cancer, diabetes, truancy, and mental illness. Secondary prevention aims to identify and treat specific problems early in their development, such as by slowing their progress or curing them, for example, helping persons with early-stage diabetes slow its progress. Tertiary prevention has the same goals as secondary prevention but for more advanced problems, such as helping someone slow the progress of relatively advanced medical or mental health problems.

Primary and secondary prevention often receive insufficient priority in health and human services. Providers are often diverted to persons with advanced and serious problems due to their sheer number and the cost and time required to help them. People often find it difficult to modify lifestyle preferences that cause them to develop social problems, such as poor diets, lack of exercise, substance abuse, and smoking. Health insurance companies often do not fund preventive services at all—or at lower rates than surgery and medications. Tobacco, industrial, and food interests have slowed or blocked bans on smoking, pollution reduction, and food labeling policies. Elected public officials often focus on short-term policies rather than long-term preventive ones. They often slash funding for prevention during budget crises and recessions.

Researchers have made considerable progress in identifying promising preventive strategies. They identify at-risk indicators that allow them to predict with considerable accuracy who will and will not develop specific problems in future years, that is, true positives. (At-risk indicators may include poor life habits like smoking, level of income, diet, obesity, lifestyle decisions such as level of exercise, abusive treatment by significant others, poor education, genetic factors, and many other variables.) Their predictions are hindered, however, when they cannot identify persons who are predicted to develop a specific problem but do not (false positives) and persons predicted not to develop a problem but who do it (false negatives).

We can distinguish between passive and active prevention. Passive prevention seeks to change the human environment so that persons are less likely to develop diseases like lung cancer without requiring persons to modify their behaviors. When pollution is reduced in the air and water, for example, persons achieve health benefits without actions on their part. Active prevention succeeds, by contrast, only when persons take specific actions, such as changing their diets or engaging in more exercise. They often have to work with other persons to prevent a problem, such as by engaging in cognitive therapy as they receive medications that help them surmount drug and alcohol addictions.

Policy advocates work on many levels to advance prevention. They convince organizations to fund and implement prevention programs. They convince legislators and government officials to develop regulations and enact statutes that promote prevention. They work with the mass media to publicize effective preventive strategies. They work with researchers to obtain evidence that specific preventive strategies are effective and cost-effective. They convince administrators and public officials to prioritize prevention—and to fund it adequately.

Social injustice occurs when members of specific vulnerable populations receive fewer preventive services than other people. Take the case of primary health care that includes regular visits of children to physicians; low-income children and children of color often have less access to primary care than more affluent, white children. Affluent children in private schools often receive assistance in a timely way for reading disorders in their schools that low-income students do not.

CORE PROBLEM 5: IMPROVING AFFORDABILITY AND ACCESS TO SOCIAL PROGRAMS

Persons often encounter specific or multiple barriers to accessing services. They often find they cannot afford them, must endure excessive waits, encounter complex and time-consuming eligibility processes, find services to be geographically distant, or possess health or mental health problems that make travel difficult. Accessibility may be hindered, as well, by a lack of advertising or publicity for specific services so that many persons do not know they exist.

Americans pay greater out-of-pocket costs for health and human services than Europeans and Canadians, where governments foot a greater share of their costs. Many other services and programs require substantial payments from consumers, including most childcare and preschool programs, many medical and mental health services, and most postsecondary education programs. The inequities are often glaring: Affluent Americans often gain greater access to services and opportunities because they can afford out-of-pocket costs unlike many low- and moderate-income persons, who often refrain from using them or discontinue them prematurely. Out-of-pocket payments take many forms including deductibles, payment for services excluded from coverage by insurance companies, fees for excluded or non-covered services, and sliding fees that adjust fees upward as personal income increases.

These fees can cause hardships for persons with low and moderate incomes, not only impeding access to services but also decreasing their ability to pay rent, purchase food, or purchase medical and other kinds of care.

Many services require excessive waits, such as emergency rooms in many hospitals, public clinics in mental health and health areas, substance abuse treatment programs, and subsidized housing. Persons who decide to seek help for substance abuse problems often encounter waits of 6 months or more. Waits often lead consumers to exit services even when they need them, such as when they prematurely leave emergency rooms.

Health and human services are not distributed equitably across the American landscape but are disproportionately located in relatively affluent areas because organizations and professionals often seek locations with persons more likely to pay their fees. Persons who cannot afford cars, cannot afford gasoline, lack public transportation, or cannot easily seek services during working hours find it particularly difficult to access services and programs not located in their communities.

Policy advocates can decrease the cost and increase the accessibility of services and programs. They can propose changes in the fee structures of public programs, such as upward changes in their eligibility levels, inclusion of excluded services, and reduction of deductibles and co-payments. They can pressure public agencies to make services more accessible in underserved areas. They can persuade agencies to establish outreach programs or storefront programs for persons in underserved areas. They can outstation services in other agencies, such as in libraries, hospitals, or schools. They can establish outreach programs to homebound persons. They can downsize their central or largest programs as they move staff and resources to smaller programs. They can publicize their programs by targeting messages to media, churches, libraries, and social media that are used by specific segments of the population. They can oppose policies that defund or repeal social programs that finance medical care for low- and moderate-income persons, such as the effort by President Trump to repeal and replace the ACA in 2017 and 2018 and to seek deep cuts in Medicare and Medicaid (Jansson, 2019).

CORE PROBLEM 6: INCREASING THE SCOPE AND EFFECTIVENESS OF MENTAL HEALTH PROGRAMS

Many Americans have serious, undetected health and substance abuse problems. These problems may be "hidden" because they are not the presenting problems when persons use medical and other agencies. Many persons do not volunteer that they have these problems due to stigma often attached to them. Mental health or substance abuse problems may be caused by other social problems, such as experiencing foreclosure, performing poorly in school, or losing work.

Mental health services are often inadequately provided in settings beyond clinics and institutions that focus upon them. Roughly 44% of males and 61% of females in federal prisons possess serious mental problems but often receive no, little, or substandard care for them. Primary care physicians provide the largest quantity of mental health services in community settings, but often have little mental health training and are not usually supervised by mental health specialists. Publicly subsidized mental health clinics exist but often have insufficient resources despite funding from local, state, and federal governments, reimbursements from private insurance, funding from Medicaid and Medicare, and clients' fees.

Unlike physicians, who are the dominant providers in the health care system, many professions provide mental health services, including primary care physicians, psychiatrists, psychologists, social workers, marriage and family counselors, and psychiatric nurses. Social workers provide more mental health services than do members of any other mental health profession.

Mental health staff are often not present, or only peripherally present, in many settings where persons present their mental health problems. Employee assistance programs (EAPs) provide mental health services in some corporate settings but reach only a small fraction of American workers. School systems employ relatively few social workers and psychologists despite the sheer number of students with depression, anxiety, autism, attention deficit disorder, and behavioral problems. Many hospitals and clinics employ relatively few social workers and psychologists. Mental health services are chronically underfunded in the United States for both outpatient care and institutional care that is still needed for suicidal persons and persons who may present a threat to others.

Access to mental health services greatly increased after the enactment of the Mental Health Parity and Addiction Equity Act of 2008, which requires employers offering group health insurance plans to their employees to include coverage for mental health services. The ACA also requires private insurance companies, as well as Medicaid and Medicare, not to discriminate against mental health services as compared with services for physical problems. Many state and federal laws protect the rights of persons with mental and substance abuse problems, such as protecting the confidentiality of patients' records, requiring legal counsel to represent persons who are subject to involuntary commitments, and requiring conservatorship for persons with serious cognitive deficits.

CORE PROBLEM 7: MAKING SOCIAL PROGRAMS MORE RELEVANT TO HOUSEHOLDS AND COMMUNITIES

Assume a person receives care for a mental or physical problem in a clinic or hospital, but her providers are unaware of her home and community environment. Unable to obtain medications and food in her neighborhood because she lacks transportation and is barely ambulatory, she dies from inadequate nutrition—a problem that might have been avoided had professionals enrolled her in services offered by visiting nurses, as well as Meals on Wheels.

Several strategies can improve linkage of services to consumer households and communities. Agencies can establish contracts or agreements with community-based agencies that provide specific services, such as from visiting nurses, case managers, transportation services, and many other agencies. Agencies can establish formal collaborations with one another to provide needed services to clients. Innovative electronic systems can be used to monitor and educate home-based persons with disabilities, chronic diseases, and other conditions.

Under the Hospital Readmissions Reduction Program, hospitals receive readmission penalties from Medicare and Medicaid if their patients return to hospitals within 30 days after receiving surgeries. This policy provides them with incentives to give patients home-based services. The Affordable Care Act (ACA) gave states license to expand their Medicaid eligibility to 138% of the federal poverty level to greatly decrease the number of low-income patients in the United States that lacked health coverage. Although most states had chosen to expand coverage, many others did not.

The Americans with Disabilities Act requires public housing and other housing agencies to provide accommodations to persons with disabilities.

RECOGNIZING THE SEVEN CORE PROBLEMS IN THE EIGHT POLICY SECTORS

We discuss how these seven core problems manifest themselves in each of the eight policy sectors in Chapters 7 through 14. In health care, for example, specific patients' ethical rights are sometimes violated when they do not give their informed consent to specific medical procedures at the level of an individual or family (Core Problem 1). Children in the child welfare sector with specific mental conditions often do not receive evidence-based care at a child or family level, such as when professional staff members do not diagnose clinical depression or anxiety (Core Problem 2). Mental health staff may fail to give specific clients culturally competent care, such as when they do not adapt their services to the cultural needs of Latinos, who often want some family members to be present during counseling sessions (Core Problem 3). Some elderly persons do not receive preventive care for early-stage chronic diseases such as congestive heart failure in the gerontology sector from health providers (Core Problem 4). Many persons cannot afford health care in the United States due to lack of insurance coverage that was partially addressed when the ACA of 2010 was enacted (Core Problem 5). The mental distress of many prisoners is inadequately addressed in many prisons and correctional facilities (Core Problem 6). Persons who receive services from agencies in the gerontology sector often receive insular care not connected to their households and communities (Core Problem 7).

JOINING THE REFORM TRADITION OF SOCIAL WORK

Social work has a social reform tradition extending back to the formation of the social work profession. Founders of social work such as Jane Addams militantly supported an array of social reforms in the Progressive era at the beginning of the 20th century, including housing codes to protect tenants, governmental inspection of food to avert illness, factory regulations to protect workers, and pensions for single mothers with children to avert dire poverty (Jansson, 2019; Wenocur & Reisch, 1989).

In succeeding eras, many social workers joined this reform tradition by working for policy reforms in local, state, and federal jurisdictions. Their work was bolstered by numerous theorists who developed a systems or environmental perspective on human behavior, arguing that social inequality, blighted neighborhoods, inadequate resources, unemployment, environmental pollution, discrimination, and economic uncertainty cause human suffering and contribute to clinical conditions such as depression and poor health (Meyer, 1970; Germain & Gitterman, 1980).

Honest differences of opinion often exist among social workers. They may disagree about the merits of specific policies. They may support different political candidates. They may draw upon conflicting research findings to support their preferred policies. Yet we are linked by a shared commitment to social justice even as we may differ about how best to advance it.

You will learn about scores of social policies in this book in eight policy sectors. You need to know about them because you will often refer clients or patients to them and would be derelict if you were not familiar with them. This book provides an empowerment and advocacy approach to social policy that facilitates your personal involvement on multiple levels. It encourages you to do the following:

- Examine your values and personal perspectives throughout the book because they shape how you—and everyone else—relate to controversies in American society about social policy. Understand NASW's Code of Ethics as a foundational statement about ethics developed by and for social workers.
- Understand and work to reduce the marginalization of many vulnerable populations in the United States and abroad by engaging in policy advocacy.
- View social policy from an empowerment perspective so that you participate in it at multiple levels, including helping specific clients and patients to obtain the rights, benefits, services, and opportunities to which they are entitled (micro policy advocacy), improving agency policies and addressing community

problems (mezzo policy advocacy), and reforming social policies in organizations, communities, and government settings (macro policy advocacy).

- Identify and support evidence-based policies.
- Seek out information on the Internet while remembering that accuracy varies from site to site.
- Learn about advocacy groups in each chapter that work toward creating more just and equitable policies. Visit their websites.

Learning Outcomes

You are now equipped to:

- Define social policies and their origins
- Define social justice and why we seek to advance it
- Advance social justice with respect to seven core issues
- Recognize the seven core problems in eight policy sectors
- Join the reform tradition of social work

References

Barria, L. & Roper, S. (2010). *The development of institutions of human rights: A comparative study.* New York, NY: Macmillan.

Brislin, R. (2000). *Understanding culture's influence on behavior* (2nd ed.). Belmont, CA: Wadsworth.

DeParle, J. (2012, January 4). Harder for Americans to rise from lower rungs. *New York Times.* Retrieved from https://www.nytimes.com/2012/01/05/us/harder-for-americans-to-rise-from-lower-rungs.html

Ehrenreich, B. (2009, August 8). Is it now a crime to be poor? *New York Times.* Retrieved from https://www.nytimes.com/2009/08/09/opinion/09ehrenreich.html

Galanti, G. A. (2008). *Caring for patients from different cultures* (4th ed.). Philadelphia, PA: University of Pennsylvania Press.

Gelb, J. & Palley, M. L. (2009). *Women and politics around the world* (Vol. 1). Santa Barbara, CA: ABC-CLIO.

Germain, C. & Gitterman, A. (1980). *The life model of social work practice.* New York: Columbia University Press.

Holland, S. (2012). *Arguing about bioethics.* New York, NY: Routledge.

Jansson, B. (2019). *The reluctant welfare state.* Cengage, 2019.

Kao, D. & Jansson, B. (2011). Advocacy to promote culturally competent health services. In Jansson, B. (Ed.), *Improving healthcare through advocacy: A guide for the health and helping professions* (pp. 179–210). Hoboken, NJ: John Wiley & Sons Ltd.

Meyer, C. (1970). *Social work practice: A response to the urban crisis.* New York: Free Press.

Perkins, J. & Youdelman, M. (2008). *Summary of state law requirements: Addressing language needs in healthcare*. Los Angeles: National Health Law Program.

Perkins, J., Youdelman, M., & Wong, D. (2003). *Ensuring linguistic access in healthcare settings: Legal rights and responsibilities.* Los Angeles: National Health Law Program.

Quigley, J. (2018, July 5). US withdrawal from the UN's Human Rights Council sends the wrong message to world. *Newsday*. Retrieved from https://www.newsday.com/opinion/commentary/un-human-rights-council-1.19620465

Tomuschat, C. (2008). *Human rights: Between idealism and realism*. New York, NY: Oxford University.

Wenocur, S. & Reisch, M. (1989). *From charity to enterprise: The development of American social work in a market economy*. Urbana: University of Illinois Press.

3

HOW POLICY ADVOCATES ADVANCED SOCIAL JUSTICE THROUGH AMERICAN HISTORY

LEARNING OBJECTIVES

In this chapter, you will learn to:

1. Understand how the American welfare state has evolved in stages from the colonial period to the present
2. Describe key policies of the following periods:
   - The colonial period
   - The Gilded Age
   - The Progressive Era
   - The Great Depression
   - The Great Society and the early 1970s
   - Conservative counterrevolutions
3. Identify how social workers engaged in policy advocacy at micro, mezzo, and macro levels in these periods
4. Describe how and why the United States lags behind many other industrialized nations with respect to many social problems including rates of incarceration, infant mortality, poor school performance, low social mobility, and rates of poverty
5. Identify how the United States can increase its revenues to fund improvements in its welfare state
6. Prepare for a new reform period

American social policy evolved in stages over roughly 270 years if we begin counting in 1750. Policy advocates have transformed the nation's social policies from relatively primitive ones at the local level in the colonial period to a substantial welfare state in 2019 that includes thousands of social policies of local, state, and national governments. Policy advocates engaged in policy advocacy when they realized that ordinary people were harmed by new problems as the nation moved from an agrarian to an industrialized nation. They initiated major periods of reform. They opposed persons and movements that sought to rescind social reforms with proven effectiveness. They fought for resources to fund social policies and against efforts to defund them. They advocated for marginalized populations when they were attacked by persons and movements. They developed movements that sought social justice such as for women's rights, civil rights, gay rights, rights of disabled persons, seniors' rights, and rights of workers.

THE COLONIAL PERIOD

The United States constructed its welfare state from the bottom up from the colonial period to the present. The initial set of 13 colonies had a primitive set of social institutions at the local level. These included almshouses (also called poorhouses) where destitute persons were placed only after they passed a test, called a means test, that revealed they had no money and were unable to find work. They engaged in manual work, such as cutting wood, ate spartan meals, and sometimes engaged in prayer. Some of them received "outdoor relief" in their homes. Public schools did not exist, so some youth attended private schools or were tutored by their parents. Some prisons existed. These colonies were overwhelmingly agricultural with only a smattering of small towns. Even Philadelphia, which hosted the Constitutional Convention, had only about 2,000 residents in 1789. The colonies relied on property taxes mostly for public roads and bridges and local administration.

When the colonists ratified the Constitution in 1787, they failed even to mention "social policy" in it because they believed social policy fell exclusively in the province of local and state governments (Jansson, 2019b). Schools, poorhouses, prisons, and other charitable institutions were ceded to local and state governments, whereas the federal government was assigned foreign policy, developing a uniform national currency, regulating trade, developing a military force to fend off foreign powers, and overseeing federal elections. It was a small federal government that lacked even a federal income tax, forcing it to rely only on excise taxes on certain products, tariffs placed on foreign imports, and several smaller taxes. State legislatures rarely met. The federal government did not participate in social policy in a major way until the Great Depression of the 1930s.

Local, state, and federal governments remained small by modern standards from 1789 to the Great Depression of the 1930s even as the nation developed mandatory public education through high school, public universities, and a sizable number of private colleges. States developed "mental asylums" but with no federal assistance after President Monroe convinced Congress in 1848 that the Constitution did not give the federal government the power to fund social programs. States and private charities developed charitable institutions for orphans and blind and deaf people and institutions for delinquent youth. Many local public and private hospitals were constructed by the end of the 19th century

(Jansson, 2019b). State-funded prisons supplemented local prisons. Many local, state, and federal regulations were established in the late 19th century and the early 20th century including ones that regulated the safety of food and water, housing regulations, and work-safety regulations.

Major setbacks took place in the colonial period, such as the legalization of slavery in the U.S. Constitution; failure to honor the rights of women to vote, own property, or join professions; and the forcible pushing of Native Americans westward. Yet the colonists enacted a Bill of Rights in 1791 that protected free speech, established local and state legislatures, and developed infrastructure. Possibly their most significant achievement was to auction land to settlers so that many persons in this agrarian society owned relatively small plots of land aside from plantation owners in the South. If we exclude slaves, Native Americans, and women, colonial society had high levels of economic equality because many white males possessed relatively small landholdings, such as 60 acres.

THE GILDED AGE

A mismatch between public policy and social problems became glaringly obvious, however, as the nation rapidly industrialized from the start of the Civil War in 1860 through the 1920s. Tens of millions of penniless immigrants from Europe and Russia flooded growing American cities and provided labor for the many factories that were constructed. Deep recessions and a depression in the 1890s devastated these immigrants, who were forced to wait in bread lines outside police stations during economic downturns. Local regulations over housing and factories hardly existed or were not enforced, leading to catastrophes like the burning of a garment factory in 1911 in New York that killed 87 female immigrant workers who jumped to their deaths from upper stories. Like most workers in factories, these factory workers were not unionized.

The period from 1865 to 1900, known as the Gilded Age, was a period of extreme economic inequality that juxtaposed these penniless immigrants with multimillionaires and billionaires, including Andrew Carnegie, John D. Rockefeller, and Henry Ford, who owned factories, railroads, and mines. Nor did either the Republican or Democratic Parties represent working-class Americans who found no allies when they engaged in strikes for better wages. Local police and federal troops forcibly attacked them when they protested their working conditions.

Nor were conditions favorable for Native Americans, African Americans, and Latinos (Jansson, 2019b). Not immune to Europeans' diseases, millions of Native Americans perished from cholera, typhoid fever, and other illnesses. Survivors were pushed westward by the incessant flow of European immigrants prior to the Civil War. The U.S. military killed many Native Americans who refused to reside on reservations after the Civil War, where many of them remain today. African Americans were emancipated during the Civil War only to suffer extreme hardships when they were not given land. Roughly 25% of them starved during and after the Civil War. Many of them became tenant farmers who rented land from former plantation owners on terms that left them nearly penniless. The Ku Klux Klan that terrorized African Americans, Catholics, Jews, and other minorities plagued them. The United States forcibly appropriated much of the lands of the southwestern part of the nation from Mexico when it invaded Mexico and forced its national government to cede this territory. White settlers took the land of indigenous Latinos by force or by claiming that their deeds under Mexican laws were bogus. Having lost their land, many Latinos became agricultural and mining laborers who were exploited by white owners. Workers in manufacturing plants were treated harshly, lacking minimum wages, ceilings on hours worked, and lack of trade unions. Social policies that we take for granted in contemporary society did not exist, such as the Supplemental Nutrition Assistance Program (SNAP), unemployment insurance, Social Security, and housing programs.

THE PROGRESSIVE PERIOD

It seemed that the extreme inequality of the Gilded Age might be broken by the first urban reform movement in the 20th century when Theodore Roosevelt became president in 1901 after the assassination of President William McKinley. Called the Progressive movement, it was led by middle-class reformers drawn from both the Republican and Democratic Parties. Roosevelt challenged the industrialists by coming to the aid of striking workers and sought to break up monopolies. Progressives' signature victories were enactment of local and state regulations over child labor, unsafe working conditions in factories, fire codes, public health, and housing codes. They enacted a small and relatively punitive "Mothers' Pensions" program for destitute single mothers.

Wanting far greater reforms, Jane Addams, a founder of the social work profession who had established the settlement house known as Hull House in Chicago, teamed with Theodore Roosevelt to form the Progressive Party in 1912 (Addams, 1907). Roosevelt and Addams hoped to enact major federal social programs but lost the 1912 presidential election to Democrat Woodrow Wilson. The primitive nature of American federal government was illustrated by its lack of Constitutional authority to levy income taxes. Without resources obtained from taxes, the United States could not fund major social programs. Even when this deficiency was removed when the 16th Amendment to the Constitution was enacted in 1913 (which allowed the federal government to levy income taxes), the federal government failed to collect major income taxes until World War II and beyond—and even today collects far lower levels of taxes as a percentage of gross domestic product (GDP) than most other industrialized nations (Jansson, 2019a).

Jane Addams and other social workers engaged in micro, mezzo, and macro policy at a time when federal safety net programs did not exist (Addams, 1961). Jane Addams's Hull House intersected with public agencies as it guided immigrants who spoke no English to educational programs, helped persons who lived in unsafe housing obtain help from city housing enforcement programs, worked with fire departments to place exits in multilevel garment factories, and helped establish juvenile courts so that children and youth would not be sent to prisons peopled by adult offenders. They engaged in mezzo policy advocacy when they developed Hull House, which provided literacy training, employment, money management, and other services to immigrants. Addams found ways to fund Hull House at a time when local, state, and federal governments had scant resources and when few public programs that we currently take for granted existed. Addams mobilized people to participate in protests about housing conditions, workplace hazards, and corruption in city government. She initiated research projects that measured the incidence and location of specific problems, such as malnourishment of children

(Addams, 1961). She engaged in macro policy advocacy when she backed reform-minded candidates in municipal and state elections as well as when she worked with Roosevelt to form the Progressive Party.

THE GREAT DEPRESSION

The Great Depression savaged the nation from the stock market crash in 1929 to American entry into World War II in December 1941. The national unemployment rate surged as high as 25% by 1933 and still remained about 15% in 1940 until it was erased by wartime industrial production during World War II. Unemployment remained far higher among African Americans, women, and Latinos. Roughly 60% of the wealth of affluent Americans disappeared in the wake of the crash of the stock market and the bankruptcy of many businesses.

Franklin Roosevelt, a rich man afflicted with polio, became the unlikely advocate for ordinary people during this national emergency. He orchestrated two major accomplishments. He proposed and enacted sweeping federal social legislation for the first time in American history (Jansson, 2019b). He attracted millions of working-class voters to the Democratic Party that he headed, making it a powerful advocate for these voters as compared with the Republican Party, which came mostly to represent relatively affluent people (Leuchtenberg, 1963). The Democratic coalition included those African Americans who were allowed to vote in the North and the South, intellectuals, Jews, and members of the working class. White conservative Southerners also joined the Democratic coalition because they had shifted to the Democratic Party in the wake of the Civil War away from the Republican Party, headed by Abraham Lincoln, who had emancipated slaves. Most Southerners belonged to the Democratic Party until the 1980s and beyond.

Franklin Roosevelt developed and nurtured this Democratic coalition as he developed his New Deal legislation to get people through the Great Depression (Patterson, 1967). Tens of thousands of persons became homeless as their houses were foreclosed on. Farmers were unable to sell corn, beef, and other products. Even conservative politicians mostly voted for New Deal programs because their constituents, too, had lost jobs and houses. New Deal programs included the following (Jansson, 2019a):

- The Federal Emergency Relief Administration (FERA) that provided grants to states allowing them to make welfare payments to destitute individuals and families. It required them to establish state commissions separate from their existing welfare programs with uniform eligibility processes throughout the state and provisions to discourage discriminatory administration. It began in 1933 and was ended when welfare programs in the Social Security Act were enacted in 1935. It ended so-called poorhouses funded by local governments, where unemployed persons were required to reside under harsh conditions.

- A variety of work-relief programs that gave recipients jobs, including the Civilian Works Administration (CWA), which hired 16 million Americans for 190,000 work projects between November 1933 and January 1934 alone; the Civilian Conservation Corps (CCC), which provided jobs to 2.5 million young men, mostly in rural areas under the administration of the Army and the Department of Interior; and the Public Works Administration (PWA), which built complex projects like airports, dams, and roads throughout the 1930s. Roosevelt replaced the CWA with the Works Progress Administration (WPA) in 1935, which funded hundreds of projects across the nation.

- The Federal Deposit Insurance Corporation (FDIC) and other programs to stabilize banks.
- Programs to combat the aggressive cutting of prices by corporations as they sought to retain sufficient market share to remain afloat during the Great Depression, including the National Recovery Act of 1933 (NRA). It helped industries in specific sectors negotiate set prices but was declared unconstitutional by the U.S. Supreme Court in 1935.
- Programs to cut foreclosures, such as the National Housing Act of 1934, that established the Federal Home Administration Act (FHA) to help banks refinance home mortgages at lower rates of interest. He averted foreclosures on farmers and home owners by having the government purchase mortgages and refinance them under the Emergency Farm Mortgage Act and the Farm Relief Act in 1933. He established the FHA to insure mortgages and home improvement loans. He established the Wagner-Steagall Housing Act of 1937 to provide low-interest loans to local authorities to build public housing.
- Programs to encourage economic growth in entire regions, such as the Tennessee Valley Authority (TVA), through a network of dams and generating stations to power fertilizer plants and to help fund reforestation and flood control projects.
- The Social Security Act that provided pensions for seniors financed from payroll deductions of employers and employees. It also provided welfare programs funded by matching federal and state funds that included Aid to Dependent Children (ADC), old age assistance (OAA), and Aid to the Blind (AB); funds for child welfare and public health programs; and unemployment insurance by appropriating funds to help states administer their unemployment funds in return for an agreement by the states to hold fair hearings to avoid discrimination in their distribution of unemployment benefits.
- The Fair Labor Standards Act of 1938, which established fair working conditions and minimum wages.
- The Wagner Act of 1935, which allowed workers to decide whether they wanted to unionize in elections monitored by the National Labor Relations Board (NLRB).
- A variety of programs that addressed nutritional, health, dental, and other needs of impoverished persons, which included partnerships among the federal government, states, and not-for-profit organizations.
- The Agricultural Adjustment Agency in 1933, which convened producers of specific crops to reduce crop surpluses that had driven their prices so low that many farmers were bankrupted—and other legislation that protected sharecroppers.

Make no mistake: When taken together, these programs helped persons in the lower economic echelons in ways that were unprecedented in the United States. They increased their resources, wages, food, housing, and employment. They increased their morale during an unprecedented decade of mass unemployment. They kept families together. They averted disease and starvation. They cut homelessness. These policies were revolutionary in a nation where virtually no social programs had been enacted by the federal government. Many social workers worked in New Deal agencies. Above all, New Deal programs gave the federal government the power to fund and administer a host of social programs as compared with state and local governments that had exclusively served these roles prior to the New Deal.

Notable exceptions to the generosity of the New Deal included Roosevelt's decision not to support federal anti-lynching legislation and his decision to incarcerate tens of thousands of Asian Americans primarily of Japanese descent at the outset of World War II. He feared that southern conservatives in the Congress would veto New Deal legislation if he supported anti-lynching legislation. He incarcerated Japanese Americans due to mass hysteria and racism against them as contrasted to white German and Italian immigrants (Jansson, 2019b).

The work relief programs of the New Deal were terminated during World War II by Republicans and conservative southern Democrats who argued they were no longer needed when the nation had returned to full employment in World War II as the United States manufactured munitions for nations that opposed Germany, Italy, and Japan (Jansson, 2001). Yet the programs of the Social Security Act remained intact as well as the federal minimum wage, housing programs, prohibition of child labor, and regulations that allowed unions to organize.

Social workers engaged extensively in policy advocacy in the 1930s. They engaged in micro policy advocacy to help destitute individuals use the New Deal work programs (Wenocur & Reisch, 1989). They engaged in macro policy advocacy to help establish and implement work relief and welfare programs for millions of people. They worked to educate local residents about the myriad services and programs established during the New Deal. Roosevelt's wife, Eleanor, worked with distressed communities during the 1930s to develop social services, housing, and businesses when most residents suffered extreme poverty. She helped find resources to rebuild their infrastructure. She helped African Americans gain access to work relief programs. Franklin Roosevelt repeatedly visited persons in the Dust Bowl in Oklahoma to find ways to improve their lives by introducing new agricultural techniques that would decrease the dust storms that ruined farmlands.

The federal government did not aggressively use its power to levy income taxes until World War II, when it had to fund American military forces at high levels. As we shall see, this power was eventually used, as well, to fund many social programs enacted in the 1960s and early 1970s even if these taxes were lower as a percentage of GDP than in most other industrialized nations. Moreover, the United States spent far more money on military forces during the Cold War from 1949 through the 1980s than other industrialized nations (Jansson, 2001).

THE GREAT SOCIETY

Lyndon Johnson succeeded John Kennedy when he was assassinated in November 1963. Inheriting Roosevelt's coalition of working-class people, African Americans, white Southerners, and intellectuals, he won a landslide victory over Republican Barry

Goldwater in 1964. He encountered massive levels of poverty, the oppression of African Americans throughout the nation, lack of medical care for seniors, and many other social problems. The Great Society included the following (Jansson, 2019b):

- Enacting the Civil Rights Acts of 1964 and 1965, which protected voting rights of African Americans, desegregated public facilities and transportation, prohibited hiring discrimination in federal contracts, gave the U.S. attorney general the right to file suits to desegregate schools—and gave rights to women through Title VII of the Civil Rights Act of 1964
- Enacting the so-called War on Poverty that came to include the Job Corps, Head Start, the Neighborhood Youth Corps, legal aid centers, health clinics, and community action programs
- Enacting Medicare and Medicaid, with the former providing health care for seniors and the latter providing means-tested medical benefits for low-income persons
- Establishing the Department of Housing and Urban Development (HUD) and enacting or expanding public housing and affordable housing programs
- Enacting the Older Americans Act, which provides Meals on Wheels and other programs
- Enacting the Elementary and Secondary Education Act (ESEA), which provided federal funds primarily to schools with large numbers of low-income children

Johnson was the first American president to fund scores of new social programs. They were often funded, however, at low levels because he enacted a massive tax cut in 1964 and made an ill-advised decision to commit 600,000 troops to the Vietnam War (Jansson, 2001). The war split the Democratic Party because many white Southerners and blue-collar voters favored the war in contrast to many liberals and Martin Luther King. Johnson chose not to run for a second full term in 1967, leaving the door open for Vice President Hubert Humphrey to oppose Richard Nixon in the election of 1968.

NIXON'S SURPRISE

Many liberals feared that Nixon's accession to the presidency would doom further social reforms because he was a staunch Republican conservative. Nixon chose, however, to initiate social reforms that rivaled ones that Johnson enacted because he wanted to rebuild the Republican Party by gaining credit for domestic reforms that had traditionally been initiated by the Democratic Party. He also wanted to draw conservative southern Democrats and white blue-collar voters into the Republican Party by supporting the Vietnam War, opposing school busing of minorities into white schools, and opposing affirmative action (Reichley, 1981). Nixon correctly predicted that many white northern Catholics and southern white Democrats, angered by Johnson's civil rights measures, would join the Republican Party. He could not accomplish this goal because he resigned from the presidency before the Congress was on the verge of impeaching him. Nixon nonetheless achieved these domestic reforms (Jansson, 2019b):

- Indexing Social Security benefits to inflation to assure regular increases
- Enacting the Supplementary Security Income (SSI) Program to provide means-tested benefits to seniors and persons with disabilities
- Enacting the Earned Income Tax Credit of 1975 to give working persons in families tax credits if they earned less than a specified income
- Expanding the Food Stamps Program by federalizing it and increasing its benefits
- Enacting the Housing and Community Development Act of 1974, which established rental subsidies for low- and moderate-income persons
- Establishing the Occupational Safety and Health Administration (OSHA) of 1970 to regulate and improve working conditions
- Enacting Title XX of the Social Security Act to fund the social service programs of states
- Enacting the Comprehensive Employment and Training Act (CETA) to train workers and provide them with jobs in public and not-for-profit agencies, with an emphasis on long-term unemployed persons; fund summer jobs for high school students; and providing full-time jobs for 12 to 24 months with the goal of providing marketable skills
- Passing the Rehabilitation Act of 1973, which prohibited discrimination on the basis of disability in programs conducted by federal agencies, in programs receiving federal financial assistance, in federal employment, and by federal contractors
- Enacting the Education for All Handicapped Children Act of 1975, which required all public schools that accept federal funding to provide equal access to education and one free meal a day for all children with physical and mental disabilities
- Establishing the U.S. Department of Education
- Approving affirmative action in the wake of the Supreme Court's decision in *Regents of the University of California v. Bakke*

When taken together, enactments of the Great Society and Nixon's presidency transformed America. Nondefense spending went from 8.1% of the gross national product in 1961 to 11.3% in 1971 and 15.6% in 1981 (Jansson, 2001). About two-thirds of this domestic budget consisted of social insurance and means-tested social programs. Total federal social spending rose from $67 billion in 1960 to $158 billion in 1970 (in 1980 dollars) and to $314 billion in 1980.

Social workers engaged in advocacy during the Great Society and Nixon's presidency. They helped persons learn about and gain access to the scores of social programs enacted in the 1960s and early 1970s including food stamps, Section 8 housing vouchers, Medicaid, and the Earned Income Tax Credit (EITC)—all new programs.

Many social workers worked in programs of the War on Poverty. They helped residents in marginalized communities develop community action councils that engaged in planning to develop new programs. They worked with welfare rights organizations (WROs) that worked to increase welfare grants for single mothers and to decrease punitive practices of some welfare departments, such as raiding homes to see if men lived in households with women on welfare.

Even with the reforms of the Great Society and the Nixon presidency, critics believed that many of them were underfunded. Welfare programs reached relatively few poor people because many states established harsh eligibility standards for Aid to Families with Dependent Children (AFDC), just as the federal government established harsh eligibility standards for food stamps. Huge sections of American cities housed persons of color in segregated communities with poor schools and housing (Jansson, 2001).

CONSERVATIVE COUNTERREVOLUTIONS

History demonstrates that policy advocates often encounter counterrevolutions in which the nation turns sharply toward conservatism. Policy advocates guided by social work's code of ethics feared that elections of Ronald Reagan in 1981 and Donald Trump in 2016 might lead to repudiations of prior social reforms because both opposed social programs widely used by marginalized populations. If Reagan repeatedly sought cuts in social programs, Trump followed suit and also attacked immigrants, African Americans, Latinos, women, and Muslims (Jansson, 2019b).

Gridlock During the Presidencies of Ronald Reagan, George H. W. Bush, and George W. Bush

In the blink of an eye, many voters bought into the arguments of Ronald Reagan, a former movie star who had changed from liberal views in the 1930s to conservative ones in the 1950s and beyond. A two-term Republican governor of California in the late 1960s and early 1970s, Reagan ran unsuccessfully for the presidency in 1976 but won a landslide victory over Jimmy Carter in the presidential contest of 1980—and won another one-sided election in 1984. He glamorized the Gilded Age even with its extreme income inequality and racism. He invented the term "welfare queen" to describe a woman in Chicago who he alleged had many children to increase her welfare grants. (Only later was it established that he had invented this woman.) He liked the theories of economist Arthur Laffer, who believed that deep tax cuts stimulated economic growth without causing deficits because economic growth caused by lower taxes would produce higher tax revenues (Roberts, 1984). Many economists and David Stockman, Reagan's budget director, believed Laffer was wrong (Stockman, 1986). Their forecasts proved to be correct: Reagan's deficits exceeded the sum total of all the deficits of all prior American presidents.

Reagan created a rival coalition to Roosevelt's that included evangelicals, southern whites, affluent Americans, and suburbanites.

Reagan's policy agenda flowed from his ideology and his coalition. He cut taxes deeply for everyone but with the largest cuts given to affluent Americans whose top marginal rates were cut from 70% to 28%. He increased military spending to the level during the Vietnam War even though the nation was not at war. He argued that the resulting budget deficits required him to cut domestic discretionary spending deeply—including food stamps, welfare, housing vouchers for poor people, and many other programs. He enacted all of this legislation in 1981 (Stockman, 1986).

Reagan ushered in a new era of income inequality that came to equal the income inequality of the Gilded Age (Jansson, 2019a). He initiated few social reforms aside from the Immigration Reform and Control Act of 1986, which gave more than 3 million undocumented immigrants a path to citizenship. He made many Americans averse to increasing taxes. He attacked the federal government and social welfare programs in general but did not cut Social Security, Medicare, and Medicaid. No major legislation was enacted in the presidency of George H. W. Bush aside from the Americans with Disabilities Act of 1990, which provided work, access to public transportation with ramps and lifts, and other accommodations in workplaces.

Social workers engaged in policy advocacy to help their clients during this conservative period. They had to help persons find assistance when myriad social programs were cut or eliminated. They helped clients use private food banks when food stamps (now SNAP) were cut. They directed clients to free clinics when mental health services were cut. Social workers worked frantically to keep social agencies afloat when they lost federal contracts in the wake of federal budget cuts. Social workers participated in coalitions in Washington, D.C., to contest President Reagan's deep cuts in social spending. They helped progressive candidates retain or gain seats in swing districts.

President George W. Bush followed Reagan's policy script during his presidency from 2001 through 2008 by cutting taxes and raising military spending even if he did not cut funding of most social programs (Jansson, 2019a). When Democrats proposed new programs, he accused them of increasing budget deficits and the national debt after he had already done so.

Gridlock During the Presidencies of Bill Clinton and Barack Obama

Two relatively liberal presidents (Bill Clinton from 1993 through 2000 and Barack Obama from 2008 through 2016) were unable to initiate new liberal eras because the federal government was gridlocked by Republicans who controlled one or both of the chambers of Congress. Recall that Presidents Franklin Roosevelt and Lyndon Johnson achieved their legislative successes because Democrats controlled both chambers of the Congress, unlike Presidents Clinton and Obama (Jansson, 2019a). Bill Clinton hoped to enact national health reform and social investments but was stymied by the Republican Congress. President Obama was able to enact the Affordable Care Act (ACA), the Stimulus Plan, and banking regulations in 2009 and 2010 because Democrats barely controlled both chambers of the Congress for most of that period. He was gridlocked, however, for the rest of his presidency as Republicans gained control of one or both chambers of Congress. He encountered numerous budget battles leading to a budget deal in 2013 that required huge cuts in domestic and military spending up to the year 2023 (Jansson, 2019a). Clinton and Obama lacked resources for major social reforms because they were burdened by the huge increases in the national debt by Presidents Reagan and George W. Bush.

The Maverick Presidency of Donald Trump

Not even Donald Trump predicted he would win the election of 2016 in his contest with Democrat Hillary Clinton. With strong support from blue-collar workers coupled with suburban voters, he achieved an upset victory. He campaigned as a populist who would help white voters in coal country and in towns and small cities where factories had moved to other nations or gone bankrupt. Trump quickly vanquished 17 Republicans in primary elections and narrowly defeated Clinton. He promised to renegotiate trade treaties with China, Mexico, and Canada as well as with other developing nations where American factories had fled. He vowed to increase tariffs on nations that impeded imports from the United States. He promised to "build a wall" on the southern border with Mexico to impede immigrants who he believed took jobs from Americans. He promised to stop the flow of refugees from the Middle East, Mexico, and Central America. He promised a large infrastructure program that would provide jobs for his base of support. He promised to "repeal and replace Obamacare" (the ACA) but offered no substitute to meet the health needs of the more than 20 million persons insured by the ACA. He promised large tax cuts particularly for the middle class. He promised not to cut Social Security, Medicare, or Medicaid (Jansson, 2019b).

He had accomplished some of these objectives by mid-2018. He renegotiated the North American Free Trade Agreement (NAFTA), although he put on hold trade agreements with many European nations. He cut taxes deeply. He oversaw a surging economy with low levels of unemployment, although it was unclear if his policies had caused this outcome because unemployment had already declined greatly during Obama's presidency.

But Trump failed to enact many of his campaign promises by late 2018. No wall had been built on the Mexican border. He had not repealed and replaced the ACA, although he had enacted some policies that weakened it. No infrastructure program had been proposed. He betrayed his white blue-collar followers in coal country and rural areas by not enacting social programs that met their employment, education, and other needs—and by seeking deep cuts in Medicaid and the ACA that they needed for opioid drug addiction and other health needs. He proposed deep cuts in SNAP and other safety net programs used by his base of support (Jansson, 2019b).

Many liberals questioned his tax cuts for several reasons. They argued that his deep tax cuts mostly favored affluent persons and corporations that received roughly 83% of their benefits as compared with only 17% for low- and moderate-income persons, including Trump's followers. They feared that he was emulating President Reagan's creation of massive budget deficits that would prevent budget increases for social programs in coming years. They contended that tax cuts that favored affluent persons and corporations would exacerbate income inequality in the United States that already exceeded levels of 20 other industrialized nations (Jansson, 2019a). They criticized his insufficient attention to opioid poisoning in 2018 that had led to 70,000 deaths in 2017.

Distressingly, Trump attacked members of many marginalized populations during his campaign and his presidency, including Muslims, Latinos, African Americans, women, and disabled persons. He attacked the press almost on a daily basis even though freedom of the press is enshrined in the Bill of Rights. He attacked leading Democrats, using words like "morons" and "clowns." He vilified Republicans who dared to question his policies. He attacked climate control, even though 97% of climate scientists have contended that the world is heading into significant temperature increases that could have major consequences like raising sea levels, health epidemics, food shortages, and conflicts among nations. These fears were registered by major climate reports issued by the United Nations that predicted these adverse outcomes were likely as early as 2040 unless carbon dioxide emissions were cut significantly, particularly by China and the United States, which ranked first and second in these emissions (Davenport, 2018).

Social workers engaged extensively in micro, mezzo, and macro policy advocacy during the first two years of Trump's presidency. They participated in marches to protect immigrants' rights, to seek policies to avert climate change, to protect women from sexual exploitation, to affirm science, and to decrease deaths of unarmed African Americans. They participated in protests at town hall meetings of conservative politicians who sought to repeal the ACA. Many women chose to run for political office, including many social workers.

Falling Behind Other Industrialized Nations

The United States had evolved a substantial set of social programs by 2018 that lifted many people from extreme poverty to higher levels of resources when cash benefits, pensions, in-kind benefits, medical services, social services, housing programs, educational opportunities, and childcare are aggregated. Many individuals and families received tax credits and deductions, such as the Earned Income Tax Credit. (I discuss these benefits in more detail in Chapter 9.)

Despite these advances, however, the United States has lagged behind 20 other industrialized nations in two ways: income inequality and many kinds of social problems (see Figure 3.1 developed by Wilkinson & Pickett, 2009). When the ratio of the income of the top 20% of the population to the bottom 20% is calculated, the United States had a ratio of roughly nine to one as compared with roughly eight to one for Portugal; seven to one for the United Kingdom; six to one for France and Canada; five to one for Germany, Belgium, and the Netherlands; four to one for Sweden, Norway, Finland, and Denmark; and 3.5 to one for Japan (see the *x*-axis in Figure 3.1). In other words, the bottom 20% of Americans find themselves far poorer than persons in the bottom 20% of 20 other industrialized nations—and often way poorer. Some social scientists contend that extreme inequality produces many ill effects for people in the bottom 20%, including higher levels of social problems measured in Figure 3.1 by the Index of Health and Social Problems (see the *y*-axis in Figure 3.1). Residents of nations with higher levels on this index have higher levels of lack of trust, low life expectancy, high infant mortality, high rates of obesity, poor educational performance, high rates of teenage births, high rates of homicide, high rates of imprisonment, and low rates of social mobility (Wilkinson & Pickett, 2009).

These findings are hardly comforting to Americans. Despite our overall affluence and success in developing a more robust welfare state, we still lead the world in maldistribution of income and in levels of a wide array of social problems. The unequal distribution of income places people in the bottom 20% in an extreme subordinate position as they witness on television and with their cell phones homes, cars, and lifestyles they cannot obtain or emulate. They also find themselves with far lower rates of health insurance, poorer schools, lower pensions, and poorer vocational training than many of the other industrialized nations—the social investments that would help them move upward in the social order. Nor can they rely on upward mobility to move them to a higher

FIGURE 3.1 ■ Rankings of 21 Industrialized Nations by Inequality and Social Problems

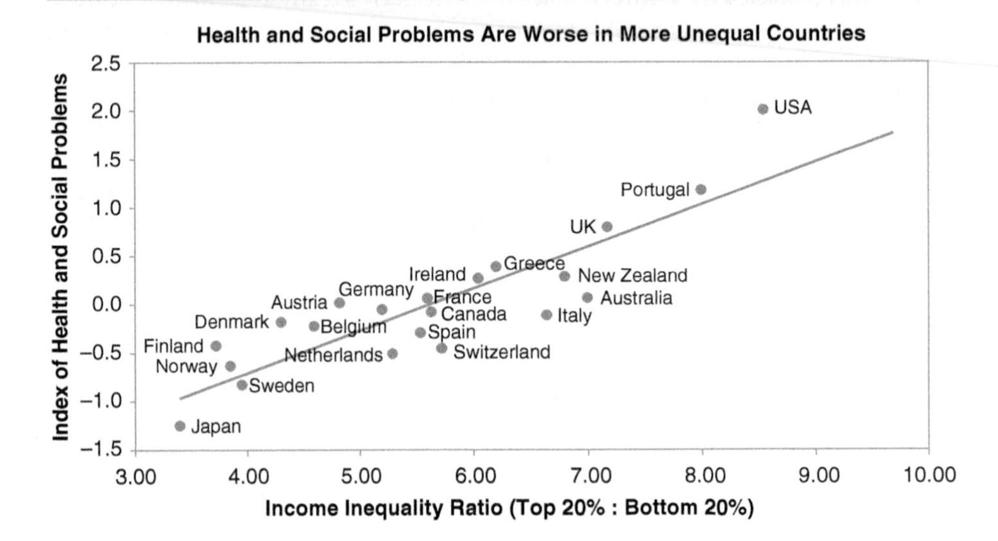

position in the nation's economy because five separate studies show that Americans in the lower economic echelons are less likely to move upward than persons in England, Germany, France, and Canada.

Policy advocates have a full agenda before them in coming years. They need to augment an array of social programs in the eight sectors discussed in Chapters 7 through 14. They need to establish or improve programs that will give persons in the lower 20% a road to upward mobility, not to mention persons in the bottom 50%, who also have restricted upward mobility and high rates of many social problems. They need to buttress safety net programs. They need to pressure public officials to work toward social justice just like policy advocates in the Progressive era, New Deal, Great Society, and early 1970s. They need to oppose deep tax cuts for affluent persons and corporations so that the nation possesses resources to help low- and moderate-income Americans receive resources and social programs commensurate with ones already in place in many other industrialized nations.

Take the case of homelessness as an example of possibilities for social reform. I witnessed almost no homeless people on the streets of Stockholm, Helsinki, and Copenhagen in the summer of 2017. With its great resources, the United States could emulate these Nordic nations—and many other industrialized nations—by constructing necessary units in and around urban areas in the United States. It is a national problem that needs leadership from a federal government whose revenues are roughly two times the aggregated revenues of the 50 states. The nation needs to greatly increase funding for construction of new units and conversion of old ones. It needs to create jobs programs. It needs to increase rent subsidies. It must resort to greater rent control in many cities were rents have skyrocketed beyond the reach of many people as witnessed by 80 million evictions from 2000 to 2018 in the United States. It needs to increase support staff, including social workers, to help homeless people cope with mental illness and substance abuse. It must be willing to spend tens of billions of dollars on this issue.

Many policy advocates work on the micro, mezzo, and macro levels to reduce homelessness as illustrated by the policy advocacy of Andy Bales, a former minister who heads the Union Rescue Mission in downtown Los Angeles, where 13,000 people receive food and lodging each day (Castellanos, 2018). Living with his father and grandfather as a child, he was homeless for extended periods of time. He gives micro policy advocacy to innumerable families with children, women, and others each week. He encourages homeless people to contact him directly through his Facebook, Twitter, and other accounts. He always interrupts his schedule when he hears that a woman is homeless, even during the early morning hours.

He engages in mezzo policy advocacy when he creates housing for communities of homeless people that takes them off the streets. He created a 77-acre campus in Sylmar in 2005 that gave permanent housing to senior women. He opened another facility for women and children in 2007 in Carson that gave them living space for up to three years combined with educational training and supportive services.

He engages in macro policy advocacy when he takes the lead in documenting the "dumping" of homeless people from hospitals' emergency rooms rather than transporting them to shelters and community programs. Using video cameras to gather evidence, he has helped authorities convict hospitals of this illegal practice that violates state and federal laws. Not only did the offending hospitals have to pay large fines, but they received adverse publicity.

He was a leading advocate for enactment of bond issues in Los Angeles City and Los Angeles County to provide funds to build many housing units for homeless people and to provide them with supportive services. Realizing these government measures and the construction of housing units will take a decade to implement, and even then will serve only a fraction of the homeless people in Los Angeles, he is initiating many "sprung-tent communities" across Los Angeles County that will each house 100 people for 6 months and provide them with case management services. He aims to create thousands of units of micro-unit housing—so-called tiny homes modeled after a program in Austin, Texas. Such housing would not require the large government subsidies needed for homes of 1,000 square feet and larger and could easily be transported to specific sites.

He perseveres despite losing a leg from three flesh-eating infections that he contracted from contact with human feces in encampments of homeless people who lacked even rudimentary facilities.

Why Social Workers Need to Engage in Policy Advocacy in 2020 and Beyond

The Positive Picture

Contemporary social workers have inherited many social policies. The United States has constructed a federal welfare state with a network of important social policies that include programs like Social Security, Medicare, Medicaid, and SNAP—called entitlements—that are automatically funded to the level of claimed benefits each year. It funds many programs annually from the federal budget—called discretionary spending—including childcare, Head Start, housing, public health, welfare, mental health programs, portions of child welfare, substance abuse, portions of education programs through the Elementary and Secondary Education Act, federal prisons, and many others. It funds eight major block grants that give funds to the states that they use for social services; community services; alcohol, drug abuse, and mental health services; maternal and child health services;

community development services; primary health services; preventive health services; and childcare and development. It shares funding with states for Temporary Assistance for Needy Families (TANF), a welfare program mostly used by single mothers.

The federal government funds an array of tax expenditures that give resources to residents not through the budget but through tax deductions or tax credits that are funded by the Internal Revenue Service (IRS). Americans receive these benefits if they apply for them on their income tax forms, such as the Child Tax Credit and the EITC.

Employers finance a considerable portion of the health care costs of their employees from deductions from their corporate taxes or from subsidies from the ACA. Some of them fund wellness programs and other health programs as well as leaves for employees for childbirth.

Expenditures of the American welfare state also include an array of programs funded by states and local governments, including roughly 90% of the nation's secondary education programs, part of the costs of eight block grant programs and TANF, and most of the costs of community colleges and public higher education, nonfederal prisons, mental health programs, and general relief for single men and women.

The private sector also funds many social programs. Charitable contributions to hospitals, universities, secondary schools, and not-for-profit agencies are considerable. Hospitals and physicians give considerable charitable care. Private donors fund a significant share of the budgets of nongovernment organizations (NGOs), such as the YWCA, YMCA, homeless shelters, and child guidance centers that are also funded by contracts with local, state, and federal agencies.

Consumers of service fund a considerable share of social programs. They often pay fees. They pay deductibles and co-insurance in many health and mental health programs. They often fund tuition of secondary schools, community colleges, colleges, and professional programs. Unfortunately, consumers' expenses and payments are often onerous for persons in the lower economic echelons—and sometimes for middle-class persons and above.

When added together, the United States possesses a significant welfare state—one that emerged over decades with considerable assistance from macro policy advocacy by public officials, advocacy groups, and private citizens. Social workers have assumed important roles in enacting and funding social programs.

The Challenge

Policy advocates must now begin a new round of policy advocacy at micro, mezzo, and macro levels to augment the American welfare state so that the United States can catch up with the 20 other industrialized nations portrayed in Figure 3.1. We should no longer encounter the following:

- Parents who have to decide who to feed in their families because they cannot afford food for everyone
- Tens of millions of Americans who are medically uninsured—and many more with health insurance that fails to cover a large portion of their medical costs
- More than 500,000 homeless people partly caused by 80 million evictions from rental housing from 2000 to 2018
- Tens of thousands of students who drop out of community colleges and colleges because they cannot afford tuition

- Tens of thousands of persons who do not receive timely assistance for opioid and other addictions
- Enhanced food distribution programs to decrease the number of families whose members are malnourished

To achieve these goals, the United States has to tap resources of affluent Americans and corporations (Jansson, 2019a). Many Americans don't realize that other industrialized nations have fewer social problems than their nation because they invest more heavily in schools, services, infrastructure, and benefits. They can make these investments for (at least) three reasons.

First, the United States taxes its affluent citizens and even its middle and lower-middle classes at lower levels than other industrialized nations. The top marginal tax rate in the United States was 39.6% in 2014—now reduced to 37.5% as compared with top marginal tax rates in Denmark (60.2%), Sweden (56.6%), Belgium (53.7%), Spain (52%), the Netherlands (52%), France (50.7%), Austria (50%), Japan (50%), Greece (49%), Finland (49%), Portugal (49%), Canada (48%), Ireland (45%), Israel (48%), Australia (47.5%), and Iceland (46.2%) (Eaton, 2014).

Second, it spends as much on its military as the next seven nations combined. Although military spending is often necessary in a world with considerable conflict, its level in the United States is viewed by many experts as excessive (Jansson, 2001, 2019a). Third, it spends two times more on its health care system than many other industrialized nations partly by ceding its management to private insurance companies rather than using the government to manage and administer it and by other inefficiencies—and has poorer health outcomes than these other nations, such as shorter life expectancy and greater infant mortality.

By increasing the top marginal tax rate on affluent Americans to from 37.5% to 40% and taxing wages, interest, dividends, employer contributions to health plans, overseas earnings, and growth in retirement accounts, the United States would gain $157 billion in tax revenues in the first year. It would gain even more ($276 billion in the first year) if the top marginal rate was increased to 45%, which would still allow this group to take home at least $1 million per year (Cohen, 2015). Nor would raising tax rates harm the super-rich since the top 0.1% of American families, who all have assets worth more than $29 million, own more than 20% of all household wealth in the nation, compared with the 1970s when it controlled only 7 percent (Cohen, 2015).

PREPARING FOR A NEW REFORM PERIOD

We have discussed the evolution of the American welfare state, including its meritorious features as well as its shortcomings. We have discussed how social workers engaged in micro, mezzo, and macro policy advocacy throughout the 20th century and into the 21st to improve social policies in the United States. We discussed prior eras of reform and conservatism. Now we need to carry policy advocacy forward at micro, mezzo, and macro levels to catch up with other industrialized nations and to produce another era of reform. Who knows? Maybe the United States can enter another reform period similar to the New Deal and the Great Society.

Learning Outcomes

You are now equipped to:

- Understand how the American welfare state has evolved in stages from the colonial period to the present
- Describe key policies of the following periods:
  1. The colonial period
  2. The Gilded Age
  3. The Progressive Era
  4. The Great Depression
  5. The Great Society and the early 1970s
  6. Conservative counterrevolutions
- Identify how social workers engaged in policy advocacy at micro, mezzo, and macro levels in these periods
- Describe how and why the United States lags behind other industrialized nations
- Identify how the United States can increase its revenues to fund improvements in its welfare state
- Prepare for a new reform period

References

Addams, J. (1907). *Newer ideals of peace.* New York: Macmillan.

Addams, J. (1961). *The spirit of youth and city streets.* New York: Macmillan.

Castellanos, C. (2018, March 1). Union rescue mission leader committed to ending homelessness. *Pasadena Outlook, 12*(9), 1, 23.

Cohen, P. (2015, October 16). What could raising taxes on the 1% do? Surprising amounts. *New York Times.* Retrieved from https://www.nytimes.com/2015.

Davenport, C. (2018, October 7). Major climate report describes a strong risk of crisis as early as 2040. *New York Times.* Retrieved from https://www.nytimes.com/2018/10/07/climate/ipcc-climate-report-2040.html

Eaton, G. (2014, January 26). Which countries have the highest tax rates? *NewStatesman.* Retrieved from http://newstatesman.com

Jansson, B. (2001). *The sixteen-trillion-dollar mistake: How the U.S. bungled its national priorities from the New Deal to the present.* New York: Columbia University Press.

Jansson, B. (2019a). *Reducing inequality: Addressing the wicked problems across professions and disciplines.* San Diego, CA: Cognella Academic Press.

Jansson, B. (2019b). *The reluctant welfare state: Engaging history to advance social work practice in contemporary society.* Boston, MA: Cengage.

Leuchtenberg, W. (1963). *Franklin Roosevelt and the New Deal: 1932–1940.* New York: Harper & Row.

Patterson, J. (1967). *Congressional conservatism and the New Deal.* Lexington: University of Kentucky Press.

Reichley, J. (1981). *Conservatives in an age of change: The Nixon and Ford administrations.* Washington, DC: Brookings Institution.

Roberts, C. (1984). *The supply-side revolution: An insider's account of policy-making in Washington.* Cambridge, MA: Harvard University Press.

Stockman, D. (1986). *The triumph of politics: Why the Reagan revolution failed.* New York: Harper and Row.

Wenocur, S. & Reisch, M. (1989). *From charity to enterprise: The development of American social work in a market economy.* Urbana: University of Illinois Press.

Wilkinson, R. & Pickett, K. (2009). Income inequality and social dysfunction. *Annual Review of Sociology, 35*, 493–511.

PROVIDING MICRO POLICY ADVOCACY INTERVENTIONS

LEARNING OBJECTIVES

In this chapter, you will learn to:

1. Define micro policy advocacy
2. Read the context
3. Decide whether to proceed
4. Decide where to focus
5. Obtain recognition that a client has an important unresolved problem
6. Analyze or diagnose why a client has an unresolved problem
7. Develop a strategy to address a client's unresolved problem
8. Implement and assess micro advocacy strategy
9. Learn skills needed by micro policy advocates
10. Learn how a social worker provided a micro policy advocacy intervention

DEFINING MICRO POLICY ADVOCACY

Curiously, social work curriculum often fails to provide students with skills to advocate for specific clients or consumers of service—an omission that is addressed in this textbook through micro policy advocacy. Highly used texts in direct practice do not discuss case advocacy at all or only briefly even though the well-being of many clients, or even their lives, depend upon timely access to programs like the Supplemental Nutrition Assistance Program (SNAP) and Medicaid (see scant attention to case advocacy, for example, in Hepworth, Larson, Rooney, & Strom-Gottfried, 2005). Micro policy advocacy helps clients or consumers obtain services, rights, and benefits that would (likely) not otherwise be received by them and that would advance their well-being. Micro policy advocacy can be provided directly to clients or through referrals, provided that micro policy advocates ascertain if their clients or consumers of service actually receive assistance.

Micro policy advocates engage in eight challenges (see Figure 1.1). They decide whether to proceed (Challenge 1). They decide where to focus (Challenge 2). They obtain recognition that a client has an unresolved problem from other staff in an agency (Challenge 3). They analyze or diagnose why a client has an unresolved problem (Challenge 4). They develop a strategy to address a client's unresolved problem (Challenge 5). They obtain support for their strategy to help resolve a client's unresolved problem (Challenge 6). They implement their strategy (Challenge 7). They assess whether their implemented strategy or policy has been successful (Challenge 8).

Advocates may decide to progress from micro to mezzo or macro policy advocacy if they decide that a client's unresolved problem is shared by other clients or persons and is caused by dysfunctional policies of agencies or communities (mezzo policy advocacy) or by dysfunctional policies in government settings (macro policy advocacy).

READING THE CONTEXT

Like with mezzo and macro policy advocacy, micro policy advocates must be acutely aware of the context that includes both opportunities and constraints as they help specific persons and families. The context includes policies, procedures, court rulings, funding sources, personnel, economic factors, and any other factor that shapes how social workers engage and work with clients and families. Our discussion will make clear why we use the term *micro policy advocacy* in this chapter, as well as the eight chapters devoted to specific sectors, because social workers and their clients are surrounded by policy and policy-related factors—and social workers have to help their clients navigate them so that they receive their benefits to the extent possible.

These contextual factors vary with different agencies and policy sectors. Social workers who work in mental health clinics, hospitals, schools, safety net programs, child welfare agencies, criminal justice programs, immigration services, and programs assisting seniors must be aware of those unique and cross-cutting policies that shape their work with clients and families.

Micro policy advocates need a strategic view of the organizations where they work so that they can identify (or find) assets that will facilitate their micro policy advocacy with specific consumers while not neglecting those contextual liabilities that will make it more difficult. For example, persons with insurance policies that exclude specific services or require considerable out-of-pocket payments may not be able to surmount them when an advocate seeks to help them contend with their health costs (a liability). An advocate who wishes to help an uninsured patient not be transferred from a private to a public hospital before he or she is medically stabilized can cite federal regulations that prohibit such transfers (an asset).

The context contains many "streams" that flow into and through specific agencies from external sources. Each of these streams may contain assets or liabilities for advocates in specific situations, including the following:

- Court rulings that require health staff to take certain actions and to avoid other actions, such as ones concerning when it is possible to withdraw food, medications, and/or hydration from a dying person
- Statutes (or legislation) that require specific actions, services, or accountability by health organizations and their staff
- Civil rights legislation that mandates that providers not violate specific rights of specific populations, such as requirements for ease of movement by persons with disabilities as required by the Americans with Disabilities Act (ADA) or ones that require agencies to provide translation services to persons with limited English proficiency (LEP)
- Flows of revenues to clients and agencies from private insurance companies, employers, Medicare, Medicaid, the State Children's Health Insurance Program (SCHIP), contracts that many not-for-profit agencies possess with public agencies, and scores of public programs
- Potential sanctions or adverse publicity that specific organizations might confront, such as allegations that specific child welfare agencies fail sufficiently to protect children's well-being in foster care placements
- Competition from other agencies in a specific policy sector, such as between not-for-profit agencies, for-profit agencies, and public and nonpublic agencies
- Accreditation standards established by accrediting agencies in each sector
- Regulations of specific local, state, and federal departments, such as ones that require schools to mainstream developmentally disabled children
- Legal litigation and inquiries of public interest attorneys and private attorneys
- Monitoring and oversight of public clinics and clinics by public authorities and funders in specific cities, counties, and states
- External advocacy groups that help consumers contest specific decisions or policies of mental health, education, and other agencies
- Resources and regulations of publicly funded programs from which consumers may be able to receive assistance, including Medicare, Medicaid, SCHIP, the Supplemental Nutrition Assistance Program (SNAP), Supplemental Security Income (SSI), insurance exchanges established by the Affordable Care Act (ACA) in many states, Social Security, unemployment insurance, rent subsidies, and scores of additional programs

- Specific research findings in the professional literature or from private or governmental sources that are widely viewed as establishing evidence-based findings based on empirical research
- Policies and procedures of specific agencies

The context includes many streams from internal sources within specific organizations that can also be assets or liabilities in specific instances of case advocacy, including the following:

- Budgets and their relative surpluses or deficits in specific health settings that establish what agency programs and functions receive priority
- Mission statements that state the priorities and goals of a provider
- Organizational culture as it favors specific activities while discouraging others—including case advocacy and policy advocacy
- Specific units that engage in advocacy or advocacy-related work, such as departments of compliance, risk management departments, bioethics committees, institutional review boards (IRBs), patient representatives or advocates, social work departments, care managers, and financial departments
- Specific officials, administrators, professionals, or staff who have expertise and clout in specific areas and programs, such as social workers in child welfare agencies

These internal factors can also be assets or liabilities in specific policy advocacy situations. Budget deficits may lead to cuts in services that can diminish the size of specific programs needed to provide services to specific consumers (a liability), but budget surpluses or increased funding from external sources can increase services (an asset).

Advocates have to update their knowledge to be effective. Laws and regulations frequently change. New and revised evidence-based findings frequently emerge in each of the eight sectors. Ethical standards change. Budget priorities of agencies often change rapidly. Policy and service preferences of the funders of services in specific agencies often change.

I place Internet sites at strategic places in this book to facilitate this updating process with reputations for accuracy. These include websites of governmental agencies, advocacy groups, and self-help groups. Policies by public agencies shift as they are modified by government agencies, regulatory bodies, and legislatures.

CHALLENGE 1: DECIDING WHETHER TO PROCEED

Social workers must ask themselves whether specific consumers might suffer harm if they did not provide them with micro policy advocacy including members of vulnerable populations, persons with high-threshold problems, persons with multiple problems, persons not receiving any care or only partial or delayed care, persons with overlooked problems, persons caught in a revolving door, persons enmeshed in destructive relationships, persons in physical or psychological jeopardy, persons who are unable to self-advocate or to receive advocacy from others, persons who are bewildered by systems of care, and persons who are currently harming themselves.

Membership in a Vulnerable Population

Vulnerable populations include consumers whose predisposing or intrinsic characteristics often expose them to one or more of the seven problems (Anderson, 1995; Jansson, 2019). They include consumers with distinctive racial or ethnic characteristics, including African Americans, Latinos/as, Asian Americans, and Native Americans, as well as specific immigrant populations from the Middle East, Africa, Eastern Europe, Russia, and Asia. They include documented and undocumented immigrants as well as persons with uncertain immigration status. They include women. They include consumers in different age-groups, such as children and adolescents as well as elderly persons. They include consumers with specific sexual orientations including the LGBTQ population. They include consumers with stigmatized health conditions, including persons with HIV/AIDS or sexually transmitted diseases and victims and perpetrators of family violence and rape. They include persons with stigmatized forms of mental illness, such as schizophrenia and clinical depression. They include persons with addictions. They include persons with eating disorders. They include persons with reduced mental capacity, such as persons with dementia, Alzheimer's disease, and developmental disabilities.

Vulnerable populations include people with enabling factors that often expose them to one or more of the seven problems (Anderson, 1995). They include consumers who lack medical insurance or who are underinsured. They include rural and inner-city consumers who live in areas with shortages of services. They include persons with limited English proficiency (LEP) and low-income consumers. They include consumers with chronic diseases including chronic obstructive pulmonary disease (COPD), congestive heart failure, heart disease, arthritis, degenerative muscle diseases such as muscular dystrophy, cerebral palsy, and genetic disorders such as Lou Gehrig's disease and Huntington's disease. They include consumers with life-threatening diseases such as some kinds of cancer and chronic diseases. They include persons with limited literacy. Many persons who belong to two or more of these vulnerable populations are at particular risk (Jansson, 2011).

Micro policy advocates should prioritize services to persons with limited income because they often possess multiple health, economic, mental health, employment, housing, disabling, and other conditions that taken together make it difficult for them to navigate the service systems in the eight sectors. They often have lower literacy skills that compromise their ability to understand written documents. They are more likely to encounter negative responses from highly educated professionals. Background factors, such as evictions, loss of employment, poor housing, and exposure to violence in their neighborhoods make it difficult for them to engage and use services. They are more likely to have high-threshold and multiple problems and to have other characteristics of persons who need priority in social workers' micro policy advocacy.

Clients With High-Threshold Problems

Clients particularly need micro policy advocacy when any of the seven core problems reach a high threshold that endangers their well-being or their ethical rights. Remember, however, that ethical issues that seem low threshold to outside observers may be important to specific consumers such as delays in service or disrespectful language (Daley, 2001). They can make this determination by using the following kinds of information:

- Information from specific consumers
- Care divergent with evidence-based findings
- Care divergent with ethical standards including informed consent
- Notations in medical records
- Direct observations
- Feedback from other professionals
- Feedback from family members

Clients With Multiple Problems

Many clients are plagued by more than one of the seven core problems that we discussed in Chapter 1. Advocates should often provide micro policy advocacy to clients with two or more of these problems. Some clients possess multiple, interacting problems that together place them in jeopardy—but receive care for only one of them. Clients who are obese, diabetic, and depressed and lack health insurance, for example, are likely to need micro policy advocacy to develop strategies for obtaining multifaceted care and for their multiple problems as well as their financial problems. Children with mental problems and poor school performance often need help from professionals from education and mental health fields.

Clients Receiving No, Partial, or Delayed Care

Some clients do not receive any care for a specific problem. Perhaps they lack transportation. Perhaps they lack childcare so they cannot use needed services. Perhaps their lives have been disrupted by evictions from their places of residence. Perhaps they lack resources to pay the fees associated with specific services. Persons of color sometimes do not seek attention for HIV/AIDS, for example, because it is stigmatized in their communities.

Clients may be deterred from seeking assistance when they find their treatments or assistance to be excessively delayed. They may be caught in long lines in public clinics. Perhaps they are placed on waiting lists.

Clients With Overlooked Problems

Many clients receive assistance for a presenting problem but find other problems to be unaddressed by specific agencies. A disabled person may receive care for his or her physical limitations but not attention for her anxiety or depression.

Clients Caught in a Revolving Door

Some clients repeatedly return to service providers. A considerable portion of medical and social services are devoted to helping the "worried well," that is, to persons who do not have serious problems but who want reassurance, attention, or companionship from providers. Some clients come to specific mental health facilities repeatedly for the same condition, such as alcoholism, substance abuse, or attempted suicide. Advocates may need to help them receive preventive, counseling, or other services to address causes of this repetition (Jansson, 2011).

Clients Enmeshed in Destructive Relationships

Persons who experience destructive relationships often need micro policy advocacy. Advocates are often required to inform authorities when specific consumers, such as women, children, or elderly persons, are subject to violent acts from family members or other persons.

Clients in Physical or Psychological Jeopardy

Social workers and other professionals sometimes engage in micro policy advocacy when they fear that specific consumers will otherwise die or suffer grievous physical injury. A nurse in a public clinic in the Bronx in New York City took it upon herself, for example, to locate consumers who had recently received biopsies revealing aggressive and advanced cancer but had not returned for follow-up visits. She persuaded many of them to return for treatment that they might otherwise have not received—and possibly saved some of their lives (Perez-Pena, 2005).

Clients Unable to Self-Advocate or to Receive Advocacy From Others

Some clients are able to advocate for themselves because they are highly skilled at navigating the service and health systems and have read about their conditions or searched the Internet diligently. They may already have received a second opinion from experts and already received advice from other persons who have experienced their condition. They are appropriately assertive, skilled communicators, and persistent. They keep records. They know about emerging treatments and tests. They are not traumatized by adverse diagnoses. Social workers may decide that these persons are able to advocate for themselves with no or minimal assistance. They should prioritize micro policy advocacy for consumers who lack these skills and knowledge, particularly when they possess serious problems that could threaten their well-being and their ethical rights (Jansson, 2011).

Some consumers have supportive families and friends who help them self-advocate or become surrogate advocates when they lack the skills, knowledge, or ability to advocate for themselves. Health care professionals need to consider providing advocacy when consumers do not have these supports.

In the case of comatose persons, persons with limited cognitive function, or persons under sedation or heavy medication, health care professionals should obtain the concurrence of family members or friends before engaging in micro policy advocacy.

Clients Bewildered by Bureaucratic and Policy Complexity

Social workers often provide services in complex bureaucratic and policy systems. Welfare, education, correctional, health, immigration, child welfare, and many other service organizations are complex institutions with multiple professions, administrative staff, internal units, and linkages with myriad external organizations. Their staff uses complicated jargon. It is not surprising that many clients find it difficult to understand them or to navigate within them, particularly when they have complex or serious problems and when they contend with medical and other professional jargon. Many providers are not good communicators.

Skilled micro policy advocates can better decide who needs advocacy by viewing these institutions from the bottom up like their clients. They can bring simple improvements to their care, such as scheduling visits so that consumers do not lose their privacy when professionals flow through their rooms at nonstop and unpredictable intervals. They can interpret procedures for them, help them locate places, facilitate family members' visits, and learn to detect clients' confusion and angst. Clients and families need this kind of assistance in each of the eight policy sectors we discuss in this book.

Clients Who Endanger Themselves

Micro policy advocates usually focus on factors that preclude clients from receiving needed services or that violate their rights. In some cases, however, they provide reverse advocacy that focuses on clients' self-destructive behaviors like smoking, substance abuse, lack of exercise, or poor diet—topics often not discussed by health and other professionals who do not want to antagonize or confront them.

Micro policy advocates can point consumers in new directions tactfully. They can ask, for example, whether anyone has informed them of support groups, Internet sites, or counseling assistance that might prove helpful in reducing their smoking, excessive use of alcohol, or other self-destructive behaviors. They can ask if they have received data that examines their health prognosis if they do not curtail or eliminate specific destructive behaviors. They can engage them in cognitive therapy, sometimes supplemented by medications, that has been found by researchers to help persons escape drug and alcohol addictions, as I discuss in Chapter 10.

CHALLENGE 2: DECIDING WHERE TO FOCUS

Advocates identify relevant policies, regulations, and organizational factors. They identify persons, programs, regulations, laws, resources, organizations, and services that are relevant to helping a client surmount an unresolved problem. Assume, for example, that a child needs a preventive preschool program, but Head Start is oversubscribed in his or her geographic area and the parents lack recourses to pay fees for private nursery schools. Also assume a parent is disabled and cannot help the child travel to other areas. A micro policy advocate might find it difficult to help this child but might nonetheless be able to locate scholarships or reduced fees to allow this child to gain entry to a church-based or other program in this area—or might find resources to empower the child's parent to help with reading or other activities that parallel ones provided in Head Start. By contrast, an advocate would have a simpler task if a local Head Start program was not oversubscribed.

The context includes key contacts of social workers. Perhaps they know a specific person in a program who can be trusted to help a client gain access. Perhaps they can counsel a client not to see a specific intake worker who has proven relatively hostile to similar clients in the past—unlike another intake worker.

Social workers need a strategic view to accomplish the navigational task not only of the programs, staff, and resources of their own agencies but also of other ones in their geographic areas. Assume, for example, that a woman with a prison record wants to have it legally expunged from the records that she gives to potential employers because she knows that it may lead to her inability to secure work and that many employers will discriminate against her no matter her actual talents and reliability. A social worker would need to know how to link this person to a legal aid attorney or other resource to help her resolve this issue.

Social workers empower persons to develop context-reading and navigational skills. They help them use the Internet. They engage them in role-playing to help them develop questions and interviewing skills. They help them become appropriately assertive and give them strategic advice.

CHALLENGE 3: OBTAINING RECOGNITION THAT A CLIENT HAS AN IMPORTANT, UNRESOLVED PROBLEM

Micro policy advocates devise strategies for placing clients' unresolved problems on the agenda of other persons so that they will provide assistance in resolving them. The advocate locates persons who have the authority and motivation to help persons with unresolved problems. They find fellow professionals at the service delivery level. They involve specific supervisors or higher-level administrators when needed.

Micro policy advocates often help assemble a team that works together to help a client resolve a problem. It may include professionals, family members, and friends. Assume, for example, that a senior citizen needs help in making his or her home accommodating to his or her physical and mental problems, including moving a bedroom downstairs, placing ramps over outside steps, remodeling the bathroom to include a downstairs shower, and removing impediments that could cause a fall. The team might include a spouse, an adult child, a contractor, a financial adviser, and a professional from a senior center. The advocate helps convince each of these persons that the issue is sufficiently important to be placed on their agendas because, absent these modifications, this senior citizen could suffer a life-threatening facture or brain injury from a fall (Jansson, 2011).

In simpler cases, the advocate may need to convince only a single person to place an issue high on their agenda. He or she might convince the senior's gerontologist to contact the executive director of a senior center funded by the Older Americans Act to assume leadership in helping this senior citizen. Advocates often involve family members and friends in their micro advocacy because their concurrence and cooperation are often needed. Without the help and encouragement of his or her spouse and children, for example, this senior citizen might be so overwhelmed at the task of remodeling the home that he or she fails to take action.

Social workers can sometimes use shortcuts. They can convince other professionals to provide micro case advocacy for and with specific persons—what I call "advocacy punting" (Jansson, 2011, p. 33)—because they have greater expertise or may already have the trust of a particular client. They can empower specific clients to self-advocate if they have the skills and motivation to be their own advocates.

CHALLENGE 4: ANALYZING OR DIAGNOSING WHY A CLIENT HAS AN UNRESOLVED PROBLEM

Before they can devise a strategy, social workers must diagnose why a client experiences one or more unresolved problems from the seven core problems that we discussed in Chapter 1.

Developing Narratives

Micro policy advocates often develop narratives of specific transactions between specific clients and their interactions with other professionals and staff. They construct a chronology of events, service transactions, motives or beliefs, the nature of communications, and key actions (Steiner, 2005). They obtain evidence from records, direct observations, feedback from other professionals, and feedback from family members, as well as their prior experiences in specific service settings.

Advocates should understand some challenges in constructing accurate narratives. They cannot assume that reports or perceptions of specific participants are necessarily accurate, such as teachers, mental health professionals, or physicians. A physician may report in the medical record, for example, that he or she had an extended discussion with a client in which he or she covered the pluses and minuses of specific treatment options, but the actual discussion may have been relatively brief and may have only covered treatment options tangentially because the physician was already committed to a specific diagnosis and course of treatment.

Clients, too, may exaggerate, distort, or even falsify events and communications. When given serious diagnoses, for example, some clients may be sufficiently traumatized that they cannot process information accurately, so they may wrongly believe they were given (or not given) specific information. A few persons may even deliberately falsify information to advance their claims for workers' compensation or to lay the groundwork for malpractice litigation.

The narrative should lead to an informed opinion about why a specific consumer needs case advocacy. Here are some recurring causes:

- Prejudice
- Narrow mind-sets
- Flawed transactions and communications between specific consumers and specific medical staff
- Poor ethical reasoning
- Inadequate use of evidence-based medicine
- Bureaucratic and organizational malfunctioning
- Flawed policies and procedures
- Causation by multiple factors

POLICY ADVOCACY LEARNING CHALLENGE 4.1

IDENTIFYING TYPES OF PREJUDICE

Identify some kinds of prejudice of professionals and staff who work in other sectors, such as education, safety net programs, mental health facilities, child welfare, gerontology, criminal justice, and immigration. Could you provide micro policy advocacy to persons who experience it?

Clients Subjected to Prejudice

Advocates need to be alert to prejudicial treatment of specific clients in specific sectors, including overt prejudice as well as subtler forms. They need skills not only in recognizing when prejudicial treatment exists but also in helping specific consumers surmount it. Prejudice can take overt form, such as excluding specific persons from services, but it has been greatly reduced by enactment of federal and state civil rights laws. Considerable evidence suggests, however, that more subtle forms of prejudice exist, such as making some persons wait for longer periods than others, discouraging some persons from using specific services when they need them, and providing better-quality care to some persons than others (Institute of Medicine [IOM], 2003).

Responding Negatively to Clients' Behaviors

Some professionals and staff may be biased against clients who are not "good" clients, such as those who are excessively compliant and non-communicative or who do not ask important questions, seek alternatives, or educate themselves about their specific problems. Unless their providers take the time to elicit their ideas and level of information, they often do not receive full explanations. Social workers should not let bothersome actions and behaviors distract them from the ethical goal of offering a person optimal care.

Narrow Mind-Sets

Professionals sometimes have excessively narrow perspectives. Many counselors focus on clients' psychological states sufficiently that they ignore realities and life pressures they experience, such as insufficient resources, lack of job training, and poor housing. Some physicians rely excessively on technological remedies. They may not refer patients with mood disorders or other mental conditions for counseling—or simply give them medications without referring them to mental health experts. They may not consider implications of specific surgeries or medications for consumers' employability or lifestyles. They may not warn consumers of side effects of specific treatments that might have led them not to agree to the treatment.

Many providers are excessively specialized, not able to see beyond the confines of their practices and specialties. Older persons may see an array of specialists, for example, who treat them for separate conditions but who are unaware that others have given them medications that adversely interact with ones they have prescribed. Specialists often do not communicate with clients' primary physicians, who often do not follow their care even when they receive surgery and other treatments in hospitals. Many of them do not refer consumers with mental health problems to psychiatrists, psychologists, or social workers.

Professionals sometimes suffer from excessive insularity that leads them not to link their services to clients' households and communities, even when their clients might benefit from community-based mental health services, job training services, alternative medicine, welfare programs, childcare services, schools, rehabilitation services, services for seniors, services for disabled persons, independent-living programs, and physical therapy programs—to name only a few (Jansson, 2011).

Inadequate Communication

Consumers often do not receive sufficient information from providers to enable them to make intelligent choices as in the case of a psychiatrist who fails to inform a patient about side effects of specific medications. They often receive information that they cannot understand because it uses terms and concepts beyond their grasp. LEP persons often cannot understand English sufficiently to comprehend mental health, medical, educational, and other options discussed with them. Providers often fail to listen to consumers.

Lack of Ethical Reasoning Skills

Some professionals possess inadequate ethical reasoning skills. They may believe they have given consumers sufficient information that they can make informed choices even when they have not. They may believe they have protected confidential information, only to divulge it to family members and others without seeking the consumer's permission. They may believe that they have honored the ethical value of honesty without having warned clients about likely side effects of treatments and medications.

Lack of Knowledge of Relevant Research

Some providers are unaware of recent empirical data that suggests that specific consumers would likely benefit from specific helping strategies. Despite in-service training and continuing education, many providers are not current with existing research (IOM, 2008). I discuss in Chapter 10 how many professionals who help clients with substance abuse refer them to Alcoholics Anonymous (AA) when recent research questions the effectiveness of AA as compared with treatment that couples specific medications with cognitive therapy.

Unwillingness to Consider New Approaches

Some providers may be aware of specific evidence-based findings pertinent to the care of specific consumers but do not want to change their customary approach. Berwick (2003) contends that "innovators" constitute a relatively small portion of the workforces of most organizations with many other staff not implementing innovations—or waiting until they see others who have adopted innovations, even years later.

Misinterpretation of Research

Professionals sometimes interpret research incorrectly. Perhaps they think it is more definitive than it is. Perhaps they do not realize that it applies to a specific population but is untested or not successful with other populations. Perhaps they do not realize that some interventions bring short-term improvements that do not last.

Bureaucratic and Organizational Malfunctioning

Social workers deliver services in relatively complex organizational settings. These settings possess not only bureaucratic structures, but also rules, protocols, budgets, and cultures that profoundly influence their provision of services. Clients, too, must navigate these bureaucracies, which often appear confusing to them.

Failure to Follow a Preferred Sequence of Events

Errors can often be reduced if providers follow a specific sequence of events when treating certain clients. Researchers have discovered, for example, that rates of injuries and deaths in intensive care units (ICUs) are markedly reduced when health professionals follow a sequence of actions (Gawande, 2010). A patient may not otherwise have been given a test, medication, or treatment at the correct time—or at all. A consumer may receive the wrong test result or even the wrong surgical procedure. Pharmacists may not have been able to read a physician's handwriting, leading to an incorrect medication. Nurses may fail to give proper medications.

Lack of Team Practice

Many persons with chronic diseases, substance abuse problems, and serious mental problems benefit from team practice that couples medical assistance, "talking therapies," and other professionals such as physical therapists, nutrition experts, and occupational therapists.

POLICY ADVOCACY LEARNING CHALLENGE 4.2

ENGAGING IN MICRO POLICY ADVOCACY—SEEING THE BROADER CONTEXT

Assume that an elderly person with early-stage dementia and some disorientation receives help from a professional in a mental health clinic. Assume, as well, that the professional focuses primarily on relieving the presenting symptoms, such as with medications provided by a consulting psychiatrist.

Learning Exercise

1. What key questions might the professional have asked during this visit that might have led to better services, including about the home situation, the presence of other family members, the possible dangers the client might encounter in her or his environment, the client's diet, and other medications that the elderly person currently receives.
2. Does this failure to ask questions parallel failure to follow a preferred sequence of events in intensive care units?

POLICY ADVOCACY LEARNING CHALLENGE 4.3

WHEN DOES A SOCIAL WORKER MOVE FROM MICRO TO MEZZO POLICY ADVOCACY?

Should a social worker, when encountering the preceding situation, move from micro to mezzo policy advocacy to develop a policy that defines the child welfare agency's obligation to children and their parents even when they do not find evidence of abuse or neglect?

Many elderly persons benefit from team practice. Many barriers frustrate team practice, however. Insurance companies and health plans often do not fund it. Some physicians do not want to share their patients—or their medical information—with other professionals. Many mental health clinics lack resources to hire or to contract with members of multiple professions.

Lack of Knowledge

Clients suffer violations of their rights or non-optimal care because specific professionals lack knowledge of specific regulations, protocols, and standards. Perhaps they did not participate in briefings or in-service training. Perhaps they received inaccurate information from other staff. Perhaps they are unaware of specific accreditation standards. Perhaps they possess a mistaken view that their professional status makes them not subject to policies and procedures, are not aware of specific court rulings and regulations, or are not current with relatively new policies. They may not realize that they could be subject to litigation and legal sanctions if they fail to adhere to specific regulations and laws.

Lack of Policy Guidance

Policies and procedures may not even exist in some situations—or may be excessively vague. Assume, for example, that child welfare staff in a particular jurisdiction conduct an investigation to see if a specific child has been abused or neglected. Not finding evidence to support this finding, they terminate the investigation but fail to refer the child and his or her family to supportive services after they document serious mental problems of the child and one of the parents.

Causation by Multiple Factors

Many factors, operating in tandem, often compromise the care of specific patients or lead to violations of their ethical rights. Perhaps a consumer is subject to prejudice, receives inadequate communication, and runs into bureaucratic snafus. The work of micro policy advocates is often made more complex as the number of causal factors increases.

CHALLENGE 5: DEVELOPING A STRATEGY TO ADDRESS A CLIENT'S UNRESOLVED PROBLEM

Social workers can accomplish the strategizing task by dividing it into steps, including deciding who will provide advocacy services at the outset, setting goals and time frames, identifying assets and liabilities, assessing motivations of decision makers, allocating responsibility as the advocacy intervention proceeds, developing a sequence of actions, and developing specific actions.

Step 1: Deciding Who Will Provide Advocacy Services at the Outset

Even when triaging suggests that specific consumers possess at-risk indicators for micro policy advocacy, health care professionals cannot always provide it for legitimate reasons. They sometimes can find advocacy shortcuts as well.

Social workers often encounter daunting workloads when their regular work is combined with paperwork and other bureaucratic requirements. The urgency of their regular work may sometimes trump case advocacy because micro policy advocacy takes time, particularly in complex situations.

Social workers sometimes decide that positive outcomes are highly unlikely in specific instances. Perhaps they have attempted case advocacy on prior occasions with a particular consumer with no success, or a consumer does not want advocacy, or they cannot gain access to specific consumers who might otherwise benefit from it.

Micro policy advocates often question the status quo, such as the failure of mental health staff to obtain informed consent, follow evidence-based practices, or attend to consumers' depression or anxiety. They may decide in some situations that they will encounter excessive risk when injecting themselves into some situations that could jeopardize their ability to engage in micro policy advocacy in future situations.

Social workers need to beware, however, of using rationalizations to refrain from engaging in micro policy advocacy, such as deciding they do not have time to do it or that they might experience adverse repercussions. They need to remember that they have an ethical obligation to increase consumers' well-being and to protect their ethical rights even when it is inconvenient or expedient not to engage in it. They may also wrongly conclude that they will suffer repercussions or wrongly decide that case advocacy takes a lot of time.

Social workers should also realize that they can use legitimate shortcuts and avoidance even in those situations in which they cannot devote much time to micro policy advocacy or fear repercussion. They can try to convince other professionals to provide advocacy to a particular client. These staff may have more time in particular situations, more expertise relevant to a specific consumer, or better standing with a particular professional or staff member—and they may be better able to fit advocacy into their workloads at a particular moment.

They may also diminish the likelihood of adverse repercussions to themselves by empowering specific consumers to self-advocate. Consumers possess considerable power in advocacy situations based on their right to self-determination and their ability to seek another provider if they are dissatisfied with their current services. Micro policy advocates can empower consumers by referring them to the Internet, educating them, coaching them, and engaging in role plays with them.

Step 2: Establishing a Goal and a Time Frame

Micro policy advocacy requires the development of a goal, a subgoal, and a time frame. Goals that contain these three components become actionable and can be monitored to see if consumers have actually achieved them (Jansson, 2011). The goals can be illustrated by using the seven core problems that consumers often confront in human services:

- *Overarching goal:* Surmount a financing problem as it adversely affects a consumer's medical care. *Subgoal:* Gain eligibility to Medicaid to allow financing of specific tests, medications, or procedures. *Time frame:* Obtain eligibility within two weeks and receive charitable care in the interim.
- *Overarching goal:* Surmount a possible threat to a consumer's quality of care. *Subgoal:* Help a consumer obtain a second opinion. *Time frame:* Obtain it within three weeks so the consumer can better decide what course to take.
- *Overarching goal:* Obtain specific preventive services. *Subgoal:* Have a diabetic consumer receive a consultation from a nutrition consultant. *Time frame:* Obtain it within one month.
- *Overarching goal:* Obtain culturally relevant care. *Subgoal:* Obtain the services of a certificated translator in Vietnamese. *Time frame:* Obtain the translator within one hour.
- *Overarching goal:* Obtain an appointment to relevant resources outside a specific agency. *Subgoal:* Obtain an appointment at a local welfare office to make applications for SNAP (food stamps) and to obtain information for filing for an Earned Income Tax Credit (EITC). *Time frame:* Obtain the appointment within two days.
- *Overarching goal:* Address excessive anxiety, depression, or other mental issues. *Subgoal:* Have a suicidal consumer at the emergency room get an appointment with a mental health provider at a mental health agency. *Time frame:* Obtain it within three days with the use of medications in the interim as well as daily phone calls.
- *Overarching goal:* Protect ethical rights of a consumer. *Subgoal:* Obtain separate consultations for a client for palliative care and hospice rather than regular care. *Time frame:* Obtain two consultations within five hours.

Micro policy advocates may decide to accomplish two or more goals in the same time frame, such as helping a consumer surmount a financial issue and obtain quality services—both within two weeks. They may decide to accomplish these two goals sequentially, beginning with one of them and progressing to another after the first one has been achieved.

Step 3: Identifying Assets and Liabilities

It is sometimes useful to make a list of assets and liabilities in the context—and their locations outside or within the site of a micro policy intervention. Assume, for example, that a social worker wants to help a client get into a smoking cessation program. Also assume that the client is uninsured and Latino—and has early-stage emphysema. Also assume that the physician who has treated him has made no effort to get him into such a program. Assets and liabilities are summarized in Table 4.1.

Even with her limited information, the advocate finds that the context contains both favorable and unfavorable factors—suggesting that a positive outcome is possible. Clinics do exist. Other physicians in the hospitals have referred consumers to them. State subsidies do exist. Empirical evidence supportive of smoking cessation is strong. Yet liabilities also exist. This physician has not made prior referrals.

The advocate realizes, too, that he or she does not know some important facts. Do culturally competent and accessible services exist for this client? Does a smoking cessation clinic exist near the home? Does it have evening or weekend hours and a short wait list? What are the costs because the consumer has no health insurance? By listing the assets and liabilities, the advocate concludes that a favorable outcome is possible.

TABLE 4.1 ■ Evaluating the Context

	Assets	Liabilities
External	1. Evidence-based research exists. 2. State-subsidized clinics at community sites exist.	1. No referral system in place.
Internal	1. Other physicians have referred persons to smoking-cessation programs. 2. Accreditation standards require referrals for persons with lung disease.	1. Client's physician hasn't made referrals in the past.

Step 4: Assessing the Motivations of Decision Makers

Advocates' effectiveness often hinges on their abilities to assess the motivations of specific decision makers. Why did the client's physician not refer this person to a smoking clinic even when early-stage lung disease is apparent? Did the physician fear offending the client? Was the doctor not aware that smoking-cessation programs exist? How likely was it that the client could be persuaded to get help from the smoking cessation program even without the physician's referral? Do subsidies from the state to the clinic allow significant or full financing of services? Micro policy advocates need to factor into their strategies ways to address these motivations effectively so that they increase the chances that they will change directions and take different actions that enhance clients' well-being.

Step 5: Allocating Responsibility

Social workers must decide who takes responsibility for advocacy intervention. Different possibilities exist. They can become the prime movers in an advocacy intervention, such as with comatose consumers, those with limited cognitive function, those with serious mental disorders, or those overwhelmed by their health conditions or external realities. Some clients may expect them to be the prime mover under these circumstances.

Social workers can empower consumers to advocate for themselves. They provide them with knowledge about resources, health conditions, treatment options, and relevant research, whether by giving them written materials or by helping them to use the Internet if they do not already have sufficient Internet skills. They may coach them, such as helping them develop questions to ask medical staff. They may help them be appropriately assertive, giving them suggestions about how to ask questions or communicate with medical staff to increase the likelihood of successful outcomes. They engage in role-play with them in which they assume the role of the medical staff.

Even when empowering consumers, social workers may wish to keep in touch with specific consumers to see if they have taken key actions. They may discover, for example, that a client fails to follow through with specific actions. Perhaps other life events intrude so that they cannot advocate for themselves.

Social workers can partner with consumers, as well as other medical staff, in a team arrangement. They can decide who does what and by when—and select someone to be the lead person who keeps track of progress. In partnering arrangements, a lead person tracks the implementation of the strategy—and troubleshoots when key actions are not taken.

Social workers can initiate a team approach in which a number of professionals and family members participate in an action system. Perhaps a nurse works with a consumer to obtain an appointment for a second opinion as a social worker helps him or her get an appointment in a community-based agency as the consumer agrees to adhere to specific medications and to keep the appointments. Perhaps progress on these three actions is monitored online, by telephone, or through a case conference to be held at a specified time. Considerable research suggests that persons with chronic diseases need to be helped by a team to obtain a full range of interventions not available to them under solo practice. Diabetics need, for example, help monitoring their weight, healthy lifestyle, insulin, occupational therapy, counseling, and participation in support groups.

Professionals can shift their roles as events unfold. Perhaps someone begins as the prime mover but discovers a family member or friend who takes more responsibility. Perhaps they discover that a client who had wanted to take responsibility for self-advocacy wants some assistance.

Step 6: Developing a Sequence of Actions

Advocates plan a tentative sequence of communications, fact finding, contacts, and meetings at the outset of their work—with some contingencies included in it. In the case of a Latino with lung disease who will not join a smoking-cessation clinic unless his physician "prescribes" it and signs a referral form to a specific smoking cessation clinic, the advocate decides to implement the following sequence of actions:

- He or she will conduct some research for the client, such as finding a smoking cessation clinic near his residence that has evening and weekend hours, provides free services, and has Spanish-speaking staff.

- He or she asks the client to take this information to his physician and ask for a medical referral to this clinic—and to let the advocate know if he has received it.
- *Contingency.* If he does not receive the referral, the advocate will work with the physician's nurse to get the referral from another physician in the clinic that serves persons with lung disease.
- He or she asks the client to get an appointment at the clinic—telling him to see a specific Spanish-speaking staff person whom the advocate has contacted by telephone—and asks him to leave a message whether he kept his appointment.
- *Contingency:* He or she will place in his medical record a notation that he failed to keep his appointment if the advocate learns this from the clinic staff person—and suggest further discussions with him about how his smoking could greatly increase the progression of his lung disease.

Different kinds of sequences exist. The simplest of them involves a single action by an advocate who helps a person who lacks skills or abilities to self-advocate. Perhaps he or she asks a physician to make a medical referral for counseling that the advocate can present to a hospital-based psychiatrist. A more complex sequence of action takes place when an advocate empowers a client to get this referral from a physician. The advocate works with the client at two points in time: coaching to get the referral from a physician and coaching to get an appointment with a psychiatrist. An even more complex sequence of action occurs when the advocate initiates a team approach that involves a physician, a social worker, a psychologist, and an occupational therapist. Assume the social worker assumes the lead role in helping a client with serious mental illness. He or she initiates a case conference to help a consumer receive medications (from the physician), counseling (from the psychologist), and job readiness and job search skills (the occupational therapist). He or she monitors the client's progress and orchestrates another case conference in three months.

Sequences also vary by the extent they include contingencies and question marks. Contingencies take place when case advocates anticipate two (or more) possible courses of action depending on the outcome of a particular meeting or event as illustrated by the referral to a smoking cessation program in which two contingencies were identified. Question marks can be placed on a sequence of actions in which the case advocate does not know what specific actions might be needed after a certain point in the sequence. Perhaps a client will have numerous options—and the advocate does not know what they will be until an interview, case conference, diagnosis, or medical procedure is complete.

Step 7: Selecting Specific Actions

Social workers possess a smorgasbord of actions they can take during their micro policy advocacy.

Using Technology

Social workers can use technology to facilitate their advocacy in several ways. They can use spreadsheets to identify the goals, subgoals, and timelines of an advocacy intervention.

They can select specific activities as well as who does them and by when. If they evolve a team approach in which different persons agree to perform specific actions, they can enter them into the spreadsheet and record when they are completed. They can e-mail spreadsheets that show what activities have been accomplished and which have not. For simpler advocacy plans, they can resort to e-mail or texts between the lead person and those clients with Internet access. Or they can use Facebook to orchestrate advocacy among two or more professionals and the client.

Empowering Consumers

Advocates can empower consumers in many ways. They can show them Internet sources that may be useful to them—and give them some instruction about how to distinguish between accurate and inaccurate information.

Advocates can refer clients to support groups that meet within a hospital or clinic or that hold meetings in community-based organizations, schools, and religious organizations. The members of support groups often give clients invaluable information as well as emotional support. They often encourage members to be appropriately assertive.

They can direct clients to reliable sources of data about specific institutions and providers on the Internet.

Advocates can help clients develop a list of questions before they see providers. They might ask about treatment options and side effects that could accompany them. They can obtain information about the skills and experience of the staff that might help them. They can inform them about their ethical rights, such as disclosure of information so that they can make informed choices. Advocates can give consumers hotlines, as well as names of advocacy groups outside their hospitals or clinics, to help them contest decisions by insurance companies or managed-care plans about the coverage and financing of specific medical procedures, tests, and medications.

Advocates can empower specific consumers by coaching them to prepare for specific encounters with agency staff. They can encourage them to be appropriately assertive. They can give them a list of specific questions they might wish to ask. They can encourage them to have a spouse, family member, or friend accompany them.

Arranging Case and Family Conferences

Social workers can arrange case conferences that include the staff at an agency and a client to discuss available services and their likely cost.

If a retirement facility is disinclined to admit an elderly person because of his or her medical condition, an advocate can orchestrate a case conference to discuss why the client was rejected—and to see if the decision can be reversed.

Case conferences can be used to arbitrate or settle disputes between a client and an agency, such as over the cost of services, the effectiveness of services, or any other issue. Denial of errors in hospitals and clinics not only harms the institution by increasing malpractice lawsuits, but often has a devastating impact on consumers and their families (Delbanco & Bell, 2007). Clients sometimes need a place where they can voice their criticisms of the services that they received even if they do not seek redress.

Family conferences can usefully expose different points of view so that they can be factored into the decision-making process. Disagreements can be brought into the open so that they can be resolved.

Locating External Advocates

Some consumers will believe that they cannot seek redress unless they use external advocates, whether advocacy groups, government personnel who regulate and oversee health care in a specific jurisdiction, private attorneys, or public-interest attorneys such as those in legal aid organizations.

Many advocacy organizations advocate for specific consumers. Some of them mediate among insurance companies, medical administrators, and consumers to solve specific reimbursement or coverage issues—often successfully. Others request that specific clients gain services or resources to which they are entitled, such as translation services, equipment funded by Medicare for older persons, and referrals to home health agencies before they are discharged.

Other advocacy organizations seek to educate consumers about services and resources that they can obtain on discharge, such as support groups for persons with breast cancer, diabetes, and many other health conditions. They sometimes visit clients in hospitals before they are discharged.

Many organizations possess internal arbitration mechanisms so that aggrieved consumers can seek redress without taking legal action. Many managed-care organizations require enrollees to sign agreements that require them to use arbitration rather than courts to seek redress.

Many private attorneys represent clients. They may threaten or take legal action against specific medical staff or specific hospitals or clinics, insurance companies, health plans, and managed-care organizations. Public interest attorneys often represent persons who lack the resources to obtain private attorneys.

When helping clients gain access to public programs like SNAP and Medicaid, advocates should not take "no" as definitive when clients are rejected. They should advise clients to ask for a supervisor's opinion. Or they should appeal an adverse decision to higher authorities within the bureaucracy because mistakes are common. Many public agencies have boards or officials designated to review decisions of lower-level staff. Or they can go to a legal aid office for advice or a private attorney.

Making Referrals

Advocates become involved in referrals when they believe that consumers will not otherwise receive them. They can coach consumers to call for appointments on repeated occasions, often early in the morning, in hopes of getting an appointment through a cancellation. Advocates can increase the odds that specific persons will receive prompt service by making direct contact with intake staff—and by getting to know them professionally.

Some clients do not keep appointments even when they have serious problems. Advocates may sometimes e-mail or phone specific clients to increase the odds that they keep the appointment—and then help them, or urge them, to reschedule them if they miss them.

Finding Resources

Clients often need to obtain eligibility to specific programs. Advocates not only inform clients of these programs and their guidelines but help them prevent adverse decisions by educating them about their requirements. They inform them about requirements that they receive eligibility information in their own language—or with special materials that are developed for persons with limited literacy. They inform them that they have the right to appeal eligibility decisions.

Using Intermediaries

Social workers often use intermediaries as advocates for their clients, such as their supervisors, highly placed nurses, administrators, or physicians. These intermediaries often have clout, information, and resources that may be helpful to a client. Perhaps a client needs a referral to palliative care or hospice—and a social worker knows that a specific nurse has oversight responsibilities for these programs. This nurse can easily initiate an action system for the client to get expedited information, such as having personnel from these programs visit the client soon after he or she calls them.

Honoring Clients' Wishes

Clients often find that some of their wishes are not honored or are marginalized by staff in a particular agency, clinic, or hospital. Advocates can state that clients and patients have the ethical right to decide what services to use as long as they are mentally competent.

Citing Regulations, Protocols, and Ethical Guidelines

Advocates can take several actions when they observe apparent violations of regulations or court rulings by other professionals, such as a specific court ruling that requires that clients give their "informed consent" to a mental health treatment. They can remind or inform persons about these regulations or court rulings. They can ask, "Are we following guidelines that mandate us to . . . ?" or "Have we heard what course of treatment the patient desires?" They can discuss possible penalties that providers can encounter if they violate specific regulations, such as when a hospital social worker informed a medical resident that he might lose his license if he persevered in discharging a homeless person who had not been medically stabilized. They can report violations to an intermediary, such as an administrator. They can report them to a relevant department, such as the compliance or risk management departments. They can inform clients where they can find specific regulations.

Steering Clients

Social workers sometimes steer consumers to specific programs and professionals because they believe they offer outstanding services and steer them away from programs and professionals that they believe offer inferior services.

CHALLENGES 6, 7, AND 8: IMPLEMENTING AND ASSESSING MICRO POLICY STRATEGY

A micro policy advocate may assist a client in developing a strategy, but the advocate and the client will succeed only if it receives support from those officials who confer resources, opportunities, benefits, or rights to a specific client. They need to implement this strategy with skill. Micro policy advocates and clients assess specific advocacy interventions to determine their relative success in improving the well-being for specific clients. If they were not successful, they may provide another advocacy intervention, possibly relying on a different strategy.

MEASURING AND PREDICTING MICRO POLICY ADVOCACY OF SPECIFIC FRONTLINE PROFESSIONALS

Some data from the research project that I discussed in Chapter 1 suggests that many frontline health professionals, including social workers, engage in relatively high levels of micro policy advocacy in hospitals as indicated in Table 1.1. (The words "patient advocacy" are synonymous with "micro policy advocacy" in the ensuing discussion.) The research team developed and validated a Patient Advocacy Engagement Scale composed of 26 items to measure the levels of micro policy advocacy of specific frontline professionals. (For more information about this Scale, see Jansson, Nyamathi, Heidemann, Duan, & Kaplan, 2015.)

The research team also developed scales that predicted levels of micro policy advocacy of specific frontline professionals. Frontline professionals who reported they had relatively high levels of competence with respect to 17 specific skills had relatively high scores on the Patient Advocacy Engagement Scale (see Table 4.2 for a list of these skills).

Data suggests that frontline health professionals who report that they possess specific skills are more likely to engage in relatively high levels of micro policy advocacy than other frontline professionals (see Jansson, Nyamathi, Duan, et al., 2015, for statistical findings that describe scales used to predict levels of micro policy advocacy of frontline professionals. The research team developed a Patient Advocacy Skills Scale that measured the respondents' estimations of their level of proficiency on 17 skills. Professionals with relatively high scores performed higher levels of micro policy advocacy than those with relatively low scores ($p < .001$).

The research team developed a Patient Advocacy Ethical Commitment Scale that measured their ethical commitment to provide micro policy advocacy. Frontline health professionals who believe that members of their profession have an ethical duty to engage in patient advocacy are more likely to have relatively high scores on the Patient Advocacy Engagement Scale ($p < .001$). Frontline professionals were asked to indicate on a scale of 1 (not at all) to 5 (a great deal) the extent they believe members of their professions are mandated by their profession's code of ethics to engage in micro policy advocacy, to inform providers of patient rights, to assist patients' decision-making about health care including understanding risks and benefits, to help patients obtain access to health care, to sometimes challenge providers' decisions on behalf of specific patients, to inform patients about benefits and rights under the ACA, to help patients gain access to additional sources of information and health opinions, and to act on behalf of those patients who cannot assert their views.

The research team developed a Patient Advocacy Eagerness Scale that asked them to indicate on a five-point scale to what extent they wanted to engage in higher levels of micro policy advocacy in coming months with respect to patients' ethical rights,

TABLE 4.2 ■

Please rate the extent you have the following skills from 1 (not at all) to 5 (high level).
I have the skill to assess why specific clients have unresolved problems.
I have the skill to develop alternative strategies to help clients resolve specific problems.
I have the skill to use influence to persuade others to help specific clients.
I have the skill to use appropriate assertiveness to help specific clients obtain assistance from others.
I have the skill to use intermediaries to facilitate micro policy advocacy.
I have the skill to negotiate and bargain on behalf of specific clients.
I have the skill to resolve conflicts that arise during micro policy advocacy.
I have the skill to empower clients to advocate for themselves.
I have the skill to coach clients to advocate for themselves.
I have the skill to advocate clients' wishes with other professionals.
I have the skill to call specific, unresolved problems to the attention of the other professionals.
I have the skill to decide and prioritize which unresolved problems warrant immediate attention.
I have the skill to faithfully represent clients' wishes when they cannot advocate for themselves.
I have the skill to question plans or actions of other professionals to improve clinical outcomes and avoid adverse events.
I have the skill to network with other professionals to facilitate micro policy advocacy.
I have the skill to refer clients to community resources.
I have the skill to use influence to persuade others to help specific clients.

quality of care, cultural content of care, preventive care, affordability of and access to care, patients' mental health conditions, and patients' community-based care. Frontline professionals with relatively high scores on this scale were more likely to score relatively high on the Patient Advocacy Engagement Scale ($p < .001$).

Other findings from the research on frontline health professionals suggest that organizational factors powerfully impact levels of their micro policy advocacy. The project validated a Patient Advocacy Organizational Receptivity Scale. Frontline professionals were asked to answer 10 questions on a five-point scale that included these: "Are you invited to participate in case conferences about patients with specific patients?," "Is there an atmosphere that invites you to question the evolution of unresolved problems with specific patients?," and "Do patients' attending or consulting physicians encourage you to inform them when you see patients' unresolved problems?" (See Jansson, Nyamathi, Duan, et al.,

2015.) Frontline professionals with relatively high scores on this scale were more likely to have relatively high scores on the Patient Advocacy Engagement Scale ($p < .001$).

The research team developed and validated a Patient Advocacy Tangible Support Scale. Frontline professionals were asked to rate levels of support in their hospitals for micro policy advocacy by responding to 10 questions on a five-point scale that included these: "Do you believe administrators are aware of unresolved patients problems?," Does your supervisor support you when your patient advocacy results in negative repercussions?," and "Is patient advocacy discussed during orientation of new staff?" Frontline professionals with relatively high scores on this scale were more likely than ones with relatively low scores to have high scores on the Patient Advocacy Engagement Scale ($p < .01$).

Two additional validated scales predicted frontline health professionals' levels of engagement in micro policy advocacy. A scale titled Belief Colleagues Engage in Advocacy measured the extent frontline health professionals believe their colleagues engage in micro policy advocacy. They answered five questions on a five-point scald that included these: "How confident are you that patients with serious unresolved problems will receive patient advocacy in this hospital?," "To what extent do the social workers engage in patient advocacy in this hospital?," "To what extent do the nurses engage in patient advocacy in this hospital?," and "To what extent do medical residents engage in patient advocacy in this hospital?" Frontline professionals who believe relatively few colleagues engaged in patient advocacy were less likely to engage in it themselves ($p < .01$).

The research team developed and validated a scale titled Belief the Hospital Empowers Patients by asking frontline professionals to indicate the extent to which their hospitals empower patients by providing them with Internet sites, support groups, evidence-based literature, professionals who can give second opinions, spiritual support, and community resources. Frontline health professionals with relatively high scores on this scale were more likely to have relatively high scores on the Patient Advocacy Engagement Scale than ones with relatively low scores ($p < .001$).

I provide this data in this chapter to demonstrate that a combination of personal factors and organizational factors facilitate or discourage frontline professionals' engagement in micro policy advocacy. It suggests that workshops, orientation meetings, training in the use of skills, and reinforcement of ethical underpinnings of micro policy advocacy may increase levels of micro policy advocacy by frontline health professionals. The data suggest, as well, that social workers need to engage in mezzo policy advocacy to make their organizations more likely to provide a supportive context for micro policy advocacy, such as by empowering patients with educational materials, highlighting micro policy advocacy of different professions within hospitals, and increasing tangible support for micro policy advocacy.

LEARN HOW A SOCIAL WORKER ENGAGED IN MICRO POLICY ADVOCACY

Read the following case study of micro policy advocacy, and then address the questions at the end of the case.

POLICY ADVOCACY LEARNING CHALLENGE 4.4

ENGAGING IN MICRO POLICY ADVOCACY—WHY IT IS PARTICULARLY NEEDED BY AND FOR LOW-INCOME PERSONS OF COLOR

Social workers often encounter families that have multiple issues with health, mental health, economic, substance abuse, and marital conflict. In such cases, they engage in micro policy advocacy scores of times over an extended period. Take the case of Jim Smith, a 45-year-old African American man living in an inner-city neighborhood of a large city along with his wife, Ava, their son, Tom, and their daughter, Lily. Jim and Ava have developed barely adequate income over many years from Jim's work as a car mechanic and Ava's as a full-time cashier at Walmart. Although sometimes scraping the bottom of the barrel, the Smiths were content with their lives, and delighted that they own their own home.

They have since experienced a series of adversities, however, that stressed their lives—and they needed micro policy advocacy for each of these adversities. In this case, please read the background factors that describe issues that the social worker observed working with them in Part A. Then proceed to Part B to see actions and strategies that the social worker used when engaging in micro policy advocacy.

Part A: Background Issues

Personal Trauma. The couple lost their son, Tom, to gang violence about five years ago when caught in a crossfire between two gangs. Jim feels that he should have prevented his son's death, that he failed to protect his son from gun violence as his father had done for him. He repressed his angst, however, for three years. He stopped talking to his wife. His health declined. He resumed smoking crack cocaine.

Inadequate Access to Key Health Services. Jim can't obtain adequate mental and physical health services because he can only afford health insurance for his daughter. His income places him just above the eligibility level of Medicaid—the prime source of funding for low- and moderate-income persons in the United States. He won't use services of a large public hospital because he does not trust non-black doctors who, he believes, don't understand his culture and are prejudiced against African Americans. He relies on a small community clinic that has had to cut its psychosocial services due to its budget shortfalls. When absolutely necessary, he uses the emergency room of a nearby hospital.

Inadequate Access to Mental Health Services. Jim has serious, under-addressed mental health problems. The small community clinic that he used for his health and mental health issues recently cut its mental health staff markedly due to budget problems. This staff was cut within months of Tom's death. It also cut its social service staff.

Adverse Consequences of Not Seeking or Receiving Help. As Jim's health, mental health, and already-difficult access to social services deteriorated—possibly related to his inability to talk with others about his son's death—he developed frequent headaches and dizziness. He was diagnosed as having high blood pressure, for which he received medications, but doctors failed to diagnose his clinical depression. Meanwhile, the small clinic he used was forced to close. Jim's depression worsened. He stopped taking medication for blood pressure. He often stayed in bed all day, and was unreliable in fixing people's cars at his family business. His use of cocaine resumed. His wife finally persuaded him to see a doctor after he fainted in the shower.

Inability to Find Preferred Services: Travel and Race. He went to a nearby public hospital, where he was referred to a mental health program that gave him help for his clinical depression and drug abuse counseling. But he soon learned that the public hospital, too, was closing. He did not want to use another public hospital due to its distance from his home. He also wanted to receive medical care from black health professionals.

Key Allies. His health improved somewhat. His depression became less serious. He was able to restrict his use of drugs to weekends. His wife continued to worry about him and feared his health problems were harming their marriage. She now carried most of the economic burden of the household due to his sporadic work. Later, the social workers enlisted help from his wife as she worked with Jim.

Ill Effects of Lifestyle Choices. Jim continued to complain about fatigue and headaches, as well as weakness in his left side; blood dripped from his nose when he sneezed. When he finally went to the public hospital, a white, male physician diagnosed him as having had a stroke caused by his use of drugs.

Negative Experiences in the Health System. Jim had negative experiences in the public hospital where he stayed for one week to recover from his stroke. He had long waits and was unable to see a physician for a full day even after going to the emergency room. Jim believed the white physician who diagnosed his condition in the emergency room was condescending and insulting—so much so that he yelled at him and accused him of being prejudiced against black people. Deciding not to argue with Jim, the doctor wrote him off as a drug addict who had poor coping skills, like those he had seen in the past.

Part B: Micro Policy Advocacy Interventions

How Good Social Workers Engage in Many Kinds of Micro Policy Advocacy. The doctor referred him to a social worker. Jim vented to the social worker, telling her all of his grievances and frustrations about his lost son, his dislike of past medical staff, his depression, and his drug habit. The social worker helped him gain perspective on his life. She complimented him for controlling his drug use. She sympathized with him with respect to his lack of medical insurance and the closures of the health clinic and the public hospital. She helped him understand that he was not responsible for his son's death because he could not stop random shooting from gang members.

She helped him obtain his prognosis from a physician, who responded that it would take six months to one year for Jim to return to his work as a mechanic. She said she would help him ascertain if he might qualify for Supplemental Security Income (SSI) and Medicaid as a disabled person partially incapacitated by a stroke. She helped Jim's wife obtain necessary paperwork after the doctor filled out documentation to verify that Jim was disabled.

The social worker provided support for Jim's wife after she told her that Jim seemed indifferent to her. She encouraged Jim and Ava to obtain marital counseling to be funded by Ava's health insurance from her work. She helped Ava file for paid leave from her job through the Family and Medical Leave Act of 1993, using the same proof of disability that Jim had already obtained from the physician to obtain SSI.

The social worker used cognitive therapy to help Jim deal with his drug addiction. She referred him to a physician who specialized in addictions to see if he might benefit from medications that ease the craving for powder cocaine. She referred him to a psychiatrist to ascertain if he might receive medications for chronic depression after Jim divulged that he wondered if it was worth living as he still grieved for his son.

The social worker helped Jim realize that he has many strengths. She was delighted that

(Continued)

(Continued)

he and Ava would receive marital counseling funded by her insurance. She helped him realize that he could receive high-level medical care in this hospital, including from white health professionals when necessary.

When Jim learned that his application for disability from SSI had been denied, the social worker helped him file an appeal with new documentation. She also referred him to legal aid to help with this appeal.

Learning Exercise

1. Compare this family with a family in the upper middle class. What economic realities frustrate this family at every turn in the road?
2. How did Jim cope with turbulence in the health and mental health system he used over time?
3. Discuss how social workers need considerable expertise in so-called safety net programs such as Medicaid and SSI.
4. Why did Jim feel far more comfortable with African American health professionals than European American ones? And why is it difficult for him and other African Americans to gain access to them? What strategies do they need to use in the meantime?
5. Identify the sheer range of micro policy advocacy interventions that the social worker employed in helping Jim and Ava. Can we view the advocate as at the center of a wheel with many spokes?

Learning Outcomes

You are now equipped to:

- Define micro policy advocacy
- Read the context
- Decide whether to proceed
- Decide where to focus
- Obtain recognition that a client has an important, unresolved problem
- Analyze or diagnose why a client has an unresolved problem
- Develop a strategy to address a client's unresolved problem
- Implement and assess micro advocacy strategy
- Learn skills needed by micro policy advocates
- Learn how a social worker provided a micro policy advocacy intervention

References

Anderson, R. (1995). Revisiting the behavioral model and access to medical care: Does it matter? *Journal of Health and Social Behavior, 36,* 1–10.

Berwick, D. (2003). Disseminating innovations in health care. *Journal of the American Medical Association, 289,* 1969–1975.

Daley, J. (2001). A 58 year old woman dissatisfied with her care. *Journal of the American Medical Association, 285,* 2629–2635.

Delbanco, T. & Bell, S. (2007). Guilty, afraid, and alive: Struggling with medical error. *New England Journal of Medicine, 22*(2), 1682–1683.

Gawande, A. (2010). *The checklist manifesto: How to get things right.* New York: Henry Holt.

Hepworth, D., Larsen, J., Rooney, D., & Strom-Gottfried, K. (2005). *Direct social work practice: Theory and skills* (7th ed.). Belmont, CA: Thomson Brooks/Cole.

Institute of Medicine (IOM). (2003). *Unequal treatment: Confronting racial and ethnic disparities in healthcare.* Washington, DC: National Academy Press.

Institute of Medicine (IOM). (2008). *Knowing what works in health care.* Washington, DC: National Academy Press.

Jansson, B. (2011). *Improving healthcare through advocacy: Guidelines for professionals.* Hoboken, NJ: John Wiley & Sons.

Jansson, B. (2019). *Reducing Inequality.* San Diego: Cognella Academic Publishing.

Jansson, B., Nyamathi, A., Duan, L., Kaplan, C., Heidemann, G., & Ananias, D. (2015). Validation of the patient advocacy engagement scale for health professionals. *Research in Nursing & Health, 38*(2), 162–172.

Jansson, B., Nyamathi, A., Heidemann, G., Duan, L., & Kaplan, C. (2015). Predicting patient advocacy engagement: A multiple regression using data from health professionals in acute-care hospitals. *Social Work in Health Care, 54*(7), 559–581.

Perez-Pena, R. (2005, October 15). At a Bronx clinic, high hurdles for Medicaid Care. *New York Times,* p. A1.

Steiner, J. F. (2005). The use of stories in clinical research and health policy. *Journal of the American Medical Association, 294*(22), 2901–2904.

5

PRACTICING MEZZO POLICY ADVOCACY INTERVENTIONS

This chapter was coauthored by Gretchen Heidemann, MSW, Ph.D.

LEARNING OBJECTIVES

In this chapter, you will learn to:

1. Define mezzo policy advocacy
2. Identify five skills needed by mezzo policy advocates
3. Read the context of agencies and communities
4. Decide whether to proceed in agencies and communities
5. Decide where to focus in agencies and communities
6. Obtain decision makers' recognition of an unresolved problem in agencies and communities
7. Analyze problems in agencies and communities
8. Develop a strategy to address problems in agencies and communities
9. Develop support for a strategy or proposal in agencies and communities
10. Implement a strategy or policy in agencies and communities
11. Address whether the implemented strategy or proposal is effective in agencies and communities
12. Analyze a case example of mezzo policy advocacy in Watts in South Central Los Angeles

DEFINING MEZZO POLICY ADVOCACY

We define mezzo policy advocacy as policy advocacy that seeks to address persons' seven core problems by changing organizations and communities. Mezzo policy practitioners also help create an organizational milieu that encourages agency staff to engage in micro, mezzo, and macro policy advocacy in the first place. Assume, for example, that agency staff are not encouraged to make suggestions about how agency services might be improved—or not encouraged to engage in policy advocacy in surrounding communities. They could engage in mezzo policy advocacy to expand the scope of their work to include organizational and community changes so that this expanded scope is not only allowed but encouraged. As one example, the executive director of an agency that served homeless persons identified macro policy advocacy actions with low, medium, and high impact, for example, getting on a mailing list of an advocacy organization that focuses on an issue relevant to your agency (low impact); calling a policy maker at local, state, or federal levels about an issue you care about (medium impact); and involving your clients in writing, calling, and visiting policy makers (Jansson, 2018).

IDENTIFYING FIVE SKILLS NEEDED FOR MEZZO POLICY ADVOCACY

Mezzo policy advocates need to have a number of skills to draw upon to be effective in changing policies at the organizational and community levels. Next, we describe several skills that a mezzo policy advocate should develop.

1. **Initiating.** Mezzo policy advocates need to initiate policy-changing interventions. This is a critical skill because failure to take the initiative often means that the wishes of other persons and officials prevail even when they lead to dysfunctional policies. Advocates need in general to be appropriately assertive. They have to place issues on the table while not unnecessarily provoking conflict.

2. **Influencing.** Mezzo policy advocates need to be able to influence other people to work with them to change specific policies. We discuss influence or power resources at more length in the next chapter.

3. **Negotiating and bargaining.** Mezzo policy advocates need to be able to negotiate and bargain with agency and community leaders to achieve their policy goals. They have to decide what outcomes or policy changes they want while realizing that policy advocates often must compromise.

4. **Mediating conflicts.** Mezzo policy advocates need to be able to mediate conflicts among stakeholder groups. They can mediate by identifying points or issues that competing groups have in common. They can help participants identify compromises. They can use process skills in task groups and committees to facilitate positive outcomes.

5. **Communicating.** Mezzo policy advocates need to be able to communicate with concerned citizens, agency heads, community leaders, public officials, and other persons who can help resolve specific issues. Communication skills of policy advocates are often different from communication skills of direct-service professionals with their clients. Policy participants often want to analyze broader issues that impact the quality and nature of services and programs. They want data that sheds light on the nature of specific issues. They want to examine policy and program options. They want to know the cost of specific proposed reforms.

A PRELIMINARY CHALLENGE: READING THE CONTEXT OF AGENCIES AND COMMUNITIES

We use the term *mezzo policy advocacy* because social workers and their clients are surrounded by policy and policy-related factors in the organizations and communities in which they work and because social workers have an ethical mandate to help resolve policy-related problems in agency and community settings with and on behalf of their clients. As with micro and macro policy advocacy, mezzo policy advocates must be acutely aware of the context that includes both opportunities and constraints as they help groups of clients in agencies or groups of community residents. Contextual factors vary with different agencies, with different communities, and with different policy sectors. Regardless of which sector they work in, social workers must be aware of the unique and cross-cutting organizational policies that shape their work with clients and families, such as organizational budgets, eligibility guidelines, and service delivery protocols. Moreover, social workers working in any agency and within any sector must be aware of community-level factors that impact their clients, such as the existence or dearth of social services or advocacy groups, the civic engagement of residents, and community demographics, such as poverty, unemployment, and crime.

In this section, we briefly discuss the policy context of agencies: their mission statements, budgets, revenue streams, organizational culture, hierarchies, and key players (or personnel). We subsequently turn our attention to the policy context of communities: their demographics, social problems, distribution of services, level of civic engagement, involvement of community leaders and advocacy groups, presence of neighborhood councils and home owners' and business associations, involvement of law enforcement, and media presence.

The Policy Context of Agencies

Policies are essential to the functioning of all agencies. Next we discuss specific types of agency policies after discussing internal versus external agency policies and formal versus informal agency policies.

Internal and External Polcies

Agencies are affected by both internal and external policies. Internal policies establish specific rules, such as intake procedures, staffing requirements, content of services, reporting mechanisms, and a general statement about the program's purposes. Some details of funded programs are left relatively vague so that units, departments, and staff must fill the gaps with their own policies. Some policies are originally external to agencies but are then internalized or adopted, like the policies agencies accept when they take funding from governmental programs. An agency that has a number of externally funded programs (not-for-profit agencies now receive the majority of their funds from the government and from foundations) will have multiple sets, or clusters, of these adopted policies. External policies, such as court rulings, shape agency policies as well. For example, the *Tarasoff* decision requires social workers to inform intended victims when a client says that he or she intends to inflict bodily injury on them.

Formal and Informal Policies

Formal agency policies include mission statements, budgets, organizational charts, and formal written policies (such as eligibility guidelines, intake procedures, and service delivery protocols as they pertain to clients as well as human resources policies, worker safety protocols, and ethics guidelines as they pertain to staff). Agencies are also influenced by many informal policies, including organizational culture, competition and collaboration from outside agencies, and values and preferences of staff. Both formal and informal policies have the power to shape staff's actions and choices at many points in their work and deliberations. Informal policies may diverge from official written policies, but they may also fill gaps when official policy is ambiguous (Chu & Trotter, 1974).

Mission Statements

Most not-for-profit agencies have a mission statement that defines their priorities and direction (Hasenfeld, 1983). Although relatively broad, mission statements establish the agency's general philosophy and help guide policies and practices related to the organization's purpose.

Written Policies

Agencies typically—but not always—have written policies that pertain to a number of interorganizational activities. These include but are not limited to service delivery policies (such as intake guidelines and referral policies), ethical practice policies, human resource policies (including hiring, grievances, and dismissals), worker safety protocols, administrative policies and procedures, meeting processes, evaluation practices, and board of directors bylaws. Human resources and worker safety policies provide structure and security to employees. Administrative policies ensure that the organization is run fairly and efficiently. Service delivery policies help clients and consumers know which services they are entitled to, how to access those services, and how long those services will last. Bylaws ensure that the board of directors operates ethically and efficiently.

These written policies provide a powerful system of checks and balances for all stakeholders within the organization. However, many small start-up nonprofits do not have such sophisticated policies and procedures in place. Moreover, even when such policies are in place, there is no guarantee that they will be followed. Indeed, examples abound of nonprofit organizations "behaving badly." Watchdog and consumer information organizations, such as CharityWatch (charitywatch.org) and Charity Navigator (charitynavigator.org), play an important role in informing the public about the practices of nonprofit organizations. CharityWatch publishes an annual "Hall of Shame," spotlighting organizations that have egregiously violated basic ethical principles or otherwise acted out of concert with widely regarded standards. Such watchdog groups play an important role in educating the public and holding nonprofits accountable. County and state legal officials serve as watchdogs, as well, such as when the New York State attorney general terminated a foundation that Donald Trump had established because its resources were used for personal gain rather than charity.

Revenue Streams

The sources from which an agency receives funding can heavily impact its policies. More than $335 billion was donated to charitable organizations in 2013 (http://www.charitynavigator.org). Of that, 15% was received from foundations. A typical request for proposals (RFP) from a foundation requires that the organization seeking funds must agree to certain policies regarding how they can use or disperse the funds, who can be served under the funded project, and in what types of activities the organization can engage. Organizations may also receive funding from city, county, state, and federal sources. This revenue is often heavily regulated in similar ways to foundation funding, which can constrict the recipient organization's activities.

Budgets

Agencies' annual budgets serve as symbolic policy statements. They shape agencies' priorities by distributing resources to programs. To fully understand an agency's policies, one must examine both its present and previous budgets to determine the agency's priorities and how they have changed (Gummer, 1990). Because service delivery is predicated on an organization having funding, mezzo practitioners should pay careful attention to the budgetary processes of their agencies so that they have a clear understanding of funding streams and expenses.

Hierarchies and Organizational Charts

Anyone who seeks to change organizational policies should develop a clear understanding of hierarchies and divisions of labor. A hierarchy is the chain of command that gives high-level executives powers to create policies, hire staff, and make budgets. The division of labor divides staff into units that focus on specific tasks. While many organizational theorists have sought ways to soften hierarchy and specialization within organizations, they remain enduring, important features of many agencies.

Hierarchy and division of labor are often reflected in an agency's organizational chart. The board of directors of nongovernmental and nonprofit agencies, which usually appears at the top of the organizational chart, makes many important policy decisions. The board establishes an agency's high-level policies (i.e., its mission); hires its executive director; oversees the development of personnel policies; examines the agency's budget; and serves as the general overseer of the organization. Executive directors (presidents or CEOs) and program directors (COOs) usually have powers that enable them to shape decisions for the entire agency. They develop the agency budget and present it to the board; participate in hiring, firing, promoting, and supervising of staff; and shape the personnel and service-delivery policies that guide the organization. Lower-level professionals and volunteers often carry out direct services to the clients. As intermediaries between management and direct-service staff, supervisors often share the perspectives of both higher- and lower-level personnel. Program directors are directly impacted by policy choices and agency budgets for their particular divisions within the agency.

An organizational chart can be misleading. Organizational charts do not tell which units or programs in an organization have considerable resources and which ones the top executives' favor. In addition, organizational charts imply that high-level persons make most general policy choices, but they do not tell who has power with respect to specific issues. Many political scientists have observed that the distribution of power often varies with an issue; persons who are exceedingly powerful concerning certain issues may have little or no power regarding other issues (Dahl, 1967). The organizational chart, moreover, does not reveal patterns of friendship and trust, enmity, or social distance that can develop among staff members. Nor does it tell us about human contacts that span different units of an organization or that cut across levels of the hierarchy, including ongoing informal clusters of persons who share knowledge and who support one another.

Organizational Culture

The culture of an organization is a powerful factor that shapes the behaviors of stakeholders, including administrators, employees, and clients. A mezzo policy advocate will need to be keenly aware of the organizational culture of the agency or setting in which they are working to advance their reforms. When surveying the agency in which they work, advocates may discover that it conforms to one of the main types of organizational cultures proposed by Cook (1987).

Constructive organizations have collaborative environments where members of a group work together toward positive goals (Cook, 1987). These organizations are effective (their members are able to achieve complex tasks), self-actualizing (members are able to realize their personal potential), humanistic (members help others grow and develop), and affiliative (members cooperate with and develop pleasant relationships with others). In a constructive organization, the advocate will likely find colleagues who are willing to partner and support their advocacy efforts. They might also find that they are able to appeal to higher-level goals as motivation for decision makers to adopt their reforms.

Passive or defensive organizations, by contrast, are characterized by social norms in which members interact with others in ways that do not threaten their own security (Cook, 1987). Members of passive or defensive cultures tend to feel pressure to please others, to avoid interpersonal conflict, and to unquestioningly follow rules and procedures. Advocates in a passive or defensive organization may find that they must appeal to their colleagues' need to please to recruit them to support proposed reforms because members of this type of organization tend to avoid interpersonal conflict. They may also have to work strategically to convey how their proposed reforms do not threaten the security of high-level officials.

In *aggressive or defensive organizations*, people tend to focus on their own individual needs at the expense of the success of the group (Cook, 1987). This culture is characterized by opposition (members tend to be critical and cynical), power (members strive for prestige and status and desire to control others), competition (members work to protect their own status by outperforming others), and perfectionism (members have extremely high standards and place excessive demands on themselves and others). In aggressive or defensive organizations, advocates might find that they have to work largely alone to advance their reforms and may have to appeal to decision makers' desire for power and perfectionism to get their reforms adopted.

Collaboration and Competition

The role of collaboration and competition in influencing organizational policies—from dictating which programs will remain in operation to who is eligible for services—cannot be understated. Nonprofit organizations are forced to compete for increasingly scarce public and private dollars to maintain and expand their programs and services. Foundations, which provide funding to nonprofit and community-based organizations, increasingly call on these organizations to collaborate on special projects of concern to the community rather than operating independently. This allows funders to apply their dollars toward one goal that is achievable through collaboration rather than spreading resources among many organizations that may address only a small piece of the problem.

Values and Preferences

Informal policies can be reflected in the values and preferences of staff members and the expression and interpretation of specific policies. For example, different staff members often have diverse and subjective notions concerning their organization's mission. Although an official mission may stress services to families with single heads of households, some staff may want the agency to redirect efforts toward serving single mother-headed households over single father-headed households. When confronting virtually identical client problems, one staff member may provide services based on traditional psychotherapy, whereas another may emphasize survival skills, advocacy, and empowerment.

POLICY ADVOCACY LEARNING CHALLENGE 5.1

MAPPING AGENCY POLICIES

Take any agency with which you are familiar and trace the following aspects:

- Mission statement
- Stated or unstated values that you can determine
- Revenue stream (i.e., primary funding sources)
- Budget priorities (i.e., how it allocates funding across its services)
- Organizational chart, including key personnel and their duties or functions
- How it collaborates and competes with outside agencies
- Any aspects of informal organizational culture that you can determine, including organizational culture type (from those described in this chapter or from any typology existing in the extant literature)

(Continued)

(Continued)

Now, identify any written policy that exists within the organization. This can include a policy related to who qualifies for services, how clients or employees file grievances, or any other written policy.

With regard to the policy you identified, discuss or respond to the following questions:

Learning Exercise

1. How is the policy shaped by the organization's mission statement?
2. How is the policy shaped by the organization's revenue streams and budgetary priorities?
3. How is the policy shaped by the organization's culture?
4. Of these, which has the most powerful influence on the particular policy you identified?
5. If you had a desire to reform or change the particular policy you identified, how might the most powerful factor you identified in Question 4 influence whether and how you proceed with your advocacy efforts?

The Policy Context of Communities

Next we discuss factors that shape community policies and form the backdrop to any policy change intervention a mezzo policy advocate may undertake.

Type of Community

Many different types of communities exist. The terms *urban*, *rural*, and *suburban* are often given to communities to describe where they exist in relation to large cities. Typically, a large amount of commercial activity occurs in urban communities, and these communities can be extremely impoverished or extremely wealthy depending on the city's tax base and revenues. Rural communities are those located far from major cities and include small towns, villages, and remote farming communities. Similar to urban communities, rural communities span the range from extremely impoverished (such as parts of Appalachia and the deep South), to those composed mostly of middle-class residents, to incredibly wealthy rural communities (e.g., those in wine country in various states). The major economic driver in most rural communities is agriculture. Suburban communities are those located on the outskirts of major cities. A map of the Chicago area, for example, lists dozens of suburbs located in 10 counties (including some in Indiana and Wisconsin) that spread out around Chicago proper. Communities that lie on the farthest outskirts of major metropolitan areas and are composed almost entirely of houses, schools, and retail locations (and are void of major commercial activity) are sometimes referred to as the "exurbs." Other types of communities that may not fit the urban-rural-suburban typology include college towns, resort towns, and mill towns.

Urban, rural, and suburban communities assume major roles in the nation's politics. In elections in Virginia in 2016 and 2018, for example, many rural areas elected Republican representatives, whereas suburban areas such as Arlington swung toward Democrats. Inner cities of Richmond mostly voted Democratic.

Demographics

Communities' demographic characteristics are described by census data, which indicates whether they are demographically heterogeneous or homogeneous. Social work advocates should use census data to inform their policy reform initiatives. Census data can be used to determine whether the community contains a dominant ethnic group or a variety of ethnicities and whether a community is predominantly low income or affluent or possesses a mix of social classes. Census data indicates whether communities are relatively integrated or racially segregated and can be used to describe the extent of poverty and unemployment, home ownership, and many other factors that are important in understanding the social problems that may exist in the community.

Land Use and Zoning

Zoning is the practice of designating land within a jurisdiction, such as a city, town, or community, for land-use purposes, such as commercial, industrial, residential, and recreational. Zoning is often used to prevent new development from interfering with existing residents or businesses and to preserve the "character" of a community. Because zoning is sometimes used for negative purposes, such as to control the social class and the ethnic makeup of neighborhoods, knowledge of these issues can be critical for the mezzo level advocate. Exclusionary zoning measures may artificially maintain high housing costs through land-use regulations and maximum density requirements, which can exclude lower-income and "undesirable" groups from a given community. The movement of a community from a low-income to an affluent community is called "gentrification," often leading to replacement of low-income persons with more affluent ones. As this process takes place, relatively affluent persons often oppose the construction of low- and moderate-income housing. As we discuss in Chapter 9, exclusionary zoning and reduction of low- and moderate-income housing contributes to increases in homelessness in many communities.

Presence and Distribution of Services

Communities can be described by the extent to which they possess social service agencies and community groups that meet the social welfare needs of their residents. Many communities have a high need for certain types of services, and yet these types of services are severely lacking. In other communities social welfare services may be abundant. Communities that suffer high rates of high school dropouts, gang violence, drug and alcohol addiction, homelessness, and incarceration are often devoid of the prevention and treatment services that would alleviate these social problems.

Civic Engagement Including Electoral Politics

The extent to which members of a community vote and are otherwise engaged in civic activities is a major factor shaping the policies that emerge from and impinge upon communities. An engaged and informed citizenry is more likely to be able to secure needed services, ensure safety, and promote a high quality of life within a community. Those communities in which large numbers of residents are disenfranchised are often the same communities with high rates of poverty, unemployment, homelessness, and other social problems. This is a factor in many low-income communities of color, where large numbers of people have been previously incarcerated and are no longer allowed to vote, believe that they are unable to vote, or believe that their vote does not make a difference.

Mezzo policy advocacy includes participating in the electoral process. Although staff in not-for-profit and public agencies cannot participate in elections as part of their working hours, they can educate clientele and community residents about specific propositions on the ballot and the positions of candidates on them. They can engage in electoral activities on their own time, such as volunteering for specific advocacy organizations (Jansson, 2018).

Leaders and Key Informants

When addressing specific issues in communities, such as where to place social service organizations or how to mobilize opposition to certain policies, policy advocates need

POLICY ADVOCACY LEARNING CHALLENGE 5.2

REGISTERING HOMELESS, INCARCERATED, AND FORMERLY INCARCERATED PEOPLE TO VOTE

The outcome of the presidential election of 2020 will be determined by outreach by staff and volunteers of both major political parties. Let's examine my outreach (Gretchen Heidemann). In the lead-up to a prior presidential election, I worked as a voter registration coordinator. My job was to register homeless, incarcerated, and formerly incarcerated people to vote. Many homeless individuals living on the streets believed that they were not eligible to vote because they did not have a permanent address. On the contrary, in Los Angeles County where I was working, a homeless person could list the park or underpass or intersection where they slept as their address on a voter registration application as long as they also provided an address where they could receive mail; this could be a relative's home, a social service agency, or a post office box. I also worked with county jail officials to register persons to vote inside Los Angeles County jails who were eligible. A large proportion of jail inmates are persons who have been charged but not yet sentenced, such as those awaiting trial. The logistics were quite complicated, but we successfully established the first-ever voter registration program inside the jail that year. I also worked to register formerly incarcerated people to vote. Although laws vary by states, in many places people who have served their sentence and are no longer on parole are eligible to vote. In many cases, I had to overcome fear and mistrust; some persons with convictions were afraid that if they completed a voter registration card, they would be sent back to prison. These were all uphill battles, but my coworkers and I believed it was important for these groups of disenfranchised citizens to participate in key decisions that would impact their lives and their communities. On election night, we rented vans, picked people up from homeless shelters and the streets, and took them to the polls to vote.

Discuss whether outreach to marginalized groups is important in the presidential election of 2020. Can social workers engage in this activity? Can they link these activities to specific candidates for political office in your specific location that seek to increase resources and services for specific marginalized populations from the lists provided in Chapters 1, 2, and 3? Can these candidates be persuaded that such outreach can increase the odds that they will be elected? Do they possess census, voting, and other data that would help social workers identify the locations of specific marginalized populations?

to discover which community residents, leaders, politicians, and institutions have traditionally focused on their issue. Community leaders might be rabbis or pastors in local, faith-based institutions, directors of social service agencies, representatives of neighborhood councils, school principals, elderly residents who have resided in the community for a long time, or any other person who knows the inner workings and politics of the community. These persons are often referred to as key informants. Many resources are available to help mezzo policy advocates identify and interview key informants whose expertise can be invaluable in policy change efforts within communities.

Resources on Key Informant Interviews

The UCLA Center for Health Policy Research provides a clear and easy-to-follow guide to conducting key informant interviews. It can be located at http://healthpolicy.ucla.edu/programs/health-data/trainings/Documents/tw_cba23.pdf.

Existing Advocacy Groups

In seeking specific reforms in communities, policy advocates often affiliate with, consult, or enlist the support of community-based advocacy groups. These groups may include civic associations and local chapters of national groups, such as the National Association for the Advancement of Colored People (NAACP). Policy practitioners sometimes form coalitions representing social agencies and community groups when they want to oppose a specific measure or establish new programs (Dluhy, 1990).

Neighborhood Councils and Watch Groups. Community residents often organize themselves into groups—either formally or informally—for various purposes. Neighborhood watch groups began developing in the late 1960s in New York as a response to the rape and murder of a young woman in a park in Queens, where more than a dozen witnesses did nothing to stop the incident or apprehend the perpetrator. Neighborhood watches spread throughout the country and were a powerful force in many neighborhoods in the 1970s and 1980s. Where they still exist, these groups are typically devoted to crime and vandalism prevention within neighborhoods. Members are trained not to intervene when they become aware of suspicious or untoward activity but rather contact law enforcement officials for assistance. The shooting death of Trayvon Martin, a black, unarmed 17-year-old in Sanford, Florida, in February 2012 by off-duty neighborhood watch captain George Zimmerman—who was eventually acquitted of murder charges—ignited a fierce social debate about the role of neighborhood watch groups. Once viewed as a positive and protective force in communities, neighborhood watch groups are now under intense scrutiny, particularly in ethnically diverse areas where racially motivated fears are high.

Neighborhood councils, on the other hand, are typically established through local governmental entities, such as city councils. Neighborhood councils do not govern neighborhoods and are designed to provide an advisory role to city officials on issues of concern. Neighborhood councils usually hold monthly meetings that are free and open to the public. Any community stakeholder, whether a resident, business owner, student, or member of a faith-based institution, can become a member of a neighborhood council. Councils also have elected positions, such as chair, secretary, treasurer, business representative, and area representative. In cities including Los Angeles and San Diego, California, and Tacoma, Washington, neighborhood councils play an active and sometimes powerful role in shaping policies within their communities.

Business Owners' Associations. Associations of business owners are also powerful entities that work actively to shape community policy. Business owners within a community will often join local chambers of commerce or form other less formal associations to protect the business interests within the community. They may work with their local government, such as the mayor, city council, or other local representatives, to develop pro-business initiatives. Mezzo policy advocates should be aware of such entities where they exist as they can be either powerful supporters or opponents of issues affecting the welfare of the community.

Law Enforcement Involvement. Although a thorough discussion of policing and its effects on communities and their residents is beyond the scope of this chapter, advocates should be aware of the powerful impact that law enforcement can have on a community. In many poor and ethnic neighborhoods, police-resident relations are strained, at best, and hostile at worst. Both historic and current examples abound of corruption and brutality on the part of law enforcement against residents of communities.

In other communities, friendlier styles of policing known as "community policing" and "restorative justice" are being practiced. Community policing focuses on building ties and working closely with members of communities for which law enforcement officials are responsible. Actual community policing practices vary widely from place to place. Restorative justice focuses on addressing the needs of both victims and perpetrators of crime instead of punishing the offender. Victims take an active role in the restorative justice process, whereas offenders are encouraged to take responsibility for their actions by apologizing, returning stolen money, or participating in community service.

Community Advocacy Groups. Specific community groups focus on issues impacting marginalized populations as well as the general public. They can include ones that seek to reduce homelessness by pressuring local government, as well as state and federal governments, to construct and finance low- and moderate-income housing, retaining or establishing rent control in specific locations, and allowing residents to construct "granny housing" on their lots. They include groups that pressure local officials to enhance the enforcement of building codes to decrease unsafe housing. They include groups that seek to redress toxic chemical sites that disproportionately exist in low-income communities and communities of color. They include groups that pressure police not to use excessive force with respect to marginalized populations. Social workers can form these community groups and can work with existing ones.

Local Media. The media, including local newspapers and television and radio stations, often assumes a pivotal role in local communities as a powerful delivery mechanism for community-level information. The media can help shape policies by playing the role of whistle-blower on local businesses, agencies, or elected officials who are operating unethically. It can also help shape policies by raising awareness about social problems within a community. Mezzo community advocates should be aware of the news outlets within their community and develop relationships with reporters who can aid their advocacy efforts.

CHALLENGE 1: DECIDING WHETHER TO PROCEED

Mezzo policy advocates engage in eight challenges (see Figure 3.1). They decide whether to proceed (Challenge 1) and where to focus (Challenge 2). They secure key agency or community leaders' attention of unresolved problems in the agency or community (Challenge 3). They analyze why the problem(s) have developed (Challenge 4). They develop a strategy or policy proposal to address the unresolved problem(s) (Challenge 5). They obtain support for their strategy or policy proposal (Challenge 6). They implement their strategy or policy (Challenge 7). Finally, they assess whether their implemented strategy or policy has been successful (Challenge 8). After discussing a preliminary challenge—reading the context—we discuss each of the eight challenges as they apply to agencies and communities.

Deciding Whether to Proceed in Agencies

Mezzo policy advocates must decide whether an issue or problem within their organization merits the development of an advocacy intervention. Social workers should consider engaging in policy advocacy within their organization when they recognize that clients possess one of the seven core problems discussed in Chapter 1 and when they can connect these problems to deficits in agency policies, budget, protocols, internal culture, mission, or other organizational factors. For example, a social worker in a county mental health facility might recognize that low-income clients from certain geographic areas are unable to regularly attend their appointments (Core Problem 5), because the agency terminated its policy of issuing bus tokens to clients in need. The social worker would need to decide whether to proceed with an agenda to reinstate the policy of bus token distribution and thereby increase access to mental health services for this vulnerable population.

Mezzo policy advocates use the ethical reasoning skills to determine whether to proceed with a policy change intervention. They consider the extent existing policies violate first-order principles of beneficence and social justice. They consider pragmatic factors, such as the relative difficulty of changing a specific policy and the amount of time it might take to change it.

Deciding Whether to Proceed in Communities

Mezzo policy advocates also must decide whether an issue or problem within their community merits the development of an advocacy intervention. The views of community residents may contribute to the decision to develop an advocacy intervention. These views may be evidenced at community forums, through focus groups, or by members of specific community groups or churches. However, communities are often divided about the merit of specific issues or problems, and so the mezzo policy advocate must carefully weigh the choice whether to proceed. Just as in organizations, mezzo policy advocates working in communities utilize an ethical reasoning process to help them decide whether to proceed.

CHALLENGE 2: DECIDING WHERE TO FOCUS

Deciding Where to Focus in Agencies

Mezzo policy advocates can seek to change an organization's mission, service-provision policies, funding sources, programs, budget priorities, strategic plans, personnel and hiring policies, relationships between units and programs, and relationships with other organizations. Given these many options, mezzo advocates have to decide where to focus and if they want to make incremental or larger shifts in policy. They have to decide whether they wish to change policies of specific units or policies at higher levels of an organization.

Mezzo policy advocates sometimes try to change the culture of specific agencies. They may seek multidisciplinary training if they want to increase cooperation across professions in an agency, such as training sessions on ethics in a hospital. They may bring new perspectives to helping certain kinds of consumers, such as persons with disabilities, by bringing in a speaker who puts less emphasis on a medical model in a rehabilitation program and more emphasis on empowerment.

Social workers are more likely to succeed in their mezzo policy interventions if they develop a strategic overview. This overview allows them to identify assets and liabilities in the policy and the organizational context that are relevant to their issue. Optimism about changing the organization's strategy to deal with a specific problem may increase as they identify regulations, statutes, or court rulings that suggest the need for change. They may also focus on changing outcomes for their organization related to resource levels, improved client satisfaction, and better outcomes for clients.

Deciding Where to Focus in Communities

Assume that strong support existed for not allowing additional liquor stores and fast-food outlets in a specific community. The community organizer would then proceed to Challenge 2, that is, helping community residents decide how to navigate community institutions that develop policies that govern whether and how many commercial outlets of specific kinds are permissible in a specific community. These institutions would include the following:

- City planning and zoning departments that decide which kinds of commercial institutions can be placed at specific locations
- Public health departments and advocates that seek to improve the health of community residents
- Members of the city council who ratify decisions of city planning and zoning departments

- The top government official of an urban area, for example, a mayor
- Specific community groups that deal with commercial interests, for example, the local chamber of commerce or business improvement district
- Associations of residents, for example, home owners' associations and neighborhood councils

Advocates to limit the placement of liquor stores and fast-food outlets in this community would have to consider the level of influence and resources of specific commercial interests that would protect their interests.

CHALLENGE 3: OBTAINING DECISION MAKERS' RECOGNITION OF UNRESOLVED PROBLEMS

Obtaining Decision Makers' Recognition in Agencies

Policy advocates work to place issues on decision makers' agendas so that they receive priority. This includes finding a propitious time to introduce policy changes to decision makers within their organizations. They ascertain whether the problem or issue has already been discussed or whether similar or related problems have already been identified. They determine if the executive director or CEO of their organization appears amenable to a specific problem or issue and if other high-level administrators might assume a pivotal role in advancing it. They work with specific stakeholders to gain their support and identify the ways in which a policy initiative enhances the financial well-being of the organization, its image in the community, or the outcomes of its clients. They also work with persons who might join a coalition or task force to address the specific issue or problem.

Timing should be considered when promoting policy changes that come with a price tag. If a problem can be solved only with considerable financial resources, advocates may want to postpone introducing it during a period of budget stringency.

Obtaining Decision Makers' Recognition of Specific Issues in Communities

Advocates place their issue on the agendas of decision makers in their community and the larger city where they are located. They decide when to initiate their advocacy by considering whether background events, such as elections of public officials, the economy, and other factors, are favorable. They might decide to join forces with several other communities to make their issue more relevant to other members of the city council and the mayor.

They should also ascertain whether the national climate and public opinion are helpful to them. The mass media can provide a mechanism for elevating interest in their issue by giving it substantial coverage and linking it to stories of national interest. Assume that a community organization decides to proceed with their quest to limit the numbers of liquor stores and fast-food outlets. They may utilize the media to draw attention to the problem, form a coalition of residents and business representatives who wish to address the problem, and set appointments with members of local zoning boards, business improvement districts, and neighborhood councils.

CHALLENGE 4: ANALYZING WHY THE PROBLEM EXISTS

Analyzing Why the Problem Exists in Agencies

Social workers seeking to address unresolved problems in agencies will need to analyze why the problem exists, including its source and factors that enable its continuance. Take the example of a nonprofit organization in which a culture of discrimination against LGBTQ clients exists. If the organization is run or funded by a faith-based entity, advocates might find that the policy is tied to a religious belief. Alternatively, they might find that the policy was put in place many decades ago before the gay rights movement put the issue of equality for people of all sexual orientations on the social radar. Advocates will need to develop a historical understanding of the problem by analyzing the history of inequality and oppression directed toward LGBTQ clients. Equipped with this information, mezzo policy advocates will be more likely to develop an effective strategy to address the problem (in Challenge 5).

POLICY ADVOCACY LEARNING CHALLENGE 5.3

ADVOCATING FOR LESBIAN CLIENTS IN A DOMESTIC VIOLENCE AGENCY

I (Gretchen Heidemann) worked as a crisis hotline counselor at a shelter for battered women early in my career. Unfortunately, I became aware of some actions on the part of staff that troubled me as I noticed that lesbian women who received services at the agency were being treated unfairly. In one instance, a staff member asserted that she would not wrap up Christmas gifts for the children of "those" women. In another, a staff member began proselytizing to lesbian clients, telling them that God viewed their behavior as sinful and that they must repent. I felt the need to advocate on behalf of the clients. I moved quickly to bring the staff's behavior to the attention of the assistant director (AD). Although the AD heard my concerns, and seemed to agree that the behavior was inappropriate and unethical, she was not quick to make changes. First, the agency did not have a formal nondiscrimination policy in place. Thus, there were no immediate grounds on which to reprimand staff or to mandate them to change their behavior. Second, there were many religious and cultural factors at play. It was not a simple matter, she informed me, to ask staff members to change their behavior on a dime. In retrospect, I wish I had done my homework before approaching the AD. Had I more knowledge about the history of the organization and about the beliefs of members of the community, I might have had a more strategic plan in place. I might have sought examples of agencies that were able to change both their organizational culture and written policies with regard to this issue while also respecting the religious and cultural beliefs of staff members. As it turned out, an informal and (in my opinion) largely ineffective conversation about "treating everyone equally" took place at a staff meeting. But a formal nondiscrimination policy was never enacted within the agency during my tenure there so staff did not feel compelled to change their behavior.

Analyzing Why the Problem Exists in Communities

A mezzo policy advocate who identifies an unresolved problem within his or her community will need to analyze why the problem exists in that community. Unlike organizations, there is no mission statement to drive a set of shared beliefs and behaviors within a community. Moreover, communities are typically part of larger entities—cities and counties—that legislate matters that influence the capacity of a community. Whereas many organizations operate as stand-alone entities with the ability to create, enforce, and change their own policies, communities must work within the complex web of city-wide ordinances, county-level law enforcement and safety (i.e., fire and emergency) services, networks of business associations, and so forth. Thus, understanding why a problem exists within a community might involve looking beyond the community itself as well as into the past to understand its historical roots.

The case description of the community of Watts at the end of this chapter describes some of the problems that the community faces. These problems cannot be understood apart from their historical context. As the case example at the end of this chapter describes, the gangs that currently exist in Watts are an outgrowth of mutual protection societies established in the 1960s by black residents who came together to protect one another from threats from the white community. Without this critical piece of information, a mezzo community advocate in Watts would miss an opportunity to work with gang members under the assumption that mutual protection (and not just crime or drug dealing) is a primary function of gang membership. Absent such a detailed analysis, the advocate's ability to develop an effective strategy to address the problem in Challenge 5 would be stymied.

CHALLENGE 5: DEVELOPING A STRATEGY OR PROPOSAL TO ADDRESS THE PROBLEM

Developing a Strategy in Agencies

Interagency strategy should include a series of activities designed to raise awareness about the problem and inform stakeholders about the proposed policy solution and its benefits. Activities might include staff briefings, publication of reports, interagency email blasts, mobilization of stakeholder groups to support the proposed reform, and face-to-face meetings with key decision makers. Each of these activities is carefully thought out with a planned timeline or sequence of events. A mezzo policy advocate should be aware of relevant timetables within the agency (e.g., board meeting schedules, licensing inspections, and major fundraisers). With this information, the effective advocate can plan activities at the most auspicious moments.

Mezzo policy advocates should also look for open policy windows to advance their causes. Open policy windows are critical events, such as administrative changes, a change in decision-maker ideology, a shift in public opinion, or the onset of a crisis or "focusing event" that reflects favorably upon the issue. They occur infrequently, so mezzo policy advocates should be prepared to "pounce" when such an opportunity arises.

There will be entities that oppose any effort at policy change, even within the smallest of agencies. Awareness of this opposition is key to survival. Opposition may come from insiders, such as staff members who prefer the status quo, ideologically opposed administrative officials, or clients who feel somehow vulnerable or threatened by the change. Outside opposition should be anticipated so that when it does arise there are talking points available to counter any hostility.

Developing a Strategy in Communities

Many of the strategic elements described apply to communities as well. Let us imagine that a social worker in the mental health clinic encounters a patient who was recently released from prison. As she works with him to develop a case plan, she realizes that this client faces obstacles and challenges beyond those of her typical client. The formerly incarcerated man not only suffers from mental illness but is unable to secure permanent housing. Moreover, the few halfway houses that exist in the community are being threatened by an ordinance that would close their doors. The social worker might consider engaging in advocacy to stop the proposed ordinance, to develop ordinances that actually address the needs of the community, or to fund alternative reentry facilities for formerly incarcerated people. Many of the strategies previously described are applicable to such a scenario.

The mezzo policy advocate wishing to change community-level policy will need to outline a series of actions, set a timetable, look for opportunities, and plan for opposition. In addition, mezzo policy advocates in communities might incorporate the use of media and social media into their strategy. Press, such as radio, television, and newspapers, can be powerful tools for any policy strategy. Social media is also highly desirable for a number of reasons. Advocates have complete control over their content, so they can target messages to certain audiences and can disseminate information quickly. Advocates can also be creative with their messaging using videos, photographs, cartoons, tweets, blogs, interactive web pages, and online petitions.

RESOURCES ON THE USE OF SOCIAL MEDIA FOR ADVOCACY

American Association of University Women:

1. Social Media 101: Getting Started With Facebook and Twitter. Go to http://aauw.org/resource/how-to-use-social-media-for-advocacy.
2. Social Media 201: Leveraging Social Media to Increase Your Visibility. Also go to http://aauw.org/resource/how-to-use-social-media-for-advocacy.

Online strategies must often be supplemented by direct contact with decision makers at the local level—or at state and federal levels when engaging in macro policy advocacy. These decision makers are often inundated with social media, so those mezzo policy advocates that talk with them in person obtain greater impact when they supplement social media with personal discussions and pressure. They can consider developing delegations to talk with local officials as well as asking questions at local hearings.

CHALLENGE 6: DEVELOPING SUPPORT FOR A STRATEGY OR PROPOSAL

Developing Support for a Strategy in Agencies

Sometimes it takes only one person to advance a policy change, but often a mezzo policy advocate will need to secure the support of individuals and groups to accomplish their agenda. The skill of persuasion is key to developing support for a proposed policy change. A range of stakeholders including clients, frontline staff, high-level or administrative staff, executives, members of boards of directors, community residents, funders, vendors and contractors, regulatory agencies, and collaborating organizations should be sought out. The mezzo policy advocate should carefully consider the interests of the stakeholder group and consider how the proposed policy will affect each group. They might wish to interview key informants from each group to better understand the positions and interests of that group, set up meetings or activities to build relationships with members of the different groups, and carefully craft messages targeted toward each group to gain support. A policy change might have obvious implications for some, but not all, of the stakeholder groups. The advocate should think about how members of each group will benefit from the change as well as why it is the best possible solution for the organization. Appealing to mission and vision statements, ethical principles, professional standards of practice, and shared moral or political beliefs can be powerful ways of winning support for policy proposals.

Developing Support for a Strategy in Communities

Mezzo policy advocates in communities appeal to a broad range of stakeholders. Residents; business owners; employees of those businesses; consumers of community-based services; members of faith-based institutions; fire, rescue, and law enforcement officials; and many other groups have vested and often competing interests in community-level policies.

Gaining support for a proposed policy agenda in a community requires much of the same strategy as discussed in relation to agencies. However, the complexity of the sheer size of some communities, when all of the many stakeholders mentioned are included, makes this challenge a daunting one for the mezzo policy advocate. In addition, residents of communities are rarely united in their beliefs about what should change and how that change should occur. Advocates must remember that it is not possible to win everyone over and that the broadest possible support for a policy is always advantageous.

CHALLENGE 7: IMPLEMENTING A STRATEGY OR POLICY TO ADDRESS THE PROBLEM IN AGENCIES AND COMMUNITIES

Mezzo policy advocacy does not end when a policy is adopted. In fact, in some ways, adoption is just the beginning. A new community ordinance, agency-wide or service-delivery policy, or interorganizational operating procedure is virtually useless unless it is fully implemented. Many things can happen to hinder the implementation of a new or modified policy. In an agency, high-level decision makers may become distracted or forget about the policy altogether. Those who are involved in the implementation, such as frontline staff, might not fully understand their new rules or duties. A lack of funding may affect full implementation of the policy. Individuals may refuse to implement the policy or work to block its implementation.

To counter these pitfalls, mezzo policy advocates should work to establish protocols and trainings that will help ensure that staff are fully aware of their duties under the new policy. They may devise creative solutions to bring resources to the agency or community to fully implement the policy. In more extreme cases, they may decide to form a watchdog group to put pressure on decision makers or administrators who are responsible for overseeing implementation.

POLICY ADVOCACY LEARNING CHALLENGE 5.4

CONDUCTING WATCHDOG ACTIVITIES TO ENSURE IMPLEMENTATION OF AGENCY POLICY

I (Gretchen Heidemann) interned at a homeless coalition that worked in the greater Los Angeles area to protect the rights of homeless individuals, ensure adequate resources for homeless services, and build grassroots support to end homelessness. The organization became aware that homeless persons were being denied benefits entitled to them by county welfare offices through the homeless assistance program. Specifically, the dozen or so welfare offices in the county were mandated, through state policy, to provide funding to homeless families for temporary shelter for up to 16 consecutive nights. We developed a protocol to hold their feet to the fire. We created a script and systematically placed phone calls to welfare offices pretending to be a family in need and requesting information about the homeless assistance program. We recorded whether the welfare worker responding to the call gave us correct, partially correct, or entirely incorrect information. We then wrote up a report from our data and used it to confront directors in the welfare offices where workers most frequently provided incorrect information. We pressed the directors to conduct trainings for their workers or, at minimum, to provide detailed materials and instructions to them so that future seekers would not be denied their entitled benefits.

CHALLENGE 8: ASSESSING WHETHER THE IMPLEMENTED STRATEGY OR POLICY WAS EFFECTIVE IN AGENCIES AND COMMUNITIES

Mezzo policy advocates will want to know whether the policy they helped pass and implement is accomplishing its intended goals. Advocates should consider using tools that will provide the data they need to determine whether or not the policy is achieving its aims. For example, they may decide to conduct in-depth interviews with agency clients who were impacted by the policy change, develop a survey to measure the extent of their satisfaction with it, or locate standardized scales if the policy is intended to have specific health, mental health, behavioral, or other measureable outcomes. It is important to understand how the policy is affecting other stakeholders and the agency or community in general. Not only do advocates seek this information to tell them whether the policy was effective, but the information they acquire when assessing the strategy will be critical if the policy is time limited and needs to be renewed.

Some of the questions mezzo policy advocates may seek to answer when assessing the policy include the following:

1. How do stakeholders perceive the policy? Subjective perceptions about the utility of the policy can be powerful tools when it comes time to renew, expand, or replicate a policy.
2. Is the policy cost-effective? Does it have any unintended financial impacts? Knowing whether the policy can be implemented without causing undue burdens on taxpayers or other revenue providers is also critical to the advocate's ability to renew, expand, or replicate it.
3. What challenges has staff faced in successfully implementing the policy? How can the lessons learned from those challenges help ensure better implementation in the future?
4. If the policy is falling short of accomplishing its intended outcomes, what ideas do members of stakeholder groups have about how to improve it?

Assessing policy is not a one-time activity. It is an ongoing process that involves the constant collection and analysis of data to understand the extent the policy is accomplishing its intended aims. Simple instruments can be developed to collect information about outcomes to help audiences understand the impact the policy is having. Some agencies may have staff or departments specifically dedicated to program evaluation that could aid in assessing policies.

A CASE EXAMPLE OF MEZZO POLICY ADVOCACY

This section presents a case study of mezzo policy advocacy in the community of Watts in Los Angeles, California. We present the history of the community and the community's current demographics. We describe the Watts Labor Community Action Committee

(WLCAC), which has been working for more than 40 years to change community-level policies to make Watts a safe and thriving place to live.

History of Watts

During the Second Great Migration in the 1940s, large numbers of African Americans moved to the West Coast, enticed by defense industry recruitment at the start of World War II. Although Los Angeles did not practice Southern-style racial segregation, the city was divided geographically by race. Minorities were expressly banned from housing in 95% of the city. As the population of the city of Los Angeles burgeoned, demands for housing skyrocketed. Developers began building new housing in undeveloped areas south of Los Angeles. Large-scale housing projects were built during this time, along with single-family homes, and by the late 1940s neighborhoods like Compton and Watts were populated by blue-collar African American families who enjoyed a middle-class lifestyle.

In the 1950s the area became a target of racial animosity. White residents burned crosses on the yards of black families, and white gang members in neighboring communities accosted black residents who passed through their "territory." So-called mutual protection clubs were formed by young black residents in response to these assaults, while white families began to flee to surrounding suburbs.

Around the same time, a new Los Angeles police chief, William H. Parker, began using military-style fear and intimidation tactics in these neighborhoods, especially against young black and Latino residents. Police brutality became widespread and erupted in violence on the night of August 11, 1965. That night, 21-year-old Marquette Frye was pulled over for driving while intoxicated near his home in Watts. His mother and brother arrived on the scene, along with several backup police officers, and an argument ensued. A physical altercation led angry resident onlookers to hurl objects at police officers. Marquette, his brother, and his mother were arrested, and an angry crowd grew. Throughout the night, police attempted to break the crowd up but were unsuccessful.

Over the next six days, a 46-squre-mile area around Watts was transformed into an all-out war zone involving looting, beatings, and mass rioting. The unrest resulted in 34 deaths, 1,032 injuries, 3,438 arrests, and more than $40 million in property damage. More than 2,300 members of the California Army National Guard were called in, along with the Los Angeles Police Department (LAPD) and LA County Sheriff's Department, and the violence was eventually contained. Yet the frustrations of community residents over the system of racial segregation and inequality, as well as biased and oppressive treatment by law enforcement, could not be contained.

Following the Watts riots, the mutual protection clubs that had developed in the 1950s transformed into the violent street gangs known today as the Bloods and Crips. Crack cocaine began to flood the streets in the late 1970s and early 1980s, while many of the living-wage manufacturing plants that employed black workers shut down or moved operations overseas. Large numbers of residents, particularly young black men, were arrested and became trapped in a cycle of incarceration, poverty, and homelessness. Many black women were forced to raise families on their own, often relying on welfare assistance when jobs were scarce and affordable childcare nearly nonexistent.

A second wave of civil unrest occurred in the area in 1992 after an incident eerily reminiscent of that which provoked the 1965 rebellion. On March 3, 1991, a driver by the name of Rodney King and his two passengers were pulled over by California Highway

Patrol officers after a high-speed chase. King was suspected of being intoxicated; a drug test later proved he was not. LAPD arrived on the scene and took the two passengers into custody. King, a parolee who knew that an arrest would send him back to prison, resisted. As they pulled him from the vehicle, five LAPD officers tasered King, kicked him in the head, and beat him with batons while he crawled on the ground. He was then tackled and cuffed. George Holliday, a resident of the community, caught the police actions on video. The video aired worldwide on nightly news stations, and national debate about police brutality against unarmed black men was ignited.

Following the incident, four of the police officers were charged with assault and excessive use of force. A lengthy trial ensued, and a jury, composed of nine whites, one black, one Latino, and one Asian, acquitted three of the officers on April 29, 1992. They could not agree on a verdict for the fourth officer.

Riots and looting began the day of the acquittal as angry mobs formed in areas of South Central Los Angeles, including the neighborhood of Watts. Innocent residents were beaten, and businesses were looted and burned to the ground in rioting that lasted for six days. Fifty-three people died, and more than 2,000 were injured. Worldwide media coverage of the unrest showed broad swaths of Los Angeles on fire. More than 13,500 military forces including the Marines and the National Guard were called in. Most of the violence subsided by early May.

Since the 1992 civil unrest the community of Watts has seen both growth and decline in relation to the violence. Despite a peace treaty signed by rival gangs in the early 1990s, more than 500 homicides were reported in Watts between 1989 and 2005, most of them gang related. In 2006, the Watts Gang Task Force (WGTF), a coalition of residents, business owners, schools, churches, city council members, and the Los Angeles Human Relations Commission was founded. Meeting on a weekly basis, the group focuses on building trust between residents and law enforcement and implementing initiatives to keep residents safe. The task force touts the creation of a safe passage program for students to get to school safely through gang territory, the reopening of a public pool that was once considered unsafe, and the creation of thousands of summer jobs for youth. Over the two years following its creation, WGTF reported that violent crime had decreased by 50% in Watts.

The community has transformed in other important ways as well, partially as a result of the work of the Watts Labor Community Action Committee (WLCAC). Founded around the time of the Watts Rebellion in 1965, WLCAC operates an array of social services in Watts, including a homeless shelter, a senior center, a WorkSource center, a childcare center, a gang reduction program, and several youth development programs. In addition, the organization promotes culture and tourism in Watts through its operation of a civil rights museum, an art gallery, and a performing arts center. The organization also hosts the Watts International Marketplace, an open-air market housing embroidery, silk-screening, ceramics, fine arts, glass blowing, woodworking, and photography studios. The organization is also at the forefront of policy advocacy initiatives to improve conditions throughout the community. WLCAC has worked in collaboration with other organizations to accomplish the following:

- Restrict the number of liquor stores operating within the community
- Remove harmful gang injunctions that profile and sweep young men of color into the criminal justice system rather than protecting the community

- Renovate Ted Watkins Memorial Park, the only green space within the community
- Establish a two-acre farm where residents can grow their own food

The organization is currently working to pass the Zero Displacement Policy to ensure that any development or construction that occurs within the community does not displace residents. It would also ensure that housing and retail developers give priority to agencies within the community to perform the construction and that they hire workers from within the community. They are proceeding with the Zero Displacement Policy (Challenge 1) and are focusing on working with entities such as the Central Area Planning Commission and the Industrial Development Authority, which will need to give approval before the policy goes to the city council for a vote (Challenge 2). They have secured the support of two important policy makers, including one city council member and the Congress member who represents the district (Challenge 3).

POLICY ADVOCACY LEARNING CHALLENGE 5.5

THE ZERO DISPLACEMENT POLICY IN WATTS

Consider the community of Watts as described and with any supplemental information you might locate. Imagine you are working with WLCAC to pass the Zero Displacement Policy. With Challenges 1 and 2 already complete as described in Figure 1.1, and Challenge 3 in progress, outline a detailed plan for completing Challenges 3 through 8 by addressing the following:

Challenge 3: Secure attention

Identify other decision makers whose support you should bring to the issue and the proposed policy. Even if they will not be the ones to ultimately decide or vote on the policy, what key figures need to be aware of the issue? Are there other persons or groups within the community or outside of it whose support you will need to secure?

Challenge 4: Analyze why the problem exists

Consider the high rates of poverty and unemployment that exist in Watts as well as the current lack of living-wage jobs. What historical and socioeconomic factors have contributed to this situation?

Consider the reasons why a developer might be unconcerned about displacing residents in the process of building new commercial structures within the community. What historical fears and mistrust might this lack of concern ignite?

Consider why a developer might choose not to hire from within the community or why it might choose to give its contracts to agencies from outside the community. What will you do to overcome this issue?

Challenge 5: Develop a strategy

Outline a strategy for getting this policy approved by the city council. How will you frame the issue? What stories will you tell and how? Who will be featured in those stories? What sorts of materials will you develop? What activities will you undertake (e.g., a press conference, one-on-one meetings, petition-gathering, or a community town hall)? What opportunities will you be looking for to advance your cause? Who is likely to be your opposition, and how will you address opposition?

Challenge 6: Develop support for the proposal

Identify the range of stakeholders that this issue impacts. Develop a strategy for gaining their support. What types of messages will you develop to target the different groups? For those who are already on your side, how will you further engage them to advance your agenda? For those who are on the fence, how might you bring them over to your side?

Challenge 7: Implement the policy

How will you ensure efficient and successful implementation of the policy? What groups of people will be involved in implementing the policy? What types of watchdog activities might you undertake to make sure that the spirit of the policy is being honored? Will you consider engaging in whistleblowing if you see that policy makers are not fully or correctly implementing the policy?

Challenge 8: Assess the policy

How will you assess whether the policy has been effective? What are the outcomes of concern to the stakeholder groups? How will you measure outcomes?

MEASURING SKILLS NEEDED FOR MEZZO POLICY PRACTICE

I discussed data drawn from the federal research project funded by the Patient-Centered Outcomes Research Institute (PCORI) in Chapters 1 and 4 as well as validated scales used to measure and predict frontline professionals' micro policy advocacy (Jansson, Nyamathi, Duan, et al., 2015; Jansson, Nyamathi, Heidemann, Duan, & Kaplan, 2015). I was unable to measure their skills in mezzo policy advocacy as systematically, however, because the expert panel decided to delete questions from the project's questionnaire that would have provided more data about their mezzo policy advocacy, whether within their hospitals or in the surrounding communities. The experts' views reflected a reality in American health care: the assignment of community health practice to the field of public health while assigning inpatient care to hospitals. In other words, health professionals within hospitals mostly or entirely work within them with notable exceptions.

The experts may have been unaware of the level of mezzo policy activity of frontline professionals in hospitals. Frontline health professionals are employees who are greatly affected by the missions of their hospitals, internal policies, cultures, budget allocations, and relationships with other hospitals and clinics. They often do relate to communities where their clients live and to which they are discharged from their hospitals. They often need to refer their clients to community agencies and programs that are discussed in Chapter 9 of this book.

Some data from the PCORI project is nonetheless relevant to measuring frontline professionals' mezzo policy advocacy. More than 75% of frontline professionals selected "somewhat, quite a bit, and a great deal" when asked if they could "influence other people to work with me to change specific policies," "mediate conflicts," and "discuss specific kinds of unresolved patient issues with hospital administrators." More than 60% of them gave these rankings to "initiate policy changing interventions," "negotiate or bargain to achieve my policy goals," "help patients become policy advocates," and "develop better coordination between units or departments of my hospital." More than 50% gave these rankings to skills to "communicate with public officials," "change policies in my hospital," and "establish multidisciplinary training sessions in my hospital."

They gave far lower rankings, however, to "make budget suggestions in my hospital" (29%) and "changing protocols or operating procedures in my hospital" (41%).

Items in a preliminary version of the PCORI questionnaire addressed some community dimensions of mezzo policy advocacy. These items included ones that asked respondents to what extent they had participated in community outreach by planning health fairs, developed joint programs with health providers in the community, identified community factors that have had negative health consequences, and worked to address community factors that have had negative health consequences. Other questions asked to what extent they had communicated with community leaders and used the mass media.

Even if hospital-based professionals mostly relate to patients and professionals within their hospitals, professionals in many other kinds of agencies routinely relate to communities, public officials, community groups, and nongovernmental organizations as revealed by the preceding case example of the WLCAC.

Further research is needed to develop scales that measure the extent frontline health professionals engage in mezzo policy advocacy in their hospitals and in the communities where their patients live. Further research is needed, as well, to develop scales that predict the extent frontline health professionals engage in mezzo policy advocacy.

Learning Outcomes

You are now equipped to:

- Define mezzo policy advocacy
- Identify key skills needed by mezzo policy advocates
- Engage in the eight challenges identified in the multilevel policy advocacy framework in Chapter 3 in agencies and organizations
- Engage in the eight challenges identified in Chapter 4 in communities
- Apply concepts from this chapter to a specific community

References

Chu, F. D. & Trotter, S. (1974). *Madness establishment: Ralph Nader's study group report on the National Institute of Health*. New York, NY: Grossman.

Cook, M. F. (1987). *New directions in human resources: A handbook*. Englewood Cliffs, NJ: Prentice Hall.

Dahl, R. (1967). *Pluralist democracy in the United States: Conflict and consent*. Chicago, IL: Rand McNally.

Dluhy, M. (1981). *Changing the system: Political advocacy for the disadvantaged*. Beverly Hills, CA: Sage.

Gummer, B. (1990). *The politics of social administration: Managing organizational politics in social agencies*. Englewood Cliffs, NJ: Prentice Hall.

Hasenfeld, Y. (1983). *Human service organizations*. Englewood Cliffs, NJ: Prentice Hall.

Jansson, B. (2018). *Becoming an effective policy advocate: From policy practice to social justice.* Boston, MA: Cengage Learning.

Jansson, B., Nyamathi, A., Duan, L., Kaplan, C., Heidemann, G., & Ananias, D. (2015). Validation of the patient advocacy engagement scale for health professionals. *Research in Nursing & Health, 38*(2), 162–172.

Jansson, B., Nyamathi, A., Heidemann, G., Duan, L., & Kaplan, C. (2015). Predicting patient advocacy engagement: A multiple regression analysis using data from health professionals in acute-care hospitals. *Social Work in Health Care, 54*(7), 559–581.

ENGAGING IN MACRO POLICY ADVOCACY

LEARNING OBJECTIVES

In this chapter you will learn to:

1. Understand the context of public policies and regulations, including the following:
    - The players in the public sphere
    - How the public sphere includes legislatures, the executive branch, elected officials, appointed officials in executive agencies, and civil servants
    - How legislatures are organized and function
    - The role of advocacy groups
    - The mind-sets of public officials
    - Connections among lobbyists, legislators, and bureaucrats
    - The important role of public opinion
    - The role of regulations
2. Understand how macro policy advocates engage the eight challenges
3. Identify specific policy advocacy skills that macro policy advocates need
4. Analyze a policy advocacy vignette to identify macro policy advocacy skills and strategies

Social workers engage in macro policy advocacy in local, regional, state, and federal governments because laws, statutes, regulations, and administrative decisions in these settings profoundly influence the well-being of their clientele. Many gaps and omissions in existing policies can be corrected only by developing macro policy interventions in these venues.

We discuss the eight challenges that social workers confront when they engage in macro policy advocacy that were depicted in Figure 1.1., that is, deciding whether to proceed, deciding where to focus, securing decision makers' attention, analyzing why a dysfunctional policy developed, developing policy proposals, securing approval or enactment of a proposal, implementing a policy reform, and assessing an implemented policy. Even before they undertake these challenges, macro policy advocates have to understand the context.

UNDERSTAND THE CONTEXT OF PUBLIC POLICIES

Necessary Information About the Executive Branch of Government

Macro policy advocates have to know how the public sphere is organized so that they can maneuver within it. They have to know the players in the public sphere. These include three kinds of elected officials—heads of government including mayors, governors, and presidents; legislators at local, state, and federal levels; and persons elected to special bodies such as local school boards. They include unelected officials, who include persons appointed to public office by heads of government such as the heads of many agencies as well as civil servants who obtain their jobs by taking exams and interviews and are charged with running government agencies at local, state, and federal levels. They have to be familiar with lobbyists and interest groups that represent a wide array of corporations, unions, professions, and nongovernment organizations (NGOs), including state and national chapters of the National Association of Social Workers (NASW). They have to understand how regulations are fashioned by civil servants as compared with statutes that are enacted by legislatures.

They have to understand the legislative process that determines how bills become laws including the existence of two chambers, the extraordinary power of the party that has a majority in a specific chamber, and the legislative calendar that determines when bills must become law or its advocates must start anew with the next legislative session.

They have to understand the executive branch. The U.S. president, as well as governors and mayors, are in charge of the executive branch of government, which comprises the myriad agencies that implement government policies. These agencies are usually called departments, such as the Department of Health and Human Services (DHHS). They usually initiate a budget, even though the legislators make many of the final budgetary choices. This initiating role represents an important power because it allows chief executives to influence priorities within the executive branch.

Chief executives usually develop a legislative agenda to which they often refer in general terms in speeches, such as the president's State of the Union speech. They have vast resources to help them fashion this legislative agenda, including personal aides in their own offices, their political appointees in the executive branch, and political allies who occupy powerful positions on legislative committees or in the party or legislature (Holtzman, 1970). Because heads of government have central positions in government and high profiles, their legislative proposals, which members of their own party or political allies introduce into the legislature, often have an advantage over individual legislators' proposals. Even with such power, the legislative proposals of many heads of government are defeated, particularly when the opposing party holds a majority in the legislature.

Chief executives can issue executive orders, which are directives to specific units of government that don't require legislative approval, such as when President Barack Obama issued executive orders pertaining to details of the Patient Protection and Affordable Care Act (ACA) of 2010 not included in the legislation.

Capitalizing on the extensive coverage the mass media usually accords them, heads of government often try to gain support for legislative measures, rally opposition against legislators who may block their policies, and educate the public about specific issues (Smith, 1988).

Finally, heads of government can veto legislation that the legislature has approved, an important power because legislatures often cannot muster the votes (such as the two-thirds of each chamber required in the federal government) to override a veto. Some governors have line-item vetoes over the budgets that legislatures have approved.

Necessary Information About Legislatures

Legislatures seem formidably complex; in fact, they are all structured rather similarly. They are usually divided into two houses, such as the House of Representatives and the Senate at the federal level, or the Senate and the Assembly at the state level, as in Wisconsin. Usually, both houses must assent to legislation or a budget before it can become operative. Many legislatures, like the U.S. Congress, convene annually, but some state legislatures convene only every two years. Moreover, the length of legislatures' sessions varies widely; Congress meets almost nonstop each year, but other legislatures convene only for several months.

The members of each house or chamber of a legislature, whether local, state, or federal, are elected by districts, whose precise shapes change over time. Districts are reapportioned as the population shifts and as the courts decide that existing district lines are unfair to specific groups, such as Latinos or African Americans. To understand specific legislators, then, we must analyze the characteristics of their constituents, whose preferences influence their positions on myriad issues. We might ask: Is the district relatively affluent or poor? Does it have a mix of ethnic and racial groups, or does a single group dominate it? Is it urban, suburban, or rural? Is it dominated by a single party or evenly divided between two parties? We can also ask whether specific legislators occupy relatively safe seats or whether they will face closely contested elections.

To understand how specific chambers of legislatures operate, we must first ask: Which party controls a majority of the legislature's members? The majority party appoints the chairs of all committees and has a majority of the members on each committee of that chamber. Moreover, the members of the majority party in each house elect the presiding officer of that chamber, such as the president of the U.S. Senate and the speaker of the House. These high-level leaders have considerable power in determining when specific measures will be debated on the floor, in mobilizing support for or against measures, and in making important parliamentary decisions at critical junctures in floor debate. Presiding officers often have the authority to establish committees, assign members to committees, and appoint chairs of committees. In addition, they often have the power to decide where to route specific bills for deliberation. This power is critical because a presiding officer can often kill a bill by insisting that it go through specific committees that are known to be hostile to it.

Each chamber has a second tier of powerful leaders: the floor leaders, which are also elected by caucuses and include the majority leader and the majority whip. Working in tandem with the presiding officer, these party leaders shepherd legislation through floor deliberations and decide which measures their party will support or oppose.[9]

A third tier of leaders, the chairs of the important committees of a chamber, are members of the majority party and have considerable power over the fate of legislation in their committees.[10]

Although at a disadvantage, a minority party often has considerable power in a specific chamber. It has its own leader, such as the U.S. House minority leader, who can mobilize support of or opposition to pieces of legislation. The minority party is allocated seats on all committees of a chamber in proportion to its share of the chamber's total membership, and its members sometimes obtain a majority vote on a committee by teaming with committee members of the majority party. Thus, in the era of Presidents Ronald Reagan and George H. W. Bush, Republicans could defeat congressional legislation that the Democrats strongly favored by joining with southern Democrats, even though the Democratic Party had a majority in the House of Representatives throughout the period and controlled the Senate from November 1986 onward. The power of the majority party is illustrated by the Republican Party during the presidency of Donald Trump. Acting in concert with Senate Majority Leader Mitch McConnell and House Speaker Paul Ryan, the Congress enacted a huge tax cut that mostly benefited affluent Americans and corporations.

Because of their size and the myriad issues they consider, legislatures are divided into specialized committees. In the federal House of Representatives, for example, the Ways and Means Committee processes Social Security, Medicare, and tax legislation, and the Committee on Labor and Public Welfare processes social programs such as Head Start.

As we have discussed, the presiding officers of each chamber often have considerable discretion in referring measures to committees. Many pieces of legislation go to multiple committees when they pose issues that cut across committee divisions. In other cases, certain kinds of legislation are automatically referred to a specific committee. The House Ways and Means Committee and the Senate Finance Committee in the U.S. Congress, for example, always consider Social Security and Medicare legislation as well as tax policy.

Each legislative committee has an internal structure. Its chairperson may be elected by the committee's members or appointed by the chamber's presiding officer. Committee chairs are usually powerful figures. Like presiding officers of the overall chamber, they can kill legislation by not placing it on the committee's agenda, by referring it to a hostile subcommittee, or by merely raising strong objections to it when the committee discusses it. Each legislative committee has subcommittees that specialize in certain issues within the whole committee's purview. Subcommittee chairs also have considerable power over issues that fall within their domain.

Legislation that presiding officers refer to a committee falls into two categories. Some of it is consigned, more or less at once, to the legislative junk heap because the committee, much less the full chamber, does not consider it seriously. The subcommittees and committees take other legislation seriously and mark it up in committee deliberations; that is, they amend it.

Most legislation, then, evolves in the course of deliberations that take weeks, months, or even years, and it can be amended on the floor of the chamber when the whole chamber decides whether to amend, accept, or defeat a bill. A bill usually starts in one chamber and progresses from committees to a floor debate and then a vote. It is then referred to the other chamber, where it follows a similar course. After the second chamber enacts its own version, representatives from each chamber seek a common version, which usually requires both chambers to make concessions. If the conference committee creates a joint version, each chamber must then ratify it before it goes to the president (or governor), who then signs it into law or vetoes it. Congress can override a veto if each chamber musters a two-thirds vote. Otherwise, the legislation dies.

Some legislative deliberations are relatively straightforward; relatively few amendments are offered, and legislators move quickly to a decision. Other legislative deliberations, particularly those concerning controversial issues, are marked by parliamentary maneuvers, such as the opponents' efforts to derail the legislation. In unusual cases, opponents in the Senate will even filibuster a bill by talking nonstop to prevent a concluding vote. Filibusters are not allowed in the House.

Necessary Information About Advocacy Groups

Advocacy groups place pressure on decision makers in city councils, on boards of supervisors, and in state and federal legislatures. They may be community groups, coalitions, think tanks, public interest groups, groups of consumers, or professional groups like the NASW. Some advocacy groups exist for long periods of time, whereas others are relatively short-lived, such as those developed to address a specific issue at a specific time. Advocacy groups vary in their size and staff: Although some are relatively small, others have substantial resources from foundations and other sources. Policy advocates often work with advocacy groups when they seek to change policies.

Necessary Information About Elected Officials

Those who engage in policy practice in governmental settings need to understand the mind-sets of heads of government and legislators. Imagine that you have just spent two years planning, fund-raising, and campaigning to obtain your job. You have narrowly

POLICY ADVOCACY LEARNING CHALLENGE 6.1

IDENTIFYING ADVOCACY GROUPS

Examine the following list of advocacy groups in the gerontology sector. Go to the websites of three of these groups. Identify their general perspectives. Are they relatively liberal, relatively conservative, or relatively neutral in their orientation? How would you estimate their likely power or influence—or is that impossible to determine from their websites? Were you desirous of improving policies for senior citizens, might you consider getting feedback from staff members of one or more of these groups? Now go to the web to find advocacy groups for a specific issue such as gun control, women's reproductive rights, or immigration rights.

Some Advocacy Groups Seeking Greater Social Justice in the Gerontology Sector

1. American Federation of Labor and Congress of Industrial Organizations (AFL-CIO)
2. Alliance for Retired Persons
3. American Association of Retired Persons
4. Center for Economic and Policy Research
5. Century Foundation: Social Security Network
6. Institute of America's Future
7. National Committee to Preserve Social Security & Medicare
8. NY Network for Action on Medicare & Social Security

defeated a determined opponent who has already pledged to prevail in the next election, which is several years away. In response to this threat, you are likely to have reelection on your mind throughout your tenure. You will look at most issues with an eye to their effect on your reelection, and you will spend hours wondering about the general public's preferences. To deduce their views, you will study the following:

- Public opinion polls
- Recent outcomes of other elections in comparable districts
- The mail you receive
- The views of subgroups within your constituency, particularly those that you believe will support you in the next election
- The preferences of state or national organizations—for example, professional associations or groups such as the American Association of Retired Persons—that might contribute funds to your next campaign

Moreover, you will nervously eye the statements and positions of potential opponents in the next election. In some cases, you will support an issue to steal the thunder of your likely opponents, and in other cases you will openly support issues that they oppose to publicize your differences from them. In districts divided between liberals and conservatives, relatively liberal candidates often support liberal issues to solidify their support among liberals. They realize that without committed support from this constituency, they may lose to conservative opponents (Arnold, 1990).

This nonstop campaigning will make you sensitive to the political ramifications of certain choices. Some issues, such as increasing funds for city parks, will cause you little or no concern. However, issues seen as more controversial by important segments of your constituency will make you hesitate before committing yourself.

If elected officials often have their ears to the ground, they are also extraordinarily busy. Assume that you are a member of several major committees and subcommittees, each of which handles many issues on which you need to brief yourself before and during policy deliberations. As you hurriedly read technical reports and briefing papers, you simultaneously try to raise funds for your reelection bid. You make weekly trips back to your district to meet with its citizens and convene regularly with lobbyists from interest groups. Constituents also come to your offices every day to speak with you.

You also are concerned about your relationships with your legislative and party colleagues. As a member of legislative committees and subcommittees, you know that the chairperson holds great power and can often determine which issues will receive a serious airing in the committee and which amendments will be enacted. You belong to a political party, as well, which meets regularly, agrees collectively to support or oppose certain measures, and parcels out rewards and penalties, such as committee assignments and campaign funds. If you are a complete renegade, you will suffer reprisals from high officials on your legislative committees and in your party.

With such a full agenda, you need to develop shortcuts. You have to rely heavily on aides to manage the bulk of your interactions with constituents, lobbyists, and others (Smith, 1988; Richan, 1991). You are not afraid to delegate much of your work to these trusted aides because, without them, you cannot function. You create a division of labor by hiring specialists in legislative matters, in handling constituent demands (such as the woman who wants help with her husband's Alzheimer's disease), in fund-raising, and in public relations. In other cases, you rely heavily on lobbyists to do technical work for you, to do reconnaissance work with other legislators, to help you draft legislative proposals, and to help you write amendments to existing legislation (Smith, 1988). Early in your term, you decide that you have to develop priorities by taking some pieces of legislation seriously and only giving glancing attention to others. You decide to expend your political capital (your power resources) liberally on issues that may bring you large political dividends when you come up for reelection or that appeal to you for other reasons.

Your decisions on specific measures will hinge on electoral considerations, personal values and life experiences, personal expertise and interests, your views of the public interest, and the political feasibility of enacting specific measures.

Our discussion of the mind-sets of elected officials suggests that policy advocates need to exercise caution in making premature judgments about legislators' choices. Many factors can compel them to support or oppose a measure and to invest energy in it.

Connections Among Interest Groups, Legislators, and Bureaucrats

Important relationships exist among lobbyists, legislators, and bureaucrats. Many lobbyists, for example, are former legislators, civil servants, or political appointees. Many civil servants are former aides to legislators, with whom they maintain important relationships, such as passing them "inside information" (McIver, 1981). So-called iron triangles sometimes link civil servants, legislators, and lobbyists (or interest groups) when the legislature considers specific issues. For instance, if legislation about child abuse goes through a state legislature, several people who know each other and have worked together in the past may become active. Such collaboration may include lobbyists associated with children's advocacy groups; an association of the directors of public child welfare agencies around the state; professional associations, such as the state chapter of NASW; a key civil servant in the state's welfare department; and an aide to a legislator with a long-standing interest in child abuse. Advocates need to be aware that these relationships exist and that they can be tapped into for technical advice or for the support of a specific measure.

Public Opinion

Politicians, bureaucrats, and lobbyists work in an environment of uncertainty. They realize that voters can end a politician's career and can bring down an administration (Arnold, 1990). Bureaucrats are also vulnerable to public opinion because scandals or unpopular decisions can ruin political appointees' and even civil servants' careers (Lynn, 1987). Legislators often try to implement programs in ways that will please their constituents, as illustrated by politicians who oppose placing mental health facilities and prisons in their districts.

Advocates must often be persistent when seeking help from civil servants, but they must also be aware that civil servants are responsible for numerous tasks (Bell & Bell, 1982).

Regulations

Regulations powerfully shape the content of specific publicly funded programs because they define program details not contained in the statutes that established them. Wanting to avoid conflict that could stymie the enactment of specific legislation, legislators often prefer to relegate some policy details to regulators, that is, civil servants and government appointees. These government regulations must be published by oversight agencies before they become official so that the public can comment on them in public hearings and through written comments. When finalized, they are officially published as administrative regulations in the Federal Register—and can be located at https://www.archives.gov/federal-register, where you can find Internet addresses for search tutorials to find key words, enabling law, issuing agency, and specific years of federal regulations. You can find them in state venues by using Google search. Policy advocates can propose possible new regulations by forwarding them to oversight agencies.

Legal Action

Advocates sometimes conclude that they can only achieve their goals through litigation, such as was illustrated by court suits against specific laws enacted by many states to allow police officers to interrogate and detain persons who they believed might be undocumented persons based upon their appearance. They can inquire about public interest law firms with established reputations in specific areas of law to determine whether legal action might be an effective strategy.

HOW MACRO POLICY ADVOCATES ENGAGE THE EIGHT CHALLENGES

Now that we have discussed the context, we can turn to the way macro policy advocates contend with the eight challenges discussed in Figure 1.1 in Chapter 1.

Challenge 1: Deciding Whether to Proceed

Policy advocates initiate a policy-advocacy intervention when they decide that an existing policy is flawed. It may be ethically flawed, perhaps because it violates ethical first-order principles or because it leads to adverse outcomes for consumers, as is discussed in Chapter 2. It may be incongruent with evidence-based practices. Policy advocates sometimes initiate policy advocacy to advance the well-being of their organizations, such as preserving revenues essential to their operations as well as consumers' needs, such as opposing cuts in a specific government program. Public officials and members of political parties often support specific policies to enhance their political fortunes. Because Latinos had voted overwhelmingly for Democrats in the presidential and congressional elections of 2012, for example, many Democratic elected officials and party members were particularly desirous of obtaining major immigration reforms to consolidate their support during the congressional elections of 2014 and the presidential election of 2016.

Policy advocates have similar motives when they decide to block specific policy initiatives. In the case of immigration reform, for example, relatively liberal policy advocates were determined to block initiatives in specific states that they regarded as anti-immigrant in nature, such as allowing the police to stop anyone who appeared to be undocumented or requiring immigrants to carry proof of citizenship with them at all times. They often received assistance from courts, including the U.S. Supreme Court, that ruled that some of these policies were unconstitutional.

Advocates often consider practical factors. They may gauge the relative difficulty of changing a specific policy, sometimes deciding that it is futile to change a policy that is supported by powerful persons and interests even when they believe the policy is flawed. Many advocates wanted, for example, to enact policies that would restrict gun ownership, outlaw possession of assault weapons, or limit the size of magazines in the wake of the

POLICY ADVOCACY LEARNING CHALLENGE 6.2

ANALYZING IDEOLOGY AND COMPROMISE

Take any major controversial policy proposal currently under consideration by a local, state, or federal legislature. Identify what its most ardent advocates would want no matter their ideology, including liberals or conservatives. Then identify some concessions these advocates might consider supporting to make it more likely that it would be enacted. (HINT: Take examples from 2017 and 2018, like national immigration reform, enactment of specific policies to decrease economic inequality, proposals to limit gun ownership, proposals to decriminalize drugs like marijuana, proposals to allow persons with severe illnesses to receive assistance to end their lives, or ways to give Deferred Action for Childhood Arrivals (DACA) immigrants a path to citizenship.) Discuss why policy advocates favor these policies. Discuss how they are likely to encounter persons with divergent positions. See if you can identify some concessions these policy advocates might need to make to get their policies adopted.

massacre of students, teachers, and administrators at Sandy Hook Elementary School in 2013, not to mention many other school massacres such as in Parkland, Florida, in 2018. They discovered, however, that the National Rifle Association (NRA) had such extraordinary power in the U.S. Congress, as well as many state legislatures, that they could not make forward progress and even lost ground in many states that passed laws enhancing the ability of persons to purchase guns, including assault weapons. We have to remember that some policy goals may even take decades to enact, such as civil rights laws that were not enacted and implemented at the federal level until a century after the Civil War. Yet many surprising policy successes take place even when they seem unrealizable to advocates at a specific point in time. President Barack Obama was warned, for example, not to seek passage of the ACA by many of his top aides on grounds that it would not be politically feasible—but the rest, as they say, is history. Some policies that seem impossible to enact can be enacted by making compromises in them, including the ACA, which made major concessions to powerful interests, including health insurance companies, hospitals, and physicians—but their defenders argue that the ACA would not otherwise have been approved by Congress.

Challenge 2: Deciding Where to Focus

We have discussed the context of public policies earlier in this chapter. Policy advocates frequently have to decide where to focus their work. Do they want to focus on local, state, or federal levels of government? Do they want to enact new legislation or to amend existing legislation? Do they seek regulatory reforms or new legislation? Which legislative committee should they approach—and which elected officials? Do they want to work with an established lobby group—and, if so, which one? Could they initiate a lawsuit to block specific legislative initiatives—and, if so, with what law firm or attorney? Do they decide to proceed with policy advocacy at the macro level or select the mezzo level?

Challenge 3: Securing Decision Makers' Attention

Policy advocates must often obtain the attention of decision makers if they wish to be effective, such as supervisors, mid-level managers, specific physicians, or high-level administrators in clinic or hospital settings; directors of community-based agencies; elected officials in legislative settings; civil servants or political appointees in government agencies; or elected public officials.

These decision makers sometimes serve as "policy entrepreneurs" in advocating for a specific policy reform. They may help give the idea a title that is appealing to others. They may couple or link it with another policy or program such as one geared to preventing obesity among Latinos who have recently moved from normal weight to early-stage obesity levels. They may negotiate and bargain with clinic program and budget officials to see if they can find resources and office space for the new program. They may assemble other sponsors and supporters of the program from within the clinic and from the surrounding community—or within a legislature in the case of proposed legislation. They may seek key endorsements, such as from the American Diabetes Association or public agencies that would implement or oversee a proposed policy.

Policy advocates sometimes persevere even when they discover opposition to specific policy reforms, such as by taking an outside-in approach in which they exert pressure on one or more decision makers. Perhaps they gain the attention of one or more advocacy groups that place external pressure on decision makers in organizational or legislative settings, such as the American Diabetes Association, which pressured principals and superintendents of public schools to provide on-site medical care for diabetic children in many states. Several movements in 2017 and 2018 placed pressure on national politicians including the #MeToo Campaign that challenges sexual harassment, Black Lives Matter that challenges the slaying of black youth by police, the National Football League (NFL protests to preserve the right of football players to kneel or lock arms en masse during the national anthem to call attention to the slaying of black youth by police), the movement to give members of DACA the right to a path to citizenship, the movement for gun control legislation in the wake of the massacre of students at Marjorie Stoneman Douglas High School in Parkland, Florida, the environment movement to enact policies to avert climate change, the movement to defend science, and the movement to prevent the separation of immigrant children from their parents.

Policy advocates work closely with decision makers in framing policy proposals so that they attract support. Imagine, for example, how less appealing Head Start would have been to many legislators had it been called "Kindergarten for Low-Income Children"—or had Medicare been called "the Medical Program for Retirees." New initiatives need to be presented in ways that appeal to top managers, boards of directors, and funders in organizational settings.

Policy advocates sometimes rely on advocacy groups to sponsor and run with an issue, such as the NASW, the American Nursing Association, the American Public Health Association, the American Diabetes Association, the American Cancer Association, or many other interest groups in the health field. (Each of these groups has national associations located in or near Washington, D.C., as well as state chapters.) They have lobbyists, researchers, media officials, and others with skills and experience in advocating for legislation and regulations in legislatures and government agencies. Some unions assume important roles in securing policy reforms or in blocking proposals.

Policy advocates use their communication skills to persuade specific decision makers, legislators, or advocacy groups to assume leadership with respect to a specific issue. They use organizing skills to develop coalitions. They use influence-using skills to assess the feasibility of getting support for a specific policy initiative and to assess the likelihood and power of potential opponents.

Policy advocates have to decide which legislators might sponsor a specific piece of legislation. They often seek legislators with proven clout, such as the chairs of key committees or relatively high-level members of specific political parties. They often must engage in a balancing act in which they seek initial support not just from die-hard supporters of specific legislation but from legislators who hold moderate positions on a specific topic or who they believe can be persuaded to become supporters.

Policy advocates also have the option to pressure legislators in their town hall meetings. Members of Indivisible, a liberal protest group, stormed town hall meetings of Republicans who favored legislation proposed by the administration of President Donald Trump to repeal the ACA or to dilute some of its provisions. It is likely that they defeated Trump's effort to repeal it by swinging several votes away from the group of Republican legislators who wanted to repeal the ACA. This tactic was also used by the far-right Tea Party in 2011 and 2012 to oppose legislation put forward by President Obama (Mai-Duc, 2017).

Challenge 4: Analyzing Why a Dysfunctional Policy Developed

Macro policy advocates must often analyze who supports existing policies and why as they develop policy proposals. Take the issue of reforming gun laws in Congress in 2013 in the wake of the massacre at the Sandy Hook Elementary School in Connecticut, where a young man, armed with multiple rifles and pistols, killed many children, teachers, and the principal. Many legislators wanted to enact multifaceted legislation that included abolition of magazines that could contain many bullets, prohibiting fast-firing weapons, tightening the checking of persons' prior criminal records and whether they had been treated for mental illness, and restriction on the number of bullets that persons could purchase. Now fast forward to the shooting at Marjorie Stoneman Douglas High School in Parkland, Florida in 2018. Proponents had to ask the following questions:

1. Did the gun lobby, headed by the NRA, possess such extraordinary power in Congress that working majorities for these reforms could not be developed?
2. What interest groups existed that could counter the power of the NRA and its allies?
3. What power could the president and vice president exert?
4. Could public opinion in the districts of legislators be mobilized so that many of them would support gun reforms?
5. What specific committees in Congress would have to support gun reforms—and how were their members likely to vote?

6. Will the memory of Sandy Hook and Parkland fade so that the urgency of enacting gun reforms ebbs?

7. Is this an issue that can only be addressed if the Democrats take control of the presidency and both chambers of Congress in 2020 or in succeeding years in light of the close alliance of many Republicans with the NRA?

These examples illustrate how even skilled advocates often cannot quickly succeed: Not only did Congress not enact far-reaching gun controls, but many states loosened their gun laws. Gabrielle Giffords, the former congresswoman from Arizona who suffered a serious brain injury from a gun-wielding assailant, concluded that gun reforms could succeed only if its proponents continued their course for years to come. Most legislative initiatives take three or four years to bring to fruition—and some take even longer—so macro policy advocates have to be ready for protracted work. They must be persistent to realize their goal, but the sheer number of enacted policies in local, state, and federal jurisdictions attests to their many successes. Imagine a nation that lacked Medicare, Medicaid, food stamps, the Earned Income Tax Credit, and many facets of the ACA, such as those that disallow health insurance companies ceasing coverage to persons with preexisting conditions.

Challenge 5: Developing Policy Proposals

Policy advocates develop proposals to reform specific policies. They typically identify a number of policy options for addressing a specific problem—and then compare the options to select the best one. They might compare the policy options with respect to criteria like cost, effectiveness, and political feasibility. They might then have to decide, on balance, which option is the best one. In a specific state, for example, advocates of gun control decide whether to seek sweeping legislation that outlaws all automatic weapons including so-called assault weapons used by the military versus legislation limiting only the size of magazines versus legislation only denying guns to persons who have been incarcerated or shown to have serious mental illness. Some advocates might favor the first option but choose to limit only the size of magazines on grounds that the first option is not politically feasible at the present time. In jurisdictions more favorable to sweeping gun reform, such as in Chicago, they might opt for a ban on all automatic weapons.

Macro policy advocates sometimes use specific analytic techniques to select policies in a process of policy analysis. They begin by identifying a policy objective that they wish to achieve, such as decreasing the number of deaths from guns in the general population (Step 1). They select a policy strategy for achieving the policy objective, such as limiting the ability of guns in the general population to injure or kill large numbers of persons (Step 2). They identify specific policy options that enable them to address this problem, such as ones we have identified in the preceding paragraph (Step 3). They develop criteria that allow them to evaluate the merits of the specific policy options—and they determine the relative weight or importance of each criterion (Step 4). They might decide, for example, that the criteria are likely cost, effectiveness (in achieving the policy objective), and political feasibility. They score each of the policy options on each of the criteria—and then sum the scores of each policy option to determine the one that they prioritize in their macro policy advocacy.

We can illustrate policy analysis in Table 6.1. We list the criteria across the top of the table—in this case, cost, effectiveness, and political feasibility. We could have chosen different or additional criteria, such as ease of implementation or ethical merit. (Ethical merit could include, for example, whether it is fair to keep weapons away from persons with prior mental illness or released felons when the vast majority of them do not kill other people with guns.) In Table 6.1 we determine the weight of each criterion by giving it a score from 0 to 1.0 and make the sum of the scores of the different criteria add up to 1, such as 0.2, 0.4, and 0.4. Here, too, we could have made different choices, such as weighting political feasibility even higher in those states where gun ownership is part of a hunting culture as compared with more urban states. We then score each of the options by each of the criterion from 1 to 10—realizing that we are making best guesses if scientific evidence does not exist. In Table 6.1, for example, we had to guess the likely cost of each option from 1 (lowest) to 10 (highest); it is hard to guess what the exact costs might be when we consider the cost to the government to monitor each of the options. We then multiply the weight of each criterion times the score we give each option—then add the total scores that are in bold print in each cell and place them as the overall score on the far right of Table 6.1. As can be seen, Option 2 (limiting the size of magazines) received the highest score at 5.8, Option 3 (limiting access to guns to mentally ill persons) received the second-highest score at 5.2, and Option 1 (banning assault weapons) received the lowest score. These scores would likely vary from state to state because the political feasibility would vary by the culture and politics of each of them.

It is important not to dwell on the details of scoring rules but on the analytic process that uses a step-by-step structure to develop a final proposal. A specific policy option rarely receives a perfect score because it likely scores relatively high on some criteria but lower on others.

As this example illustrates, many other options can be identified to address this problem, including requiring persons who purchase or own military-style weapons to obtain licenses from military personnel, ban hand guns because they are not used in hunting but

TABLE 6.1 ■ Policy Analysis to Compare Options for Gun Reform

Criteria Weight	Cost (0.2)	Effectiveness (0.4)	Political Feasibility (0.4)	Total Scores
Option 1 Ban Assault Weapons	Score = 4 Final Score (4 × 0.2 = 0.8)	Score = 6 Final Score (6 × 0.4 = 2.4)	Score = 2 Final Score (2 × 0.4 = 0.8)	Overall Score = 4
Option 2 Limit Size of Magazines	Score = 7 Final Score (7 × 0.2 = 1.4)	Score = 5 Final Score (5 × 0.4 = 2.0)	Score = 6 Final Score (6 × 0.4 = 2.4)	Overall Score = 5.8
Option 3 Limit Access to Guns for Mentally Ill Persons	Score = 6 Final Score (6 × 0.2 = 1.2)	Score = 2 Final Score (2 × 0.4 = 0.8)	Score = 8 Final Score (8 × 0.4 = 3.2)	Overall Score = 5.2

often lead to deaths of persons, or (the most radical proposal) limit gun ownership to the police and the military as in many other nations.

Challenge 6: Securing the Approval or Enactment of a Proposal

Policy advocates have to develop an effective strategy if they wish to secure policy reforms. They have to establish policy goals, hone their proposals, establish a style, and select and use influence resources.

Developing Policy Goals

Policy advocates have to decide whether they want incremental change or basic change. They often have to downsize a proposal in the give-and-take of the political process. Policy advocates sometimes decide they want basic (or major) policy changes even if they encounter substantial opposition. They often develop fallback strategies if they cannot obtain fundamental changes.

Honing the Proposal

Policy advocates often develop an initial proposal but fine-tune it as they proceed. When asked to identify his most difficult challenge, a top aide to President Bill Clinton said that he and the president constantly had to decide when to hold fast to specific policy proposals and when to accept compromises in them in the push-and-pull of the political process.

Developing a Style

Policy advocates sometimes use a high-conflict strategy that may include emotion-laden language and use of the mass media, particularly if they wish to appeal to relatively liberal or conservative constituents with strong ideological views. They often use a low-conflict strategy that emphasizes the technical details of a proposal and shared goals of different legislators. They select a style that they believe will be most effective in a specific situation. They may change their style as deliberations proceed. Stylistic options include the following:

- Focusing on technical details versus underlying principles
- Inviting substantive changes in the proposal versus resisting changes
- Expanding the number of persons who participate versus limiting participation
- Seeking rapid resolution versus encouraging an extended process

Implementing and Assessing the Strategy

Even wise strategy choices come to naught, however, if advocates do not implement them skillfully. They need to be flexible, such as by revising their strategies as events change. These revisions may include expanding the number of persons that they consult, developing or expanding a coalition, or making concessions to others as they shape their policy proposal. They sometimes harden their initial positions if they believe they cannot attract others to their cause or if they fear negative responses from members of their own party or their constituencies.

Challenge 7: Securing Implementation of a Policy Reform

Enacted policies sometimes come to naught because they are not implemented at all, such as when they fail to receive needed budgetary resources, or are poorly implemented, such as when executives and staff are not committed to them. Sometimes government agencies decide they need to monitor implemented programs more stringently, such as when they think their resources are used unwisely. Perhaps implementing staff need more training to be able to achieve positive results or to be true to the provisions of a statute.

Challenge 8: Assessing Implemented Policies

Policy innovations have to be evaluated to ascertain if they improve consumers' well-being (effectiveness) and if they achieve these gains at acceptable costs (cost-effectiveness). Rigorous evaluations are often not conducted, however. They take considerable expertise to conduct as well as considerable resources—and their findings may not emerge for several years.

Evidence-based policies emerge only as implemented policies are evaluated. Gold-standard evaluations require an experimental design to ascertain if persons who receive services or benefits from a specific program benefit from them when compared with a control group that does not receive these services and benefits. In many cases, evaluators have to rely on evaluations with less rigor due to lack of necessary resources or expertise—or because agency directors block evaluations of specific programs. An array of evidence-based policies can be identified by going online, including to sites like the Campbell Collaboration, which provides a searchable list of reviews of social research in the United States and Europe (http://www.campbellcollaboration.org).

POLICY ADVOCACY LEARNING CHALLENGE 6.3

GETTING TO THE TRUTH

Go online to see if the following assertions are true that are often made in the public arena by public figures:

1. Few users of safety net programs work, including those who use SNAP and Medicaid.
2. Increases in Medicaid or welfare programs in one state will attract low-income persons from another state to emigrate to it.
3. Tax increases for millionaires will cut the nation's economic growth.
4. Requiring Medicaid and SNAP beneficiaries to work before they receive benefits will greatly reduce the cost of these programs without denying assistance to persons who truly need their services and benefits.
5. Select any other assertion made by a public figure, and inquire whether it has been shown to be true by use of research. Use the Internet to inquire how many times President Trump made claims not supported by evidence. Can you obtain data about other public officials in both political parties by going to the website of PolitiFact at https://politifact.com/?

Many policy assertions are made by public figures as if they have been subjected to evaluations when this is not true. Mitt Romney, the Republican candidate for the presidency, asserted in 2012, for example, that most spending by the federal government on public programs, like the Supplemental Nutrition Assistance Program (SNAP), housing vouchers, and Medicaid, have "massive overhead [so that] very little of the money that's needed by those that really need help . . . actually reaches them." In fact, administration costs of these programs range from 1% to 8% of total federal spending on them (Greenstein, 2012).

Improvising When Engaging in the Eight Challenges

Policy advocates rarely engage in the eight policy advocacy challenges in a sequential process. They may bypass a challenge. They may engage in a challenge only to return to it at a later point. They improvise as events change.

IDENTIFYING MACRO POLICY ADVOCACY SKILLS

Developing Overarching Strategy

Macro policy advocates have to develop a political strategy, that is, a sequence of actions and verbal exchanges that will increase the likelihood that a proposal will be enacted. Their political strategy contains seven steps.

1. They have to organize a team or coalition that will spearhead their project with leadership provided by an existing advocacy group (such as Planned Parenthood in a specific state) or by a new group.
2. They have to establish policy goals, such as whether they want incremental change or basic change. They may decide in some cases that they want to block someone else's proposal if they believe it to be contrary to their values and preferences.
3. They have to specify a proposal's content—and move quickly to find early sponsors. They often write a policy brief that identifies a policy issue, summarizes why existing policy is inadequate, discusses available options, and selects a preferred option.
4. They have to establish a style such as opting for a nonconflict process or a higher-conflict approach that may include considerable publicity.
5. They have to decide who does what, including meeting with specific persons, making certain presentations, doing research, and developing lists of supporters.
6. They have to implement their strategy with use of influence and power resources.
7. They have to revise their strategy as they proceed in light of changing events.

Diagnosing Who Stands Where—and Recruiting Persons to a Cause

Macro policy advocates seldom have the luxury of unanimous support for a specific policy proposal. They often have to determine who stands where by talking with legislators, members of key interest groups, public officials, and others. Some persons are intractable opponents. Others are enthusiastic supporters. Others are undecided but potentially persuadable—and these persons often swing the balance.

Advocates have to be skilled in recruiting persons to their cause. They have to decide what provisions of a policy can be altered or added to their proposed legislation to gain support. They have to identify legislative leaders who are willing to persuade other legislators to support it.

Framing Issues to Gain Support

Macro policy advocates have to use words and concepts that elicit support for a policy among those legislators who are critical to its enactment. They often appeal to cultural symbols that attract support, like prevention, improving health, increasing opportunity, increasing fairness, saving public resources, increasing upward mobility, helping children, increasing transparency, decreasing overhead, and decreasing fraudulent use of public funds.

Networking and Coalition Building

Successful macro policy advocates have remarkably inclusive networks that they cultivate over an extended period. They can sometimes decrease opposition from likely opponents by finding commonalities with them, such as shared experiences. They can sometimes establish personal relationships by sending them thank-you notes or get-well notes. As social work policy advocate Melissa Bird said, "Most legislators are used to getting negative comments from their constituents for not supporting a specific measure

POLICY ADVOCACY LEARNING CHALLENGE 6.4

IDENTIFYING AND USING CULTURAL SYMBOLS

Take a stab at identifying cultural symbols that might prove adverse to securing support for a legislative measure, particularly with relatively conservative legislators. Might President Barack Obama have considered a different title for the Affordable Care Act to enhance support for it among conservatives, and, if so, what might a better title have been? Or would they have mostly opposed this legislation no matter what he called it? Many Democrats in the run-up to the 2018 congressional elections decided to attack Republican candidates by citing their effort to attack the ACA provision that disallows private health insurance companies from denying eligibility for preexisting conditions such as diabetes or heart conditions. Polling data led them to this conclusion. Discuss why this issue was so compelling with many voters as compared to other possible issues such as gun control.

or for supporting another measure. They really like positive communications, such as thank-you notes or congratulatory notes or even birthday cards.

Managing Conflict

Conflict often occurs but often needs to be managed, whether upward or downward. Conflict sometimes enhances a policy advocates' quest to enact a specific piece of legislation, such as when it draws attention to a specific measure from opponents of the persons who create conflict by attacking it. Liberal and moderate legislators may, for example, believe a measure must be meritorious if conservatives attack it!

We have already discussed how many Democrats and some Republicans prevented the termination of the ACA by the Trump administration by noting how it gave health insurance to many people who otherwise would not have coverage. By sitting in town meetings of many Republican officeholders, advocates of the ACA obtained extensive coverage from the news media in their districts—the very persons who benefited from the ACA, including many Republicans.

Assume that many Democrats favor enactment of Medicare for All in the aftermath of the 2020 presidential and congressional elections. Also assume that this would phase in coverage to all Americans over a five-year period, including persons currently receiving private insurance and persons currently receiving Medicaid. Now imagine what position private insurance companies that currently serve more than 80% of the American population might take when this proposal is advanced. Imagine, too, how pharmaceutical companies, medical device companies, the American Medical Association, and the American Hospital Association might respond to this proposal.

Realize that many of these groups prosper under the current health arrangement. Private insurance companies depend on the purchasing of their policies by many employers. Pharmaceutical companies and medical device companies depend on payments from private insurance companies, Medicare, and Medicaid.

Also assume that many consumers would fear that their costs might rise under Medicare for All as compared with their current costs for private insurance. Realize that their current costs are significantly underwritten by their employers that receive huge tax write-offs for the cost of the insurance policies they provide for their employees.

Recall, too, that opponents of Medicare for All already are calling it a "socialist scheme." They often cite long waits for surgeries in Canada, which is akin to Medicare for All. Many private insurance companies may side with these opponents. Judging by their opposition to the relatively small ACA, most Republicans will likely oppose Medicare for All, even when its supporters make a compelling case that it will save the nation and health consumers tens of billions of dollars each year by eliminating costs paid to private insurance companies. Its supporters will also argue that Medicare for All will cover everyone unlike the current health system that leaves more than 20 million persons without coverage—and possibly far more persons if Republicans repeal the ACA or dilute its provisions.

We can anticipate high levels of conflict if Democrats initiate Medicare for All in 2021 and beyond. Stake out a position in this possible debate:

Strongly support Medicare for All. Discuss the likely level of conflict and partisan division that will occur.

Seek a compromise position that will leave the current health system intact, but seek some incremental additions to the nation's health system that might eventually lead to Medicare for All. These include allowing persons at age 55 to obtain Medicare in 2021—and persons over 45 to obtain it by 2025.

Support the status quo while recognizing that it leaves at least 20 million with no health insurance and is highly costly to the federal government when the tax deductions for employers' health insurance, Medicare and Medicaid costs, and emergency room costs of uninsured persons are tallied. Also realize that health consumers expend a lot of money on deductibles and co-payments in their private health insurance plans—and that private health insurance companies expend huge amounts of money to market and administer their insurance policies or far more money than Medicare and Medicaid expend for managing their health coverage.

Discuss likely levels of conflict in this debate if Democrats rally behind Option 1. Discuss what size majorities Democrats would likely need in the House and Senate even if they elect a Democratic president. Recall that President Obama barely enacted the ACA in 2010 when Democrats had only a slim majority in the Senate. Extreme conflict can sabotage a specific public policy. Discuss whether Option 2 has merit. Would some Republicans support it? Would Option 2 split the Democratic Party if many Democrats favored Option 1? What would be drawbacks of Option 3? Read the following article to obtain an overview of the debate about Medicare for All: Rosenthal, E., & Luthra, S. (2018, October 19). Don't get too excited about Medicare for All. *New York Times,* Opinion Section. Retrieved from https://www.nytimes.com/2018/10/19/opinion/.../medicare-single-payer-health-care.html.

Using Procedural Power

Macro policy advocates often use procedural power to increase the chances that a particular measure will be enacted. They convince prominent and influential legislators to sponsor their legislation. They try to route their measures to those legislative committees and subcommittees that will be favorable to them. They try to get hearings for their policy proposals before legislative committees because few measures are enacted that do not receive hearings in a timely way.

Developing and Using Personal Power Resources

Consider power transactions to be transactional in nature. Person A interacts with Person B in a way that makes Person B adopt a position or take actions that she or he would not otherwise take. Assume, for example, that Person B is predisposed to oppose a specific policy but changes her mind after Person A discusses it with her. If Person B would not otherwise have changed her mind, Person A has successfully used her personal power sources, such as her communication skills, to induce Person B to support legislation she would otherwise have opposed.

Each of us has a considerable number of personal power resources in specific situations but only to the extent that other persons respond to them. We possess expertise based upon our professional and other knowledge. We possess reward power to the extent others believe we have and will use inducements such as money, recognition, friendship, favors, or increased support from their constituents. We possess coercive power to the extent that other persons believe we can withhold inducements or impose penalties, such as excluding them from deliberations, harming their reputations, decreasing support for them from their constituents, firing them, or otherwise harming them. We possess the power of authority to the extent others believe we hold powerful or influential positions that might ultimately help or harm them. We possess charisma to the extent that other persons respond positively to our requests because of our personal qualities. We develop substantive power as other persons come to see us as able to improve policy proposals or to make changes in them that are viewed as correct or advisable.

We develop these and other power resources as others come to view us in positive terms as they interact with us or learn about us. Persons who come to be viewed as "straight shooters" often benefit from this reputation in legislative settings because legislators want and need accurate information and truthful interactions. Persons come to be seen as credible as they develop reputations, often over many years, for providing accurate information.

Developing Credible Proposals

Macro policy advocates have to develop credible proposals. They have to demonstrate that they will be effective in addressing specific problems or issues. They have to have the support of recognized experts or be supported by researchers. They have to show that they can be funded by existing resources.

Negotiating Skills

Macro policy advocates have to be effective negotiators. They need to make concessions that appear reasonable. Yet they cannot be perceived as pushovers who will always back down. Negotiators often "test the waters" by floating relatively minor concessions—and then expanding them or withdrawing them as circumstances dictate.

Developing Power by Changing Legislators' Environment

Legislators do not live in cocoons but in systems that expose them to multiple interactions with the external world. Macro policy advocates can often impact legislators' decisions by changing their external environment. They can place pressure on them by using social media that leads many persons to send them messages on Facebook, Twitter, and other social media—messages that may urge them to take or refrain from specific actions. Working with advocacy groups, macro policy advocates can pressure legislators to take certain actions, not only through lobbying but by encouraging their constituents to take specific actions. They can take direct action, such as flooding their town halls with protesters as members of Indivisible, an advocacy group, employed to block Republicans' attempts to repeal the ACA in 2017. Macro policy advocates can place pressure on legislators by obtaining stories in the mass media that give them positive or negative coverage for specific positions or actions that they have taken.

Macro policy advocates can place pressure on specific legislators by participating in electoral politics. They can elicit support for them by engaging in political campaigns or can work to defeat them. They can organize specific parts of legislators' constituencies, such as homeless persons, low-income persons, residents of public housing, and other constituencies. They can raise funds for specific candidates.

Developing Personal Connections

It is easy to forget that public officials are people with the same problems and issues as everyone else. Policy advocates need skills to forge personal relationships with public officials, even getting to know names of their family members, learning their hobbies, discovering mutual friendships, and seeing them at a number of occasions. Nor does it hurt to empathize with them about the significant challenges public life brings to them, such as loss of anonymity, hectic working schedules, and exposure to criticism in the mass media.

POLICY ADVOCACY LEARNING CHALLENGE 6.5

KEY SKILLS AND PERSONAL ASSETS

Read the vignette that follows to identify key skills needed by macro policy advocates like social worker Dr. Melissa Bird, who successfully persuaded members of the legislature of the State of Utah to enact a specific piece of legislation. Identify some of the skills needed by macro policy advocates in the preceding discussion—but also identify some additional skills not discussed in preceding pages. Discuss which of these skills you think you possess—and which ones you would like to further enhance. Can you imagine becoming a macro policy advocate for an issue that interests you, whether like Dr. Bird or in some other capacity, for example, working on a political campaign or helping an advocacy group?

A SOCIAL WORKER PERSUADES THE UTAH LEGISLATURE TO ADVANCE WOMEN'S REPRODUCTIVE RIGHTS

Melissa Bird, MSW, Ph.D.

In my second year of my MSW program at the University of Utah, I engaged in research on homeless youth. It was my intent to discover how many homeless youth identified as lesbian, gay, bisexual, transgender, or questioning (LGBTQ). At the time, there was no homeless youth shelter, and I partnered with Volunteers of America—Homeless Youth Resource Center (HYRC), a nonprofit agency in Salt Lake City that provided drop-in services to homeless youth under the age of 25. Estimates by advocates indicated that more than 1,000 young people were homeless, oftentimes forced to sleep in subzero temperatures. As my working relationship with HYRC evolved, I became aware that the law in Utah stated that youth could not be sheltered for longer than eight hours without parental consent or emancipation, and there was no emancipation statute in the state of Utah.

This policy situation was untenable to me as a social welfare advocate. Armed with the policy advocacy skills I had learned in my MSW program and the data I had collected, and with the knowledge that hundreds of young people were sleeping in the streets, I decided to tackle the issue of emancipation. Using the principles of social work and the science of social change, I sat at my dining room table and wrote my first piece of legislation. Building upon the emancipation statutes from 26 other states, I crafted my bill (six more policies would be drafted over the next eight years, five of which were passed into law). As I combed through the policies from other states, it became clear that if I was to be successful in my endeavor to pass legislation, I would have to tailor my words carefully so that they would be palatable in a conservative political environment. After my draft was written, I contacted former Utah representative Roz McGee and asked her if she would look at my bill. Within minutes of reading it, Representative McGee said not only that it was an incredible document, but that she would be honored to sponsor it, thus HB77 Emancipation of a Minor was born.

The number one social work principle that has carried me throughout my macro practice career has been working with people where they are, not where I want them to be. Working with politicians, stakeholders, nonprofit advocates, state agencies, and lobbyists can be an incredibly complicated and daunting task. I honed my skills and learned the legislative process with patience and an open mind. I understood that emancipation of youth was a complicated subject, especially in a conservative state in which family and the rights of parents are paramount above all else. There is a saying that laws are like sausages; you don't want to see how either one is made. Being a social welfare advocate and a professionally trained social worker meant that I was ready and able to deal with the messiness of the Utah legislature.

Honing My Skills

Representative McGee worked with me closely as I honed my skills as a macro-level practitioner. Not only did I need to learn how a bill becomes a law, but I needed to understand the fundamental difference between the words "may" (permissive) and "shall" (mandate), why budgets are important (the fiscal note had to be minor to get it through the process), who the people in power were (in Utah the conservatives held the keys to the kingdom), and how to deal with conflicts and setbacks in a professional (read unemotional) manner. Strategically, you must get to know your legislature. Members of the legislature can be broken into three categories: saints (the legislators who support you 100% of the time), sinners (the legislators who never support you 100% of the time), and savables (the legislators who are on the fence). When developing macro policy strategy, you want to target the savables; those are the legislators whom you can move to your side, and they are the ones who help you win.

Building a Coalition

As the person who brought the issue of emancipation to Representative McGee, coalition building became my main task after the final draft of the bill was introduced. I went to the usual sources (Health and Human Services—Division of Child and Family Services and Juvenile Justice Services) but also got creative. The best example of this is my cold calling the attorney general of Utah, Mark Shurtleff. From the local media, I was aware of his special interest in the "lost boys" of polygamous communities throughout Utah. Young men were being kicked out of their homes and forced into homelessness by their parents, and it was my hope that his passionate interest in this issue would give me the backing I needed from the attorney general's office. Not only did I gain the valuable support of the attorney general, by my willingness to be candid and honest with him in our initial meeting led to a professional relationship that lasted years even though we were on different spectrums politically and philosophically.

Learning to Deal With Conflict

When engaging in macro-level policy you will encounter an unusual amount of conflict, not just from people who you know are your enemies (those who will vote or work against you 100% of the time) but also from people who you would think would be inherent allies. Despite this conflict your success as a macro policy practitioner is dependent on being able to work with everyone, even people you don't think you can. I cannot begin to tell you how many times I was shocked when meeting with peers and fellow social workers to

(Continued)

(Continued)

hear things like "You will never pass this legislation," "We cannot support you; it would be detrimental to our agency," and "This is too controversial, and you won't be successful anyway." If I had listened to everyone who told me this was impossible, hundreds of youth in Utah would still be sleeping in the snow. The person who became my main source of conflict was a conservative lobbyist named Gayle Ruzicka. She is the head of the Eagle Forum and someone who was dead set on killing my emancipation bill. Gayle could have been a sinner, and she was successful in killing the bill the first year it was introduced. In 2006, HB 30 Emancipation of a Minor was reintroduced. This time I worked with Gayle to insert one crucial line about parents' fundamental rights over their children. This ensured that the bill passed, and we both ended up feeling like it was a good piece of legislation that would help children and their families.

I remember asking my policy professor if there were macro-level jobs for me in Utah; she told me no. I proved her wrong when my experience as a citizen lobbyist, a macro-level social worker, and a professional that works with people in a nondivisive manner led to a job as the executive director of Planned Parenthood Action Council and vice president of public policy for Planned Parenthood of Utah. Never did I believe that one piece of legislation would lead to six. I continue to attribute my success as a lobbyist to successfully making the connection with the missing half of advocacy—bringing case advocacy skills together with policy advocacy skills to improve the lives of women and children throughout Utah.

- Identify specific policy advocacy skills that Melissa Bird used in this vignette.
- What attributes does she possess that contribute to her success in enacting legislative proposals? Can you emulate her when you engage in macro policy advocacy in organizations, communities, and legislatures?

FACILITATING MACRO POLICY PRACTICE AMONG FRONTLINE HEALTH PROFESSIONALS

The research project that surveyed roughly 300 frontline health professionals in eight hospitals that was discussed in Chapters 1 and 4 developed and validated a Policy Advocacy Engagement Scale that measured the extent frontline health professionals have engaged in macro policy advocacy during the prior six months (Jansson, Nyamathi, Heidemann, Duan, & Kaplan, 2015). Respondents were asked to indicate how often they had engaged in macro policy advocacy under these circumstances: When patients' ethical rights may be at risk, their quality of care may be at risk, their cultural content of care may be at risk, their preventive care may be lacking, the affordability and accessibility of care may be problematic, their mental health conditions may be lacking, and their community-based health care may be lacking. The Policy Advocacy Engagement Scale has acceptable levels of content validity, construct validity, and reliability, making it the first scale that measures levels of macro policy engagement by frontline health professionals.

The research team developed six additional scales that predicted levels of policy advocacy engagement among acute-care health professionals (Jansson et al., 2016). The Eagerness Scale contains seven questions that asked respondents to indicate how often they wished they had been able to engage in more policy advocacy during the past six months when patients' ethical rights may have been at risk, patients' quality of care may have been at risk, patients' culture may not have been respected, patients' preventive care may have been lacking, patients' affordable or accessible care may have been problematic, patients' mental health conditions may have been lacking, and patients' community-based health care may have been lacking. Respondents with relatively high scores were more likely to have relatively high scores on the Policy Advocacy Engagement Scale than ones with relatively low scores ($p < .001$). It appears that frontline health professionals who engage in macro policy advocacy not only provide more of it than their frontline peers but are more likely to believe they were remiss in not providing even more of it during the prior six months.

Respondents were asked to rate the extent they had 13 specific skills that the research team hypothesized were often needed for engagement in macro policy practice. These included these skills: the skill to influence other people to work with themselves to change specific policies, to initiate policy-changing interventions, to negotiate or bargain, to mediate conflicts, to talk with community leaders, to communicate with public officials, to help patients become policy advocates, to discuss specific kinds of unresolved patient problems with administrators, to change hospital policies, to establish and develop multidisciplinary training sessions, to develop better coordination among different units, to make budget suggestions, and to change protocols of operating procedures. Ones who scored relatively high on this scale were more likely to score relatively high on the Policy Advocacy Engagement Scale ($p < .01$).

Respondents were asked to indicate they received "tangible support" from their hospital to engage in macro policy advocacy on a scale from not at all (1) to a great deal (5) with respect to five items. These items were the following: "Policy advocacy is part of my job description," "My supervisor encourages me to engage in policy advocacy," "My supervisor supports me when I experience negative repercussions resulting from my policy advocacy," "I believe my supervisor will come to my defense if I am criticized for doing policy advocacy," and "I believe administrators are aware of unresolved policy problems." Respondents with relatively high scores were more likely to score relatively high on the Policy Advocacy Engagement Scale ($p < .01$). It is likely that frontline professionals perceive these tangible supports as an organizational mandate to engage in macro policy advocacy in hospitals whose staffs are overwhelmingly committed to daily patient care.

Frontline professionals were asked to gauge whether policy advocacy is effective in improving health care policies from not at all (1) to a great deal (5) in organizational, community, and government settings. Respondents who believed macro policy engagements were relatively effective were more likely than other respondents to score high on the Policy Advocacy Engagement Scale than other respondents ($p < .05$). Because policy changes are often difficult to achieve, these positive perceptions of their effectiveness may contribute to their willingness to seek policy changes.

Respondents were asked to indicate the extent their hospitals were "receptive" to macro policy advocacy. They were asked, for example, to what extent they had been excluded from discussions about unresolved patient problems and invited to participate in case conferences about patients with specific unresolved problems. They were asked to what extent they experienced hostile behaviors from other professionals and whether sufficient discussion is devoted to patients' unresolved problems by health care professionals, supervisors, and administrators. They were asked whether attending and consulting physicians ask them to gather information from patients relevant to specific unresolved problems—or to gather information from family members, friends, or significant others that is connected to patients' unresolved problems. Respondents with relatively high scores on the Organizational Receptivity Scale were more likely to score relatively high on the Policy Advocacy Engagement Scale ($p < .05$). It may be that frontline health professionals who believe they are not restricted to relatively minor roles by higher-level health care professionals, supervisors, and administrators may also use this perceived autonomy to engage in macro policy advocacy. As with our discussion of micro policy advocacy at the end of Chapter 4, frontline health professionals' macro policy advocacy is strongly influenced by a variety of factors that include personal orientations ("eagerness"), belief that they possess key macro policy practice skills, belief that they possess considerable autonomy in their professional work, belief that macro policy engagements are effective, and the perception that they receive tangible support for their macro policy advocacy from administrators and supervisors.

One could imagine a combination of training programs and policy reforms in hospitals could elicit higher levels of macro policy advocacy among frontline professionals who are well positioned in proximity to patients, unlike higher-level administrators. It would be worth the investment in developing and implementing training programs and policy reforms if frontline professionals identified problems in addressing the seven core issues among hospital staff and if they sought effective solutions. Hospitals could survey their frontline health professionals at regular intervals to ascertain if their training programs and policy changes have led to higher levels of macro policy advocacy among them.

Learning Outcomes

You are now equipped to:

- Use macro policy advocacy skills
- Analyze the context of macro policy advocacy
- Discuss how macro policy advocates engage the eight challenges
- Use a macro policy advocacy vignette to review content in this chapter

References

Arnold, R. D. (1990). *The logic of congressional action.* New Haven, CT: Yale University Press.

Bell, W. & Bell, B. (1982). Monitoring the bureaucracy: An extension of legislative lobbying. In M. Mahaffey and J. Hanks (Eds.), *Practical politics: Social work and political responsibility.* Silver Spring, MD: National Association of Social Workers.

Greenstein, R. (2012, April 17). Testimony before House Budget Committee hearing on strengthening the safety net. Center on Budget and Policy Priorities. Retrieved from http://www.cbpp.org/cms/index.cfm? fa=view&i=3745

Holtzman, A. (1970). *Legislative liaison: Executive leadership in the Congress.* Chicago: Rand McNally.

Jansson, B. (2018). *Becoming an effective policy advocate: From policy practice to social justice.* Boston: Cengage Learning.

Jansson, B., Nyamathi, A., Heidemann, G., Bird, M., Ward, C., Brown-Saltzman, K., Duan, L., & Kaplan, C. (2016). Predicting levels of policy advocacy engagement among acute-care health professionals. *Policy, Politics, & Nursing Practice, 17*(1). Retrieved from https://www.ncbi.nlm.nih.gov/pubmed/27151835

Jansson, B., Nyamathi, A., Heidemann, G., Duan, L., & Kaplan, C. (2015). Validation of the policy advocacy engagement scale for frontline healthcare professionals. *Nursing Ethics, 54*, 24–34.

Lynn, L. (1987). *Managing public policy.* Boston: Little, Brown.

Mai-Duc, C. (2017, February 6). Noisy town hall protests show how the left is using Tea Party tactics to fight Trump. *Los Angeles Times.* Retrieved from https://www.latimes.com/politics/la-pol-ca-indivisible-protests-trump-congress-20170206-story.html

McIver, J. W. (1981). *Tribes on the Hill.* New York: Rawson Wade.

Richan, W. (1991). *Lobbying for social change.* New York: Haworth Press.

Smith, H. (1988). *The power games: How Washington works.* New York: Ballantine Books.

7

BECOMING POLICY ADVOCATES IN THE HEALTH CARE SECTOR

LEARNING OBJECTIVES

In this chapter you will learn to:

1. Develop an empowering perspective toward persons with unaddressed health problems
2. Analyze the evolution of the health care system in the United States
3. Analyze how health disparities are powerfully linked to economic inequality in the United States
4. Describe the political economy of the health sector, including powerful players and interests as well as underrepresented ones—and identify some key advocacy groups in the health sector
5. Analyze seven core problems in the American health system, and identify some examples of Red Flag Alerts with respect to each of them
6. Identify some background information about the causes and scope of each of the seven core problems
7. Identify some important regulations, statutes, and programs that are important resources for advocates
8. Drawing upon vignettes in this chapter, discuss how advocates can move from micro to mezzo policy advocacy and to macro policy advocacy
9. Learn about some breaking news in the health care system
10. Discuss a major proposal for reforming the American health care system

The health sector will likely consume as much as 20% of the gross domestic product (GDP) of the United States soon—or nearly one in five dollars spent by Americans. Roughly 12.2% of American adults lacked health insurance at the end of 2017. Even medically insured Americans have to pay large out-of-pocket costs, such as deductibles and co-insurance. Many low- and moderate-income people have too much income and property to qualify for Medicaid and lack insurance coverage. Americans have poorer health outcomes than many other industrialized nations, such as in rates of infant mortality, longevity, and rates of specific chronic diseases. Poisoning from opioid drugs killed roughly 72,000 people in 2017. Low-income Americans, Native Americans, and low-income African Americans have poorer health outcomes than affluent white people.

DEVELOPING AN EMPOWERING PERSPECTIVE TOWARD PERSONS WITH UNADDRESSED HEALTH PROBLEMS

Wide health disparities exist in the United States. As compared with low-income Americans, affluent Americans live longer, have fewer disabilities such as diabetes and heart disease, and are more likely to have health insurance. (Low-income Latinos, particularly recent immigrants, are a notable exception.) Affluent people are more likely than low-income people to have private health insurance. They often have healthier food. They do not suffer stress that accompanies poverty. They live in neighborhoods where they can safely walk and work. Low levels of violence exist in their neighborhoods. They are more likely than low-income people to live in neighborhoods with relatively low levels of air pollution, toxic waste dumps, and violence. They are more likely than low-income people to have access to gyms, pools, and physical trainers. They are less likely to smoke, possibly the single most harmful cause of poor health and short longevity.

Roughly 40 million Americans suffer from disabilities that restrict their movement, their ability to work, and their cognitive functioning capabilities (Pew Research Organization, 2017). Disabilities include advanced forms of chronic diseases like diabetes and congestive heart failure, mental conditions like schizophrenia, birth defects such as Down syndrome, degenerative disease associated with aging, genetic conditions such as Huntington's disease, and injuries from gunshots, automobile accidents, and warfare. Persons with disabilities are a marginalized population because they sometimes cannot be self-sustaining, cannot ward off attackers, require special assistance to enter and use public transportation, suffer discrimination, and are disproportionately poor. Many Americans have unhealthy health habits whether because they eat excessive amounts of meat; eat relatively few vegetables and fruit; consume high levels of fat, salt, and sugar; watch television for long periods; and are excessively sedentary (Knight, 2004).

Policy advocates need to work on at least six fronts. They have to enact and fund public health projects to increase health literacy and health prevention. They have to equalize access to health care among social classes and between whites and members of marginalized groups. They have to provide services, housing, and protections for persons with disabilities. They have to develop strategies to prevent smoking, opioids, and other toxic drugs; unhealthy diets; and lack of exercise. They have to provide civil rights for persons with disabilities and other health conditions in the workplace and in the broader society. They need to cut the cost of medical care while improving its effectiveness.

ANALYZING THE EVOLUTION OF THE AMERICAN HEALTH CARE SYSTEM

The evolution of the American health care system from the colonial period to the present is depicted by the following timeline:

- The United States develops a two-tiered health system in the 19th century that consists of private, fee-paying persons and low-income persons who use public institutions.
- A national network of private not-for-profit hospitals emerges in the early 20th century that are typically started by Jewish, Protestant, and Catholic religious groups.
- States enact licensing laws of physicians and hospitals with the emergence of the American Medical Association (AMA) as the most powerful lobbying group in the states in the early part of the 20th century.
- Some states reject proposals to enact state health insurance plans funded by public resources from 1915 to 1920.
- Public systems of health care emerge in the late 19th century and early 20th century in many big cities that primarily serve low-income persons who cannot afford private care.
- Franklin Roosevelt decides not to include national health insurance in the Social Security Act because he fears opponents, led by the AMA, will block the act itself.
- Private health insurance begins in the 1930s and early 1940s but rapidly expands during and after World War II, the 1950s, and the 1960s so that a vast majority of Americans receive private health insurance from their employers.
- The Veterans Administration is established after World War II—and soon establishes a national network of hospitals.
- The National Institutes of Health (NIH) is established after World War II.
- Congress fails to enact President Harry Truman's proposal for national health care in 1949.
- A national system of health benefits for federal civilian employees and their dependents is established after World War II that covers about 9 million persons by 2014.

- Congress enacts Medicare in 1965 to help finance episodes of acute care among seniors. Medicare provides inpatient care to persons over 65 for relatively short-term health conditions (Part A), which is financed by payroll deductions of employers and employees. It also provides hospital care (Part B), which is financed by monthly premiums from seniors and from general revenues. (Other groups were added to Medicare in subsequent years, such as persons with kidney disease who need dialysis and kidney transplants.)
- Congress enacts Medicaid in 1965 to serve two populations: low-income, non-elderly persons who do not have private health insurance or whose medical costs bankrupt them or who are on welfare rolls and seniors who become medically bankrupted when their medical costs exceed Medicare's benefits and their private savings. It is financed by a combination of federal payments and state payments. It covers many previously uninsured persons.
- Congress fails to enact national health insurance proposals put forward by Presidents Richard Nixon, Jimmy Carter, and Bill Clinton.
- The Children's Health Insurance Program (CHIP) was established in 1997 to finance health care for children not eligible for Medicaid.
- The Medicare Prescription, Improvement, and Modernization Act of 2003 added prescription drug coverage to Medicare (Medicare Part D) beginning on January 1, 2006.
- President Barack Obama enacted the Affordable Care Act (ACA) in 2010. We discuss its provisions subsequently in this chapter as well as its implementation in succeeding years.
- Senator Bernie Sanders (D, VT) campaigned for "Medicare for All" in his presidential campaign of 2016.
- President Donald Trump and congressional Republicans fail to repeal and replace the ACA in 2017 and early 2018 as well as to sabotage its implementation. They failed but succeeded in ending the mandate for young adults to get coverage, thus increasing premiums of insurance companies when relatively healthy persons were removed from coverage.
- House Republicans sought to impose a work requirement for Medicaid but were rebuffed by a federal judge in June 2018. However, some states imposed this policy.

IDENTIFYING HEALTH PROBLEMS CAUSED BY ECONOMIC INEQUALITY

Low-income consumers and consumers with low levels of education have higher rates of morbidity and shorter lives than other people (Kawachi, Daniels, & Robinson, 2005). Health disparities are caused by a combination of economic and social factors including lack of nutrition and adequate housing, lack of stable employment, lack of health knowledge, community violence, fatalism, mental disorders linked to poverty, inaccessible health care, inability to finance health care, substance abuse, and unstable support systems (Jacobs, Kohrman, Lemon, & Vickers, 2003).

The nation has made remarkable progress in lengthening life spans and improving public health during the last century, but health disparities remain unacceptably high in the United States when we compare white Americans with specific racial groups and when we compare persons in the top fifth of economic distribution with persons in the bottom fifth (Kawachi et al., 2005; Satcher et al., 2005). If relatively affluent African Americans have better health than low-income members of their races, even they often have poorer health outcomes than relatively affluent whites (Kawachi et al., 2005).

Low-income Americans are more marginalized economically and educationally than their counterparts in Canada, the UK, and Japan. They are less educated and lack strong supports from safety net programs (California Newsreel, 2008). They have lower rates of civic participation with voting rates of only 49.1% as compared with 56.1% in Canada and 83.2% in Sweden (Jackson, 2002).

Some researchers have contended that Latinos appear to have escaped the nexus between race, poverty, and poor health in the so-called Mexican American paradox. Mexican Americans have a life expectancy of 77 years for men and 83 years for women, for example —and this longevity, they contend, extends even to low-income Mexican Americans (Lee & McConville, 2007). They conjecture that social factors, such as cohesive families and religiosity, as well as diet, may have caused these positive effects. Other researchers argue, however, that sampling errors may account for some of these results. They note that immigrants to the United States have better health than migrants who have resided in the United States for a considerable period, which would suggest that Latinos' health will erode as more of them have been residents for longer periods. They also note that Latinos have been plagued by an epidemic of obesity and diabetes that already is eroding these positive health outcomes. Whereas 6.4% of Caucasians are diabetic, 11.1% of Latinos are now afflicted with this disease (Office of Minority Health, 2009). Latinos also suffer from increasing rates of HIV/AIDS and heart disease (Lee & McConville, 2007).

Poverty is associated with shorter lives and greater illness for several reasons. Poor persons are more likely than affluent persons to lack insurance and to lack accessible health practitioners and services. Some theorists contend that poverty causes adverse biological changes (Barr, 2008). Poverty is associated with poorer diets and less exercise as well as higher rates of smoking. Some theorists contend that inequality in American society marginalizes American poor people as compared with poor persons in other nations because they have less generous safety net programs, poorer schools, and lower levels of upward mobility (Jansson, 2019).

ANALYZING THE POLITICAL ECONOMY OF THE AMERICAN HEALTH SYSTEM

The enactment of the ACA during the Barack Obama presidency, as well as efforts to replace and repeal it in 2017 and early 2018, reflect the extraordinary conflict between supporters and opponents of government health programs (Brill, 2015). Recall that

President Obama entered his presidency in 2009 when roughly 45 million Americans lacked health insurance. It is important to understand why many Americans lacked health insurance. Although the federal government granted significant tax write-offs to corporations that provide health insurance in 1948, it made coverage optional. So many corporations chose not to provide it, particularly smaller companies and non-unionized ones, who contended they could not afford it. President Obama provided many incentives in the ACA to induce more companies to provide health insurance, such as offering federal subsidies to small companies. Obama also offered Medicaid coverage to millions of people whose incomes fell just above the existing eligibility levels in states, although the U.S. Supreme Court subsequently ruled that each state could decide whether to accept this offer. As of February 2019, 37 states and the District of Columbia had expanded Medicaid eligibility levels, whereas 14 had not. The ACA developed insurance exchanges in the states that offered private insurance to uninsured people. It also required young people to purchase health insurance in a so-called mandate so that insurance premiums would be lower as young people are less likely to use health care than older Americans. It made youth who failed to purchase health insurance pay a penalty. Arguing that a mandate on youth was unfair and also wanting to sabotage the ACA, Republicans repealed this ACA penalty in their tax legislation in late 2017. To keep premiums down, some states such as Maryland hope to enact their own version of a mandate. These policies succeeded in reducing levels of lack of health insurance markedly from 2010 to 2018.

President Obama also wanted to reduce the cost of health care in the United States because it consumed almost 20% of the nation's GDP by 2010—or far more than health costs in other industrialized nations. This high cost partly stemmed from the nation's reliance on private health insurance companies that charged roughly 25% extra to fund their advertising and management costs as compared with a government-run program like Medicare. They also established unethical rules like declining coverage to persons with "preexisting conditions" when they applied for coverage, such as diabetes and heart disease. Although the ACA successfully outlawed this practice, the Trump administration moved in court to allow insurance companies to resume this practice. Democrats made retention of coverage of persons with preexisting conditions their centerpiece in the congressional elections of 2018. Other factors contributed to the high cost of American medicine, including excessive use of scans, medications, and surgical interventions. Its costs rose as well because the American health system relies on a fee for service that reimburses physicians and hospitals separately for each procedure, medical test, and medication. Fee for service provides an economic incentive to provide more services as compared with national health systems in many other nations that hire physicians and own hospitals. American hospitals conduct, for example, far more scanning tests of patients than other health systems at enormous cost.

This American health system erred as well in its emphasis on curative services rather than (also) preventive services. Physicians and hospitals focused on curative services because they would bring huge reimbursements as compared, say, with smoking cessation services, exercise programs, or dietary programs. The sheer cost of medical care has important negative consequences. Consumers have shouldered many of the costs through out-of-pocket payments such as for deductibles, exclusions, and co-insurance. Many states spent as much as 25% of their budgets on Medicaid alone. Medicaid and Medicare, taken together, constitute a large part of the federal budget.

Despite an overall cost that was roughly double the per-capita cost of many other health systems in other industrialized nations, the American health system achieved poorer health outcomes than other industrialized nations. Americans live shorter lives and higher rates of infant mortality, higher levels of some chronic diseases, and longer periods of absence from jobs due to illness (Jansson, 2019).

The two largest public programs enacted in the United States in 1965—Medicare and Medicaid—had problems of their own. Medicare gave seniors health coverage when they had none, and Medicaid gave poor people health coverage and coverage of seniors when they had virtually none. But Medicare covered only "acute" medical conditions that required no more than 90 days of coverage while neglecting seniors who developed chronic health conditions that often continue for years. Seniors could then switch to Medicaid but only after "spending down" their resources to the eligibility level of their state—a demeaning process that forced them to give up their assets. Medicaid became the funder of most of the nation's nursing homes by the early 1970s. Many physicians refused to accept Medicaid patients, however, because their states funded their services at much lower levels than Medicare and private insurance. The ACA had several provisions that Obama hoped would cut the cost of American medicine including commitments from physicians and hospitals to cut fees.

President Obama prioritized health reform because he heard from thousands of people at campaign stops that they had no insurance, had to foot a huge share of their medical costs, or could not afford medications. Obama hoped that congressional Republicans would work with him to develop a bipartisan health plan but discovered in 2009 that virtually none of them supported health reform. Relying entirely on support from Democrats, he got it enacted in early 2010 with seven prongs: (1) cover roughly 30 million uninsured Americans, (2) increase the coverage and quality of private health insurance, (3) decrease the rate of cost increases in medical services, (4) increase primary care and health prevention, (5) raise the eligibility levels of Medicaid in states so that millions of persons could join the program, (6) mandate young people to purchase insurance to balance out seniors' high medical costs, and (7) raise states' eligibility levels of their Medicaid plans to allow millions of people with incomes just above them to enroll in Medicaid at no cost to themselves as the federal government would foot their additional costs.

Obama secured the enactment of the ACA primarily by convincing powerful interests that they would benefit from its passage, including pharmaceutical companies, the AMA, hospitals, medical-device manufacturers, and private insurance companies (Brill, 2015). He asked in return that they promise to cut their fees to reduce the costs of the ACA. Armed with their backing, President Obama and Democratic allies narrowly secured the support of the ACA in early 2010. Despite some early implementation problems, the ACA achieved its major goals from 2011 through 2016. Opinion polls showed that it was and is highly popular. It did insure roughly 24 million people and had accomplished many of its other goals by 2017.

The implementation of the ACA proved to be difficult (Brill, 2015). Conservatives succeeded in getting cases before the Supreme Court that proposed to end the mandate that young people get insurance and the requirement that states raise the eligibility levels of their Medicaid program. In a one-vote decision, the Supreme Court upheld the mandate that young people buy health insurance. In another one-vote decision, it struck down the ACA's requirement that states raise the eligibility level of their Medicaid plans, but gave them the *option* to do so instead. Fourteen states decided by late 2018 not to raise Medicaid eligibility, whereas 37 states and Washington, D.C. did. This expansion of Medicaid has been one of the ACA's great successes because it has provided Medicaid coverage to tens of millions of low-income Americans (Kaiser, 2018). Medicaid expansion was possibly the most important achievement of the ACA, particularly because it induced many red states to expand Medicaid, including Virginia in late May 2018.

Infused with hard-right members from the so-called Tea Party, House Republicans adopted the pledge to repeal and replace the ACA as their major goal even by late 2010—and voted to repeal and replace it on many occasions from 2011 through 2016. Donald Trump demanded the repeal and replacement of the ACA during scores of campaign appearances in 2015 and 2016 (Jansson, 2019b). When he achieved an upset victory over Hillary Clinton in 2016, the stage was set for a titanic battle between Democrats and Republicans. The stakes were high for both sides: Trump feared loss of support from many of his followers if he failed to keep his campaign promise to repeal and replace the ACA, whereas Democrats fought for the ACA's survival. Trump contended that the ACA had caused rising medical costs and had caused many private insurance companies to abandon the ACA's state exchanges through which many people bought their insurance, whereas Democrats contended that the ACA insured roughly 24 million people and had accomplished many of its other objectives. Whereas Republican wanted to replace the ACA with private health accounts but with minimal funding from the government, Democrats wanted to fix it by subsidizing private insurance companies that were unable to make a profit on the state exchanges (Jansson, 2019b).

When Republican Paul Ryan introduced an alternative to the ACA in early 2017, he encountered a firestorm of protest from moderate House Republicans and many Republican voters, not to mention unanimous opposition from congressional Democrats. They flooded town hall meetings, shouting their opposition to his plan because it would cut federal subsidies that allowed them to purchase health insurance from the ACA's state exchanges. They demanded that states continue to have the option to raise their Medicaid eligibility levels, not just to give them medical coverage but to allow them to continue to receive treatment for poisoning from opioid drugs that had killed roughly 72,000 people in 2017. The Congressional Budget Office (CBO) estimated that Ryan's version would cause 23 million persons to lose insurance within 10 years. Democratic activists joined Republican opponents. Ryan was unable to get sufficient votes to repeal and replace. When he proposed moderate provisions, hard-right Republicans were furious. When he added proposals favored by Republican hard-right politicians, moderate Republicans were furious. A similar process took place in the U.S. Senate when Mitch McConnell, the Republican Senate majority leader, put together a proposal that the CBO estimated would cause 20 million people to lose health insurance. As in the House, efforts to add moderate features outraged far-right Republicans (Jansson, 2019b).

Unable to repeal and replace the ACA in either chamber, President Trump tried to sabotage it by drastically cutting federal funds for advertising the state exchanges. They wrote into their budget legislation in early 2018 a provision that rescinded the requirement that young people obtain health insurance, hoping that insurance companies would be forced to raise their premiums sufficiently to reduce the number of persons

TABLE 7.1 ■ Advocacy Groups Seeking a Just American Health System
American Medical Student Association
American Nurses Association
American Public Health Association
California Nurses Association/National Nurses Organizing Committee
Families USA
Kaiser Family Foundation
Medicare Rights Center
National Association of Social Workers
National Physicians Alliance
Robert Woods Johnson Foundation

purchasing health insurance on the state exchanges. The ACA survived these strategies in 2017 and early 2018. Much would depend on the outcome of the congressional elections in November 2018 and the presidential election in November 2020 because control of either or both chambers of the Congress would give Democrats the ability to block efforts to repeal the ACA or to weaken it.

Democrats hoped that they could expand government health programs if they regained control of the House, the Senate, and the presidency in the congressional elections of 2018 and 2020 and the presidential election of 2020. Some wanted to push for national health insurance such as Sanders's proposed Medicare for All plan, which would mostly eliminate private insurance. Others favored incremental reforms, such as expanding Medicaid's eligibility level yet further and allowing people to become eligible for Medicare at age 55 rather than age 65. Many advocacy groups seek an American health system that is in accord with ethical principles of social justice. (See Table 7.1.)

ANALYZING THE SEVEN CORE PROBEMS

Core Problem 1: Engaging in Advocacy to Promote Ethical Rights, Human Rights, and Economic Justice— With Some Red Flag Alerts

Many ethical issues arise in the American health system. Patients have to make difficult choices often involving life-and-death issues. Health systems collect extensive data from patients about sensitive issues. Patients and parents have to make difficult decisions about preserving and ending lives. Decisions must often be made in a brief time span. Medical errors frequently occur, including between 100,000 and 400,000 preventable deaths from hospital errors per year (Consumer Reports, 2014). A power imbalance often exists between physicians and their patients that can lead to poor communication.

- **Red Flag Alert 7.1.** A patient does not give informed consent to a procedure because health professionals fail to discuss the possible benefits, dangers, or side effects of treatment options—or do not inform them about other treatment options, including the option of not using invasive procedures or specific medications.
- **Red Flag Alert 7.2.** The confidentiality of a patient's medical information is not preserved because it is divulged to other persons without the patient's consent.
- **Red Flag Alert 7.3.** Medical professionals perform surgery or provide medications to a patient who is not competent to give informed consent, such as a person with limited cognitive function.
- **Red Flag Alert 7.4.** A patient is not informed of his or her right to sign an advance directive that tells health professionals what kinds of treatments he or she does not want under specified circumstances, such as if he or she is comatose, cannot swallow, or has a terminal health condition such as incurable cancer. Advance directives also allow patients to name persons who will act in their behalf if they are unable to express their own wishes. Health professionals sometimes do not honor patients' advance directives.
- **Red Flag Alert 7.5.** A patient is given misleading information or insufficient information so that he or she cannot give informed consent as required by specific regulations and court rulings.
- **Red Flag Alert 7.6.** Inequitable or inferior treatment of specific patients takes place against members of specific vulnerable populations, such as members of racial minorities, women, LGBTQ persons, homeless persons, persons with mental health or substance-abuse conditions, disabled people, and low-income persons.
- **Red Flag Alert 7.7.** The health system is financed in ways that lead to inferior care to persons who cannot afford its care.
- **Red Flag Alert 7.8.** Health professionals fail to disclose conflicts of interest to patients, such as referring them to nursing or convalescent homes that they wholly or partly own.

Background

We do not have data about the frequency of violations of patients' ethical rights in the United States. In a recent survey of 300 frontline health professionals (social workers, nurses, and medical residents), many respondents reported that they had seen considerable numbers of patients with the seven kinds of unresolved problems during the prior two months, as I discussed in Chapter 1 in Table 1.1.

Resources for Advocates

Health advocates can draw upon many existing policies that protect patients' ethical rights at the micro level, including accreditation standards, statutes and constitutional rights, state and federal regulations, court rulings, and standards of state-level boards that license health professionals.

Accreditation Standards

A chapter of the Joint Commission's accreditation standards for hospitals is devoted to "rights and responsibilities of the individual" (Joint Commission, 2009). It asks hospitals to develop written policies on patient rights and to disseminate them to consumers, including their right to refuse care, treatment, and services in accordance with laws and regulations. It requires patients be given informed consent. It requires that patients be informed about advance directives including their right to forgo or withdraw life-sustaining treatment and to withhold resuscitative services. It identifies patients' rights to participate in care decisions and to be treated in a dignified and respectful manner. It discusses patients' rights to privacy including their right to decline being filmed or recorded. It requires that patients receive caring treatment, including knowing the names of physicians or other practitioners that care for them. It discusses patients' rights to participate in end-of-life decisions. It says that consumers have the right to receive information in a manner that they understand because "communication is a cornerstone of patient safety and quality care," including the right to translation services and communication with visual and hearing impairments, children, and persons with cognitive impairments. It asks hospitals to respect consumers' cultural and personal beliefs and preferences—as well as their right to religious and other spiritual services. It asks hospitals to allow consumers access to information disclosures of their health information. It requires a hospital to "put its respect for the patient's rights into action by showing its support of these rights in the ways that staff and caregivers interact with the patient and involve him or her in care, treatment, and services" (Joint Commission, 2009). It asks hospitals to involve surrogate decision makers when consumers are unable to make their own decisions. It asks them to involve consumers' families in making health decisions to the extent permitted by specific patients or surrogates. It requires hospitals to inform consumers or their surrogates of unanticipated outcomes of care, treatment, and services. It asks hospitals to respect consumers' rights during research, investigation, and clinical trials. It requires hospitals to give consumers copies of some of its policies, including ones about informed consent and advance directives.

These accreditation standards ask hospitals to develop measures to determine the extent to which they implement many of these policies, whether through documentation (such as in medical records) or consumer surveys. The accrediting team grades the hospital as achieving insufficient compliance, partial compliance, or satisfactory compliance with these ethical standards. It can recommend loss of accreditation or probationary status for hospitals that violate ethical standards—a decision that could lead to loss of Medicare and Medicaid funding and make it difficult to attract qualified staff.

Statutes and Constitutions. Many court rulings, as well as the federal Patient Self-Determination Act of 1990, give consumers the right to make their own medical decisions as long as they are not minors and are competent. They can decline treatment even when informed that they will suffer harm, or even death, without it. The Patient Self-Determination Act requires medical personnel to educate consumers about so-called advance directives that allow them to state what medical procedures they want, or do not want, if they are unable to communicate their wishes. The 14th Amendment to the federal constitution gives persons a right to privacy and their right to informed consent to medical procedures (Stein, 2004).

Congress enacted the Health Information Portability and Accountability Act (HIPAA) of 1996 (Annas, 2003). It protects individually identifiable health information from disclosure to other persons without their consent, which it calls "protected health information, or PHI," including patient information identified by name, address, telephone number, and e-mail address in "any form or medium" and including "any information, oral or recorded, relating to the health of an individual, the health care provided to an individual, or payment for health care provided to an individual" (California Hospital Association [CHA], 2008, p. 16.4). Providers must obtain advance authorizations from consumers to release their PHI in advance and in "plain language." Some exceptions to HIPAA requirements include the release of information to researchers that is not personally identifiable, the release of HIV test results to public health officials in many states, and psychotherapy notes used by students, trainees, or mental health practitioners to develop their treatment skills. HIPAA allows release of information with respect to victims of abuse, neglect, or domestic violence to specific authorities.

Health advocates can advance social justice by helping patients gain access to federal programs that have improved access to health care for tens of millions of patients including Medicare, Medicaid, CHIP, and programs established by the ACA.

Regulations

Many state and federal regulations protect consumers' ethical rights covering diverse topics such as determining when medical interventions can be terminated for dying persons, advance directives, safeguards for consumers who participate in research on human subjects, privacy rights, and informed consent. Persons who do not heed regulations can be subject to fines and criminal sanctions.

Regulations sometimes allow release of patients' information to appropriate authorities under specific circumstances. Physicians, hospitals, and other health care providers must report all AIDS cases, HIV infections, and viral hepatitis infections to the local health officer or other designated agency within specific time frames, including ones caused by transfusions.

Consumers who have been subject to sexual and other assaults often visit emergency rooms, clinics, and other health providers. Health practitioners are required to make reports to local law enforcement when they treat persons with specified injuries from possible assaults. "Assaultive or abusive conduct" includes (in California) murder; manslaughter; mayhem; aggregative mayhem; torture; assault with intent to commit mayhem, rape, sodomy, or oral copulation; administering controlled substances or anesthetic to aid in commission of a felony; battery; sexual battery; incest; throwing corrosive materials with intent to injure or disfigure; assault with a deadly weapon; rape, spousal rape; procuring any female to have sex with another man; child abuse and endangerment; abuse of a spouse or cohabitant; sodomy; lewd and lascivious acts with a child; oral copulation; sexual penetration by a foreign object; elder abuse; and attempts to commit any of these offenses (CHA, 2008). Persons who are victims of abuse or domestic violence must be informed that a report has been or will be made unless (in California) the providers fear a report could place them in serious harm or fear that they would be informing a personal representative who was responsible for the abuse, neglect, or other injury, which could result in injury to the victims (CHA, 2008). Health practitioners often enter into the medical record comments by the injured person regarding past domestic violence or the names of persons suspected of inflicting injuries or assaultive or abusive conduct as well as a map of the injured persons' bodies showing where wounds were inflicted and a copy of the law enforcement reporting form (CHA, 2008). Failure to report is a misdemeanor punishable by fines, imprisonment, or both. Persons suspected or accused of inflicting injuries, and their attorneys, cannot be allowed access to the injured person's records (CHA, 2008).

Regulations in one state are often different from those in other states—so you need to find experts to help you navigate ones in your state. Take the case of regulations about HIV tests. Many states require consent to HIV tests, such as California, which requires physicians to inform patients that the test is planned, provide information about the test, inform consumers of treatment options if they test positive, and advise them they have the right to decline the test. The medical care provider has to note in patients' medical records if they decline the HIV test in some states. Different procedures exist for minors 12 years and older and persons under age 12 as well as for criminal defendants and inmates. Some states declare results of HIV tests to be confidential as do federal privacy regulations. However, disclosures can be made to specific organizations or persons including health personnel who may have had contact with persons who test positive. Physicians who order HIV tests in California may, but are not required to, disclose confirmed positive test results to patients' spouses, persons reasonably believed to be a sexual partner, persons who shared use of hypodermic needles, or local health officers—but they must first discuss the results with patients, counsel them, and attempt to obtain their voluntary consent to notify their contacts. (When contacts are notified, the physicians must refer them for appropriate care.) Improper disclosures can bring civil penalties including fines (CHA, 2008). First responders and health care personnel who have experienced significant exposure to patients' blood or other potentially infectious materials can ask for HIV testing under certain conditions (CHA, 2008).

Parents face difficult choices when they have infants who have serious medical conditions that may terminate their lives. Using information from their physicians, parents make final decisions unless they are incompetent or can't agree with one another. Nonetheless, court rulings state that physicians should always act in the best interests of the child and presume life-sustaining treatment should be provided unless and until a court resolves the dispute. In general, withdrawal or withholding of life-sustaining treatment for newborns should not occur merely because they have a disability such as Down syndrome or because their medical care will be costly. When parents and members of the health care team can't agree, they should generally turn to a multidisciplinary hospital ethics committee before resorting to the courts, with the committee not making treatment decisions or deciding whether to disqualify parents but facilitating communication and providing advisory guidance when ethical conflicts exist (CHA, 2008). Health staff often need legal advice when making these decisions because the Federal Child Abuse Amendment of 1984 only allows withholding of life-sustaining treatment if infants are comatose and if treatment merely prolongs dying and would be medically futile (CHA, 2008, pp. 5–15).

Court Rulings. The U.S. Supreme Court, as well as lower courts, have ruled on diverse topics such as the right of a state to enact legislation allowing physicians to prescribe lethal drugs to terminally ill patients and allow the use of medical marijuana for patients in pain, women's right to abort a fetus, informed consent, and the right of physicians to discontinue medical care for comatose patients. Many liberals feared that the appointment of Neil Gorsuch to the Supreme Court in 2017 and Brett Kavanaugh in 2018, after the retirement of Justice Anthony Kennedy, could lead to rulings that might even declare abortion to be illegal.

Consumers can sue physicians, hospitals, and clinics for violating their ethical rights as established in constitutions, statutes, regulations, and accreditation standards. Failure to obtain proper consent to treatment can result in battery, professional negligence (malpractice), and/or unprofessional conduct against the physician or other health care provider for even the simplest of procedures (CHA, 2008). Battery is defined legally as "the intentional touching of a person in a harmful or offensive manner without his or her consent"—and can be filed against any health care provider who performs a medical procedure without a patient's consent or one that exceeds the scope of the consent even when the physician has no wrongful intent (CHA, 2008). Physicians who fail to disclose the risks and alternatives open to consumers can be sued for malpractice. They must inform consumers of potentially conflicting interests, such as research or financial interests (CHA, 2008). Courts have frequently ruled that patients have the right of self-determination and even the right to decline treatment against physicians' advice.

Regulating Professionals' Conduct. Physicians, nurses, and social workers are licensed and regulated by state boards or commissions. They have the right to place them on probation or to prohibit them from practicing if they violate specific ethical standards, such as engaging in fraudulent behavior, using drugs, battery, malpractice, or sexually abusing their patients.

Core Problem 2: Engaging in Advocacy to Promote Quality Health Services—With Some Red Flag Alerts

- **Red Flag Alert 7.9.** A specific patient does not receive evidence-based care—or does not even discuss evidence-based options with those health professionals who care for her.
- **Red Flag Alert 7.10.** A specific patient does not inquire about a health professional's track record with respect to treating specific kinds of medical problems.
- **Red Flag Alert 7.11.** A specific patient does not use online and other resources to find information about performance indicators of specific professionals or providers who wish to care for him or her.
- **Red Flag Alert 7.12.** Specific patients do not seek second opinions for nonroutine medical procedures.
- **Red Flag Alert 7.13.** Specific patients do not use the Internet to gain information about nonroutine medical problems.
- **Red Flag Alert 7.14.** Specific patients do not know how to search for information about clinical trials for their medical conditions.

POLICY ADVOCACY LEARNING CHALLENGE 7.1

CONNECTING ADVOCACY TO PROTECT PATIENTS' ETHICAL RIGHTS AT MICRO, MEZZO, AND MACRO LEVELS

Mary had been in the hospital for weeks. The nurse reported that there were reports that Mary had specifically stated that she did not want dialysis or life-sustaining measures if her health worsened. Yet Mary was on dialysis. She had been transferred through institutions far from her home in San Diego. Her caseworker was still in San Diego and called me yesterday morning. The caseworker, Joanne, sounded flabbergasted: "She's on dialysis? I've been her worker for over five years. I haven't been able to get there, but I'm coming this week, and we'll settle this." Mary's wishes had been blatantly disrespected despite the voices of advocates who had attempted to bring attention to Mary's wishes and how her current level of care contradicted those desires.

Joanne arrived. Furious, she had already put motions into action to obtain a court order to stop the heroic measures that were artificially keeping Mary alive against her will and her previously stated desires. Joanne received her court order and came directly to the hospital with it. Joanne and I stood in the room as the nurse stopped the dialysis and ventilator, which were supporting Mary's weak last moments of existence. Just then, the doctor who had been opposing Mary's wishes and who was adamant about keeping Mary alive walked in, equally angry and demanding that the heroic measures continue. We observed as Joanne, Mary's micro policy advocate, showed the physician the court order and assured him that Mary's wishes and needs were being met by being allowed to pass with dignity as she had wanted. The physician was angry but finally left as Mary died.

Learning Exercise

1. What preventive strategies might have averted this outcome at the micro policy level?
2. Why does this clinic rely on a Good Samaritan nurse to perform this life-preserving function—and what are some disadvantages of this informal practice?
3. It is not obvious how a social worker might initiate a macro policy intervention, but see if you can identify an option.

- **Red Flag Alert 7.15.** Specific patients have been subjected to medical error but do not know how to proceed.
- **Red Flag Alert 7.16.** Specific patients do not know how to contest premature discharge from hospitals.

Background

A study from the RAND Institute discovered that roughly half of physicians in a national sample failed to use evidence-based medicine (EBM) findings to treat patients with asthma and depression—and even more of them failed to use an array of preventive measures (Adams et al., 2003). Researchers reached similar conclusions when surveying providers for Medicaid enrollees in a broad array of hospitals and when examining ambulatory care for children (Mangione-Smith et al., 2007).

Persons of color are often less likely than white persons to receive gold-standard medical care; for example, African Americans who present themselves in emergency rooms with acute chest pain are less likely to be admitted or triaged into the coronary care unit (Council on Ethical and Judicial Affairs, 1990; Institute of Medicine [IOM], 2003; McBean & Gornick, 1994). More African Americans receive the poorest quality of care for congestive heart failure, acute myocardial infarction (AMI), pneumonia, or stroke than European Americans and are less likely to undergo angioplasty and bypass surgery even after accounting for patient refusals of treatment—and receive poorer care for coronary heart disease (Fincher et al., 2004; IOM, 2003; Kahn et al., 1994; Paschos et al., 1994; Taylor et al., 1998). Physicians spend less office time with Hispanic patients than Caucasians (Hooper et al., 1982). Summarizing considerable research, Beach et al. (2006) contend that provider behaviors and practice patterns contribute to health disparities—and that African Americans and other minority patients often receive differential care than Caucasians (Cooper-Patrick et al., 1999).

Females often don't respond to specific medications or medical procedures that are effective, respectively, with males and younger persons (Wartik, 2002). They sometimes don't receive state-of-art care as compared with males, such as for diagnosing myocardial infarctions (Willingham & Kilpatrick, 2005). Medical care is often insensitive to many needs of older women (Mayo, Nasmith, & Tannenbaum, 2003). Older women often receive contradictory and uncoordinated care (Lawrence, 2003). Women often lack access to birth control measures due to the reduction of abortion clinics in many states such as Texas as well as ongoing attacks on Planned Parenthood by Republican legislators (Jansson, 2019b).

Many critics contend that Americans often receive suboptimal health care. *Consumer Reports* (2014) maintains that between 98,000 and 400,000 patients suffer preventable fatalities each year from medical errors and infections in American hospitals. Considerable evidence suggests, as well, that specific physicians and hospitals provide poorer health care than others as revealed by federal data that compares rates of mortality and injury of persons that use them (Consumer Reports, 2014; IOM, 2000).

Researchers have discovered that medical errors substantially decline in intensive care units and elsewhere when staff follow checklists when providing specific medical treatments. In intensive care units (ICUs), for example, staff provide mechanical ventilators, insert tracheotomy tubes, conduct dialysis, use aortic balloon pumps, feed patients through tubes, provide direct infusions into the bloodstream, and care for open wounds—providing 178 individual actions per patient per day. If they use checklists for these procedures, death rates plummet by as much as 66% in ICUs (Gawande, 2010).

Considerable research suggests that persons with chronic diseases, obesity, end-of-life conditions, and some other health issues benefit from care from multidisciplinary teams (Brumley et al., 2007; Lin et al., 2000). Many health providers who are used to solo practice need to be retrained to view new practitioner roles that facilitate partnerships, such as the mental health integration model (MHI) developed for helping persons with depression (Reiss-Brennan, 2006). Team-based chronic disease management (CDM) models have widely evolved in health settings, even if many health providers still rely on solo-based care (Bower & Gask, 2002).

Resources for Advocates

Federal Agencies That Disseminate EBM. The Agency for Healthcare Research and Quality (AHRQ) was established in 1997 to promote EBM as a tool for using scientific standards of evidence to discover which clinical practices were most effective in preventing and treating specific medical problems using outcome measures such as mortality and morbidity rates, numbers of infections, numbers of readmissions, adverse drug events, and costs (Lefton, 2008). The National Guideline Clearinghouse, which is supported by the AHRQ, now supports more than 2,200 guidelines. The volume of medical research has greatly increased with more than 500,000 articles indexed by MEDLINE annually in recent years (IOM, 2008). Many other organizations establish clinical guidelines and recommendations, including the American Heart Association, the American College of Physicians, the American Diabetic Association, the American Society of Clinical Oncology, and the National Heart, Lung, and Blood Institute. Many other organizations synthesize evidence collected by medical scientists, including the AHRQ, the Blue Cross and Blue Shield Association Technology Evaluation Center, the Cochrane Collaboration, the ECRI Institute, and Hayes Inc. (IOM, 2008).

The ACA created an Innovation Center in the Centers for Medicare and Medicaid Services (CMS) in 2011 as well as a national quality improvement strategy to improve the delivery of health care services and advance patient health outcomes.

Evaluations of specific physicians, clinics, and hospitals have increasingly been released to the public by state departments of public health, Medicare, state Medicaid programs, and state departments of health. These have included results of patient satisfaction surveys, overall mortality rates, mortality rates for specific kinds of cancer and heart disease, and rates of complications after specific kinds of surgeries. They have evaluated the extent to which specific managed care plans provide preventive services. They also have evaluated the extent to which specific providers use outdated equipment or equipment that isn't properly maintained.

The federal government, as well as many states, place on the Internet data about health outcomes of specific hospitals and physicians (see https://www.medicare.gov/hospitalcompare/search.html? for comparisons of hospitals), including the extent of adverse events that cause consumers to develop serious disabilities including surgeries performed on the wrong body parts or on the wrong patient; a surgical procedure to which the patient hadn't given informed consent; death up to 24 hours after surgery of a healthy patient; and disability from contaminated drugs, devices, or biological agents. Some states place data online about the extent specific physicians have caused adverse events as well as data about the extent specific physicians have been disciplined by state boards that regulate medical practitioners.

Consumers' Litigation. Consumers sometimes take matters into their own hands when they believe that they have been provided medical care that diverges from accepted norms, particularly if they believe they have suffered injuries because of subpar care. Millions of them have engaged private attorneys in past decades to litigate issues such as alleged malpractice or violation of their ethical rights. They have often sued hospitals, as well, when they believe they lacked sufficient quality controls in their procedures, hiring of staff, in-service training, and monitoring of the quality of care provided. Consumers

POLICY ADVOCACY LEARNING CHALLENGE 7.2

CONNECTING MICRO, MEZZO, AND MACRO POLICY INTERVENTIONS TO ADVANCE QUALITY OF CARE

Assume that you work in a neonatal intensive care unit (NICU) that currently lacks a protocol for deciding how to treat infants born before 23 weeks of gestation. Upon examining evidence-based research, you discover that babies born before 23 weeks rarely survive, but their survival rate increases from 29% to 65% between 23 and 25 weeks. You further discover that two-thirds of infants born at 23 weeks have some form of functional disabilities when they reach two to three years of age—and one-third of these assessed survivors have a severe disability. Infants born at 25 weeks have much better outcomes, with only one-third of them possessing functional disabilities and 13% with a severe disability at two to three years of age (Singer, 2007). (No babies born before 26 weeks survived without entering a NICU.) You want to be a micro policy advocate for traumatized parents with babies born before 23 weeks.

Learning Exercise

- Would you refer these traumatized parents to evidence-based literature or Internet sites that gave them these probabilities at the micro policy advocacy level as well as economic, marital, and mental health issues related to premature births?
- Might you want to consult the hospital's legal counsel to determine whether and when it is legal to allow some infants to perish by not giving them medications and other advanced medical treatment when the parents favor this policy at the mezzo advocacy level?
- Can you think of possible macro policy advocacy interventions at the regulatory or government level?

have often litigated against pharmaceutical companies when they believe they failed to disclose potential side effects of medications or adverse drug interactions—or when they overstated likely benefits from them. Consumers have sued manufacturers of medical devices that they believed malfunctioned.

The extent of consumer' litigation is considerable but should not be exaggerated. Its elimination would only modestly cut American health costs. Consumers have directed their litigation against a relatively small number of physicians. Some states have placed limits on the size of allowable malpractice awards, such as $250,000 in California. Many managed care plans require enrollees to sign agreements when they enter their plans in which they agree to forgo litigation through the courts and to rely, instead, on the findings of internal administrative boards and the awards they suggest, which are often limited to a specific ceiling.

Core Problem 3: Engaging in Advocacy to Promote Culturally Competent Health Services—With Some Red Flag Alerts

- **Red Flag Alert 7.17.** A specific patient's culture is not honored in interactions with health professionals.

- **Red Flag Alert 7.18.** A specific patient with limited English proficiency (LEP) is not given appropriate translation services.
- **Red Flag Alert 7.19.** A patient with limited health knowledge or literacy fails to receive health information that he or she can understand.
- **Red Flag Alert 7.20.** Health professionals do not communicate effectively with a patient with cultural perceptions of health and disease that are different from theirs.
- **Red Flag Alert 7.21.** Health professionals fail to honor a patient's specific religious and spiritual practices.
- **Red Flag Alert 7.22.** Health professionals fail to honor a patient's desire to use complementary and alternative medicine (CAM).
- **Red Flag Alert 7.23.** A member of a specific vulnerable population is not treated with respect.
- **Red Flag Alert 7.24.** Persons from a specific cultural group are provided poorer services than other persons.

Background

It is not surprising that many patients receive culturally incompetent care in a nation formed by successive migrations of persons from other cultures. Physicians are disproportionately white and male as compared with patients whom they see despite recent progress in diversifying health professionals. Many health professionals, too, fail to make sufficient accommodations for persons with limited health literacy.

Persons with low levels of health literacy also face health-related challenges. Evidence strongly shows that health literacy is significantly related to both health and health care (DeWalt et al., 2004). Persons with low health literacy are more likely to self-report fair or poor health, compared with those with higher health literacy (Kutner, Greenburg, Jin, & Paulsen, 2006). This health disparity may be linked to their lower levels of knowledge about diseases and risk factors, such as smoking, contraception, HIV/AIDS, diabetes, and asthma. It may also be linked to underuse of preventive services such as mammography. Parents with low literacy may be less able to diagnose health problems of their children. Consumers with low literacy may be less likely to adhere to physicians' recommended treatments (DeWalt et al., 2004).

Racial discrimination and intolerance remain widespread but now assume more subtle forms. Minorities who face racial discrimination may be less likely or willing to access the health care system and tend to exhibit worse health outcomes, such as self-reported health or chronic conditions (Williams & Mohammed, 2009). In its 2003 report on racial health disparities, *Unequal Treatment*, the Institute of Medicine argued that real or perceived racial discrimination can shape the expectations, attitudes, and behaviors of minority patients toward the health care system and health providers. African Americans tend to have less trust in their health care providers than do Caucasians (Halbert, Armstrong, Gandy, & Shaker, 2006). Minority patients who are treated by a doctor of the same race or ethnicity (as in a racially concordant patient-physician relationship) tend to be more satisfied with the services they receive, possibly due to perceived personal or ethnic similarities (Street, O'Malley, Cooper, & Haidet, 2008). Relatively few minority physicians practice in many settings in the United States (Reede, 2003).

Compared with the U.S.-born population, immigrants and their children are less likely to have access to health services (Brown et al., 1999; Huang, Ku, & Matani, 2001; Leclere, Jensen, & Biddlecom, 1994; Stella & Ledsky, 2006). Some qualitative studies suggest that culture may powerfully shape interactions among many immigrant and refugee families and the U.S. health system. Ngo-Metzger et al. (2003) found, for example, that Chinese and Vietnamese immigrants often encountered negative reactions regarding their use of traditional practices and commonly felt disrespected and devalued by their physicians.

The inability to speak or understand English can also lead to less access to health services and worse health outcomes. LEP persons are less likely to have a usual source of care, utilize fewer preventive services, and tend to be less satisfied with their health services (Carrasquillo, Orav, Brennan, & Burstin, 1999; Ponce, Ku, Cunningham, & Brown, 2006). Consumers' lack of linguistic and cultural competency can also lead to medical errors. In a study of six randomly selected accredited hospitals, for example, Divi, Koss, Schmaltz, and Loeb (2007) found that almost 50% of adverse events with LEP patients resulted in some physical harm (and in some cases, death), compared with only 30% of those with English-speaking patients. The majority of these adverse events involved communication errors, such as questionable documentation, inaccurate or incomplete information, questionable assessment of patient needs, and questionable advice or interpretation.

LGBTQ persons have significant problems in accessing health care (Mirza & Rooney, 2018). Court rulings and an administrative rule promulgated by President Obama protect LGBTQ persons from discrimination due to gender identity and sex stereotypes. Some conservatives and the Trump administration have worked to ease rules that protect LGBTQ persons from discrimination in the health care system on grounds medical providers should be able to not care for patients whose sexual orientation or identity are offensive to their religious views. Discrimination includes excessive waits or denial of health service, failure to provide HIV medications to gay males, and hostile treatment. Discrimination, in turn, leads some LGBTQ persons to avoid or delay medical care. In a survey of LGBTQ persons in 2017, 8% said a health provider had refused to treat them, 6% said a health provider refused to given them care related to their actual or perceived orientation, 7% said a health provider refused to recognize their child or partner, 9% said health providers used harsh or abusive language, and 7% said they experienced unwanted physical contact from a health care provider such as fondling or sexual assault. The 2015 U.S. Transgender Study found that nearly one in four transgender persons avoided seeking needed health care in the prior year due to fear of discrimination or mistreatment (Mirza & Rooney, 2018). Nor are there easy answers. Roughly one-fifth of LGBTQ persons believe it is difficult or impossible to find better services—and far more so in rural areas. Transgender people report even higher rates of poor or hostile services—and are even more pessimistic that they can find better services. Relatively few areas have LGBTQ community health centers.

Resources for Advocates

Race, color, and national origin are protected under the nation's anti-discrimination laws. Title VI of the 1964 Civil Rights Act provides the foundation for these protections:

> No person in the United States shall, on the ground of race, color, or *national origin*, be excluded from participation in, be denied the benefits of, or be subjected to discrimination under any program or activity receiving Federal financial assistance. (42 U.S.C. § 2000d, italics added for emphasis)

"National origin" has been interpreted in the courts to also include an individual's primary language so that no one should be excluded from federally funded programs because of their inability to speak, read, or understand English (Perkins, Youdelman, & Wong, 2003). President Bill Clinton issued Executive Order 13166, which reinforced the protections afforded to LEP individuals under Title VI and instructed all federal agencies and federally funded programs to provide "meaningful access" to LEP persons. Because most health providers receive some federal funding (e.g., in the form of Medicare or Medicaid), hospitals, clinics, and health providers have a legal obligation to provide linguistically appropriate services to LEP persons. However, the implementation of Title VI regarding LEP persons varies greatly across health providers. The actual enforcement of Title VI is largely driven by lawsuits or complaints, so the burden often falls on those who are facing discrimination.

The Department of Health and Human Services (DHHS) Office of Civil Rights' LEP Policy Guidance—first published in 2000 under the Clinton administration but later finalized in 2003 under the Bush administration—requires that interpreter services be offered at no cost, that they be offered to persons with the most frequently spoken languages in specific areas, and that interpreters should be qualified with demonstrated proficiency (Perkins et al., 2003). Many state-level policies also govern the linguistic accessibility of health care services in specific states.

Most statutes, regulations, and litigation focus upon provision of translation and interpretation services rather than provision of culturally competent health services. The DHHS Office of Minority Health's 2001 Culturally and Linguistically Appropriate Services (CLAS) Standards are an exception (see Figure 7.1).

FIGURE 7.1 ■ National Standards on Culturally and Linguistically Appropriate Services (CLAS)

Standard 1:	Health Care organizations should ensure that patients/consumers receive from all staff members effective, understandable, and respectful care that is provided in a manner compatible with their cultural health beliefs and practices and preferred language.
Standard 2:	Health Care organizations should implement strategies to recruit, retain, and promote at all levels of the organization a diverse staff and leadership that are representative of the demographic characteristics of the service area.
Standard 3:	Health Care organizations should ensure that staff at all levels and across all disciplines receive ongoing education and training in culturally and linguistically appropriate service delivery.

Source: U.S. Department of Health and Human Services Office of Minority Health.

Accreditation standards may have the greatest influence in how hospitals, clinics, and other health care provider organizations actually deliver their services. The Joint Commission has released requirements dealing with cultural competency that can be obtained from the Joint Commission's website. The National Committee for Quality Assurance (NCQA), a not-for-profit organization that assesses and accredits health organizations and plans, has multicultural health care standards that apply to health plans. The ACA mandated enhanced collection and reporting of data on race, ethnicity, gender, primary language, disability status, and underserved rural populations in 2012.

POLICY ADVOCACY LEARNING CHALLENGE 7.3

CONNECTING MICRO, MEZZO, AND MACRO POLICY ADVOCACY FOR PATIENTS' UNRESOLVED PROBLEMS REGARDING CULTURALLY COMPETENT CARE

As I sat with my mother in the crowded tiny clinic waiting room, I wondered why she had made me drive her for forty-five minutes to receive medical services at a clinic that appeared dirty and required a lengthy wait. I finally asked my mother why she had dragged me down to the middle of Los Angeles for her to see a doctor when there were perfectly good clinics and hospitals in the region where we lived. She quickly reminded me that there were few Spanish-speaking doctors and nurses in the hospitals near our home and the costs were much too high for someone with no health coverage. Although my brother and I had frequently played the role of translator when it came to my mother's communication between her and doctors, she felt the need to speak directly to the doctor to truly convey what she was feeling. She expressed that she was tired of having to communicate her ailments through someone else and believed it was a major reason why she had not received the appropriate medical treatment that she needed. As our wait continued we suddenly heard my mother's name being called by the nurse who was admitting clients. We quickly gathered our belongings and made our way toward the nurse. While we maneuvered through the crowd toward the nurse, she began to verify the symptoms that had brought my mother into the clinic out loud in front of the other patients in the waiting room. Upon hearing the information being disclosed, I could see my mother shake her head with a look of embarrassment to confirm the reasons that had brought her into the clinic. She looked around to scan the crowd to see if the other patients had heard. It was clear that everyone had heard what was ailing my mother. Upon approaching the nurse I asked her if she was aware of her responsibility to maintain client confidentially, and I openly verbalized my anger at the manner in which she had just violated my mother's privacy. My mother pulled me aside and asked that I stop challenging the nurse's authority as she did not want to upset her. I explained to my mother that I had every intention to file a formal complaint against the woman so that this would never happen to any other individual again. My mother asked that I do no such thing as she felt it would make it impossible for her to return to this clinic, which allowed her to pay a fee that was affordable while having access to Spanish-speaking doctors. I respected my mother's wishes and never filed a formal complaint. Upon leaving the clinic that afternoon I had a sense of helplessness and frustration at my inability to protect my mother's right to adequate, affordable, and culturally competent health services.

Learning Exercise

1. How does this vignette demonstrate that case advocates have to begin with wishes of consumers rather than proceeding without heeding their wishes?
2. How might the social worker consider mezzo policy advocacy at the organizational level? (HINT: Recall that actual policies, such as HIPAA, are not actualized until they are implemented, so policy advocates can focus on strategies to bring this result.)
3. Can you think of a macro policy advocacy intervention you could initiate?

Core Problem 4: Engaging in Advocacy to Promote Preventive Health Services—With Some Red Flag Alerts

- **Red Flag Alert 7.25.** A person has not been helped to identify personal at-risk factors through family history and diagnostic tests.
- **Red Flag Alert 7.26.** A person has not been tested for a possible chronic disease or diseases at a possible early stage.
- **Red Flag Alert 7.27.** Persons fail to receive assistance in addressing medical conditions such as obesity, lack of physical activity, poor nutrition, and smoking.
- **Red Flag Alert 7.28.** Adults and children have not been given help in receiving vaccines.
- **Red Flag Alert 7.29.** A person lives in an inner-city area and is not given assistance in obtaining better nutrition and more exercise.
- **Red Flag Alert 7.30.** Specific members of vulnerable populations are not given access to preventive care.

Background

Corporations have often blocked enactment of important prevention measures by lobbying politicians. Tobacco companies, fast-food companies, food manufacturers, automobile companies, and manufacturers and purveyors of alcoholic beverages have often resisted regulations such as warning or information labels, safety devices, and restrictions on sale of their products. Corporations have often opposed regulations to combat pollution by toxic chemicals, whether in the air, water, or soil. In places like Fresno County, California, air pollution has led nearly one in three of its children to contract asthma (Anderson, 2007).

Considerable resistance to preventive care exists among health practitioners. Only 19% of primary care physicians discuss exercise with consumers, 22% discuss diet, and 10% discuss weight reduction. Only 4% encourage consumers to stop smoking, only 14% refer overweight persons to dieticians, and only 1% recommend exercise to overweight persons (Knight, 2004; Centers for Disease Control and Prevention, 1998).

Only 11% of the contact time of primary care physicians is devoted to disease prevention—or seven minutes per patient per year on average (Knight, 2004; Rafferty, 1998). Only 50% of physicians follow the National Cholesterol Education Program guidelines even with patients with high-risk coronary heart disease (Knight, 2004; Frolkis, Zyzanski, Schwartz, & Suhan, 1998). A survey of medical practice discovered that physicians give consumers only 54.9% of recommended preventive care (Knight, 2004; McGlynn et al., 2003).

Many experts hoped that managed-care plans would provide more preventive services than traditional medical arrangements for two reasons: They require consumers to access their plans through gatekeeper primary care physicians, and they cut their overall costs by emphasizing prevention rather than treatment. Physicians in managed care plans are no more likely, however, to provide many preventive services than other physicians (Pham, Schrag, Hargraves, & Bach, 2005).

American culture sometimes adversely affects health. Americans work harder than residents of many other nations except for Japan, taking only 10 days of vacation per year as compared with 17 days for Canadians, 24 days in the UK, and 8 days in Japan (Expedia.com, 2009). If Americans consume 97 pounds of beef per year, Canadians consume 69 pounds, UK residents consume 38 pounds, and the Japanese consume 21.3 pounds (Department of Agriculture, 2006; Red Meat Industry Forum, 2007). Forty-one percent of Americans eat at least once a week in fast-food restaurants, including 59% of 18- to 29-year-olds (Pew Research Center, 2006). Many Americans have become addicted to television, with adults watching it for eight hours per week compared with only three hours for adults in Canada, three hours in the UK, and 3.5 hours in Japan (Ramsay, 2008). Although 5.2 Americans die from homicides per 100,000 each year, only 1.7 Canadians, 1.4 persons from the UK, and 0.5 Japanese die from them (United Nations Office on Drugs and Crime, 2010). Although 626 Americans per 100,000 persons are injured in automobile accidents each year, only 312 persons in England are injured per year per 100,000 people (Economist, 2009). Roughly 30% of young people 18 to 24 years old still smoke, along with significant increases in the female population since 1950. About 87% of the 159,000 lung cancer deaths in 2008 were linked to smoking—with 444,000 Americans dying from the effects of tobacco in 2004 (American Cancer Society, 2009).

More persons have chronic diseases such as diabetes in the United States that are closely linked to lifestyle factors than in Canada, the UK, or Japan. Whereas 9.3% of persons in the United State have diabetes with 30% of these cases undiagnosed, roughly 3.6% of persons in the UK, and 5% of persons in Japan have this disease (International Diabetes Federation, 2010).

The United States has a higher infant mortality rate than any of the other 27 wealthy nations. African American women have high rates of infant mortality and maternal death (Ingraham, 2014). One in eight babies are born prematurely in the United States, partly because many women do not receive sufficient prenatal care (March of Dimes, 2009).

The high poverty rate in the United States also contributes to relatively high rates of sickness and mortality in the United States as we discussed earlier in this chapter.

Many uninsured Americans receive less preventive care than insured Americans. They often use emergency rooms for health care rather than possessing ongoing relationships with primary care physicians. They receive less health care per year than insured consumers—or $583 per year as compared with $3,915—and self-fund 35% of it as compared with self-funding by insured persons of 17% (Hadley Holahan, Coughlin, & Miller, 2008). Between 35,000 and 45,000 uninsured persons die unnecessarily each year (Wilper et al., 2009).

Resources for Advocates

Resources and Data. The Centers for Disease Control and Prevention (CDC) has many programs within it such as the National Center for Chronic Disease Prevention and Promotion, the Division for Heart Disease and Stroke, the Division of Cancer Prevention and Control, the Diabetes Prevention Program, the National Breast and Cervical Cancer Early Detection Program (B&C), the Office of Public Health Genomics, the Division of Reproductive Health, and the Division of Adolescent and School Health. (Each of these programs has a website that can be accessed through http://www.cdc.gov.) These programs not only conduct research about the incidence of specific diseases but mobilize coalitions in different states and tribes to promote and fund screening programs that are funded and implemented by states (Collins, Koplan, & Marks, 2009). All states have screening and wellness programs in areas such as tobacco use, diabetes, breast and cervical screening, and comprehensive cancer control, as well as the Behavioral Risk Factor Surveillance System, which surveys consumers and collects information on health risk behaviors, preventive health practices, and health care access primarily related to chronic disease and injury.

The CDC often issues reports geared to mobilizing action, such as *A Public Health Action Plan to Prevent Heart Disease and Stroke* as well as *Healthy People 2000*, *Healthy People 2010*, and *Healthy People 2020*, that respectively established health and prevention goals for specific decades. It helps fund a network of 33 prevention research centers that fund collaborative research of community, academic, and public health partners with use of participatory research that seeks implementation of public health programs, such as the Enhance Fitness Program to increase physician recommendations for exercise by senior citizens at 300 sites in 26 states. It develops evidence-based prevention strategies, such as a study that found that type 2 diabetes can be prevented or delayed with moderate weight loss, improved nutrition, and greater exercise (Collins et al., 2009). It promotes programs such as the Health Communities Program, which mobilizes action in local communities, including schools, worksites, and health care settings, to increase physical activity, improve nutrition, and curtail smoking to prevent chronic diseases. It targets women between ages 50 and 64 who are uninsured or underinsured in its Well-Integrated Screening and Evaluation for Women Across the Nation (WISEWOMAN) to enhance lifestyle changes as well as monitoring of blood pressure and cholesterol. The CDC promotes early detection of public health issues in different states by helping them implement surveys of citizens that probe the extent they have problems such as obesity and specific mental health problems.

The DHHS has developed many prevention programs, not just its Medicare and Medicaid programs but many specific divisions and programs. The Center on Medicare and Medicare Services (CMS) administers Medicare and Medicaid. Medicare promotes prevention by covering immunizations; screening for cancer, cardiovascular disease, glaucoma, and diabetes; bone density measurement; and smoking cessation programs. It will cover additional prevention programs under the ACA of 2010. Medicaid promotes prevention by covering smoking cessation, preventive health and dental care, prescription drugs, laboratory tests and X-rays, and family planning and prenatal care. Coverage and services vary by state. Public health staff are out-stationed in sexually transmitted diseases (STDs), tuberculosis (TB), HIV-AIDS, women and infant children (WIC), and prenatal clinics that are scattered throughout urban and some rural areas but that often aren't closely linked with primary care clinics or hospitals.

The federal Agency for Healthcare Research and Quality (AHRQ) oversees and funds considerable research on prevention. It contains the U.S. Prevention Services Task Force, which grades specific preventive interventions from A to D depending on their cost-effectiveness in preventing specific health conditions—a task force that will have expanded roles under the ACA.

Regulations. Many local jurisdictions ban smoking in public places. Federal regulations require food labeling on many products and require fast-food establishments to post the ingredients in their food. Some local jurisdictions ban the use of partially hydrogenated oils in cooking French fries as well as ending use of trans fats in fast-food and other food establishments. The ACA requires chain restaurants and vending machine companies to post nutritional content and calories of their food.

Congress enacted legislation that defined tobacco as a drug in 2009, which gave the Food and Drug Administration (FDA) the power to regulate it by establishing, expanding, and monitoring local and state regulations that prohibit smoking advertisements and vending machines near schools, sales to minors, giving of free samples, and smoking in public places. Forty-three states currently allow employers not to hire persons who smoke. Many jurisdictions prohibit smoking in public places or in places at work where others will breathe secondhand smoke. Scientists contend that electronic smoking leads to smoking actual cigarettes (National Institute of Drug Abuse, 2018).

Some local jurisdictions have regulated the number of fast-food outlets in specific places where they are disproportionately located, such as low-income neighborhoods. Some have placed restrictions on food provided in local schools, requiring lunches with lower fat content and more vegetables. Some have prohibited vending machines in schools that sell soda and candy. Some jurisdictions place limits on the number of bars in specific communities as well as locations near schools. Many jurisdictions take away driver's licenses from persons caught driving while intoxicated (DWI)—and may give them prison sentences.

The Occupational Safety and Health Administration (OSHA) establishes specific regulations to protect workers' safety, and state chapters administer them and establish some of their own regulations. It regulates emissions of toxic chemicals within plants, requires the use of safety practices and equipment, and requires employers to monitor hazards and maintain records of workplace injuries and illnesses.

Statutes. The ACA has a Prevention and Public Health Fund to modernize disease prevention and improve access to clinical preventive services in schools as well as Medicare annual wellness programs. It funds many community health centers that came into creation when Senator Bernie Sanders placed them in appropriations bills. It will fund health prevention programs in occupational settings. It will subsidize many not-for-profit, community-based clinics.

Prevention in Places of Work. Many corporations have instituted wellness programs in their work sites, such as rail company CSX Transportation, Johnson & Johnson, and Coors Brewing (Brink, 2008; CSX, 2005). Many technology companies, such as Google and Microsoft, have wellness programs as well as exercise facilities at their work sites. The ACA permits employers to offer employees rewards of up to 30%, increasing to 50% if appropriate, of the cost of coverage for participating in a wellness program and meeting certain health-related standards in 2014. It will establish state pilot programs in 10 states to offer similar rewards to consumers in wellness programs who purchase individual policies.

The Lurking Danger of Epidemics. Numerous epidemics have killed remarkable numbers of persons. Examples include 50 million people killed in 541 A.D. in the Middle East, Asia, and the Mediterranean basin; 25 million people killed by the Great Plague of London that started in China and wiped out entire cities; and 11,300 people killed by the Ebola epidemic in West Africa in 2014. Experts fear that "a great flu pandemic" similar to the so-called Spanish flu that killed between 30 million and 50 million people worldwide (including 675,000 Americans) is highly likely within the next 30 years (CNN, 2004). The Ebola epidemic swept Liberia and other African nations, soon followed by the Zika epidemic in Central America and Florida. Experts warn that budget cuts by the Trump administration have cut the number of scientists at the CDC and elsewhere who can develop vaccinations capable of killing mutated viruses emerging in Asia from viruses in chicken and pig farms.

POLICY ADVOCACY LEARNING CHALLENGE 7.4

CONNECTING MICRO, MEZZO, AND MACRO POLICY ADVOCACY WITH RESPECT TO PREVENTION

Clinicians at Children's Hospital, Boston, were concerned about the sheer number of asthma attacks among low-income youth in Boston's inner city. They developed an innovative prevention strategy that included the following:

- Having nurses visit parents after their children's discharge to educate them about adherence to medications and follow-up visits to their pediatricians
- Home inspections for mold and pests
- Provision of free vacuum cleaners to parents who lacked them
- Funding costs of these interventions from the hospital's budget because insurance covered only the cost of a prescribed inhaler

This strategy was so successful that hospital readmissions of these children dropped by more than 80%—and costs of treating dropped precipitously. This intervention threatened to bankrupt the hospital, however, because it had depended on revenues from public and private insurances and programs for the many beds that had been occupied by these children.

Source: Case is drawn from Gwande, A. (2010, April 5). Now what? *New Yorker*, p. 22.

(Continued)

(Continued)

Learning Exercise

1. If you worked in this hospital and wanted to be an advocate for this creative program, what conflict of interest would you confront —and what ethical issues might you face?
2. How might you have to shift from micro policy advocacy to mezzo policy advocacy?
3. What macro policy advocacy might you consider launching?

Core Problem 5: Engaging in Advocacy to Promote Affordable and Accessible Health Services—With Some Red Flag Alerts

- **Red Flag Alert 7.31.** A person does not receive help in selecting an insurance plan; gaining eligibility to Medicaid, CHIP, or portions of Medicare—or in obtaining Medicare Supplemental Insurance—or becoming dually eligible for Medicare and Medicaid. A person receives no assistance for accessing insurance on the ACA's insurance exchanges or from specific employers.
- **Red Flag Alert 7.32.** A person is not given assistance in accessing charitable health funds in those clinics and hospitals that have them.
- **Red Flag Alert 7.33.** A person is not given assistance in contesting specific decisions made by Medicare, Medicaid, CHIP, Supplemental Security Income (SSI), Social Security, or specific private health insurance plans.
- **Red Flag Alert 7.34.** An uninsured person is not helped to find free clinics, public clinics and hospitals, or private hospitals that serve uninsured persons through the Medicaid Disproportionate Share Hospital (DSH) program.
- **Red Flag Alert 7.35.** A person does not receive help in obtaining disability health benefits.
- **Red Flag Alert 7.36.** A person does not receive assistance in obtaining in-home supportive care from Medicare, Medicaid, or private insurance.
- **Red Flag Alert 7.37.** An undocumented immigrant is not given help in where to find health care.
- **Red Flag Alert 7.38.** A senior citizen who has exhausted her Medicare benefits is not helped to understand his or her options.
- **Red Flag Alert 7.39.** Persons are not given help in financing their out-of-pocket costs.

Background

A combination of private insurance, government programs, corporations, and out-of-pocket payments fund American health care—a chaotic system that has left many consumers without insurance or access to public programs. The ACA had sought to decrease medically uninsured persons from 45 million to 16 million by 2020—an ambitious goal by employing a series of strategies. The ACA's framers wanted to decrease the number of uninsured persons by using the following strategies in tandem:

- Requiring private insurance to cover persons with preexisting conditions—and allowing parents to keep children on their policies until age 26
- Making uninsured persons pay penalties if they do not obtain insurance
- Providing tax credits to small businesses to encourage them to fund health insurance for their employees
- Providing subsidies to individuals who fall beneath relatively high income levels to induce them to purchase private health insurance
- Establishing insurance exchanges in the states—whether under the aegis of specific states or the federal government
- Raising the eligibility level of Medicaid and expanding its coverage to single persons as well as families—and requiring Medicaid to raise its reimbursements of providers to the same levels as Medicare
- Greatly expanding federal funding of Medicaid

We have already discussed reasons why the ACA did not bring health insurance to millions of people even as it helped others gain it. Some states refused to raise eligibility levels of Medicaid as allowed under the ACA. State insurance exchanges sold fewer private insurance policies than anticipated because many private insurance companies withdrew their products from the exchanges on grounds their products were not profitable. It took the ACA several years to get its programs implemented. The Trump administration and congressional Republicans tried to repeal and replace the ACA in 2017 but had only succeeded in rescinding several of its provisions by the summer of 2018.

Even these gains are important because lack of coverage adversely affects health outcomes. Roughly 30,000 uninsured persons were estimated to lose their lives due to lack of coverage each year prior to the enactment of the ACA in 2010. Lack of insurance causes the following:

- Delays in care when consumers don't initiate recommended care because they fear it will bankrupt them or make them destitute
- Interruption of care when consumers discontinue care because they fear it will bankrupt them or make them destitute

- Specialists becoming unavailable to low-income patients because Medicaid pays at such low levels
- Inadequate primary care because consumers use emergency rooms excessively due to lack of insurance
- Hostile or shunning actions by health providers because they don't want to serve uninsured or underinsured persons
- Increased angst from uncertain or inadequate health coverage
- Patients not taking prescriptions or splitting pills because they can't afford expensive drugs
- Increases in life-threatening diseases

Poor people disproportionately have poor health coverage because their employers rarely give them health insurance. Lack of coverage is particularly problematic for low-income people because poverty itself causes shortened lives and higher levels of chronic diseases. Women born in the lowest tenth of the economic distribution, for example, die 12 years earlier than their counterparts in the top tenth (Jansson, 2019a).

Core Problem 6: Engaging in Advocacy to Promote Care of Health Consumers' Mental Distress—With Some Red Flag Alerts

Mental health and physical health are often linked to one another. The opioid crisis, which killed roughly 60,000 people in 2016, for example, involved both physical and mental health policies and programs. Persons with opioid poisoning primarily receive services funded by Medicaid—a medical program that provides both mental health and physical health services. I discuss the opioid crisis in Chapter 10.

Core Problem 7: Engaging in Advocacy to Promote Health Care Linked to Households and Communities—With Some Red Flag Alerts

- **Red Flag Alert 7.40.** A clinic or hospital fails to promote referrals of patients to community-based agencies.
- **Red Flag Alert 7.41.** A hospital fails to commit sufficient staff to help patients move into assisted-living and nursing homes.
- **Red Flag Alert 7.42.** A clinic or hospital fails to monitor community nursing homes and convalescent homes so that it does not refer patients to ones with poor quality.
- **Red Flag Alert 7.43.** A clinic or hospital fails to link its patients to community-based preventive services.
- **Red Flag Alert 7.44.** A clinic or hospital lacks community workers.

Background

Many physicians view health as mostly confined to the physiological realm, failing to factor culture, social class, community realities, fiscal constraints, family dynamics, emotions, and patients' mental conditions into diagnoses and treatments (Glasgow et al., 1999). Many physicians assume, as well, that most consumers will adhere to treatment regimens rather than exploring financial, family, mental, cultural, and community factors that might contravene them (Glasgow et al., 1999). Health consumers live in ecosystems that profoundly shape their health as well as their use of health systems (DuBois & Miley, 2002). They may live in housing that causes or exacerbates asthma (Bell & Standish, 2005). They may live in neighborhoods with relatively spartan health services so that they are less likely to use them. They may lack transportation services to health programs. They may reside in housing that is not designed for persons with disabilities or be able to access health services because of physical limitations. The medical problems of consumers who come to emergency rooms are often linked to mental trauma, family violence, malnutrition, exposure, drug overdoses, gang violence, homelessness, and other social and economic problems that cause or exacerbate their medical problems.

Effective health care systems need to have information about their patients' ecosystems—and have the ability to provide outreach services. The term *hard to reach* is used in the disability literature to define disabled consumers who won't typically be reached unless a dedicated effort is made to engage with them, such as women whose visual, auditory, or mobility limitations make it difficult for them to leave their homes (Smeltzer et al., 2007). They need to have resources and personnel to locate persons who don't return for appointments when they have serious health conditions.

Many emergency room physicians, nurses, and social workers engage in referral, brokerage, and liaison services, including with law enforcement, mental health, child welfare, substance abuse, assisted-living and nursing home agencies, and shelters. They often help consumers gain eligibility to specific safety net programs such as Medicaid, CHIP, food stamps, and Section 8.

Resources for Advocates

Social workers are uniquely equipped to link health services to communities because many have worked in the network of community-based nongovernmental organizations (NGOs). Several developments should speed these linkages. The ACA requires hospitals to establish "medical homes" for patients where they receive a broad array of outpatient and inpatient services and where hospitals provide them with those services needed to prevent illness and to slow the development of chronic diseases. The ACA has moved reimbursement of health providers from the current fee-for-service approach to so-called capitated care, in which they receive flat fees for each patient. Such fees give hospitals incentives to cut their costs by helping patients prevent health problems by working with an array of community agencies. Medicare reduced regular reimbursements to hospitals for inpatient stays for specific patients who were readmitted within 30 days—reductions of up to 3% for readmissions of patients with problems such as chronic lung disease, heart attacks, hip and knee replacements, and pneumonia. These penalties encouraged better care of these patients in the hospital, as well as better follow-up for them in their communities with problems that contribute to their early readmission including diet, the ability to obtain needed medications, poor life habits like substance abuse, poor housing, mental illness, and failure to adhere to suggested treatments (Kaiser Family Foundation, 2018). The ACA established a national Medicare pilot program to develop and evaluate a bundled payment for acute inpatient hospital services, primary care services, outpatient hospital services, and post-acute care services for an episode of care. It will evaluate whether this approach will decrease medical silos and improve outcomes. The CMS in the Trump administration emphasized voluntary bundled payments rather than mandatory ones that the Obama administration had favored (Castellucci, 2018).

Health professionals can promote community-oriented care for consumers by establishing links between community agencies that can include mental health, family, substance abuse, shelter, income assistance, health insurance, public health, housing, employment, physical therapy, speech and language, and other health-related problems. These links can take place among autonomous agencies (linkage), can include more systematic relationships (coordination), or can involve "joint goals, very close and highly connected networks . . . and high degrees of mutual trust and respect" (integration) as discussed by Glendinning (2002). Technology allows providers to link providers with consumers in their homes and communities including video conferencing, electronic monitoring, and video cameras.

Several policies give hospitals incentives to relate to their communities. The federal Emergency Medical Treatment and Active Labor Act (EMTALA) established procedures that apply not only to consumers who are transferred from emergency rooms to other locations but to all transfer cases (CHA, 2008). Hospitals are legally responsible "for the safety, appropriateness, and monitoring of the transfer process and protocol" (CHA, 2008). Medicare, Medicaid, and the Joint Commission add additional requirements. EMTALA prohibits transfers before patients are medically stabilized. Some states prohibit transfers of consumers for financial reasons from specific vulnerable populations, such as race, disability, or sexual orientation, because they are suspected of lacking resources to fund their medical care without determining that they are actually medically indigent. Some states and federal law also require emergency services to be given to consumers without first investigating to see if they can fund them even though payment can be sought after they have been treated—and prohibit discrimination in providing emergency services on the basis of race, gender, and other characteristics (CHA, 2008).

Transition planning is often inadequate because hospitals and clinics don't hire sufficient social workers and discharge nurses to provide effective interventions. These professionals are often pressured to speed the release of patients from hospitals to save them resources when they are reimbursed with capitation or with prospective payments such as Medicare's diagnosis related groups (DRGs). Some populations may be most subject to poor discharge planning. Many states have specific regulations regarding the discharge of homeless persons, for example, that have been inappropriately transported to skid row areas in cities such as Los Angeles, sometimes in their hospital gowns, and without ensuring they receive a place in shelters.

The development of a national health program, such as Medicare for All that was proposed by Senator Bernie Sanders in his presidential campaign in 2016, could make community-based models of health care more feasible. Assume, for example, that all persons with diabetes received health insurance from a single, national source rather than fragmenting the population among persons with different health

POLICY ADVOCACY LEARNING CHALLENGE 7.5

CONNECTING MICRO, MEZZO, AND MACRO POLICY ADVOCACY TO HELP FUGITIVE PATIENTS OBTAIN NEEDED SERVICES

A Good Samaritan Nurse

A nurse in a health clinic used primarily by low-income consumers in the Bronx decided to follow up with patients who had received biopsies that indicated they had cancer at relatively advanced stages but who failed to keep appointments to receive their biopsy results. She often assumed the role of a detective because many of them lacked telephones and frequently changed addresses. She was not asked to undertake this assignment by clinic staff, but felt ethically impelled to locate these fugitive consumers in after-work hours. She was able to locate many of them and to convince them come to the clinic so that they could begin treatments.

Learning Exercise

1. How does this vignette illustrate a tendency to define medical practice as confined to the four walls of specific clinics and hospitals?
2. Why does this clinic rely on a Good Samaritan nurse to perform this life-preserving function—and what are some disadvantages of this informal practice?
3. Could a social worker in this clinic have engaged in mezzo policy advocacy to persuade clinic administrators to hire an outreach worker with this assignment?
4. Can you think of a possible macro policy intervention at the state or federal level?

insurance plans. One could imagine the formation of community groups of diabetics in every town and city in the United States that could support one another in dietary, exercise, education, and other activities geared toward managing and preventing this disease. They could develop macro policy advocacy projects to reduce the use of high-sugar soda drinks and many other initiatives. Similar community-based programs could evolve for other chronic diseases—both to help patients with these diseases and to prevent them.

THINKING BIG AS ADVOCATES IN THE HEALTH CARE SECTOR

Several major proposals for reforming social policy currently exist in health care. Discuss these proposals as you think they do or do not have merit.

1. Work to enact Medicare for All. Medicare is currently a popular and respected program for insuring health care of seniors but could be reformed to include the entire population. This reform would bring the United States into conformity with most other industrialized nations. These nations have better health

outcomes than the fragmented American system, including longer longevity, lower infant mortality, lower levels of obesity, and lower levels of specific diseases. They cost roughly half as much as the American health care system.

2. Work to enact incremental reforms that might eventually lead to universal health care. These include lowering eligibility for Medicare from age 65 to age 55 and raising eligibility levels of Medicaid to even higher levels than were achieved by the ACA.

3. Move the health system toward a wellness model by requiring insurance companies, Medicare, and Medicaid to place greater emphasis on wellness programs and services. If the nation fully endorses a wellness model, what policies might it consider regarding fast-food outlets, lack of places to exercise for many Americans, and new approaches for balancing work and recreation? How might particular assistance be given to low-income and other vulnerable populations to engage in wellness programs? How could the wellness programs and services be funded?

Learning Outcomes

You are now equipped to:

- Identify stages in the evolution of the American health system in seven eras
- Identify key interest groups in the political economy of the American health system as well as groups that are not well represented
- Analyze why income inequality in the United States exacerbates health disparities
- Identify seven major problems encountered by consumers of American health care in their policy and regulatory contexts
- Develop Red Flag Alerts at the micro, mezzo, and macro levels
- Develop micro, mezzo, and macro policy advocacy initiatives
- Discuss and analyze a major proposal for reforming the health care system

References

Adams, J., Asch, S. M., DeCristofaro, A., Hicks, J., Keesey, J., Kerr, E. A., & McGlynn, E. A. (2003, June 26). The quality of healthcare delivered to adults in the United States. *New England Journal of Medicine, 348*(26), 2635–2645.

American Cancer Society. (2009, December). *Tobacco-related cancers fact sheet.* Retrieved from http://www.cancer.org/docroot/PED/content/PED_10_2x_Tobacco-Related_Cancers_Fact_Sheet-asp?sitearea=PED

Anderson, B. (2007, December 12). Fresno is state's asthma capital. *Fresno Bee.* Retrieved from http://www.fresnobee.com/2007/12/12/263218/fresno-is-states-asthma-capital.html

Annas, G. (2003). HIPAA regulations—a new era of medical-record privacy? *New England Journal of Medicine, 348*, 1486–1490.

Barr, D. (2008). *Health disparities in the United States: Social class, race, ethnicity, and health.* Baltimore, MD: Johns Hopkins University Press.

Beach, M. C., Gary, T. L., Price, E. G., Robinson, K., Gozu, A., Palacio, A., & Cooper, L. A. (2006). Improving healthcare quality for racial/ethnic minorities: A systemic review. *BioMed Central Public Health, 6*, 104. Retrieved on April 10, 2010, from http://www.biomedcentral.com/1471-2458/6/104

Bell, J. & Standish, M. (2005). Communities and health policy: A pathway for change. *Health Affairs, 24*(2), 339–342.

Bower, P. & Gask, L. (2002). The changing nature of consultation-liaison in primary care: Bridging the gap between research and practice. *General Hospital Psychiatry, 24*(2), 63.

Brill, S. (2015). *America's bitter pill: Money, politics, and the fight to fix our broken healthcare system.* New York: Random House.

Brink, K. (2008, April 29). *Wellness programs at work.* Retrieved from http://www.asociatedcontent.com/article/724387/wellness_programs_at_work_pg3.html?cat=5

Brown, E. R., Wyn, R., Yu, H., Valenzuela, A., & Dong, L. (1999). Access to health insurance and healthcare for children in immigrant families. In D. J. Hernandez (Ed.), *Children of immigrants: Health, adjustment, and public assistance* (pp. 126–186). Washington, DC: National Academy Press.

Brumley, R., Enguidanos, S., Jamison, P., Seitz, R., Morgenstern, N., Saito, S., . . . Gonzalez, J. (2007). Increased satisfaction with care and lower costs: Results of a randomized trial of in home palliative care. *Journal of the American Geriatrics Society, 55*(7), 993–1000.

California Hospital Association (CHA). (2008). *Consent manual—a reference for consent and related healthcare law.* Sacramento, CA.

California Newreel (Producer). (2008). *Unnatural causes . . . is inequality making us sick?* [Television series]. Washington, DC: Public Broadcasting Service.

Carrasquillo, O., Orav, E. J., Brennan, T. A., & Burstin, H. R. (1999). Impact of language barriers on patient satisfaction in an emergency department. *Journal of General Internal Medicine, 14*(2), 82–87.

Castellucci, M. (2018, January 9). CMS launches new voluntary bundled-payment model. Modern Healthcare. Retrieved from https://www.cms.gov/newsroom/...releases/cms-announces-participants-new-value-bas...

Centers for Disease Control and Prevention (CDC). (1998). Missed opportunities in preventive counseling for cardiovascular disease—United States, 1995. *Journal of the American Medical Association, 279*, 741–742.

CNN. (2014). Deadly diseases: Epidemics throughout history. Retrieved from http://www.cnn.com/interactive/2014/10/health/epidemics-through-history/

Collins, J. L., Koplan, J. P., & Marks, J. S. (2009). Chronic disease prevention and control: Coming of age at the Centers for Disease Control and Prevention. *Preventing Chronic Disease, 6*(3), A81.

Consumer Reports. (2014, May). Survive your stay at the hospital, pp. 44–46.

Cooper-Patrick, L., Gallo, J. J., Gonzales, J. J., Vu, H. T., Powe, N. R., Nelson, C., & Ford, D. E. (1999). Race, gender, and partnership in the patient-physician relationship. *Journal of the American Medical Association, 282*(6), 583–589. doi:10.1001/Journal of the American Medical Association.282.6.583

Council on Ethical and Judicial Affairs. (1990). Black-white disparities in healthcare. *Journal of the American Medical Association, 263,* 2344–2346.

CSX. (2005). Safety is a way of life. Retrieved from http://www.csx.com/?fuseaction=about.safety

Department of Agriculture, Foreign Agricultural Service. (2006). Beef: Per capita consumption summary selected countries. Retrieved from http://www.fas.usda.gov/dlp/circular/2006/06-03LP/bpppcc.pdf

DeWalt, D. A., Berkman, N. D., Sheridan, S., Lohr, K. N., & Pignone, M. P. (2004). Literacy and health outcomes. *Journal of General Internal Medicine, 19*(12), 1228–1239.

Divi, C., Koss, R. G., Schmaltz, S. P., & Loeb, J. M. (2007). Language proficiency and adverse events in US hospitals: A pilot study. *International Journal for Quality in Health Care, 19*(2), 60–67.

DuBois, B. & Miley, K. K. (2002). *Social work: An empowering profession* (4th ed.). Boston, MA: Allyn and Bacon.

Economist. (2009). *The Economist pocket world in figures 2010 edition*. London, UK: Profile Books.

Expedia.com. (2009). *2009 international vacation deprivation survey results.* Retrieved from http:/ media.expedia.com/media/content/expus/graphics/promos/vacations/Expedia_International_Vacation_Deprivation_Survey_2009.pdf

Fincher, C., Williams, J. E., MacLean, V., Allison, J. J., Kiefe, C. I., & Canto, J. (2004). Racial disparities in coronary heart disease: A sociological view of the medical literature on physician bias. *Ethnicity & Disease, 14*, 360–371.

Frolkis, J. P., Zyzanski, S. J., Schwartz, J. M., & Suhan, P. S. (1998). Physician noncompliance with the 1993 national cholesterol education program (NCEP-ATPII) guidelines. *Circulation, 98*(9), 851–855.

Gawande, A. (2010). *The checklist manifesto: How to get things right*. New York, NY: Henry Holt.

Glasgow, R. E., Wagner, E. H., Kaplan, R. M., Vinicor, F., Smith, L., & Norman, J. (1999). If diabetes is a public health problem, why not treat it as one? A population-based approach to chronic illness. *Annals of Behavioral Medicine, 21*(2), 159–170.

Glendinning, C. (2002). Breaking down barriers: Integrating health and care services for older people in England. *Health Policy, 65*, 139–151.

Hadley, J., Holahan, J., Coughlin, T., & Miller, D. (2008). Covering the uninsured in 2008: Current costs, sources of payment, and incremental costs. *Health Affairs, 27*(5), w399–w415.

Halbert, C. H., Armstrong, K., Gandy, O. H., Jr., & Shaker, L. (2006). Racial differences in trust in health care providers. *Archives of Internal Medicine, 166*(8), 896.

Hooper, E. M., Comstock, L. M., Goodwin, J. M., & Goodwin, J. S. (1982). Patient characteristics that influence physician behavior. *Medical Care, 20,* 630–638.

Huang, Z. J., Stella, M. Y., & Ledsky, R. (2006, May). Health status and health service access and use among children in US immigrant families. *American Journal of Public Health, 96*(4), 634–640.

Ingraham, C. (2014, September 29). Our infant mortality is a national embarrassment. *Washington Post.* Retrieved from https://www.washingtonpost.com/.../2014/.../29/our-infant-mortality-rate-is-a-national...

Institute of Medicine (IOM). (2000). *To err is human: Building a safer health system.* Washington, DC: National Academy Press.

Institute of Medicine (IOM). (2003). *Unequal treatment: Confronting racial and ethnic disparities in healthcare*. Washington, DC: National Academy Press.

Institute of Medicine (IOM). (2008). *Knowing what works in healthcare*. Washington, DC: National Academy Press.

International Diabetes Federation. (2010). *IDF diabetes atlas*. Retrieved from http://www.diabetesatlas.org/content/regional-data

Jackson, A. (2002). Canada beats USA—but loses gold to Sweden. *Women, 81*(79.2), 81–83.

Jacobs, E. A., Kohrman, C., Lemon, M., & Vickers, D. L. (2003). Teaching physicians-in-training to address racial disparities in health: A hospital-community partnership. *Public Health Reports, 118*, 349–355.

Jansson, B. (2019a). *Reducing inequality: Addressing the wicked problems across professions and disciplines*. San Diego: Cognella Academic Press.

Jansson, B. (2019b). *The reluctant welfare state: Engaging history to advance social work practice in contemporary society* (9th ed.). Boston: Cengage.

Joint Commission. (2009). *Comprehensive accreditation manual for hospitals: The official handbook: Refreshed core*. Oakbrook Terrace, IL: Joint Commission Resources.

Kahn, K. L., Pearson, M. L., Harrison, E. R., Desmond, M. S., Rogers, W. H., Rubenstein, L. V., & Brook, R. H. (1994). Healthcare for black and poor hospitalized Medicare patients. *Journal of the American Medication Association, 271*, 1169–1174.

Kaiser, H. J. (2018, March 1). Where are states today? Medicaid and CHIP eligibility. Retrieved from *https://www.kff.org/.../state.../medicaid-and-chip-income-eligibility-limits-for-children...*

Kaiser Family Foundation. (2018). The Medicare hospital readmission reduction program, 2018. Retrieved from https://www.kff.org/.../aiming-for-fewer-hospital-u-turns-the-medicare-hospital-readmiss

Kawachi, I., Daniels, N., & Robinson, D. E. (2005). Health disparities by race and class: Why both matter. *Health Affairs, 24*(2), 343–352.

Knight, J. A. (2004). *A crisis call for new preventive medicine: Emerging effects of lifestyle on morbidity and mortality*. World Scientific Publishing Company Incorporated.

Ku, L. & Matani, S. (2001). Left out: Immigrants' access to health care and insurance. *Health Affairs, 20*(1), 247–256.

Kutner, M., Greenburg, E., Jin, Y., & Paulsen, C. (2006). *The health literacy of America's adults: Results from the 2003 national assessment of adult literacy* (NCES 2006-483). National Center for Education Statistics.

Lantos, J. D. (2007). The edge of the known world. *Health Affairs, 26*(2), 510–514.

Lawrence, D. (2003). My mother and the medical care ad-hoc-racy. *Health Affairs, 22*(2), 238–242.

Leclere, F. B., Jensen, L., & Biddlecom, A. E. (1994). Health care utilization, family context, and adaptation among immigrants to the United States. *Journal of Health and Social Behavior, 35*(4), 370–384.

Lee, H., & McConville, S. (2007). Death in the Golden State: Why do some Californians live longer? *California Counts: Population Trends and Profiles, 9*, 1. Retrieved from http://www.ppic.org/content/pubs/cacounts/CC_807HLCC.pdf

Lefton, R. (2008). Reducing variation in healthcare delivery. *Healthcare Financial*

Management: Journal of the Healthcare Financial Management Association, *62*(7), 42.

Lin, E. H., VonKorff, M., Russo, J., Katon, W., Simon, G. E., Unützer, J., . . . Ludman, E. (2000). Can depression treatment in primary care reduce disability? A stepped care approach. *Archives of Family Medicine*, *9*(10), 1052.

Mangione-Smith, R., DeCristofaro, A., Setodji, C., Keesen, J., Klein, D., Adams, J., Schuster, M., & Glynn, F. (2007). Quality of ambulatory care delivered to children in the U.S. *New England Journal of Medicine, 357*(15), 1515–1523.

March of Dimes. (2009, October). *PeriStats: Born too soon and too small in the United States.* Retrieved from http://www.marchofdimes.com/peristats/pdflib/195/99.pdf

Mayo, N., Nasmith, L., & Tannenbaum, C. B. (2003) Understanding older women's healthcare concerns: A qualitative study. *Journal of Aging & Women, 15*(1), 3–16.

McBean, A. M., & Gornick, M. (1994). Differences by race in the rates of procedures performed in hospitals for Medicare beneficiaries. *Healthcare Financial Review, 15*, 77–90.

McGlynn, E. A., Asch, S. M., Adams, J., Keesey, J., Hicks, J., DeCristofaro, A., & Kerr, E. A. (2003). The quality of health care delivered to adults in the United States. *New England Journal of Medicine*, *348*(26), 2635–2645.

Mirza, A. & Rooney, C. (2018, January 18). Discrimination prevents LGBTQ people from accessing health care. Center for American Progress. Retrieved from https://www.americanprogress.org/issues/lgbt/.../discrimination-prevents-lgbtq-people-...

National Institute on Drug Abuse (NIDA). (2018, June 6). Electronic cigarettes (e-cigarettes). Retrieved from https://www2.drugabuse.gov/publications/drugfacts/electronic-cigarettes-e-cigarettes

Ngo-Metzger, Q., Massagli, M. P., Clarridge, B. R., Manocchia, M., Davis, R. B., Iezzoni, L. I., & Phillips, R. S. (2003). Linguistic and cultural barriers to care. *Journal of General Internal Medicine*, *18*(1), 44–52.

Office of Minority Health. (2009, December). *Diabetes and Hispanic Americans.* Retrieved from http://minorityhealth.hhs.gov/templates/content.aspx? ID= 3324

Paschos, C. L., Normand, S. L., Garfinkle, J. B., Newhouse, J. P., Epstein, A. M., & McNeil, B. J. (1994). Trends in the use of drug therapies in patients with acute myocardial infarction: 1988 to 1992. *Journal of the American College of Cardiology, 23*, 1023–1030.

Perkins, J., Youdelman, M., & Wong, D. (2003). *Ensuring linguistic access in health care settings: Legal rights and responsibilities.* National Health Law Program.

Pew Research Center. (2006, April). Eating more, enjoying less. Retrieved from http://pewresearch.org/pubs/309/eating-more-enjoying-less

Pew Research Center. (2017, July 7). 7 facts about Americans with disabilities. Retrieved from pewresearch.org/fact-tank/2017/07/27/7-facts-about-americans-with-disabilities/

Pham, H. H., Schrag, D., Hargraves, J. L., & Bach, P. B. (2005). Delivery of preventive services to older adults by primary care physicians. *JAMA: The Journal of the American Medical Association*, *294*(4), 473–481.

Ponce, N. A., Ku, L., Cunningham, W. E., & Brown, E. R. (2006). Language barriers to health care access among Medicare beneficiaries. *Journal Information*, *43*(1), 66–76.

Rafferty, M. (1998). Prevention services in primary care: Taking time, setting priorities. *Western Journal of Medicine, 169*(5), 269.

Ramsay, N. (2008, May 24). *TV viewing figures vs. IQ ranking by country.* [Web blog post]. Retrieved from http://www.longcountdown.com/2008/05/24/tv-viewing-figures-vs-iq-ranking-by-country/

Red Meat Industry Forum. (2007, May). *Beef and veal consumption and choosing the right cut.* Retrieved from http://www.redmeatforum.org.uk/supplychain/BVConsumption.htm

Reede, J. Y. (2003). A recurring theme: The need for minority physicians. *Health Affairs, 22*(4), 91–93.

Reiss-Brennan, B. (2006). Can mental health integration in a primary care setting improve quality and lower costs? A case study. *Journal of Managed Care Pharmacy, 12*(2), 14.

Satcher, D., Fryer, G. E., McCann, J., Troutman, H., Woolf, S. H., & Rust, G. (2005). What if we were equal? A comparison of the black-white mortality gap in 1960 and 2000. *Health Affairs, 24*(3), 459–464.

Singer, P. (2007, January 3). Early births fade to grey. *The Australian*, p. 10.

Smeltzer, S. C., Sharts-Hopko, N. C., Ott, B. B., Zimmerman, V., & Duffin, J. (2007). Perspectives of women with disabilities on reaching those who are hard to reach. *Journal of Neuroscience Nursing, 39*(3), 163–171.

Stein, T. (2004). *Role of law in social work practice and administration.* New York, NY: Columbia University Press.

Street, R. L., O'Malley, K. J., Cooper, L. A., & Haidet, P. (2008). Understanding concordance in patient-physician relationships: Personal and ethnic dimensions of shared identity. *The Annals of Family Medicine, 6*(3), 198–205.

Taylor, H. A., Jr., Cano, J. G., Sanderson, B., Rogers, Q. J., & Hilbe, J. (1998). Management and outcomes for black patients with acute myocardial infarction in the reperfusion era. National Registry of Myocardial Infarction 2 Investigators. *American Journal of Cardiology, 82*, 1019–1023.

United Nations Office on Drugs and Crime. (2010). *Homicide statistics, criminal justice, and public health sources—trends 2003–2008.* Retrieved from http://www.unodc.org/unodc/en/data-and-analysis/homicide.html

Wartik, N. (2002, June 23). Hurting more, helped less? *New York Times*, pp. 1, 6.

Williams, D. R. & Mohammed, S. A. (2009). Discrimination and racial disparities in health: Evidence and needed research. *Journal of Behavioral Medicine, 32*(1), 20–47.

Willingham, S. & Kilpatrick, E. (2005). Evidence of gender bias when applying the new diagnostic criteria for myocardial infarction. *Heart, 91*, 237–238.

Wilper, A. P., Woolhandler, S., Lasser, K. E., McCormick, D., Bor, D. H., & Himmelstein, D. U. (2009). Health insurance and mortality in US adults. *Journal Information, 99*(12), 2289–2295.

Woolhandler, S., Campbell, T., & Himmelstein, D. (2003). Costs of healthcare administration in the United States and Canada. *New England Journal of Medicine, 349*, 768–775.

BECOMING POLICY ADVOCATES IN THE GERONTOLOGY SECTOR

This chapter was coauthored by Dawn Joosten, Ph.D.

LEARNING OBJECTIVES

In this chapter you will learn to:

1. Develop an empowering perspective toward seniors
2. Analyze problems of seniors created by economic inequality
3. Analyze the political economy of the gerontology sector
4. Analyze seven core problems in the gerontology sector
    - Identify barriers in remedying each of them
    - Recognize important policies, regulations, and organizational factors that provide the context for each of them
    - Understand some Red Flag Alerts for each problem
    - Understand some connected micro, mezzo, and macro policy interventions by social workers seeking to ameliorate the seven problems
    - Understand gerontology-related initiatives
5. Think big in the gerontology sector

Note: Clinical Professor Dawn Joosten, Ph.D. at the Suzanne Dworak-Peck School of Social Work, made invaluable contributions to this chapter in the first edition of this book. Bruce Jansson was the sole author of the second edition, editing and updating contributions from each original contributor.

DEVELOPING AN EMPOWERING PERSPECTIVE TOWARD SENIORS

Most persons over age 65 live in their own homes, whether as renters or home owners. Only roughly 5% live in institutional settings, like nursing homes or congregate retirement settings. We will move between these two worlds frequently in this chapter.

Seniors contribute to the nation in many ways such as by working, volunteering, participating in civic activities, mentoring youth, engaging in civic associations, engaging in politics, or participating in whatever fulfilling activity seniors wish to pursue.

Yet seniors are a marginalized group in United States for many reasons. They are often portrayed as takers rather than makers because they make heavy use of public programs to address their financial, medical, and living expenses. They are accused of consuming a large part of federal and state budgets due to their heavy use of Medicare, Medicaid, and Social Security—entitlements that constituted more than 50% of the federal budget in 2019. They are often equated with subgroups of the elderly population that have degenerative diseases like Alzheimer's, not to mention heart disease, strokes, diabetes, and other maladies.

These negative perceptions of seniors ignore basic facts. Most seniors have been employed for decades prior to their retirements. They have to use public programs to meet their human needs unless they have sufficient pensions. Many seniors rely on Social Security for resources—and many of them have pensions of less than $25,000 per year. The aging process is part of the human condition. Everyone who ages needs medical care during their retirement years because they usually lose coverage from private insurance from their employers when they retire, forcing them to rely on their own resources, Medicare, and Medicaid. Some seniors develop mental health problems, such as depression, at some point during the aging process. Many seniors develop health problems during the natural process of aging. Seniors often need help finding employment to make ends meet even after they have retired. They need social services that link them to existing programs.

The nation benefits from funding seniors' services sufficiently in many ways. It saves resources by preventing medical problems, mental health problems, and substance abuse services. It reduces the burden of seniors' children who help their aging parents. It reduces the cost of nursing home care. It allows many seniors to enter the workforce.

The United States often neglects its seniors by providing inadequate Social Security pensions, inadequate social services, inadequate home health care, and insufficient and affordable continuing care for seniors who cannot live independently. It also trains negligible numbers of geriatricians, not to mention social workers and nurses with expertise in aging. Institutional settings, such as nursing homes, are often like hospitals rather than places where people lead fuller lives.

Most seniors live in the community rather than in institutions. Many of them live alone, including one-third of them between ages 65 and 85—and half of those over age 85. Although many of these seniors are connected to family and other networks, others experience isolation that public health researchers have discovered is equivalent to smoking 15 cigarettes a day and has greater adverse health effects than the current obesity epidemic (Seegert, 2017).

ANALYZING PROBLEMS OF SENIORS CAUSED BY ECONOMIC INEQUALITY

Nonaffluent seniors not only often live in poverty but have greater difficulty in negotiating the aging process than affluent seniors. This harsh treatment of nonaffluent seniors is a form of ageism—ageism that also includes discrimination in employment, condescending or deriding actions, negative stereotypes, and hostile actions.

The baby boom and increased life expectancies have contributed to a dramatic demographic shift among older adults in the United States. The number of adults 65 and older will increase from 12.4% of the total population in 2004 to roughly 20% in 2030 (Administration on Aging, 2005b). The percentage of adults 85 and over, as a proportion of adults 65 and over, will increase from 12% in 2000 to roughly 24% in 2050 (Administration on Aging, 2004) and from 4.6 million in 2002 to roughly to 9.6 million in 2030 (Administration on Aging, 2005b). This dramatic increase in the population of persons over 65 is partly due to the increased longevity of seniors from roughly 68 and 72 for males and females in 1960 to roughly 78 and 83 for males and females in 2017.

Many seniors encountered severe financial straits in 2019. Roughly one-half of them retired with no assets like a house, savings, stocks, a private pension plan, or a government pension that many Americans receive from their places of work (Jansson, 2019). Many have considerable credit card debt. Many seniors rely on Social Security benefits that averaged $15,528 in 2014 for single persons and $25,332 for retired couples, when poverty levels were $11,173 for single persons and $14,095 for couples (Jansson, 2019). Pensions for seniors are twice as large or more in many other industrialized nations than the United States (Jansson, 2019). Millions of seniors spend down their assets and income to become eligible for Medicaid to pay for chronic health problems when they have exhausted their Medicare coverage or when they have dual coverage from Medicare and Medicaid—a demeaning process that deprives them of hard-won assets and that makes it difficult to pass resources to family members when they die. Many affluent people benefit from private pensions that they receive when they retire, such as pensions that are funded by payroll deductions and employers' contributions (Jansson, 2019). Tens of millions of low-income and moderate-income people work in companies, often relatively small ones, that do not fund private pensions, so they must rely exclusively on modest Social Security payments during their retirement years. Elderly women are often poor because they worked in jobs with no private pensions and at low salaries.

Extreme poverty has tragic effects on many seniors. It places them under stress when they cannot meet daily food, housing, and other expenses. Women who spent their lifetimes in the lower 20% of the economy live about 12 years shorter lives than women who spent their lifetimes in the top 10% (Tavernise, 2016). If affluent people can afford units in residential continuing-care settings, low-income people may suffer from the toxic effects of isolation in rental units or owned homes. Increasing numbers of seniors die alone in hospitals with no visitors. Growing numbers of seniors are homeless for brief or extended periods. Many live in unsafe or undesirable areas because they cannot afford rentals in safer and more desirable areas (Angel & Hogan, 2004, as cited in Joosten, 2008). Many seniors cannot afford deductibles and coinsurance of Medicare and Medicaid.

POLICY ADVOCACY LEARNING CHALLENGE 8.1

TALK ABOUT YOUR AGING RELATIVES

All of us have aging relatives or ones who have recently died. Select one or several of them. If they were relatively affluent, discuss some advantages that they have or had as compared with relatively poor seniors. Identify some of the sources of their income. Discuss how their access to economic resources facilitated their ability to cope with the aging process. If they were relatively poor, identify some problems they encountered to meet their daily needs or to find housing. Did their relative poverty contribute to burdens of their children and relatives? Do you think it negatively impacted their health and mental health?

ANALYZING THE POLITICAL ECONOMY OF THE GERONTOLOGY SECTOR

Our discussion reveals a complex political economy of aging. A two-sided political system exists for seniors. On the one hand, many advocacy groups and public officials support policies that are beneficial for seniors. Hospitals, convalescent and nursing home facilities and companies, pharmaceutical companies, and other interest groups influence policies for seniors. On the other, some interest groups and public officials seek deep cuts in programs that help seniors such as Medicare and Medicaid as reflected by budget priorities of the Trump administration.

Older persons have had many advocates over the past century who helped enact and defended Social Security, Medicare, Medicaid, and Supplementary Security Income (SSI), nursing-home regulations, the Older Americans Act, civil rights for seniors, Meals on Wheels, and other programs (see descriptions of these programs in Chapter 9). Once enacted, Social Security, Medicare, Medicaid, SSI, and other beneficial programs created huge constituencies favorable to seniors. Indeed, the Democratic Party made many electoral gains in the 1980s and beyond when they attacked efforts by leading Republicans to cut or privatize Social Security, Medicare, and Medicaid. Interest groups catering to hospitals, nursing homes, and convalescent homes, as well as retirement homes, often fought cuts in Medicare and Medicaid that funded their operations. Growing numbers of younger persons came to defend seniors' programs as they realized that their aging parents and grandparents rely on these government programs to meet their basic needs. (See Table 8.1 for advocacy groups that seek greater social justice in the gerontology sector.)

No longer a national organization, the Gray Panthers still exists in many cities, such as Sacramento and Detroit. Formed in 1970 by Maggie Kuhn, who was angry that she was forced to retire at age 65, the Gray Panthers militantly sought greater rights for seniors. Kuhn established a collection of local networks that used advertising and demonstrations to seek universal health care, lower cost medications, greater patient rights, affordable housing for all, and a patient bill of rights. Kuhn believed that ageism was a potent ideology that restricted elderly persons' participation in society. She favored house sharing for seniors as well as intergenerational living as compared with nursing homes.

TABLE 8.1 ■ Some Advocacy Groups Seeking Greater Social Justice in the Gerontology Sector

1. American Federation of Labor and Congress of Industrial Organizations (AFL-CIO)
2. Alliance for Retired Persons
3. American Association of Retired Persons
4. Center for Economic and Policy Research
5. Century Foundation: Social Security Network
6. Institute of America's Future
7. National Committee to Preserve Social Security & Medicare
8. New York Network for Action on Medicare & Social Security

Support for policies that help seniors is bolstered, as well, by increased understanding of the biology of aging. Many Americans have a denial complex about the aging process, believing healthy diet, exercise, medications, and other remedies can end the natural biological processes that occur during aging. It is true: Seniors currently live longer lives than their predecessors, and longevity has risen markedly from 1960 to the present. Statins and other medications have markedly reduced cholesterol and blood pressure, the rate of smoking has declined, and more seniors exercise regularly. Yet people inexorably age as tissues, organs, and eyes decline. Brains get smaller, bones get thinner, and teeth deteriorate. The risk of falling increases with age, whether from loss of balance from medications, weakened muscles, or poor balance. Healthy lifestyles and medications slow the aging process; they don't stop it. Many older people experience medical emergencies and recover from successive ones over periods of time but at successively lower stages (Gawande, 2014). The aging process should not suggest that many seniors do not lead fulfilling and active lives. Many are supported by excellent health care, supportive services, friendship networks, and economic resources from their savings, pensions, and assets.

Older adults already are a strong political force. Seniors are more likely to vote than younger people, particularly since 1994 (Kuhn, 2010). A Gallop Poll of registered voters for the 2012 presidential election suggests "age is a predictor of voter turnout" as 87% of registered voters ages 60 to 69 reported that they definitely would vote compared with 59% of registered voters ages 19 to 29 (Newport, 2012, para. 8). They will constitute a massive part of the American electorate—as much as 25% in some jurisdictions. A majority of seniors has voted for Republican candidates in the last several national elections, but this may change as more elderly persons perceive that they lack adequate living standards, health, and long-term care. Many younger Americans also favor enhanced programs for seniors because they often find it difficult to bear the burden of supporting their parents—a burden that disproportionately falls on daughters or other female relatives who sometimes have to resign from their jobs to help aging parents or relatives.

Greater funding of programs for seniors, such as greater pensions, better medical care, enforcement of policies that forbid discrimination in places of work, and greater supportive services can reduce current costs of caring for them. These policies can allow more seniors to remain independent in their places of residence. They can reduce their medical and mental health costs. They can reduce emergency medical care. They can reduce depression among seniors (Gawande, 2014).

Opponents of policies that help seniors include Republicans' failed efforts to convert Medicare into a program of medical savings accounts during the presidencies of Bill Clinton and George W. Bush to be funded by the federal government at such low levels that they would have placed many seniors in financial jeopardy. Republicans have repeatedly proposed to devolve Medicaid to the states rather than having the federal government share its costs. States cannot afford bearing all of Medicaid's costs when many states already spend 25% of their budgets (or more) on Medicaid even with the federal government absorbing roughly half of the cost of Medicaid. Many members of Congress and presidents want to reduce the costs of Medicaid, Medicare, and Social Security on grounds that they absorb far more than half of a federal budget that has had huge deficits for decades. Why not, many advocates for seniors ask, eliminate or decrease these budget deficits by raising additional revenues by increasing taxes on affluent people, cutting medical costs by moving to a Medicare for All program, and decreasing military costs? Why not fund evidence-based social programs for seniors that allow them to live independent lives, such as geriatric teams that give seniors multidisciplinary assistance for their full range of problems including medical, mental health, disability, and home health ones (Gawande, 2014)?

The following brief chronology of policies reflects both policy successes and policy failures during American history.

- Christian Morality and Civil War, 1800–1860: Elders occupied a high position in society in the colonial period based on Puritan beliefs that elders should be revered as their spirituality and wisdom peaked in old age and as they provided moral leadership to younger generations (Quadagno, 2008).
- By 1820, veneration of elders had declined as reflected by the fact that they no longer held the best seats in churches. Legislatures established requirements for public officials to retire between the ages of 60 and 70 (Quadagno, 2008).
- The cultural value of a youthful society began to emerge as elders' manner of dress became designed to make them appear younger.
- The attitude of older adults as unproductive emerged and retirement became a practice, which further reinforced negative stereotypes of older adults. Poverty rates among older adults rose (Quadagno, 2008).
- Post–Civil War Degradation, 1860–1920: Manufacturing employers focused on mass production with use of assembly lines to increase profits. Ageist stereotypes well embedded in American culture by this period often led employers to perceive older adults as less productive than younger workers, further degrading the value of the older adult in the new industrial workforce.
- The demographic profile of the population in the United States in 1905 resembled a pyramid shape consisting of a small proportion of older adults over age 65, reflecting mortality rates, which were high. The population contained a larger proportion of infants and children, reflecting high fertility rates (Aldwin & Gilmer, 2004).

- Addressing destitution in the New Deal and its aftermath: Stereotypes of older adults as frail, abandoned by their children, and impoverished became deeply embedded in the attitudes of Americans toward older adults (Quadagno, 2008).
- President Franklin Roosevelt sought to alleviate widespread poverty among the elderly by enacting the Social Security Act in 1935, which contained Social Security that gave pensions to most seniors regardless of income. It also contained the Old Age Assistance Program, which gave financial assistance to low-income seniors.
- Poverty rates of older adults 65 and older were slightly more than double those in the 18 to 64 age-group, 35.2% and 17%, respectively, in 1959 (U.S. Census Bureau, 2011).
- The private nursing home industry began to grow during the 1950s and 1960s, taking the place of traditional almshouses. This growth was caused by decisions of many hospitals to divert seniors to nursing homes to make room for younger patients and by the funding of nursing home stays by Medicaid after its enactment in 1966.
- Growth of public social services, personal rights, and social programs, 1960 to 1980: The proportion of older adults 65+ below the poverty rate continued to decline during this time period from the high of 35.2% in 1959 to a low of 15.7% in 1980 (U.S. Census Bureau, 2011).
- A new stereotype of older adults emerged in society as the economic condition of older adults improved: They were now viewed according to Binstock as "a prosperous, selfish, and politically powerful group that is gobbling up scarce societal resources" (as cited in Quadagno, 2008, p. 103).
- The Kerr-Mills Act of 1960 was a small means-tested program to enhance medical care of some low-income seniors (Quadagno, 2008).
- Medicare and Medicaid as Titles XVIII and XIX, respectively, of the Social Security Act revolutionized health care for seniors in the years after their enactment in 1965. Whereas Medicare provided acute medical care for all seniors regardless of income, with treatments usually lasting fewer than 90 days in a specific episode, Medicaid provided care to low-income seniors for chronic health conditions. Medicaid became the major funder of nursing homes from 1966 onward. Many patients became dually eligible for both Medicare and Medicaid, receiving funding for primary care and brief episodes of treatment from Medicare and care for chronic conditions and nursing home care for low-income patients (Midgley, Tracy, & Livermore, 2000).
- Seniors increasingly received private pensions from those employers (mostly large companies) that chose to co-fund them with employees. Millions of seniors who mostly worked in smaller companies did not and do not currently receive private pensions.
- The means-tested SSI was created in 1972 to provide income to low-income seniors as well as low-income non-elderly persons with disabilities and blind persons (Torres-Gil & Villa, 2000). It replaced Old Age Assistance.

- The National Aging Network was established to meet the needs of non-institutionalized older adults through the enactment of the Older Americans Act (OAA) on July 14, 1965. It funded a variety of programs including Meals on Wheels.
- The OAA established the Administration on Aging (AoA), which accomplished its mission through local and state Area Agencies on Aging (AAAs) (Angel & Hogan, 2004).
- Devolution of federal programs to states and individuals in the presidencies of Ronald Reagan and George W. Bush: President Ronald Reagan enacted Social Security amendments in 1983 intended to keep Social Security solvent through reforms that included "advancing the age of eligibility for benefits, increasing the federal withholding tax, including federal and nonprofit employees in the program, and a 6 month delay in benefit increases" (Midgley, Tracy, & Livermore, 2000, p. 145).
- The Americans with Disabilities Act (ADA, P.L. 101-336) was enacted in 1990 during George H. W. Bush's presidency, creating a national mandate of anti-discrimination for disabled individuals in the workplace, the community, public places, and public and transportation services (Hayden, 2000).
- Social reform in a polarized context during the presidencies of Bill Clinton and Barack Obama: President Bill Clinton enacted the Balanced Budget Act of 1997 seeking to end the budget deficit in five years and promised to "divert any federal budget surplus resulting from the budget balancing to the Social Security trust fund so that baby boomers would be assured of benefits on their retirement" (Stoesz, 2000, p. 149). This promise remained unfulfilled when President George W. Bush greatly increased federal deficits and debt.
- Medicare provided health insurance to 38 million older and/or disabled adults and Medicaid to 36 million low-income adults in 1998 (Pear, 1998, as cited in Midgley, Tracy, & Livermore, 2000).
- Political and public debates took place during the Clinton administration surrounding concerns about the sheer cost and future of Medicare, Medicaid, and Social Security, but Republicans like Newt Gingrich were unable to privatize Medicare, turn Medicaid back to the states, or replace Social Security with private savings accounts (Jansson, 2018).
- President Obama released his plan for reducing the federal deficit in September 2011—a plan that included spending caps and other cuts in Medicare and Medicaid for programs used by seniors (Office of Management and Budget, 2011, p. 1).
- These cuts came on top of the $424 billion in cuts in Medicare over a 10-year period to offset some of the costs of the Patient Protection and Affordable Care Act (ACA) that was enacted in 2010 (Brill, 2015).

POLICY ADVOCACY LEARNING CHALLENGE 8.2

IDENTIFY RESOURCES AND SERVICES USED BY A RETIREE THAT YOU KNOW

Identify a specific retiree that you know. Ask him or her what publicly funded resources or services have been helpful to him or her in the wake of retirement. Which ones were most helpful? Which ones were not helpful? What changes in current public policies for seniors would he or she support?

- The ACA funded free wellness exams for seniors.
- Medicaid funds most of the nation's nursing home care in the United States. Many low-income seniors are dually eligible for both Medicare and Medicaid, which gives them care for acute-care medical conditions, primary care, and nursing home care.

ANALYZING SEVEN CORE PROBLEMS IN THE GERONTOLOGY SECTOR

Core Problem 1: Engaging in Advocacy to Promote Ethical Rights, Human Rights, and Economic Justice for Older Adults—With Some Red Flag Alerts

- **Red Flag Alert 8.1.** Older persons do not receive adequate information to make an informed decision about the options for end-of-life care such as palliative care, hospice, and legal options, including advance directives and/or living wills.
- **Red Flag Alert 8.2.** Seniors' right to self-determination is violated when competent seniors are not included, or sufficiently included, in decision-making about their care.
- **Red Flag Alert 8.3.** The wishes of elderly persons with advance directives are not honored by health professionals.
- **Red Flag Alert 8.4.** Seniors have the right to resources sufficient to give them a decent standard of living, yet many seniors unnecessarily live in poverty partly because Social Security pensions are far lower than ones in other industrialized nations and because low-income seniors are less likely to have private pensions to augment their Social Security pensions.
- **Red Flag Alert 8.5.** Nursing homes, physicians, and family members are sometimes overly protective of seniors with disabilities that are not life-threatening. They forget that seniors often want to be self-directing even when they might occasionally fall. The ethical norm of self-determination suggests

that seniors need a sense of purpose like caring for a spouse, helping others, volunteering, and remaining mobile to the extent possible (Gawande, 2014).

- **Red Flag Alert 8.6.** Seniors need help in obtaining and retaining employment if they want to remain in the workforce but often do not receive it from caregivers and professionals.
- **Red Flag Alert 8.7.** A patient's right to self-determination is violated when older adults with competence are not included in decision-making about their care or preferences for care.
- **Red Flag Alert 8.8.** A resident of a nursing home, board and care, or assisted living facility experiences a violation in the Residents' Bill of Rights.

Background

Refer back to the discussion of ethics in Chapter 1 that discussed ethical norms of self-determination, confidentiality, honesty, and equality as well as ethical dilemmas where decisions require balancing of two or more ethical norms. We make a distinction between "older adults 65+" and "oldest-old 85+" in the ensuing discussion. We also discuss some ethical issues that seniors encounter in residencies as well as in retirement homes and nursing homes.

Most seniors live in non-institutional settings. Most of them are relatively healthy, join community groups, and attend places of worship. A growing number of seniors are employed on a part- or full-time basis. Many of them have lived in their neighborhoods for extended periods of time.

Home-based seniors often encounter health, social, and economic problems as they age. Affluent elderly people have distinct advantages over low-income elderly people. They can hire helpers to maintain their homes. They are more likely to have purchased private insurance that supplements Medicare. They can hire home services and health care teams to help them with daily health problems or seek funding from Medicare and Medicaid. However, even wealthy seniors need help to deal with daily activities. Seniors can become isolated if they lose the ability to travel around their neighborhoods. Seniors who lose a spouse and who do not have children or relatives in close proximity are particularly likely to become isolated. Isolation often causes social and mental problems.

The decision about whether to leave their residence to go to a retirement home or a nursing home is a critical one. Affluent seniors can afford private retirement facilities that can cost as much as $500,000 to enter and $3,000 per month. These facilities may have units for independent living, other units for persons who are quasi-independent, and an infirmary unit for persons who have serious health problems. These facilities often have numerous clubs and activities as well as high-level food services. Less affluent seniors have far fewer choices: They may have to go to a nursing home funded by Medicaid that is usually like a hospital with minimal activities and with many residents who have serious health problems.

Ethical issues arise as seniors decide whether to remain in their houses or rental units or to move into congregate retirement places or nursing homes. They need to decide their course of action. Some seniors are determined to remain in their homes or rental units to the end. As they have increasing problems with daily activities, they may find it increasingly difficult to clean and maintain their homes. Persons with scant resources realize that many nursing homes—their only option if they cannot live with their children—are not desirable places because they are like hospitals (Gawande, 2014). Seniors develop serious health and mental health problems, or even die, during the transition.

Older adults 65+ accounted for 55.8% of all intensive care unit (ICU) stays in the United States in 2007 with rates rising by 2017 with the graying of America (Bala, Casey, & Happ, 2007). The oldest old (ages 85+) have the highest rates of functional limitations and hospitalizations. The prevalence of chronic diseases is greatest among older adults 65+ with 88% having at least one chronic disease (Centers for Disease Control and Prevention [CDC], 2011). Each year in the United States 70% of deaths are due to chronic diseases, with prevalence rates for heart disease at 31.7%, cancer at 26.9%, and diabetes at 17.1% from 2008 to 2009 (CDC, 2011).

Many ethical issues arise during end-of-life situations in which seniors must make difficult decisions, including whether to sustain support or treatments such as mechanical ventilation, dialysis, nasal or gastric feeding tubes, and cardiopulmonary resuscitation. They often must make these decisions when they have diminished decision-making capacity. When they are declared to be mentally incompetent by courts, surrogates must make these decisions, such as spouses or adult children.

Social workers, nurses, physicians, attorneys, family members, and patients often have to make difficult ethical choices about initiating specific medical treatments or ceasing them. They have to decide whether patients are sufficiently cognitively competent to make these choices and, if not, who should make them. Every hospital has a bioethics committee composed of members of different health professions as well as laypeople.

Patients, family members, and health professionals can bring cases to them for discussion and recommendations. They often act on behalf of those with diminished capacity or empower proxy decision makers when patients lack competence. Here are some examples:

- Are specific patients cognitively competent to make medical decisions as determined by their cognitive function as measured by specific tests?
- When should limits be placed upon patients' activities because of possible injuries they may sustain, such as whether they can drive or walk unassisted? When should limits be placed upon health professionals' recommendations when they override patients' wishes?
- When have health professionals not provided specific patients with sufficient information about the benefits and dangers of specific treatments that is required when giving them informed consent?
- Will specific recommended treatments not prove to be effective in light of evidence-based literature—and should patients seek other treatments under these circumstances?
- Do members of specific ethnic groups, specific genders, specific sexual orientations, age, or any other personal characteristics receive treatments that are inferior to ones given to other patients (Csikai & Chaitin, 2006)?

Many factors can impede ethical treatment of patients. Family members may want medical options not effective with elderly persons. Time pressures may lead physicians not fully to consult with patients and their families. The availability of medical technology may skew decisions toward their use even when patients do not want them. Some physicians impose their judgments on patients or are unable to communicate effectively with them. Some medical choices are made to avert medical costs. Some family members may seek to influence medical choices because they want to inherit money and other goods from a parent or relative.

Social work advocates are in a unique position to intervene on behalf of vulnerable older adults and their family members to ensure that their ethical rights are not being violated. Advocates trained in advance care planning, bioethics, and end-of-life care ensure that terminal older adults and those facing life-threatening illnesses are informed about the range of options available.

Older adults' ethical rights can be violated if they fail to document their preferences for end-of-life care. The federal Patient Self-Determination Act, enacted in 1990, requires that patients admitted to hospitals be encouraged to fill out an advance directive or a durable power of attorney that states their preferences, such as whether they want "heroic treatments" when they have terminal illness, whether they want a feeding tube to be placed in the abdomen when they cannot swallow, and whether they want to donate organs when they have died. It was estimated in 1999 that only between 2% and 15% of adults in the general population and 55.7% of patients with terminal cancer had advance directives in the United States (Ott, 1999), but more seniors have advance directives due to better enforcement and implementation of federal laws, accreditation requirements of the Joint Accreditation Commission of the American Hospital Association, pressure by state health departments, and diligence of the staff of hospice and palliative care units.

More palliative care physicians who specialize in end-of-life care need to be trained. Only 86 new palliative care physicians completed accredited training for the subspecialty annually in 2014. Only 300 geriatricians graduated from medical schools in 2013—and many graduating physicians have not had training in geriatrics (Gawande, 2014).

Resources for Advocates

A study sponsored by the Robert Wood Johnson Foundation that sought to understand the prognosis and preferences for outcomes and risks of treatment (SUPPORT) by looking at 5,000 dying patients in American hospitals concluded that most Americans "die alone in institutions, in pain, and attached to machines against their wishes" (Berzoff & Silverman, 2004, p. 7). Three major alternatives seek to avert these negative outcomes: advance directives, hospice, and palliative care.

Advance Directives. The Patient Self-Determination Act of 1991 requires hospitals to provide patients with advance directives in which they indicate whether they want specific medical procedures if they become terminally ill and are not able to request them due to their medical conditions. They might decide, for example, not to have a feeding tube inserted into the stomach if they are unable to swallow. They can also decide they want to designate a person, such as a spouse or an adult child, to make medical choices for them if they are unable to make their own choices—often called "durable power of attorney."

Federal courts have generally supported patients' right to make end-of-life choices—and the right of relatives to make them for them when they are unable to make these choices due to their medical conditions. In the landmark case of *Cruzan v. Director, Missouri Department of Health*, for example, the court upheld a request by Nancy Cruzan's parents to have her feeding tube removed due to her expressed preference not to have life-supportive measures prior to the automobile accident that left her in a vegetative state.

Five states and the District of Columbia had enacted "dignity laws" that allowed physicians to engage in physician-assisted dying by April 2018, including California, Colorado, Oregon, Vermont, and Washington. These ballot measures stated many procedural safeguards to ensure ethical decisions including the following:

- Medical determination that persons are not depressed and do not have other mental conditions that might make them suicidal
- A stated preference to physicians on multiple successive occasions
- A prescription by physicians for the lethal substances

Researchers have discovered that relatively few persons choose physician-assisted dying in these states.

Physicians are allowed in all states to issue do not resuscitate orders (DNRs) for patients with end-of-life conditions rather than have physicians provide medically futile treatments.

Hospice Care. Hospice care is defined by the Institute of Medicine as a program that provides dying individuals and their families with supportive and medical services at a specific cite of care with an emphasis on a philosophy of care that incorporates spiritual, clinical, social, and metaphysical principles (as cited in Reese & Raymer, 2004). Hospice seeks to reduce some of these problems by giving patients the choice not to receive non-heroic treatment for terminal conditions while allowing them, as well, to select heroic treatments. It allows them to die in their homes if they wish. It provides them with social work and nursing services geared toward helping them avoid pain. Medicare covers inpatient care at skilled nursing facilities, hospice services, and home health services for Medicare beneficiaries in need of long-term, end-of-life, and intermittent care. Hospice care is funded by Medicare as well as by many private health insurance plans for patients under age 65. The Medicare benefit for hospice services was established in 1982 as part of the Tax Equity Fiscal Responsibility Act—and became a national guaranteed benefit under the administration of President Bill Clinton (National Hospice and Palliative Care Organization, 2011). Patients must have a physician's order that confirms they have a terminal illness with six or less months to live to qualify for hospice coverage, but hospice can be extended beyond this time limit. They can receive the benefit at home, hospital, or skilled nursing settings at no charge with a co-payment of up to $5 for medication that manages their pain and symptoms (Centers for Medicare and Medicaid Services [CMS], 2011). Hospice patients originally could not use heroic treatments, but they are now given this option as well. These same provisions apply, as well, to hospice care funded by private insurance for patients under age 65. The Community Health Accreditation Program (CHAP) accredits hospice programs.

Palliative Care. The Dana-Farber Cancer Institute (2015) describes palliative care as a type of care that focuses on the whole person (mind, body, and spirit) to allow individuals to be more comfortable during medical treatments and to reduce pain and suffering in the later stages of disease. Unlike most hospice programs, palliative care is not defined by

POLICY ADVOCACY LEARNING CHALLENGE 8.3

CONNECTING ADVOCACY TO PROTECT PATIENTS' ETHICAL RIGHTS AT MICRO, MEZZO, AND MACRO LEVELS

Social workers are often the health care professional that family and patients reach out to when making end-of-life decisions. Social workers have an ethical obligation to advocate for the rights of self-determination among cognitively competent patients. They need to help patients write clear advance directives and living wills indicating their preferences and instructions for withholding and withdrawal of care as well as the surrogate decision makers they appoint, or power of attorney, in the event they become cognitively impaired.

Mr. K was an 80-year-old white male who resided at his home with his spouse. He had advanced Parkinson's disease and was dependent upon a ventilator/tracheotomy and artificial hydration/nutrition through a feeding tube.

Mr. K was bedbound and required total care, was alert and oriented times four (person, time, place, and situation), but was nonverbal. His only way of communicating was by writing on a whiteboard. Mr. K had an active spouse and two adult children who were professionals. Mr. K received long-term home health care and had a visiting nurse. Mr. K had the financial resources to pay for around-the-clock care. Mr. K identified his religious view of an afterlife as a source of hope and his ability to cope. To Mr. K the ability to communicate with loved ones and visitors via the whiteboard helped define his quality of life. He stated that once he could no longer write, he wanted life support discontinued. Mr. K had an advance directive that designated his spouse as his health care proxy/decision maker. In Mr. K's advance directive he requested that his life not be prolonged if he had "an incurable and irreversible condition that would result in his death within a relatively short time." About two weeks after the social worker met with Mr. K and his spouse, the spouse contacted the social worker indicating the time had come that Mr. K could no longer write on the whiteboard; he wanted to be taken off life support according to his instructions in his advance directive, and he wanted to die at home surrounded by his family.

Learning Exercise

1. How should the social worker proceed with a micro policy intervention?
2. Does Mr. K have a right to die at home?
3. What are the legal requirements or programs available?
4. What are the goals and timeline to execute the actions?
5. Should an interdisciplinary team meeting, bioethics consult, and/or psychiatric evaluation be requested to establish a plan of care and patient rights, legality of the patient's wishes, and whether the patient still has the mental capacity to make such a decision?
6. How can the end-of-life options be best explored with the patient and family?
7. Assume that a physician insisted that all medical means be used to prolong this patient's life. How might a social worker initiate a macro policy intervention at the organizational level to decrease the likelihood of incidents like this?

a patient's life expectancy because it can be provided at the time of diagnosis and for years for individuals managing chronic diseases (Morrison & Meier, 2011). Both programs, however, provide a team of professionals that include physicians, nurses, social workers, and chaplains to work with patients and their families.

Palliative care is a board-certified specialty that is similar to hospice care. Patients are often contacted in hospitals where they have come for treatment of chronic and terminal conditions. They confer with palliative care physicians and other health professionals including social workers as they make choices about how to have dignified deaths under terms and conditions that they select. The Joint Commission accredits hospitals, including their palliative care and intensive care programs.

Core Problem 2: Engaging in Advocacy to Promote Quality Services and Programs for Seniors—With Some Red Flag Alerts

- **Red Flag Alert 8.9.** Seniors receive treatments that are less effective among them than among younger populations.
- **Red Flag Alert 8.10.** Competing or conflicting values among older adults, family members, and/or health care providers can result in violations of an older adult's right to autonomy and self-determination. Older adults who are cognitively competent should be empowered to make medical decisions even when others disagree with them.
- **Red Flag Alert 8.11.** Persons with disabilities are sometimes placed in institutions in violation of the *Olmstead* decision made by the U.S. Supreme Court in July 1999, which prohibits unnecessary institutionalization of individuals with disabilities.
- **Red Flag Alert 8.12.** Elderly persons often receive medical care that falls short of gold-standard care, including excessive numbers of medications or overly strong medications, fragmented care where physicians fail to communicate with one another, and excessive use of surgical remedies before less invasive strategies are attempted to see if they work.

Background

The Geriatric Social Work Initiative was funded through the Hartford Foundation to provide training and establish competencies for geriatric social workers in 1999 to meet the biopsychosocial needs of the growing older adult population. The need for geriatric social workers and physicians will continue to grow as the population of older persons expands. If there was only a ratio of one physician per 4,400 older adults 65+ in 2011, a ratio of one geriatrician for every 9,833 older adults will exist by 2050 if training rates remain the same (U.S. Census Bureau, 2011).

Seniors are less likely to receive evidence-based care when physicians treat them with no training in geriatrics. They have health conditions that are less common than in younger populations. They often respond differently than younger people to specific medications or require different doses. They often need assistance with daily activities whether they live at home or in institutional settings. They have to deal with life-threatening conditions more often than younger people.

Resources for Advocates

The Joint Commission's Long Term Care Accreditation program, created in 1966, is responsible for the accreditation of more than 15,000 skilled nursing facilities and 1,000 eligible long-term care organizations in the United States. It requires these facilities to undergo an on-site survey every three years to maintain accreditation (Joint Commission, 2011). The Joint Commission developed long-term care standards in the following performance areas: environment of care; emergency management; human resources; infection prevention and control; information management; leadership; life safety; medication management; national patient safety goals; record of care, treatment, and services; rights and responsibilities of the individual; and waived testing (Joint Commission, 2011).

Inpatient care at skilled nursing facilities, hospice services, and home health services are covered benefits for Medicare beneficiaries in need of long-term, end-of-life, and intermittent care. Medicare Part A covers medically necessary intermittent care for community dwelling beneficiaries with a stay-at-home disability, including home nursing; physical, occupational, and speech therapies; personal assistance from certified home health aides; and psychosocial services from social workers at no charge (CMS, 2011). Once patients have spent three days in an acute care hospital, Medicare covers skilled needs such as intravenous antibiotics or physical therapy with no co-pays for the first 20 days and coinsurance for days 21 to 100 when Medicare coverage ends (CMS, 2011).

The Long-Term Care Ombudsman Program, begun under the OAA in 1972, provides advocacy for the vulnerable older adult residents in skilled nursing facilities, board and cares, and assisted-living facilities (Administration on Aging, 2011b). Each state has a Long-Term Care Ombudsman Program that investigates and resolves complaints to ensure that residents' ethical rights are not being violated. The top five resident complaints in skilled nursing facilities in 2010 were

> unanswered requests for assistance such as when residents with functional dependence require toileting or assistance with transferring out of the bed to a chair or the toilet; inadequate or no discharge/eviction notice or planning; lack of respect for residents, poor staff attitudes; and resident conflict, including roommate to roommate when residents were not transferred to a separate room or not provided conflict mediation. (Administration on Aging, 2011b, para. 2)

Ombudsmen nationwide resolved 74% of all complaints at nursing homes and 39% of all complaints at assisted living and board and care facilities in 2010 (Administration on Aging, 2011b).

Congress passed the Nursing Home Reform Act in 1987 that required nursing homes to give residents the right to choose activities and schedules and to promote residents' quality of life. The Nursing Home Reform Act established a Residents' Bill of Rights, such as the rights to exercise self-determination; communicate freely; privacy; participate in care; and be free from abuse, neglect, and physical restraints (Klauber & Wright, 2001). States have also enacted a variety of bills to ensure the protection of residents' rights in residential care facilities such as the Nursing Home Resident's Rights in California in 2006 (California Advocates for Nursing Home Reform, 2011).

POLICY ADVOCACY LEARNING CHALLENGE 8.4

CONNECTING MICRO, MEZZO, AND MACRO POLICY INTERVENTIONS FOR QUALITY SERVICES FOR SENIORS

Mrs. C is an 89-year-old resident in a skilled nursing facility where she has resided for three years. Prior to entering the nursing home, she lived in her home with a private, 24-hour caregiver. Once she required the assistance of a Hoyer lift, she decided she could not afford the extra caregiver to assist with transfers, and she made the decision to relocate to enter a nursing home. She is alert and oriented, suffers from no psychological disorders, has the capacity to engage in all decision-making, and signs all of her own consents. She is obese and had her left leg amputated four years ago due to complications with diabetes. She suffers from severe diabetic neuropathy and cannot ambulate without an assistive device and cannot transfer without assistance. Since entering the nursing home, she now requires three attendants to assist with the use of a Hoyer lift to transfer her from her hospital bed to her wheelchair. You are the new social worker at this facility and are meeting with each resident to update psychosocial assessments and treatment plans. In your assessment interview Mrs. C states that she would like to eat her meals in the dining room but is often brought her meals to eat alone in her room. She also reports that she would like participate in the social activities such as Bingo and music but reports that when she asks the attendants, they say "I am too busy. I will come back." And they "never come back, so I sit in my room watching television by myself."

Learning Exercise

1. How should the social worker proceed with a micro policy intervention?
2. Does the long-term care ombudsman need to be notified?
3. What are the legal requirements for the nursing home?
4. What resident rights have been violated?
5. How can the resident's rights be best addressed?
6. What policies may need to be updated at the organizational level to reflect state and/or national policies ensuring protection of residents' rights?
7. Could state regulations governing nursing homes be modified to limit seniors' isolation?

Core Problem 3: Engaging in Advocacy to Promote Culturally Competent Services for Seniors—With Some Red Flag Alerts

- **Red Flag Alert 8.13.** An older adult's culture is not honored in interactions with health and/or other professionals.
- **Red Flag Alert 8.14.** An older adult with limited English proficiency (LEP) is not given appropriate translation services.

- **Red Flag Alert 8.15.** An older adult with low literacy fails to receive health information that he or she can understand.
- **Red Flag Alert 8.16.** Older adults from a specific cultural group are provided poorer services than other persons.
- **Red Flag Alert 8.17.** Specific cultural norms are not honored, such as inclusion of husbands and fathers in medical decisions with Latino patients and reluctance to disrobe among women among some Muslim patients.

Background

We have an increasingly diverse population of seniors, yet many of them fail to receive culturally competent care—an omission that could endanger their safety and health. The population of persons ages 65 and older will grow to 81 million in 2050 from 37 million in 2005.

The elderly population will be considerably more diverse. If white persons constituted 82% of seniors in 2005, they will constitute only 63% by 2050. Hispanic seniors will increase from 6% to 17% from 2005 to 2050. African Americans will increase from 8% to 12% during this time span—and Asian seniors will increase from 3% to 8% (Passel & Cohn, 2008).

Membership in a minority group is often a risk factor for specific diseases. African American males are 30% more likely die from heart disease, and American Indians are 1.2 times more likely to be diagnosed with heart disease than Caucasian elderly persons (Minority Health, HHS, 2011). African American females are 34% more likely to die from breast cancer than non-Hispanic whites. Non-white Hispanic females are twice as likely as white females to be diagnosed with cervical cancer. Asian & Pacific Islanders are 2 to 2.6 times more likely to die from stomach cancer than whites (U.S. Department of Health and Human Services, 2011).

According to the LGBT Older Adult Coalition (2011), there are approximately 2 million lesbian, gay, bisexual, transgender, and questioning (LGBTQ) people in the United States. Risk factors for LGBTQ older adults include limited caregiver assistance and/or families have excluded them because of their sexual orientation (nine out of 10 do not have a child who can help), fears of mistreatment by caregivers due to sexual orientation exacerbates isolation in times of need (seven out of 10 live alone), and a lifetime of workplace discrimination and partner benefit access are contributing factors of poverty. In comparison with heterosexuals, older LGBTQ persons are three times more likely to live in poverty. Policy advocates must be aware of the myriad social, economic, and societal factors that create barriers to accessing health care, housing, and social services for this population. In an LGBT Older Adult Summit, 75 LGBTQ advocate participants identified the following areas as the greatest needs for LGBTQ older adults: isolation (38%), health care affordability (34%), and LGBTQ culturally incompetent care (13%) (LGBT Older Adult Coalition, 2011). Advocates recommend improving access to culturally competent care for LGBTQ older adults through competency training for providers in residential, community, and health care settings as well as through education to individual care providers and families.

Resources for Advocates

The first investigation of health disparities that compared minorities with Non-Hispanic whites began in 1984 when the Secretary of Health and Human Services established the Task Force on Black and Minority Health, which examined mortality rates for six leading causes of death: stroke and cardiovascular disease, cancer, diabetes, infant mortality, chemical dependence, and accidents and homicides (National Institute of Minority Health and

POLICY ADVOCACY LEARNING CHALLENGE 8.5

CONNECTING MICRO, MEZZO, AND MACRO POLICY ADVOCACY FOR CULTURALLY SENSITIVE CARE FOR SENIORS

Mr. A is a 66-year-old male referred to home health services for crisis intervention and discharge planning by social work; physical therapy for a new DME (walker), strengthening and endurance, and gait training due to a fracture sustained to his hip when he fell; and nursing to monitor vitals and medication management and compliance. Mr. A was admitted via ambulance to the emergency room upon being found unconscious in a park. He flatlined in the ambulance on the way to the emergency room but was brought back with cardiopulmonary resuscitation (CPR) and chest compressions. He was admitted and transferred to the ICU where he was on life support for three days; he was successfully weaned from the ventilator and transferred to medical floor until stabilized. The hospital social worker discovered Mr. A had just been released from prison the day before he was brought to the emergency room. A Medicaid application was not completed at the hospital. Mr. A could not go to a shelter due to the change in his functional health. The hospital discharged Mr. A to a board and care where his basic needs were met for 30 days, fully paid through a charity program by the hospital. He had no income, housing, insurance, transportation, ID, clothing, or adequate social support. He was a janitor for 15 years on and off in between times he was not in prison. Mr. A reveals to the home health social worker that he missed his parole officer meeting and has a diagnosis of schizophrenia. He asks the social worker for help in changing his life circumstances: He tells his life story leading to his present state caught in a cycle of committing crimes, using drugs, going to prison, and returning to the same environment to repeat the cycle. One of the home health providers labels Mr. A as a "drug dealer" unlikely to change. The hospitalist refuses to provide pain medication beyond the discharge prescription, requesting that the patient find a primary care physician.

Learning Exercise

1. How might the social worker intervene as a case advocate?
2. What might the social worker's plan of care look like?
3. What programs, services, and benefits is the client eligible for and in need of advocacy for coordination and implementation?
4. What state and national policies relate to this case?
5. How might the social worker advocate for changes in policy at the organizational level?

Health Disparities, n.d.). The task force discovered that 80% of deaths among minorities were due to these six causes of death (National Institute of Minority Health and Health Disparities, n.d.). From the 1990s on, a multitude of initiatives sponsored by both the National Institutes of Health and National Institute on Aging have followed. In 1999, at the 25th anniversary of the National Institute on Aging, the director requested an evaluation of its past and current research efforts on minority aging and developed new initiatives to study minority aging research, training, and outreach activities (National Institute on Aging, 2010). The U.S. Department of Health and Human Services has created a special link under the Health Resources Services Administration website (http://www.hrsa.gov/culturalcompetence/index.html), which provides resources and training materials for culture, language, and health literacy for organizations.

The ACA promotes cultural competence training for health care providers and improved collection and analysis of data on health disparities—and on disability status, primary language, race, sex, and ethnicity of persons who use health care (U.S. Department of Health & Human Services, 2011). Many states have passed their own laws and regulations to require and promote culturally competent health care and mental health care.

Core Problem 4: Engaging in Advocacy to Promote Preventive Services and Programs for Seniors—With Some Red Flag Alerts

- **Red Flag Alert 8.18.** An older adult has not been helped to identify personal at-risk factors through family history and diagnostic tests.
- **Red Flag Alert 8.19.** An older adult has not been tested for a possible chronic disease or diseases at a possible early stage.
- **Red Flag Alert 8.20.** An older adult with a relatively sedentary lifestyle has not been given help in increasing exercise or is not given modifications in exercise programs based on functional health and chronic diseases.
- **Red Flag Alert 8.21.** An older adult has not been given help in improving his or her nutrition.
- **Red Flag Alert 8.22.** An older adult has not been given help in not smoking.
- **Red Flag Alert 8.23.** An older adult has not been given help in receiving vaccines.
- **Red Flag Alert 8.24.** A person lives in an inner-city area and is not given assistance in obtaining better nutrition and more exercise.
- **Red Flag Alert 8.25.** An older adult is not given access to prevention.

Background

Health promotion is particularly important to older adults as health behaviors, both to modify negative factors such as obesity, alcohol abuse, and smoking and to promote factors such as avoiding tobacco, engaging in exercise, and drinking alcohol in moderation (Aldwin & Gilmer, 2004). Prevention activities such as immunizations, well checkups, screenings, and good nutrition are extremely important (Aldwin & Gilmer, 2004).

A myriad of barriers contribute to the overlooked need for health prevention among older adults. These include the high prevalence of chronic disease and functional limitations of older adults, use of preventive interventions that are designed for middle or younger adults, lack of recognition and financing of prevention initiatives by Medicare, and overreliance on medications rather than emphasis on proper nutrition and exercise (Richardson, 2006). Many professionals wrongly believe that prevention programs primarily help younger persons rather than older persons (Aldwin & Gilmer, 2004).

Development of preventive programs for seniors markedly increased after 2000 (Albert, 2004, p. 16). Prevention activities for older adults now include primary prevention such as vaccinations for flu and shingles, medication therapies such as prescription of medications to decrease cholesterol and blood pressure, counseling, and prosthetic devices. They include secondary prevention to detect diseases in their early states, such as screenings for cognitive deficits, diabetes, hypertension, and osteoporosis. They include tertiary prevention such as management of advanced diabetes and chronic congestive pulmonary disease. Primary care or geriatrician physicians can coordinate care by using telemedicine with which they communicate with patients electronically (Albert, 2004).

Recommended health promotion and disease prevention activities in nursing homes include:

> immunization against common pathogens; development of infection control policies, minimizing adverse drug events, screening to measures for disease at an early stage (eg, depression, tuberculosis); chemoprophylaxis for common illnesses such as osteoporosis; counseling for patients and their families; and management of pressure ulcer and fall risk. (McElhone, Limb, & Gambert, 2005, pp. 24–31)

Many residents are still not provided opportunities for exercise and improved nutrition (Kayser-Jones, 2009). Prevention among geriatric persons is slowed because "many geriatric disorders have multifactorial risk factors, interventions, and expected outcomes; older adults are often not represented in clinical trials; and important outcomes may not be measured and reported in ways that are conducive to evidence synthesis and interpretation" (Leipzig et al., 2010).

Self-rated health is a greater predictor of health and mortality among older adults than physician ratings, even those with chronic illnesses. Most seniors rate their health as excellent, very good, or good (Aldwin & Gilmer, 2004). Functional health is also impacted by mental health as low self-esteem and depression can lead older adults to neglect their health through poor diet, lack of exercise, and isolation from a support network (Aldwin & Gilmer, 2004). Older adults can participate in disease prevention and health promotion activities through goal setting, progress monitoring, and rewards for behaviors or steps (Leventhal, Rabin, Leventhal, & Burns, 2001). Health promotion activities must emphasize sociocultural aspects of change, including the older adult's community, age, culture, and social factors as well as their prior life events (Aldwin & Gilmer, 2004).

Resources for Advocates

The Evidence-Based Prevention Program was initiated by the AoA in 2003 to ensure that older adults have access to evidence-based interventions to reduce their risk for disability, disease, and injury (Administration on Aging, 2010). As part of the Aging Network, the

Department of Health and Human Services collaborates with the AoA to ensure that effective programs are implemented in older adults' community settings. The Aging Network also includes collaboration with 30+ private foundations, the Agency for Healthcare Research & Quality, the CDC, the CMS, the Substance Abuse & Mental Health Services Administration, and the Health Resources & Services Administration (Administration on Aging, 2010). Evidence-based health programs include physical activity, fall prevention, smoking cessation, medication management, diabetes, chronic disease management, and nutrition programs (Administration on Aging, 2010). AoA has funded evidence-based programs in many states, including Enhance Fitness, Matter of Balance, Healthy IDEAS, and Stanford University Chronic Disease Management Program. These and other programs are provided to older adults in settings such as senior housing, faith-based organizations, senior centers, and nutrition programs (Administration on Aging, 2010).

POLICY ADVOCACY LEARNING CHALLENGE 8.6

CONNECTING MICRO, MEZZO, AND MACRO POLICY ADVOCACY FOR PREVENTION FOR SENIORS

Mrs. G is a 61-year-old LEP female patient hospitalized after fainting. She has high blood pressure, a family history of type 2 diabetes and a history of high cholesterol, is overweight, and does not exercise; her educational level is sixth grade. You are asked by the hospitalist (an English-speaking-only physician) to identify a clinic that will accept Mrs. G's new preferred provider organization (PPO) insurance she received after September 23, 2010 (the date on which the ACA required new insurance plans to provide free preventive services for all new plans) for primary care follow-up of her hypertension. Mrs. G is instructed by the hospitalist to start exercising and modify her diet to a low sodium diet (1,200 mg or less). You are brought in right at the time she is being discharged. You ask whether she has been screened for diabetes, and she replies that she was not. She cannot recall her cholesterol levels from a test when she was 40. She indicates she does not know what type of exercise or diet to follow.

Learning Exercise

1. If you worked in this hospital and wanted to be an advocate for diabetes screening for this patient before discharge, what conflicts might arise, and how might you best mitigate them?
2. What procedures or protocols at the organization level should be in place to prevent a patient with risk factors for diabetes not receiving screening tests before discharge?
3. How might you as the social worker best coordinate the patient's linkage to an outpatient physician and/or other services necessary to ensure the patient has access to a physician in her community?
4. What policies or programs exist in the community you work in for older adults with diabetes or hypertension? How might you link an older LEP patient to such services?
5. How would you advocate for policy change at the organizational level?

The ACA has several initiatives that benefit older adults and make prevention services more available. It funds improved access to primary care, making prevention activities such as well checks and screenings more available to at-risk and underserved populations (U.S. Department of Health & Human Services, 2011). It funds programs for preventive care for persons with disabilities including ones to prevent chronic disease (U.S. Department of Health & Human Services, 2011). Yet other programs funded by the ACA give seniors greater access to primary prevention health services through physical exams and free preventive care, as do Medicare B deductibles or co-payments for screenings such as colorectal, cervical cancer, cholesterol, cardiovascular, mammogram, and prostate.

The National Prevention Strategy released in June 2011 by the National Prevention, Health Promotion and Public Health Council establishes 10-year targets to improve screening of adults ages 50 to 75 for colorectal cancer from 54.2% to 70.5% and to increase the percentage of adults 65+ vaccinated annually against influenza from 67.0% to 90.0% (Office of the Surgeon General, 2011). The National Prevention Strategy seeks to promote injury and violence-free living through home modifications such as improved lighting, grab bars, and railings; strength and balance exercise programs geared specifically to older adults; monitoring of polypharmacy to reduce side effects that contribute to falls; and improving access and linkage between prevention programs that are based in health and community-based settings (Office of the Surgeon General, 2011). The National Prevention Strategy also prioritizes increased support for those older adults who prefer to age in place and remain in their community as a means of promoting their emotional and mental health (Office of the Surgeon General, 2011).

Core Problem 5: Engaging in Advocacy to Promote Affordable and Accessible Services for Seniors—With Some Red Flag Alerts

- **Red Flag Alert 8.26.** Older adults are not informed of financing options for home and community-based services and programs to allow them to remain in their community and out of an institution.
- **Red Flag Alert 8.27.** An older adult is incorrectly informed about his or her eligibility for long-term Medicaid and receives services through In Home Supportive Services, not knowing about the estate recovery aspect.
- **Red Flag Alert 8.28.** A person is not linked to programs and services in his or her community that can help subsidize services he or she cannot afford, such as housekeeping and errands.
- **Red Flag Alert 8.29.** An older adult is unaware of the costs associated with programs and services as their functional health changes, and he or she requires more assistance in the future to fund these costs.
- **Red Flag Alert 8.30.** An older adult needs insurance counseling about how to finance out-of-pocket costs related to services he or she may use, such as deductibles and coinsurance.

Background

To be effective advocates, social workers should be aware of the variety of funding options, subsidized services through programs in communities, and experts who can assist the older adults with financial planning. Access to programs hinges not only on financial considerations, but on seniors' "functional level" as measured by ADLs and IADLs (Senior Planning Services of Santa Barbara at http://www.seniorplanningservices.com/files/2013/12/Santa-Barbara-ADL-IADL-Checklist.pdf):

- Activities of daily living (ADLs) are activities in which people engage on a day-to-day basis to manage their personal care. These include bathing, dressing, grooming, mouth care, toileting, transferring from bed to chair, walking, climbing stairs, and eating. Each of these ADLs is measured on a scale that includes the following levels of competence of each ADL: "independent," "needs help," "dependent," and "cannot do." Social workers need this information to decide what kinds of help specific persons need to care for themselves and to maintain independence.
- Instrumental activities of daily living (IADLs) are activities related to independent living. They are useful for evaluating persons with early-stage disease "both to access the level of disease and to determine the person's ability to care for himself or herself." These include shopping, cooking, managing medications, using the phone and looking up numbers, doing housework, doing laundry, driving or using public transportation, and managing finances. Each of these activities is measured on a scale that includes "independent," "needs help," "dependent," and "cannot do."

Advocates must be aware of the multiple options available for funding for services to allow older adults to remain in their communities and out of institutional facilities. Low-income older adults are at a higher risk for premature institutionalization when they have utilized all public programs available to finance their care and still require additional personal assistance with ADLs and IADLs to remain safely in their homes. Both ADLs and IADLs identify specific middle-income older adults with limited income and savings who own a home and are at risk for having unmet needs when their limited incomes cover only their monthly expenses and they are overqualified for public means-tested programs and services.

Resources for Advocates

Medicare does not finance services for older adults to help them stay safely in their homes when changes in their functional health undermine their ability to perform tasks of daily living in their homes. Medicare will finance Home Health Services for Medicare beneficiaries with a stay-at-home disability that prevents them from leaving their homes without the assistance of other persons, provided that they have a need for skilled nursing or rehabilitative services such as physical or occupational therapy. Under these circumstances, a certified home health aide for personal care assistance can be provided under Home Health Services for an intermittent period. However, for those who do not regain their prior level of functioning and still require assistance, few options exist to help them finance services for assistance with tasks of daily living. Some older persons are eligible for means-tested personal assistance services through Medicaid. Such persons may also qualify for In-Home Supportive Services offered by Medicaid in some areas as well as Multipurpose Senior Services and Integrated Care Management Programs and Adult Day Care, but these programs have been severely cut in many areas and may not exist in others so social worker advocates have to ascertain what programs exist in their areas and states for low-income seniors who need personal services. Many other middle-income older adults, such as those with Social Security whose income is too high to qualify for Medicaid, must finance these personal assistance services themselves or with help from family members. Unfortunately, many of these persons cannot afford the expenses associated with privately financed services such as caregiver agencies, which charge between $12 and $20 per hour (or more) in major urban areas, not to mention the cost of transportation services and home-delivered meals.

The OAA funds many services for older persons (age 60+) through the Administration on Aging. They include the following:

- Title III allocations to states for community programs on aging including supportive services, congregate meals, home meals, preventive services, and the National Family Caregiver Support Program (NFCSP). The budget allocated to these services increased from $1.1 billion in 2006 to $1.2 billion in 2011 (Administration on Aging, 2011a).
- Title III-NSIP funds the Nutrition Services Incentive Program. The budget allocated to these services increased from $143.5 million in 2006 to $154.2 million in 2011 (Administration on Aging, 2011a).
- Title VII funds the Vulnerable Elder Rights Protection Activities. The budget allocated to these services increased from $19.9 million in 2006 to $ 21.8 million in 2011 (Administration on Aging, 2011a).
- Title VI provides the Tribal Organization Allocation for Native Americans, including the Nutrition Services Incentive Program. The budget allocated to these services increased from $33.7 million in 2006 to $ 37.1 million in 2011 (Administration on Aging, 2011a).
- Higher proportions of the budget are allocated to states with large populations of older adults such as Florida, California, Massachusetts, Pennsylvania, and New York because "money is divided up on the basis of the number of people in the state who are over the age of 60" (Day, n.d.).

Medicare provides funding for 100 skilled nursing days per year, covers intermittent home health services, and covers hospice services for those with a terminal diagnosis of six months or less to live. With trends among several presidential administrations to control costs of aging programs over the past several decades, long-term care benefits provided under Medicare have faced cuts. In particular, the Medicare Home Health benefit implemented in 1975 under Title XX of the Social Security Act over the last 20 years has been dramatically cut to reduce costs as at "one point home care was the fastest growing segment of the Medicare budget, reaching almost 10 percent of the Medicare budget and

POLICY ADVOCACY LEARNING CHALLENGE 8.7

FINDING THE LOCATIONS OF STATE PROGRAMS FUNDED BY THE FEDERAL GOVERNMENT

Imagine that you have taken a position working with seniors in your state. You want to know where federally funded programs exist in your state. See Figure 8.1, the Aging Integrated Database.

To find AoA programs in your state, go to the Aging Integrated Database (AGID) to the online query system https://agid.acl.gov/StateProfiles/ that presents you with Figure 8.1. Click on your state. Find "state profiles" once you click on your state that identifies population. Now click successively on "Long Term Care Ombudsman Program," "Title VI Grants in State," "Title III Clients," "Focal Points/Senior Centers," "Service Profile," "State Unit Staffing," and "AAA Staffing (Total State)," as well as other items. Do note that you can also click on "AAA Staffing," "Expenditures," and "Providers." Are you surprised that so many programs for seniors exist in your state? Were you working in an agency that serves seniors, might you use Figure 8.1 to contact experts on senior programs to consider starting a new federally funded program in your geographic district?

FIGURE 8.1 ■ Aging Integrated Database

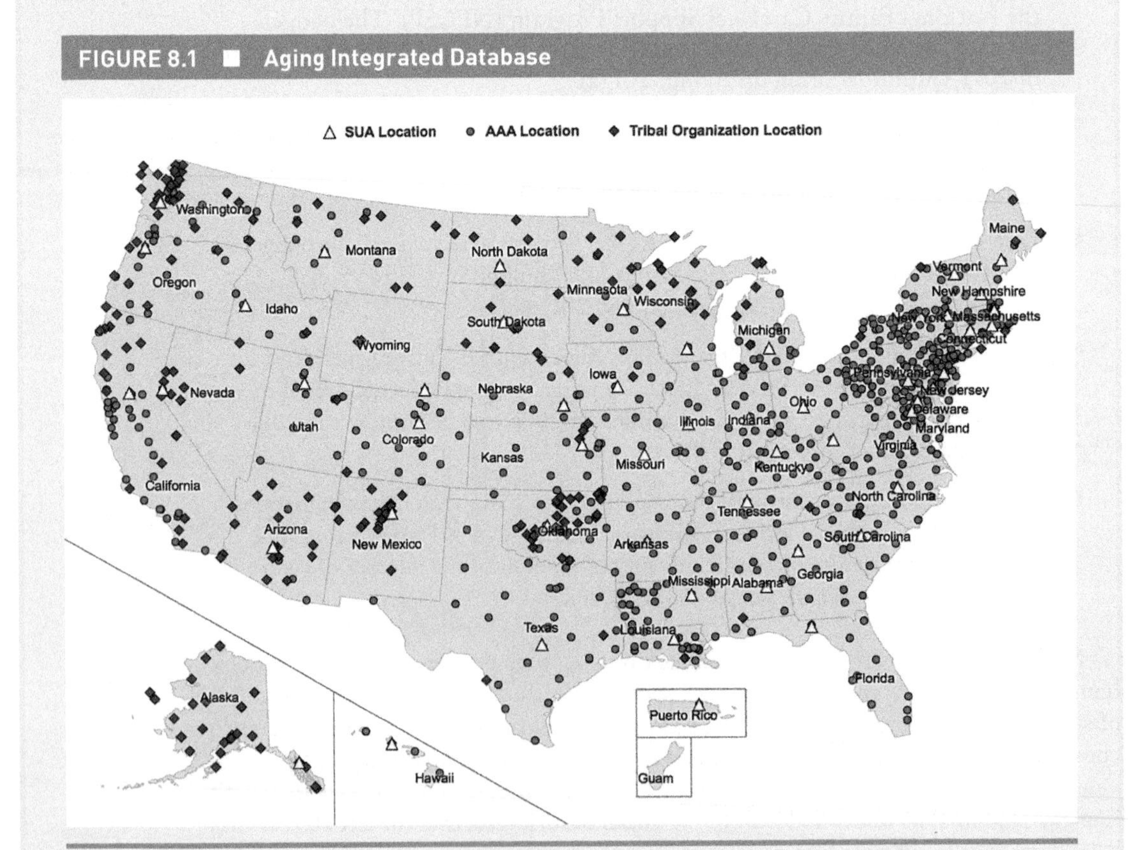

Source: US Administration on Aging.

increasing four-fold in 10 years" (Ferrini & Ferrini, 2008, p. 457). Title XX under the Social Security Act ensured federal matching with states for some social services (Ferrini & Ferrini, 2008). Medicaid provides a long-term care benefit that subsidizes custodial care at a skilled nursing facility for older and disabled adults. In an attempt to recover money states spend on long-term Medicaid programs, in 1993, the Federal Budget Reconciliation Act permitted states to begin claiming assets of long-term Medicaid beneficiaries through Estate Recovery Programs, provided that the recipient and the recipient's spouse are both deceased and/or if there is not a surviving child or spouse that is either 18 or under, blind, or disabled (Ferrini & Ferrini, 2008).

Long-term care insurance has also been an option to those who have the resources to purchase it. These plans cover all or part of the cost of personal care services and skilled nursing-home costs as well as some other costs. Seniors must, however, begin paying premiums years before they use resources to make monthly premiums affordable, such as in

POLICY ADVOCACY LEARNING CHALLENGE 8.8

CONNECTING MICRO, MEZZO, AND MACRO POLICY ADVOCACY FOR AFFORDABLE SERVICES FOR SENIORS

An affluent couple in a rural area ran into financial problems when they encountered health problems that meant they could not perform many daily activities. The female member of this couple (Mary) became, in effect, the head of household because her husband (John) was severely handicapped by a stroke. They found a woman who had cared for other persons who became, in effect, the head of their home health team (Joan). Although a caring person, she had only a high school diploma and was often remiss in keeping the house clean or helping Mary and John clean themselves because they could not ascend steps to the second floor to the bathroom with a shower. This woman recruited other younger women who also had not progressed beyond high school. She frequently had to recruit new persons as others left this job due to the treatment they received from Mary, who viewed them as intruders into her house. Joan screened applicants by placing a 10-dollar bill on the floor near John's bed. If they picked it up and did not give it to Joan, she did not hire them. Little did they know that it was a fake 10-dollar bill that was inscribed "Take Jesus Into Your Everyday Life" on the side facing the floor. Almost bankrupted by the cost of this home health team, Mary had to endure the visit of a realtor to her home, organized by her children, to appraise its value in the event that she exhausted her savings and could not obtain Medicaid financed home health services.

Learning Exercise

1. Millions of additional seniors will need home health teams in coming years. What does this situation tell us about relative preparedness of the United States for this situation?
2. If a relatively affluent couple encountered this situation, what additional hardships would many less affluent couples encounter?
3. What macro policy initiatives might social workers consider in the state's capitol to shape programs offered by the state's Medicaid program?

their 60s. Moreover, many of these plans raise the level of premiums in the intervening years so that their cost becomes exorbitant—and some of the companies go out of existence before seniors can claim benefits. Considerable care should be given to researching specific plans before committing to one of them.

The ACA would have allowed persons to contribute to a long-term care insurance benefit, but this feature was deleted due to its likely high costs.

Many seniors need financial planning so that they can preserve their estates in the event that they run into situations like Mary and John. Advocates should be aware of where such services are in the communities of consumers they work with to ensure they have access to this specialized financial planning in advance. They need to be aware that Medicaid authorities will "go after" their resources and estates if older persons try to transfer their resources to their children shortly before they need help from Medicaid for home health costs in some states.

Advocates who seek increases in funding of programs for seniors sometimes encounter opposition from persons who fear that seniors' programs divert funds from children and youth as well as other programs. Social Security expenditures currently represent one-fifth of the federal budget at a cost of $733 billion, with 82% allocated specifically to Old Age Survivors Insurance (OASI) and 18% to Disability Insurance (DI) (Congressional Budget Office, 2011). However, many of these costs are offset by payroll taxes of employers and employees.

Core Problem 6: Engaging in Advocacy for Services to Help Seniors With Mental Distress—With Some Red Flag Alerts

- **Red Flag Alert 8.31.** An older adult fails to receive needed clinical interventions to address diagnosed mental health disorders (counseling and medication, medication, and counseling).
- **Red Flag Alert 8.32.** An older adult's physician discounts his or her depression, calling it normal, due to contextual changes associated with aging.
- **Red Flag Alert 8.33.** An older adult with chronic diseases presents with depression, anxiety, or other mental disorders but is not referred to a geriatric specialist for care.

Background

According to the National Institute of Mental Health more than 7.5 million older adults 55+, or one in five, have a diagnosable mental health illness each year—a number projected to increase to 15 million older adults by 2030 (Ferrini & Ferrini, 2008). Depression, anxiety, sleep disorders, substance use, dementia, and delirium are the most common mental disorders among older adults (Ferrini & Ferrini, 2008). For community-dwelling older adults rates of depression or symptoms of depression range between 8% and 20% of the population (Ferrini & Ferrini, 2008). Research indicates that older adults whose depression is treated have improvements in their medical conditions and respond well to traditional best treatment practices that use either medication, medication and psychotherapy, or psychotherapy alone (National Institute of Mental Health, 2011).

Estimates of the prevalence of anxiety disorders among older adults in the United States range between 3.2% and 14.2%. Diabetes, hyperthyroidism, and gastrointestinal medical conditions are often associated with anxiety among older adults (Wolitzky-Taylor et al., 2010). Anxiety disorders among some older adults present serious risks for their health and well-being. Anxiety increases the risk for mortality among older adults post heart surgery, and cardiovascular morbidity along with mortality have been linked to panic attacks among older adults (Wolitzky-Taylor et al., 2010). Comorbidity prevalence estimates of chronic obstructive pulmonary disease (COPD) and anxiety among older adults range from 18% to 50% and 40% to 43% for Parkinson's disease and anxiety (Wolitzky-Taylor et al., 2010). Five percent to 21% of older adults with dementia have an anxiety disorder—and cognitive decline and anxiety disorders occur frequently (Wolitzky-Taylor et al., 2010). Anxiety has been identified as a likely risk factor for developing dementia for older adults with mild cognitive impairment (MCI) as 83% of older persons with anxiety went on to develop Alzheimer's disease three years later in comparison to only 40.9% of individuals with MCI only and 6.1% of cognitively intact individuals. Among individuals with both MCI and anxiety, the relative risk of developing Alzheimer's almost doubled with each anxiety symptom, from 1.8 to 2.7 per symptom (Wolitzky-Taylor et al., 2010).

Undetected mental disorders among older adults can have life-threatening consequences. Older adults have higher rates of suicide than any other age-group, and older adult males have the highest suicide rates. In 2004, older adults 65+ accounted for 16% of deaths by suicide, and non-Hispanic white males had the highest suicide rates of all age and ethnic groups, with 49.8 deaths per 100,000 for those ages 85+ (National Institute of Mental Health, 2010). Even more startling is the fact that up to 75% of older adults who commit suicide visited their primary care physician shortly before their suicide (National Institute of Mental Health, 2010). Mental health professionals and social workers are needed to detect mental disorders among older adults in primary care settings.

Older adults who require home health care have depression rates estimated at 13.5% (National Institute for Mental Health, 2010). For older adults who cannot leave their homes without assistance, they may actively grieve the loss of their previous independence, engagement in work and community, lifestyle, youth, and leisure activities. Depression risk increases with functional dependence and poor health. Prevalence rates of hospitalized older adults are 11.5% (National Institute for Mental Health, 2010). Definitions of what constitutes normal mental health may not provide the best fit for older adults with functional dependence because the general definition of mental health is "the ability to engage in productive activities and fulfilling relationships and to cope successfully with change and adversity" (Ferrini & Ferrini, 2008). Among the difficult tasks for older persons are accepting the multiple changes they experience and readjustment of what it means to them in their current life stage to engage in productive activities and fulfilling relationships, recognizing how it may be different or similar from prior periods. Models of successful aging may not provide the best fit for older adults with unmet mental health needs and chronic diseases causing functional dependence. For example, three conditions characterize successful aging, according to Rowe and Kahn (1998): "(1) positive self-attitude, (2) high cognitive and physical functional capability, and (3) active engagement in life, including maintaining personal relationships and sustaining productive activity" (in Ferrini & Ferrini, 2008, p. 193). Depression can be a normal reaction to the multiple losses older persons encounter, making it difficult to detect mental disorders among older adults (National Institute for Mental Health, 2010).

Older adults must adjust to myriad changes, including changes in their functional health due to chronic diseases and subsequent needs for assistance (physical losses), changes in their cognition or facing their mortality (psychological losses), shrinking social support networks, surviving a spouse or child (interpersonal losses), relocating to a residential care facility or home of a family member, loss of ability to drive (social losses), and retirement and living on a fixed income (economic losses).

Barriers exist within the medical profession for differential diagnoses of mental disorders among the older adult population. Differential diagnosis and detection of mental health disorders are complicated by the comorbidity of medical conditions among older adults. Their primary care physicians most frequently treat older adults for mental disorders, and differentiating between medical conditions and mental disorders can be difficult (Ferrini & Ferrini, 2008). Social workers should advocate for geriatric assessment and/or geropsychiatric assessment to ensure proper diagnosis of underlying mental health disorders among older adults. The scarcity of medical and mental health professionals specializing in geriatric care, assessment, and psychiatry impedes detection of mental disorders among older adults.

Resources for Advocates

Lack of parity in coverage among insurance companies for treatment of mental disorders in comparison with physical conditions has historically existed and has created barriers to mental health services for older adults. The Mental Health Parity Act of 2007 has eased this problem.

Changes to mental health services coverage under Medicare have been made with the implementation of the ACA. Medicare Part A covers inpatient hospital care for a mental health disorder; Part B covers outpatient mental health services such as visits with a psychiatrist or physician, clinical psychologist, social worker, nurse specialist, nurse practitioner, or physician's assistant, and laboratory tests ordered by the physician; and Part D covers medications prescribed to treat mental disorders (Centers for Medicare and Medicaid Services, 2009). Although certain deductibles and coinsurance may apply, Part B covers the following outpatient mental health services: group and individual psychotherapy, family counseling, testing, evaluation by a psychiatrist, management of medications, occupational therapy, education and training, partial hospitalization if a physician certifies that the patient would otherwise require inpatient treatment, diagnostic tests, and depression screening during the one-time physical exam under "Welcome to Medicare" benefit (for beneficiaries within the first 12 months of Part B enrollment) (Centers for Medicare and Medicaid Services, 2009). Parity was further addressed under the ACA, ensuring that Medicare beneficiaries' coinsurances for mental health services must be similar to coinsurances for medical care services, with the coinsurance rates for outpatient treatment falling to 20% in 2014 (Centers for Medicare and Medicaid Services, 2009). Medicare beneficiaries pay 20% of the accepted Medicare rate for diagnosis of their mental health conditions (Centers for Medicare and Medicaid Services, 2009). Medicare Part A covers 190 lifetime days of inpatient psychiatric care at a psychiatric hospital rather than a general acute care hospital (Centers for Medicare and Medicaid Services, 2009). At a general hospital a deductible of $1,100 applies for benefit periods of one to 60 days, $275 coinsurance for days 61 to 90 of the benefit period, and $550 coinsurance per day after the 90th day of the benefit period (Centers for Medicare and Medicaid Services, 2009).

Medicaid is the largest funder of mental health services in the United States, paying $26 billion and representing 26% of all expenditures in the United States for mental health in 2003 (Substance Abuse and Mental Health Services Administration, 2007). Medicare made up 7% of all mental health expenditures in the United States in 2003 (Substance Abuse and Mental Health Services Administration, 2007).

Under the ACA, access to mental health services will be implemented through two key provisions: integration of mental health services across the continuum of care within the health care system and the provision of annual well visits, which provide risk assessments for cognition and health along with a plan for health promotion tailored to the individual (American Association for Geriatric Psychiatry, 2011).

POLICY ADVOCACY LEARNING CHALLENGE 8.9

CONNECTING MICRO, MEZZO, AND MACRO POLICY ADVOCACY FOR SENIORS' MENTAL HEALTH

Anna Gorman contends in a *Los Angeles Times* article from September 5, 2011, that "ERs are becoming costly destinations for mentally disturbed patients: Budget cuts are creating added safety risks at hospitals and placing a burden on already crowded emergency rooms." She notes that many hospitals discharge patients with unaddressed mental health disorders into local communities who are either a danger to themselves or to others. (By contrast, hospitals in affluent areas often have available psychiatric services.) The outcome is that many mentally ill persons are released homeless into communities without their mental health needs addressed, posing a danger to themselves and the local community.

Imagine you are a social worker who works in the emergency room of the local hospital, and there is not a staff psychiatrist. Mr. Smith is a 70-year-old homeless male who is brought to the emergency room by a police officer after he threatened to harm a street vendor in the local community. The police officer reported that he displayed "bizarre behavior" and threatened to harm the police officer as well. The patient presents in the emergency room with auditory hallucinations, is dishevelled in appearance, and has poor hygiene. The patient is able to report his name, the date, the city he is in (oriented times three). The attending physician requests a social work consultation. You identify a county hospital that can take this patient because the patient cannot be admitted to the hospital. The patient has an identification card and no insurance. You find out that the patient has been homeless for more than 15 years. The patient states he does not want to go to a shelter and denies that he needs help with his psychiatric symptoms.

Learning Exercise

1. How might this social worker intervene to provide case advocacy to address the mental health needs of this patient?
2. What policies and/or procedures/protocols exist in the agency you are in for intervening with homeless older adults with unaddressed mental health needs?
3. What programs and services (inpatient vs. outpatient) are available in the community you work in that can take uninsured older adults?
4. What local, federal, or state policies can help inform policy changes at the organization to improve services to older adults with unaddressed mental health needs?

Core Problem 7: Engaging in Advocacy for Home- and Community-Based Services (HCBS) for Seniors—With Some Red Flag Alerts

- Red Flag Alert 8.34. Seniors lack a range of choices in specific communities other than nursing homes, including the community-based village program or small group homes or more programs that allow seniors to remain in their own places of residence.
- Red Flag Alert 8.35. Seniors may not receive adequate information to make an informed decision about the options for long-term care available across the continuum of independent living and residential care.
- Red Flag Alert 8.36. A senior with changes in functional health or a new chronic disease is not provided with referrals to home- and community-based services at the time of discharge from an acute care setting.
- Red Flag Alert 8.37. An older adult who inquires about HCBS at a primary care setting is often unable to obtain information about relevant services due to a lack of social services at primary care settings.
- Red Flag Alert 8.38. Although social isolation exists across generations, it is particularly important among seniors, where one in three persons 65+ live alone—and 50% of seniors 85+ live alone.
- Red Flag Alert 8.39. Older adults are unaware of the range of HCBS that may be relevant to them over the remaining period of their life course as their functional and residential status changes.
- Red Flag Alert 8.40. An older adult searches the yellow pages but cannot find relevant HCBS such as hot delivered meals or transportation with disabled access.
- Red Flag Alert 8.41. Hospitals or health care settings that serve older adults do not allow seniors sufficient time, prior to discharge, to receive information from a social worker so that they can make informed decisions.
- Red Flag Alert 8.42. Residents of nursing homes often become depressed due to lack of activities or participation in decision-making in the institution.

Background

Most older adults live in traditional communities with only slightly over 5% of the 65+ population in nursing homes, congregate care, assisted living, and board-and-care homes. In other words, most seniors live independently, yet insufficient attention has focused on hardships of live-alone seniors. Nor has the nation constructed and funded a network of alternatives to nursing homes.

Realize that many seniors reach age 65 not owning their homes but renting. Those seniors who do not have private pensions to supplement Social Security have to find housing at rental levels they can pay without approaching or entering poverty. Growing numbers of seniors have joined the ranks of homeless people for brief or long periods because they cannot afford rents, do not have housing in public senior housing buildings, and cannot find units subsidized by Medicaid or SSI.

So a mad search takes place for places to live. Take the example of Illinois, where a range of living options exist for seniors who "are at risk of living alone but do not require nursing care" (Nursing Home Law Center Staff, 2009). They include shared housing establishments (SHEs) with 16 or fewer residents 55+, assisted-living facilities (ALFs) with residence of at least eight unrelated adults mostly 55+, supportive-living facilities (SLFs) that combine apartment-style housing with personal care and other services, and residential care facilities for the elderly (RCFs) that provide nonmedical care to seniors in a group living arrangement along with meals, laundry, and housekeeping. These small group homes offer alternatives to nursing homes for persons who do not require nursing care but do not want to live alone. Small group homes are not panaceas. They won't accept applicants who need extensive help with ADLs and IADLs. Mistreatment of residents by staff can take place. Yet congregate care in smaller units, as compared with large nursing homes, offers many benefits. If nursing homes are like hospitals, smaller units can give residents less anonymity and less isolation. Social workers need to identify the mix and kinds of congregate living situations that exist in their states, their cost, and the extent to which Medicare, Medicaid, and SSI fund them so that they can refer seniors to them and engage in macro policy advocacy to change the mix and kinds of congregate living situations.

At the higher end, affluent persons can rent or purchase units in not-for-profit congregate living organizations. They have to pay a steep fee just to enter, such as $150,000 and up, as well as monthly rents of $3,000 and up. Some have independent living units, semi-independent units, and infirmary units. Some of them allow residents to purchase their units and to place them in their estates.

Many seniors cannot afford these options, often because they cannot afford alternative housing. They often lose a spouse who moves to a nursing home or who dies. They often do not have relatives in close proximity because Americans are highly mobile. They sometimes have to migrate from rental or housing options in relatively expensive and desirable areas to areas of relative poverty. They may have difficulty finding groceries and other amenities in these areas.

If they cannot meet ADLs and IADLs and lack resources, they may have to migrate to a nursing home. Here, too, a wide range exists. Some charge rents above the level that Medicaid accepts so many seniors lack resources to select them. Other nursing homes accept Medicaid and SSI reimbursements.

Life in nursing homes is often not pleasant. Most of the residents are elderly women. Some are bedridden, whereas others are mobile. Some have few planned activities and few support staff, whereas others are more resident friendly.

Social workers who work with seniors need to help seniors navigate the range of congregate housing and nursing home options with an eye to higher quality ones. They can help seniors find ones that have not been cited for poor conditions or service or that have been subjected to lawsuits by finding appropriate sites on the Internet. They need to know the managers or owners of housing and nursing home options to gauge whether they place residents' needs above profits and whether they implement resident-friendly activities.

Social workers need to be adept at micro policy advocacy in a complex system of services and housing and medical care for seniors. Many seniors living in traditional community settings require assistance with IADLs, ADLs or a combination of both ADLs and IADLs. Older adults with ADL and IADL limitations may need a variety of HCBS to compensate for functional losses so that they can remain safely in their homes.

Eligibility criteria for HCBS create barriers too for older adults attempting to access services. Advocates must be informed of the eligibility criteria for programs and services in older adults' communities. Age and functional dependence are traditional criteria used to determine eligibility for long-term care assistance. For example, Meals on Wheels programs typically target adults age 60+ with physical limitations that affect their ability to shop or prepare meals on a daily basis. Specialty transportation services base eligibility on age and functional dependence. However, criteria related to functional dependence are much more stringent. For example, an older adult who uses a motorized scooter or wheelchair is ineligible for door-to-door wheelchair accessible transportation. Such a client is considered capable of accessing public transportation services with wheelchair-accessible ramps (Joosten, 2008).

Eligibility criteria for services should be modified to reflect the diverse needs of the aging population. Yet adding more criteria is a simplistic solution that can inadvertently create barriers to service access. For example, Meals on Wheels programs frequently do not serve disabled adults under age 60, even though they meet functional dependence criteria. An additional barrier pertains to living arrangements. Meals on Wheels programs with long waiting lists often exclude older adults, for example, who live with a family member, regardless of whether the family member works full time or may need respite care (Joosten, 2008). A wider range of food options had developed by 2018, including packaged meals from private companies that offered a greater variety of full food options than Meals on Wheels.

The ability to identify programs and services in a community can also present challenges. Services for older adults are fragmented at community levels. Caregiver agencies, home-delivered meals, and transportation services are not combined at one location where an older adult can access all three services simultaneously. Each service requires a separate phone call and a separate application—and each has its own criteria for eligibility. Locating services to meet geographical needs of an older adult can create barriers to service access as well. Some transportation programs have service areas, for example, that do not go outside of the city where they operate, so seniors in these areas cannot access their programs.

Community-dwelling older adults who receive long-term care from a home health agency have access to a clinical social worker that refers them to services that they need. Persons without access to a social worker often do not know how to identify services in their community—or even realize that they need them, as in the case of persons with chronic health conditions. Older adults often do not know what types of services they will need until they return to their home environment from an acute care center. Older adults comprised 37% of all hospital discharges, and older adults accounted for more than 50% of the 1 million discharges related to fractures in 2007 (Hall et al., 2010). More research is needed to see if the increasing use of nurses rather than clinical social workers as discharge planners impedes needed referrals for complex crisis intervention cases (Joosten, 2008).

Resources for Advocates

The sheer number of programs and agencies is revealed in the following discussion that includes many abbreviated terms. The complex hierarchy of agencies exists to oversee, fund, and administer community programs for older persons:

- Funding for AoA programs is allocated to each State Unit on Aging (SUA) and is based on each state's number of older adults 60+ and for the National Family Caregiver Support Program (NFCSP) the number of older adults age 70+ (Administration on Aging, 2011a).
- The funding of programs is administered by each SUA to Area Agencies on Aging (AAAs), which in turn designate funds within states to planning and service areas (PSAs) (Administration on Aging, 2011a).
- The AAAs determine the specific service needs of older persons in the PSA and work to address those needs through the funding of local services and through advocacy (Administration on Aging, 2011a).
- The program service areas in a state's aging network customize program services to the specific needs of the older adults in the communities they serve (Administration on Aging, 2011a).

Yet another layer of complexity exists. Each state can request a "waiver" from the federal CMS for their Medicaid program that allows them to obtain funding for a program innovation. They have to demonstrate that providing waiver services to a target population is not more costly than the cost of services these individuals would receive in institutions such as hospitals or long-term care institutions. Currently, roughly 287 waiver programs exist in the United States that provide a wide array of community-based programs to decrease the extent older persons are institutionalized (Center for Medicare and Medicaid Services, 2011, para. 7).

A 1999 ruling of the U.S. Supreme Court in *Olmstead v. L.C.* increased greater emphasis on community-based programs when it ruled that segregation of disabled adults in institutions violates Title II of the Americans with Disabilities Act (Centers for Medicare and Medicaid Services, 2011). It led to development of waiver programs funded by CMS to serve disabled persons in the "least restrictive environment" possible. Roughly 1 million disabled persons, including many seniors, now receive community-based programs funded by the federal government.

A program called Money Follows the Person Demonstration (MFP) has allowed thousands of persons to move back into the community from institutions such as nursing homes. The idea is simple: Increase the use of HCBS, use Medicaid funds to allow persons to get long-term care of their choice, and put procedures in place to provide quality assurance and improvement of HCBS. This program was expanded by the ACA with additional appropriations because it was successful. If only 62% of its enrollees were satisfied with their living situations before they moved back to the community, for example, 92% were satisfied with their new living situation in the community, and it saved $978 million in Medicare and Medicaid expenses in its first year of operation. Unfortunately, President Trump ended the program by not funding it in his budget (Schubel, 2018).

School-based intergenerational programs promote positive perceptions of older adults among youth and enhance positive relationships between the generations (Bales, Ekland, & Siffin, 2000). The benefits of intergenerational programs are mutual. Older adults who mentor at-risk youth report both greater meaning in their lives and opportunities to enhance positive emotions (Larkin, Sadler, & Mahler, 2005).

Social workers can work with and help to form "villages" to decrease isolation among seniors and to help them live at home as long as feasible. Started in Beacon Hill in Boston in 2001, a national network of not-for-profit and community-based organizations of seniors has spread across the United States (Scharlach, 2012). Each of them has a governing board. Each of them has one or more staff persons funded by grants from foundations and other donors as well as memberships. Each organizes programs, such as hiking clubs, lectures, meetings, potluck meals in members' homes, and other events. Members

POLICY ADVOCACY LEARNING CHALLENGE 8.10

CONNECTING MICRO AND MACRO POLICY ADVOCACY FOR SENIORS' COMMUNITY SERVICES

You are a home health social worker who is referred by the primary care physician for information on community resources for your 62-year-old single female patient who has just undergone a mastectomy and will be receiving chemotherapy for three months. She was working full time up until the time of her surgery, and her doctor has informed her that she will be unable to work while she is receiving chemotherapy. She was working as an engineer at a large aerospace organization and has no children. She does not have affiliations with religious or social supports beyond her coworkers as she reports she worked 60 hours per week. She was not seen by a social worker at the hospital and has no help coming into the home with the exception of the nurse, physical therapist, and you the social worker. She is now ambulating with a walker due to postoperative weakness and a prolonged hospital stay of five weeks as she underwent testing, surgery, and postoperative complications. There was a delay in the home health referral causing the social worker not to make the first home visit until two weeks after she was already discharged. She has been ordering cabs to get to her chemotherapy and reports she fell getting out of the cab coming into her home because the cab was too low. Now that she is home, she reports she can barely make it to the bathroom, she has difficulty preparing food and has a low appetite, and she cannot drive to run errands, go shopping, or make it to her chemotherapy and doctor's appointments. She ran out of food yesterday and has prescriptions that need to be picked up from the pharmacy.

Learning Exercise

1. How does the social worker intervene?
2. What HCBS might this patient need?
3. What programs are relevant for this patient?
4. What state or federal benefits might this patient be eligible for?
5. How would you advocate for policy change at the hospital organization that discharged her without seeing a social worker?

volunteer services to other members who need help with specific tasks in their homes or need help in traveling to health appointments. An online site, the Village to Village Network, provides information about how to start a village (https://www.brookings.edu/blog/usc...on.../how-villages-help-seniors-age-at-home/).

THINKING BIG AS POLICY ADVOCATES IN THE GERONTOLOGY SECTOR

The United States clearly needs a comprehensive plan to deal with the tens of millions of persons—so-called baby boomers—who will reach age 65 in the next three decades. We currently have a disjointed set of programs and policies as our discussion in this chapter clearly reveals. Take a stab at trying to identify some components of a multifaceted policy proposal called Addressing the Social, Medical, Community, and Resource Needs of Baby Boomers. Part of this plan should include a strategy for preventing the inhumane policy of spend-down, where millions of seniors have to deplete their assets and savings to a welfare level to be eligible for long-term care financed by Medicaid. Yet other parts should involve far more alternatives to traditional nursing homes, including small residency facilities. Yet other parts should involve rigorous enforcement of civil rights for aging workers to prevent employers from firing them to obtain lower-cost, younger workers. Other parts should include greatly expanding the size of Social Security pensions that are currently less than half the size of those in many other industrialized nations. Other parts could include funding dental care for seniors as well as cutting the cost of their medications by allowing them to shop for them in Canada, where medications are often less than half the cost of American medications. Other parts should include helping homeless seniors obtain places to live.

Learning Outcomes

You are now equipped to:

- Analyze the evolution of the gerontology sector in seven eras
- Discuss the political economy of the gerontology sector
- Identify some injustices in the gerontology sector
- Analyze seven core problems of the gerontology sector, including the following:
    1. Barriers to resolving or mitigating them
    2. The policy and regulatory context of each of them
    3. Red Flag Alerts in micro, mezzo, and macro policy
    4. Making connections among micro and macro policy interventions
- Write your own policy alert as well as connected micro policy and macro policy interventions
- Develop a multifaceted policy proposal

References

Administration on Aging. (2004). Statistics on the aging population. Retrieved from http://www.aoa.dhhs.gov/prof/Statistics/online_stats_data?AgePop2050.asp

Administration on Aging. (2005a). Census 2000 data on aging. Retrieved from http://www.aoa.gov/prof/Statistics/Census2000/census2000.asp

Administration on Aging. (2005b). Statistics on the aging population. Retrieved from http://www.aoa.gov/prof/Statistics/statistics.asp

Administration on Aging. (2010). Evidence-based disease and disability prevention program. Retrieved from http://www.aoa.gov/AoA_programs/HPW/Evidence_Based/index.aspx

Administration on Aging. (2011a). Aging integrated database. Retrieved from http://www.agidnet.org/DataGlance/SPR

Administration on Aging. (2011b). Office of long-term care ombudsman programs. Retrieved from http://www.aoa.gov/

Albert, S. M. (2004). *Public health and aging: An introduction to maximizing function and well-being.* New York, NY: Springer.

Aldwin, C. M. & Gilmer, D. F. (2004). *Health, illness, and optimal aging: Biological and psychosocial perspectives.* Thousand Oaks, CA: SAGE.

American Association for General Psychiatry. (2011). Legislative & regulatory agenda. Retrieved from http://www.aagponline.org/index.php?submenu=pub_reports_submenu&src=gendocs&ref=LegislativeRegulatoryAgenda&category=pub_reports_submenu

Balas, M. C., Casey, C. M., & Happ, M. B. (2007). Geriatric nursing protocol: Comprehensive assessment and management of the critically ill. In M. Boltz, E. Capezuti, T. Fulmer, D. Zwicker, & A. O'Meara (Eds.), *Evidence-based geriatric nursing protocols for best practice* (4th ed.). New York, NY: Springer. Retrieved from http://consultgerirn.org/topics/critical_care/want_to_know_more/

Bales, S., Ekland, S., & Siffin, C. (2000). Children's perceptions of elders before and after a school-based intergenerational program. *Educational Gerontology, 26*(7), 677–689.

Berzoff, J. & Silverman, P. (2004). *A handbook for end-of-life practitioners.* New York: Columbia University Press.

Brill, S. (2015). *America's bitter pill.* New York: Random House.

California Advocates for Nursing Home Reform. (2011). Twenty-eight years of advocacy. Retrieved from http://www.canhr.org/legislation

California Pan-Ethnic Health Network. (2011). SB 853: The Health Care Language Assistance Act. Retrieved from http://www.cpehn.org/sb853.php

Centers for Disease Control and Prevention. (2011). Healthy aging: At a glance 2011. Retrieved from http://www.cdc.gov/chronicdisease/resources/publications/AAG/aging.htm

Centers for Medicare and Medicaid Services (CMS). (2009). Medicare and your mental health benefits [CMS Product No. 10184]. U.S. Department of Health and Human Services.

Centers for Medicare and Medicaid Services (CMS). (2011). HCBS waivers—Section 1915(c). Retrieved from http://www.cms.gov/MedicaidStWaivProgDemoPGI/05_HCBSWaivers-SEction1915(c).asp

Congressional Budget Office (CBO). (2011). CBO's 2011 long-term projections for Social Security:

Additional information. Retrieved from http://www.cbo.gov/ftpdocs/123xx/doc12375/08-05-Long-TermSocialSEcurityProjections.pdf

Csikai, E., & Chaitin, E. (2006). *Ethics in end of life decisions in social work practice*. Chicago, IL: Lyceum Books.

Dana Farber Cancer Institute. (2015). Adult palliative care. Retrieved from http://www.dana-farber.org/Adult-Care/Treatment-and-Support/Treatment-Centers-and-Clinical-Services /Pain-Management-and-Palliative-Care.aspx

Day, T. (n.d.). About the national aging network. National Care Planning Council. Retrieved from https://www.longtermcarelink.net/eldercare/area_agencies_on_aging.htm

Ferrini, A. & Ferrini, R. (2008). *Health in the later years* (4th ed.). New York, NY: McGraw-Hill.

Gawande, A. (2014). *Being mortal*. New York: Macmillan.

Hall, J. H., DeFrances, C. J., Williams, S. N., Golosinsky, A., & Schwartzman, A. (2010). *National hospital discharge survey: 2007 summary* [National Health Statistics Reports, No. 29]. Hyattsville, MD: National Center for Health Statistics.

Hayden, M. F. (2000). Social policies for people with disabilities. In J. Midgley, M. Tracy, & M. Livermore (Eds.), *The handbook of social policy*. Thousand Oaks, CA: SAGE.

Jansson, B. (2018). *The reluctant welfare state: Engaging history to advance social work practice*. Boston: Cengage.

Jansson, B. (2019). *Reducing inequality: Addressing the wicked problems across professions and disciplines*. San Diego: Cognella Academic Press.

Joint Commission. (2011). Facts about ambulatory care accreditation. Retrieved from http:// www.jointcommission.org/assets/1/18/Ambulatorycare_1_112.PDF

Joosten, D. (2008). *Aspects of clinical social workers' decision-making with older adult clients with unmet psychosocial and/or physical needs: Outcomes, patterns, and processes of referrals for services*. (Doctoral Dissertation, University of California, Los Angeles).

Kayser-Jones, J. (2009). Nursing homes: A health promoting or dependency-promoting environment? *Family & Community Health, 32*(1), S66–S74.

Klauber, M. & Wright, B. (2001). *The 1987 Nursing Home Reform Act*. Public Policy Institute (February 2001). Retrieved from https://www.aarp.org/home-garden/livable-communities/info-2001/the_1987_nursing_home_reform_act.html

Kronenfeld, J. J. (2000). Social policy and health care. In J. Midgley, M. Tracy, & M. Livermore (Eds.), *The handbook of social policy*. Thousand Oaks, CA: SAGE.

Kuhn, D. P. (2010). The senior wave: Older voters set for historic turnout. *RealClearPolitics*. Retrieved from http://www.realclearpolitics.com/articles/2010/10/18/the_senior_wave_older_voters_set_for_historic_turnout_107608.html

Larkin, E., Sadler, S., & Mahler, J. (2005). Benefits for older adults mentoring at-risk youth. *Journal of Gerontological Social Work, 44*(3–4), 23–37.

Leipzig, R. M., Whitlock, E. P., Wolff, T. A., Barton, M. B., Michael, Y. L., Harris, R., . . . Siu, A. (2010). Reconsidering the approach to prevention recommendations for older adults. *Annals of Internal Medicine, 153*, 809–814.

Leventhal, H., Rabin, C., Leventhal, E. A., & Burns, E. (2001). Health risk behaviors and aging. *Handbook of the Psychology of Aging, 5*, 186–214.

LGBT Older Adult Coalition. (2011). *Report: The needs of LGBT older adults in metro Detroit.* Retrieved from http://lgbtolderadults.files.wordpress.com/2012/02/report-needs-of-older-adults-6-12-11.pdf

LGBT Older Adult Coalition. (n.d.). LGBT older adults—a population at risk. Available at http://lgbtolderadults.files.wordpress.com/2012/02/lgbt-older-adults-a-population-at-risk1.pdf

McElhone, A., Limb, Y., & Gambert, S. (2005). *Retooling for an aging America: Building the health care workforce*. National Academies Press. Retrieved from https://www.ncbi.nlm.nih.gov/books/NBK215402/

Mendes, E. (2010). In US, health disparities across incomes are wide-ranging. *Gallup*. Retrieved from http://www.gallup.com/poll/143696/Health-Dispariies-Across-Incomes-Wide-Ranging.aspx

Midgley, J., Tracy, M., & Livermore, M. (2000). *The handbook of social policy.* Thousand Oaks, CA: SAGE.

Morrison, S. & Meier, D. (2011). The National Palliative Care Research Center and the Center to Advance Palliative Care: A partnership to improve care for persons with serious illness and their families. *Journal of Pediatric Hematology/Oncology, 33*, S126–S131. doi: 10.1097/MPH.0b013e318230dfa0

National Hospice and Palliative Care Organization. (2011). History of hospice care. Retrieved from http://www.nhpco.org/history-hospice-care

National Institute of Mental Health (2010). Older adults: Depression and suicide facts (fact sheet). Retrieved from http://www.nimh.nih.gov/health/publications/older-adults-depression-and-suicide-facts-fact-sheet/index.shtml

National Institute of Mental Health (2011). NIH senior health: Depression. Retrieved from: http://www.nihseniorhealth.gov/depression/research/o1.html

National Institute on Aging. (2010). Research Goal E: Improve our ability to reduce health disparities and eliminate health inequities among older adults. Retrieved from http://www.nia.nih.gov/about/living-long-well-21st-century-strategic-directions-research-aging/ research-goal-e-improve-our

Newport, F. (2012). In presidential election, age is a factor only among whites. *Gallup.* Retrieved from http://www.gallup.com/poll/154712/Presidential-Election-Age-Factor-Among-Whites.aspx

Nursing Home Law Center Staff. (2009, December 6). Are group homes a viable alternative to nursing homes? *Nursing Home Law News.* Retrieved from https://www.nursinghomelawcenter.org/news/nursing-home-abuse/are-group-homes-a-viable-alternative-to-nursing-homes/

Office of Management and Budget. (2011). *Living within our means and investing in the future: The president's plan for economic growth and deficit reduction*. Washington, DC: U.S. Government Printing Office.

Office of the Surgeon General (2011). National prevention strategy. Retrieved from http://www.surgeongeneral.gov/initiatives/prevention/strategy/

Ott, B. B. (1999). Advance directives: The emerging body of evidence. *American Journal of Critical Care, 8*(1). Retrieved from https://www.ncbi.nlm.nih.gov/pubmed/9987550

Passel, J. & Cohn, D. (2008, February 11). *U.S. population projections: 2005–2050.* Washington, DC: Pew Research Center.

Quadagno, J. (2008). *Aging and the life course* (4th ed.). New York: McGraw Hill.

Reese, D. & Raymer, M. (2004). Relationships between social work services and hospice outcomes: Results of the National Hospice Social Work Survey. *Social Work, 49*(3). Retrieved from https://www.ncbi.nlm.nih.gov/pubmed/15281696

Richardson, J.P. (2006, May 17). Considerations for health promotion in disease prevention in older adults. *Medscape News Today.* Retrieved from http://www.medscape.com/viewarticle/531942

Rowe, J. W., & Kahn, R. L, (1998). *Successful Aging*, New York: Pantheon.

Scharlach, A. (2012). Creating aging-friendly communities in the United States. *Ageing International, 37,* 25–38.

Scharlach, A., Graham, C., & Lehning, A. (2011, August 25). The "village" model: A consumer-driven approach for aging in place. *The Gerontologist*. doi: 10.1093/geront/gnr083

Schubel, J. (2018, February 23). Trump should have proposed "Money Follows the Person" funding. Center on Budget and Policy Priorities. Retrieved from https://www.cbpp.org/.../trump-should-have-proposed-money-follows-the-person-fun...

Seegert, L. (2017, March 16). Social isolation, loneliness negatively affect heath for seniors. *Covering Health*. Retrieved from *https://healthjournalism.org/.../social-isolation-loneliness-negatively-affect-health-for-s...*

Stoesz, D. (2000). Social policy: Reagan and beyond. In J. Midgley, J. Tracy, & M. Livermore (Eds.), *The handbook of social policy.* Thousand Oaks, CA: SAGE.

Substance Abuse and Mental Health Services Administration. (2007). *National expenditures for mental health services and substance abuse treatment 1993–2003.* [Publication No. SMA 07-4227]. U.S. Department of Health and Human Services.

Tavernise, S. (2016, February 12). Disparity in life spans of the rich and the poor is growing. *New York Times.* Retrieved from https://www.nytimes.com/2016/.../disparity-in-life-spans-of-the-rich-and-the-poor-is-g...

Torres-Gil, F. & Villa, V. (2000). Social policy and the elderly. In J. Midgley, M. Tracy, & M. Livermore (Eds.), *The handbook of social policy.* Thousand Oaks, CA: SAGE.

U.S. Census Bureau. (2011). *Poverty.* Social, Economic, and Housing Statistics Division. Retrieved from http://www.census.gov/hhes/www/poverty/data/historical/people.html

U.S. Department of Health and Human Services. (2011, April). HHS action plan to reduce racial and ethnic health disparities. Washington, DC. Retrieved from https://minorityhealth.hhs.gov/omh/browse.aspx?lvl=3&lvlid=102

Wolitzky-Taylor, K., Castriotta, N., Lenze, E. J., Stanley, M. A., & Craske, M. G. (2010, January 22). Anxiety disorders in older adults: A comprehensive review. *Depression and Anxiety, 27*(2), 190–211.

9

BECOMING POLICY ADVOCATES IN THE SAFETY NET SECTOR

LEARNING OBJECTIVES

In this chapter you will learn to:

1. Develop an empowering perspective of people who need and use safety net programs
2. Analyze the evolution of the American safety net sector
3. Analyze how defects in safety net programs often exacerbate income inequality in the United States
4. Analyze the political economy of the safety net sector
5. Analyze important safety net programs including ones related to income-enhancing programs, regulations over credit-providing agencies, job-related programs, nutrition-enhancing programs, shelter-enhancing programs, and asset-creating programs
6. Engage in policy advocacy to address seven core problems in the safety net sector
7. Think big to develop reforms in the safety net sector

The United States has an array of safety net programs that address the basic living needs of tens of millions of persons. You need to know about these programs because you will refer thousands of your clients or patients to these programs during your career. You will engage in micro policy advocacy for them—and this advocacy will allow some families to escape malnutrition, escape extreme poverty, and escape homelessness. You will engage in mezzo policy advocacy when you educate persons in specific neighborhoods about programs such as the Supplemental Nutrition Assistance Program (SNAP), subsidized winter heating programs, senior discounts for public transportation, Social Security benefits, and many others. You will engage in macro policy advocacy when you work with others to head off cuts in safety net programs in your state or at the federal level.

The nation's safety net sector is underfunded so that it can provide meager food allotments only for very poor families. Large numbers of Americans are homeless, have inadequate nutrition, live under federal poverty levels, and cannot meet their survival needs at minimum-wage levels in many states. We cast a wide net when discussing safety net programs because you will be surprised to find how relevant many of them are to your specific clients.

DEVELOPING AN EMPOWERING PERSPECTIVE TOWARD PERSONS WHO NEED SAFETY NET PROGRAMS

Almost everyone needs assistance from government during their lives, whether seniors, parents, adolescents, or children. Some people experience unexpected tragedies like illness or unemployment. Some people find their homes demolished by natural disasters such as hurricanes, tornadoes, floods, or earthquakes. Some people experience illness, disabilities, chronic diseases, or acute medical emergencies. Some people develop acute or chronic mental problems. Some people have unanticipated accidents such as car crashes. Some people are victims of violent acts such as when mass killings take place or attacks happen on the street. Some people lose their jobs—or cannot find work—such as during layoffs, recessions, or depressions. Some people earn wages not sufficient to meet their daily needs. Many people are evicted from rentals when they lose work, develop health problems, and cannot afford ever-rising rental levels.

People often don't know in advance when they might need assistance. Earthquakes, accidents, and illness, for example, often occur suddenly. Nor do people know, in many cases, the extent or duration of harm they may suffer. In the case of the Great Recession of 2007 to 2009, for example, hundreds of thousands of families came to realize that many banks were in economic trouble but did not realize the banking crisis would soon lead them into foreclosure proceedings that would put their housing and finances in jeopardy, possibly for years.

Scores of safety net programs help persons cope with these needs. These small programs often have millions or tens of millions of recipients. Many safety net programs are funded, as well, by the private sector, such as from food banks. Most safety net programs are funded by budgets of local, state, and federal governments, but others are funded through the tax system through tax credits or tax deductions. (We discuss the Earned Income Tax Credit [EITC] and the Children's Tax Credit [CTC] subsequently.)

Many persons and families use combinations of safety net programs, such as a low-income family that used SNAP, the EITC, the CTC, rent subsidies, and school lunches in 2019. When used in combination, these programs help millions of people rise above survival levels of living.

Some Americans have reservations about these programs on the grounds that they discourage work or burden taxpayers. This argument ignores several facts. Many retirees, disabled people, and veterans cannot work. Wages of most workers have stagnated during the past three decades, so they need safety net benefits and tax credits to meet their basic needs. Many families would become homeless if they did not receive rent subsidies or public housing. Recipients of the EITC, Unemployment Insurance (UI), and Temporary Assistance for Needy Families (TANF) are required to work or to seek work. Nor do safety net programs provide extravagant benefits as we discuss subsequently.

Critics of safety net programs often fail to consider what would happen if they did not exist. Already the most unequal industrialized nation, inequality in the United States would become even more pronounced if SNAP, the EITC, UI, rent subsidies, Social Security pensions, and other programs were deeply cut. Indeed, it can be argued that the safety net system has already been excessively cut during the past 40 years when its programs have been slashed in numerous federal budget battles. Recall from Chapter 3 that American history is strewn with efforts to limit use of safety net programs, whether by creating stringent means tests that limit use of these programs to very poor people or by imposing harsh treatment on them, such as manual labor in the thousands of poorhouses constructed in the 19th and early 20th centuries. Conservative critics of safety net programs, such as Republican House Speaker Paul Ryan in 2018, want "work tests" and reduced benefits and services for SNAP and Medicaid even though many of these recipients are physically disabled, lack transportation to places of employment, are unemployed, or earn low wages—and even though many experts believe work tests will worsen health outcomes as low-income families often do not use Medicaid either because they cannot prove they are working or because they cannot prove that they cannot work due to disabilities or lack of job openings (Katch, Wanger, & Aron-Dine, 2018).

These critics of safety net programs also forget that affluent Americans have their own subsidies through the tax system. They benefit from many tax loopholes and deductions not available to most Americans. They often receive large pensions from their corporate

POLICY ADVOCACY LEARNING CHALLENGE 9.1

EXAMIINE YOUR FAMILY'S USE OF SAFETY NET PROGRAMS

Examine your own family to better understand safety net programs. To what extent have your immediate family members used safety net programs—or more distant relatives? To what extent do they view them in a positive or negative light? To what extent have they helped family members—and, if so, how? As you read this chapter, can you find ones that you or your family members do not use because they do not know about them? Did they not use them even when they knew about them—and, if so, why? Drawing upon your family's experience, do you think American safety net programs are underfunded or overfunded?

employers unlike many low-income people who receive only Social Security pensions. They do not experience discrimination by banks when they apply for mortgages or loans unlike many persons of color.

Most people need assistance from government at some point in their lives. We discuss universal programs that are provided to most Americans, such as UI and Social Security pensions, as well as means-tested programs that are provided to persons who meet specific income, employment, and other criteria. We discuss government regulations over banks, other lenders, purveyors of credit cards, and landlords that prevent some practices that negatively impact low- and moderate-income persons. We discuss federal government programs such as Social Security, Medicare, SNAP, and Supplemental Security Income (SSI); federal and state government programs such as TANF and Medicaid; state programs such as those that give specific kinds of assistance to specific populations like immigrants or general assistance; local government programs such as specific shelters, and private-sector programs such as food banks run by not-for-profit or religious organizations. Realize that regulations vary by state for many of these programs.

ANALYZING THE EVOLUTION OF THE AMERICAN SAFETY NET SYSTEM

Figure 9.1 provides an overview of the evolution of the American safety net system, from the colonial era to the present. It demonstrates that the federal government had little role in the safety net system through the 19th century and up to the start of the Great Depression in 1929. It shows how that role expanded in the New Deal, Great Society, and the early part of the 1970s—only to contract substantially in the decades

FIGURE 9.1 ■ Evolution of the American Safety Net Programs

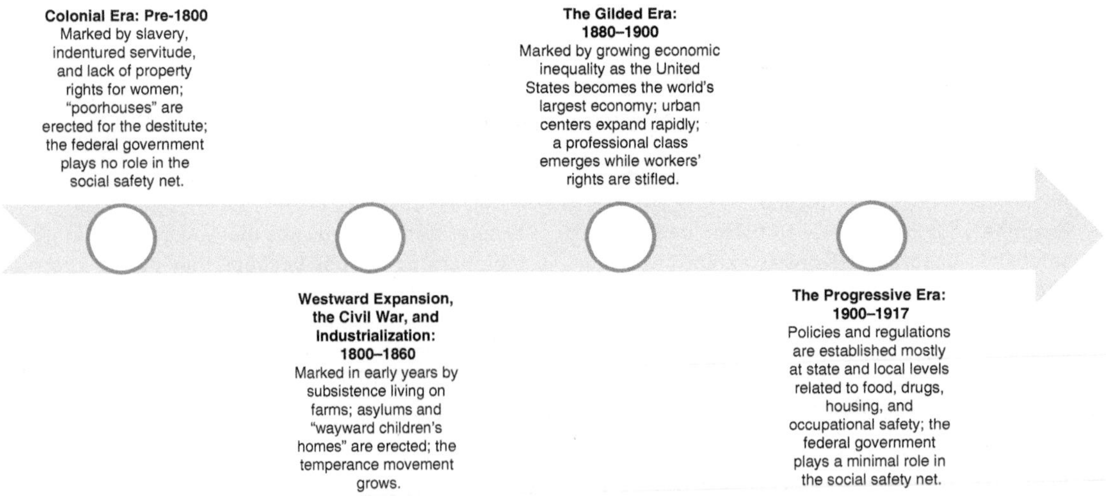

Colonial Era: Pre-1800
Marked by slavery, indentured servitude, and lack of property rights for women; "poorhouses" are erected for the destitute; the federal government plays no role in the social safety net.

Westward Expansion, the Civil War, and Industrialization: 1800–1860
Marked in early years by subsistence living on farms; asylums and "wayward children's homes" are erected; the temperance movement grows.

The Gilded Era: 1880–1900
Marked by growing economic inequality as the United States becomes the world's largest economy; urban centers expand rapidly; a professional class emerges while workers' rights are stifled.

The Progressive Era: 1900–1917
Policies and regulations are established mostly at state and local levels related to food, drugs, housing, and occupational safety; the federal government plays a minimal role in the social safety net.

The Great Depression and the New Deal: 1929–1942
In response to massive unemployment and poverty following the stock market crash of 1929, President Franklin D. Roosevelt establishes numerous federal programs to put millions (mostly men) back to work, creates public housing, creates a progressive federal income tax, establishes a federal minimum wage, and provides huge subsidies to states to finance their welfare programs; federal government begins playing a major role in the social safety net.

The Social Security Act of 1935
The backbone of the U.S. social safety net system is established; two means-tested programs—Aid to Dependent Children and Old Age Assistance—are established, along with two universal programs—Unemployment Insurance and Social Security.

Modifications to Social Security in the 1950s
Social Security becomes a family-oriented program that provides benefits to widows and children of beneficiaries; Social Security Disability Insurance (SSDI) is added in 1956 for persons with "an impairment of mind or body" that precludes gainful occupation.

Lyndon B. Johnson's "Great Society": 1960–1969
Aimed at eliminating poverty and racial injustice; President Johnson initiates a "war on poverty," enacts the Civil Rights Act, the Economic Opportunity Act, the Food Stamp Act, and the Voting Rights Act and establishes Medicare and Medicaid.

Safety Net Expansion: 1969–1980
Republican President Richard Nixon enacts the Employment and Training Act of 1973, expands the food stamps program, establishes SSI, enacts the EITC, and establishes Section 8 Housing; Democratic Presidents Ford and Carter subsequently do little to expand the safety net.

Era of Devolution: 1981–1992
Presidents Ronald Reagan and George H. W. Bush enact "supply side" economic policies, such as tax cuts for the wealthy, cuts to food stamps and EITC programs, and restrictions on union organizing; many federal programs are devolved to the states through block grants.

Era of Budget Surplus: 1992–2000
President Bill Clinton balances the national budget while keeping many safety net programs intact; the Children's Health Insurance Program (CHIP) is enacted, and the EITC program is expanded.

Personal Responsibility and Work Opportunities Reconciliation Act of 1996
President Clinton "ends welfare as we knew it" by removing entitlement status, implementing lifetime limits and work requirements, allowing states to develop their own eligibility standards, and barring immigrants from a range of safety net programs.

Era of Budget Deficits: 2000–2008
President George W. Bush vastly increases the national debt by slashing taxes and increasing military spending; funding for pharmaceutical benefits through Medicare is enacted.

The Great Recession of 2007–2009 and beyond
President Barack Obama inherits the economic devastation wreaked by a banking crisis resulting from massive foreclosures in the housing market; millions are unemployed; "bailouts" for large institutions are enacted; despite the recession, the Democratic Congress successfully enacts the American Recovery and Investment Act of 2009 and the Affordable Care Act of 2010.

following 1980 with cuts in many programs and welfare reform, including passage of the Personal Responsibility and Work Opportunities Reconciliation Act (PRWORA) of 1996, otherwise known as Welfare Reform, that transformed the largest income-supporting safety net program, Aid to Families with Dependent Children (AFDC), from an entitlement program with no time limits to a time-limited means-tested program called TANF. It also added work requirements and instituted bars to many safety net programs for immigrants. The American Recovery and Investment Act of 2009 (often called the Stimulus Program) greatly expanded funding of many safety net programs during the presidency of Barack Obama but only for several years. Deep cuts in SNAP and other safety net programs took place in the Obama administration due to a gridlocked budget process. Some of them received additional funding in the first two years of the presidency of Donald Trump when Republicans acceded to increases in safety net programs in return for large increases in military spending. However, other safety net programs received major cuts during the Trump presidency.

Social workers need to engage in policy advocacy with respect to safety net programs in each of the eight policy sectors discussed in this book, including the following:

a. Medicaid and Medicare as well as tax policies that allow some families to deduct medical expenses help persons in the health sector

b. Medicaid, TANF, SNAP, subsidized rent and housing programs, and tax incentives for families that adopt foster children in the child and family sector

c. Mental health services of Medicaid and Medicare, income supports for persons with mental disabilities from SSI, TANF, and SNAP in the mental health and substance abuse sector

d. Medicare, Medicaid, Social Security, SSI, and senior citizen housing in the gerontology sector

e. General relief, shelters, Medicaid, and SSI in the corrections sector

f. SNAP, TANF, and school lunch programs in the education sector

g. General relief and Medicaid in the criminal justice sector

Safety net programs cut across age-groups. Children often benefit from SNAP, TANF, Medicaid, school breakfast and lunch programs, food banks, and subsidized housing. Working persons benefit from Medicaid and SNAP. Unemployed adults benefit from General Assistance and UI. Elderly persons benefit from Medicare, Medicaid, SSI, Social Security, senior housing, Meals on Wheels, and deductions of mortgage interest from income. Each of these age-groups benefits from subsidized rent programs, public housing, and temporary shelters.

Safety net programs also cut across the many marginalized populations identified in Chapter 1 as a few examples demonstrate. Many white persons need help dealing with opioid addiction as well as help from SNAP and private food banks. African Americans need federal assistance to purchase housing. Women need assistance in finding birth control services. Persons with disabilities need help purchasing equipment that allows them to be mobile. Seniors need access to group homes when they are unable to live independently. Many low-income persons need SNAP.

ANALYZING HOW DEFECTS IN AMERICAN SAFETY NET PROGRAMS OFTEN EXACERBATE INCOME INEQUALITY

Poverty is the most important social problem in the United States for multiple reasons. The federal poverty level is $20,780 for families of 3, $16,460 for families of two, and $12,140 for single persons in 2018. Persons near and beneath poverty possess far higher levels of physical and mental illnesses and possess high levels of chronic disabilities and substance abuse (Jansson, 2019a). Many experience chronic or frequent unemployment. Many of them do not graduate from high school and have low levels of literacy. Many are exposed to violence in their households and communities. Poverty causes stress not only from economic hardships but because many of its victims are and feel marginalized as they see more affluent Americans in their daily lives, as well as on the mass media. Millions of families live on the edge of deep poverty—and fall into it when losing only a single paycheck.

Many safety net programs have helped people meet their basic needs in contemporary society. Indeed, income from safety net programs reduced the poverty rate in 2010 from 29% of the population when income from safety net programs was not included to 15% when it was included. This means that safety net programs "cut poverty nearly in half compared to what it would otherwise (have) been" (Greenstein, 2012).

The positive effects of safety net programs were demonstrated during the Great Recession of 2007 to 2009 and beyond. We would have expected sharp increases in poverty rates when unemployment rose to more than 12% during the early part of this recession, yet poverty rates remained stable at 15.3% for 2007, 15.7% for 2009, and 15.5% for 2010 according to a poverty measure of the Bureau of Census (Greenstein, 2012). Safety net programs automatically trigger benefits during economic downturns as more people become eligible for means-tested programs as their income declines—and as most unemployed persons become eligible for UI. Moreover, the time-limited American Recovery and Reinvestment Act (also called the Stimulus Program) temporarily or permanently augmented benefits of some safety net programs during the Great Recession. The U.S Census Bureau estimates that the EITC, the CTC, and SNAP respectively lifted 9 million, 5 million, and 4 million low-income working families out of poverty (Greenstein, 2012).

Even when we lack data that demonstrates the effectiveness of safety net programs, such as the impact on persons lifted from poverty, we can support them on ethical grounds. We have lauded good Samaritans for centuries who help persons obtain food, shelter, and income when they might otherwise suffer malnutrition, illness, and exposure—and who relieve parents of angst when they cannot meet their children's needs.

The safety-net sector is rife, however, with problems. Tens of millions of Americans remain in poverty, and many experience hunger. Nor have safety net programs redressed growing economic inequality in the United States from the early 1980s to the present, when economic inequality has risen to historic levels. More robust safety net programs would give low- and moderate-income persons more income or income substitutes, like food and subsidized shelter. The American tax system is inequitable as many affluent Americans pay income taxes at rates of 15% or lower—unlike many other industrialized nations that require affluent persons to pay income taxes at 50% or higher. Billionaires Warren Buffett and Bill Gates—each with assets exceeding $60 billion—have pleaded with public officials to raise their tax rates—with Buffett noting that his secretary's tax rate is higher than his tax rate. The tax legislation enacted by the Trump administration in late 2017 and early 2018 made far greater cuts in taxes of affluent Americans as compared with middle-class and low-income Americans. Other industrialized nations not only tax affluent people at higher rates than Americans but also have more generous safety net programs as well. Their higher income taxes and more liberal safety net programs cause them to have less economic inequality than the United States (Jansson, 2019a, 2019b).

The positive benefits of safety net programs would be even larger if eligible persons used them. Roughly half of persons eligible for SNAP (food stamps) and the EITC do not use them, for example. Many single heads of households are sufficiently intimidated by restrictions on use of TANF that they do not even apply for it, even when some of them resort to illegal activities, doubling up, and selling blood to afford survival needs for themselves and their children. Many persons with permanent and total disabilities do not apply for SSDI or SSI. Many persons do not update their eligibility for specific safety net programs because of the complexity of their application forms.

The most radical proposal for reducing inequality in the United States would be to provide a "guaranteed income" to most Americans. A trial of this idea was tested in the 1960s and it was discovered that it did not cause people to stop working. Switzerland and Canada are currently testing this policy.

Senator Bernie Sanders floated a policy idea for a safety net program that might markedly reduce income inequality in 2018. Why not, he asked, consider enacting a "jobs guarantee" that would fund hundreds of projects across the nation that addressed infrastructure, health care, the environment, education, and other areas? Why not pay $15 per hour along with health, pension, and other benefits of federal jobs? Why not include job training when needed? Why not use the 2,500 federal job training centers and employment offices that already exist around the country to implement this program? Supporters of the proposed initiative argue that it would drive up wages in the U.S. economy because employers would have to compete for workers' guaranteed wages of $15. They argue it would reduce racial inequality by cutting unemployment of African Americans that is double the size of unemployment of whites (Stein, 2018).

ANALYZING THE POLITICAL ECONOMY OF THE SAFETY NET SECTOR

Liberals and conservatives have debated the merit of increasing funding for safety net programs from the Great Depression onward. Many liberals have lauded the positive economic effects of the safety net sector, including presidents Franklin Roosevelt, Harry Truman, Lyndon Johnson, Bill Clinton, and Barack Obama, as well as Senator Chuck Schumer. Although some conservatives have supported specific safety net programs, such as SNAP that increases markets for farmers in the Midwest and South, many conservatives have viewed safety net programs in relatively negative terms. They blame them for increasing budget deficits at federal and state levels even as they have traditionally supported other budget and tax policies that have increased deficits, such as military spending and low levels of taxation on affluent Americans. They argue that safety net programs erode the work ethic of many Americans. They believe many people make fraudulent use of some safety net programs.

Liberals counter that evidence does not support the argument that safety net programs cause persons to stop working. Experts from the National Bureau of Economic Research found that increased spending on SNAP and other safety net programs during the Great Recession did not impact decisions about whether to work (Ben-Shalom, Moffitt, & Scholz, 2011). Other experts contend that imposing a work test upon applicants for SNAP or Medicaid in 2017 and 2018 would not increase employment of recipients for many reasons. Few persons make fraudulent use of safety net programs—some experts also suggest that more people cheat on income tax filings that cheat on safety net programs.

The divide between liberals and conservatives was dramatically exposed during the presidential campaign of 2012. Romney told a roomful of affluent supporters in May 2012 that:

> 47 percent of the people will vote for (Obama) no matter what . . . who are dependent upon government, who believe they are victims, who believe that government has a responsibility to care for them, who believe they are entitled to health care, to food, to housing, to you name it. . . . These are people who pay no income tax. So our message of low taxes doesn't connect. (Corn, 2013)

Unbeknownst to Romney, someone videotaped his remarks and placed them on YouTube, remarks discovered months later by Jason Carter, the grandson of former president Jimmy Carter, who divulged them to David Corn, who published them in *Mother Jones*. Romney later admitted that these remarks were a devastating blow to his campaign. Already viewed as unsympathetic to ordinary people, they suggested that he viewed those who used programs widely needed by many Americans to survive harsh economic realities as lazy persons who are dependent on government programs—and who support Obama only because he gives them government benefits. Critics promptly noted that many working Americans have such low wages that they do not pay federal income taxes but do pay payroll taxes and sales taxes in amounts that usually exceeded Romney's federal tax rate of 14%. Many other Americans do not pay federal taxes because they are retired. Many disabled veterans do not pay income taxes because they cannot work. Romney compounded his comments by saying, "My job is not to worry about those people." (He later said that he had misspoken but soon attributed the Democrats' and Obama's electoral victory in 2012 to "gifts" they gave to poor people, college students, and others, such as SNAP and college scholarships.) Conflict emerged during the presidential race of 2016. If Democrats Bernie Sanders and Hillary Clinton wanted to raise the minimum wage significantly, Republican Donald Trump wanted hardly to raise it at all.

Conflict over tax policies and funding of safety net programs surfaced during the presidency of Donald Trump in 2017 and 2018 led by House Majority Speaker Paul Ryan. According to Krugman (2018), "Everything Ryan did and proposed was to comfort the comfortable while afflicting the afflicted," whether repealing the Affordable Care Act (ACA) or seeking deep cuts in safety net programs to reduce budget deficits he helped cause by cutting the taxes of affluent people. Cohn and Delaney (2018) document how he spent most of his political career "trying to shed America's safety net so that literally tens of millions of Americans would lose supports they use to get food, healthcare and pay their most basic bills." Trump's election gave him his chance to attack safety net programs in the aftermath of tax cuts he helped enact that created a $1.5 trillion deficit. This tax legislation made some cuts in taxes for low- and moderate-income persons, but the Tax Policy Center estimated that the legislation would "harm many middle class and low-income families in the short term and the vast majority of families in the long term." It estimated that 48% of the tax cuts benefited the top 1% of the population even with some modest cuts of taxes for affluent Americans (Leonhardt, 2017; Delaney, 2018).

Ryan hoped to make deep cuts in SNAP, Medicaid, Medicare, and many other safety net programs in early 2018—and to impose a work requirement on SNAP and Medicaid. He planned to use the argument that the $1.5 trillion deficit he and fellow Republicans had created by engineering huge tax cuts and increasing military spending by $70 billion required spending cuts from safety net programs. It appeared in May 2018 that he would not succeed because he announced his impending retirement from politics just as many pundits predicted a "blue wave" that would allow Democrats to retake control of the House and possibly the Senate in congressional elections of 2018 (see Policy Advocacy Learning Challenge 9.2). Polls suggested that turnout by Democrats would surge in 2018 and 2020 if millennial, female, African American, and Latino voters vote in historic numbers as compared with Republicans. Elections in Alabama, Virginia, and Pennsylvania in 2017 and 2018 suggested, as well, that many college-educated suburban voters were moving toward Democrats from Trump. Were these electoral shifts to occur, national budget priorities might favor greater funding of many safety net programs.

Join this great debate between liberals and conservatives by reading Policy Advocacy Learning Challenge 9.2.

Websites of key advocacy groups that specialize in safety net programs are listed in Table 9.2. Each of them have websites that provide excellent data about these programs.

POLICY ADVOCACY LEARNING CHALLENGE 9.2

CONNECTING MICRO, MEZZO, AND MACRO POLICY ADVOCACY INTERVENTION TO ADDRESS ETHICAL ISSUES IN THE SAFETY NET SECTOR

The State of Florida enacted a law that required applicants for welfare to obtain drug tests. Its backers believed it would reduce applications and catch significant numbers of drug users. State data revealed that it accomplished neither of these goals as only 2.6% of welfare applicants failed this drug test during a four-month period (Alvarez, 2012). In fact, Florida may have lost money on this experiment because it reimbursed passing applicants $30 each for their cost for the test for a total of $118,140. Advocates of the law insisted that it be retained because the "drug testing law was really meant to make sure that kids were protected (and that) our money wasn't going to addicts, that taxpayer generosity was being used on diapers and Wheaties and food and clothing." The American Civil Liberty Union (ACLU) of Florida sued the state for unconstitutional invasion of applicants' privacy as protected by the Fourth Amendment to the U.S. Constitution. The Southern Center for Human Rights intends to sue the State of Georgia once it begins to implement a similar law that it just enacted.

Learning Exercise

1. Discuss why backers of this law in Florida were confident that it would catch many drug offenders.
2. Can you think of public programs used by middle- and upper-income families that would require drug tests?
3. Do similar views exist with respect to users of SNAP, the EITC, Section 8 Housing, or other safety net programs?
4. Is the Florida law consonant with social workers' values as stated in the National Association of Social Workers (NASW) Code of Ethics?
5. Discuss whether requiring users of SNAP and Medicaid to demonstrate they hold jobs should be added to their eligibility requirements. Do remember that many disabled and elderly persons cannot work.

Also realize that tens of millions of workers work part time and only receive a low minimum wage that has not increased for three or more decades. Also some consumers are unemployed. Yet others lack transportation, making it difficult for them to work in distant locations. Some people lack childcare. Others have disabled children or spouses that they help in their households.

TABLE 9.1 ■ Important Safety Net Programs Discussed in This Chapter

Income-Enhancing Programs

1. Temporary Assistance for Needy Families (TANF)
2. General Assistance or General Relief (GA or GR)
3. Supplementary Security Income (SSI)
4. Social Security for Retirees and Social Security for Permanently and Totally Disabled Persons (SSDI)
5. Social Security for survivors of deceased beneficiaries
6. Income-enhancing provisions of the federal tax code
    a. Earned Income Tax Credit (EITC)
    b. Other tax provisions relevant to low- and moderate-income families in the federal tax code
    c. Tax provisions in state and local jurisdictions relevant to low- and moderate-income persons
7. Unemployment Insurance
8. Workers' Compensation
9. Minimum wage and living wage

Regulations Over Credit-Providing Agencies

1. Regulations of lending practices of banks, payday lenders, and other providers of loans, mortgages, and credit cards

(Continued)

TABLE 9.1 ■ Important Safety Net Programs Discussed in This Chapter (Continued)

Job-Related Programs

1. Job-finding and placement programs
2. Job-training programs
3. Job creation programs
4. Transportation and childcare programs to help persons work

Nutrition-Enhancing Programs

1. Supplemental Nutrition Assistance Program (SNAP), formerly the Food Stamps Program
2. Women, Infants, and Children Program (WIC)
3. Meals on Wheels
4. Food banks run by not-for-profit and religious organizations
5. School breakfast and school lunch programs

Shelter-Enhancing Programs

1. Section 8 Subsidized Rental Housing
2. Continuum of Care (COC) and other programs for homeless persons funded by the McKinney-Vento Homeless Assistance Act
3. Public housing
4. Shelter for veterans
5. Tax incentives to build affordable housing

Asset-Creating Programs

1. Private and public pensions
2. Home ownership programs
3. Individual Development Accounts (IDAs)

TABLE 9.2 ■ Advocacy Groups That Support Income-Enhancing Programs

1. **Center on Budget and Policy Priorities (CBPP).** The CBPP conducts wide-ranging research on a wide variety of safety net programs. It has been rated as the most effective lobbying group in its area of specialty in Washington, D.C. It has close links with similar groups in many states in the United States as well as in other nations.
2. **Urban Institute (UI).** The UI conducts research on welfare, health, nutritional, housing, and many other domestic programs.
3. **Brookings Institution (BI).** The BI conducts research on a wide range of domestic programs.
4. **Center for American Progress.** Its fellows engage in wide-ranging research on safety net programs.
5. **Brookings-Urban Institute Tax Center.**

USING AMERICAN SAFETY NET PROGRAMS TO IMPROVE CONSUMERS' WELL-BEING

You will now analyze the safety net programs in Table 9.1. Remember this: Even though it can be tedious to learn about these many programs and policies, you can remarkably improve the well-being of specific families and persons by using micro policy advocacy to connect them with these programs—and you can engage in mezzo and macro policy advocacy to improve them. Many technical details are discussed because the policies that you and your clients will navigate are detailed in nature—and you sometimes will advocate for clients who are wrongly declared to be ineligible. You can use websites of key advocacy groups that support these programs, such as the ones listed in Table 9.2.

Income-Enhancing Programs

Temporary Assistance for Needy Families (TANF)

TANF is the nation's major welfare program for families that President Clinton and the Congress approved in 1996 as the replacement for AFDC. Under pressure from Republicans to live up to his campaign promise of 1992 "to end welfare as we know it," Clinton decided to enact welfare reform with many provisions that Republicans supported in return for some concessions from them. Although he got Republicans to keep Medicaid as an entitlement, he signed the legislation over the vehement opposition of some cabinet members and top civil servants who produced data that predicted that it would cast millions of children into poverty. TANF consists of nine titles or sets of provisions that cover welfare, SSI, immigrants, childcare, child nutrition, and food stamps (now SNAP).

TANF replaced AFDC as a block grant to be funded until 2002 at roughly the annual level of federal expenditures for AFDC in the year preceding the enactment of TANF—but Congress could fund it at any level it desired after 2002. Liberals feared that conservative Congresses would slash federal funding and involvement considerably or entirely in future years, leaving welfare entirely in the hands of states. They feared a "race to the bottom" would then occur as generous states would fear that their high welfare benefits would be a magnet to destitute low-income persons—saddling them with higher welfare costs and forcing them to raise their individual and corporate tax rates higher as they left low-benefit states.

TANF stipulated that the states must ensure that adult recipients participate in work or work-related activities after receiving two years of cumulative benefits, with 25% of the single-parent family caseload participating by 1997 and 50% by 2002. It stated that recipients could receive cash assistance for a maximum of five years over a lifetime with limited exceptions for no more than 20% of caseloads. It prohibited the use of federal funds for minor parents under 18 not participating in school activities or living in an adult-supervised setting.

The legislation gave states far more latitude than the defunct AFDC program. They could eliminate cash aid entirely if they chose, replacing it with any combination of cash and in-kind benefits; deny assistance to teen parents or other kinds of recipients; establish even more severe time limits; provide benefits to new residents at the same level as the state from which they emigrated for up to one year; or deny aid to persons convicted of a drug felony after August of 1996 unless they participated in a rehabilitation program. Nor did the legislation require uniform statewide standards. The legislation gave local welfare workers enormous discretion in deciding whom to cut off the rolls as recipients faced time limits and as they tried to comply with work requirements—discretion that had been greatly reduced in the AFDC program by legislation and court rulings. TANF families were no longer automatically eligible for Medicaid, even though persons meeting the old AFDC income standards could often receive it based on income. Food stamp benefits were reduced by not allowing families to deduct more than 50% of their rent or housing costs from their income to determine the amount of stamps they could receive. The maximum food stamp benefit level was reduced by 3%—and severe restrictions were placed on benefits of childless, able-bodied individuals.

TANF initiated other harsh policies. The welfare reform legislation restricted eligibility for SSI for children with behavioral disorders even though Congress agreed in 1997 to continue Medicaid benefits to children who lost their SSI eligibility. (It also disqualified adults whose primary disability is substance abuse or alcoholism.) It decreased funding for meals of children in family day-care homes as well as cut reimbursements for summer food programs and eliminated start-up and expansion funding of the School Breakfast Program. The welfare reform legislation eliminated the guarantee of childcare for welfare recipients trying to move into employment, leaving it to individual states to determine whether and for how long former recipients received subsidized childcare. It did, however, consolidate federal childcare programs into the existing Child Care and Development Block Grant and increased funding through a new childcare block grant. It also allowed states to transfer up to 30% of their TANF block grant funds to the Child Care and Development Block Grant and the Title XX Social Services Block Grant.

The intent of the legislation was to move the nation toward a work-based safety net by making TANF sufficiently harsh that many recipients would migrate to the labor force. Many of them assumed that TANF recipients would earn sufficient monies to make them as affluent as when they were on AFDC. They assumed that TANF would decrease childhood poverty.

When coupled with an incentive to work spurred by marked expansion of EITC eligibility and benefits, as well as rapid economic growth in the 1990s, TANF appeared initially not to harm former welfare recipients. Vast numbers entered the workforce and increased their incomes (Grogger, 2003). As welfare rolls declined by two-thirds or more in 32 states in the decade after enactment of TANF, most families remained "near poor," and few social or educational benefits were given to children in these families (DeParle, 2012). TANF met its most stringent test during the Great Recession from 2007 to 2009 and beyond. Cash rolls barely rose from 2007 through 2011 despite the most severe economic downturn since the Great Depression (DeParle, 2012). Only one in five poor children received cash aid—or the lowest level in 50 years. Sixteen states cut TANF rolls after 2007, often using the TANF grants they received from the federal government for other programs such as foster care or adoptions—or using cuts in TANF to cut their budget deficits (DeParle, 2012).

The ineffectiveness of TANF to cut poverty by increasing jobs of low-income women continued from 2009 to 2019. TANF benefits fell by more than 20% below 1996 levels when adjusting for inflation in most states (Floyd, 2017). The number of TANF beneficiaries declined. TANF cash benefits fell to (at least) 20% of their 1996 levels in 35 states plus the District of Columbia after adjusting for inflation (Floyd, 2017). These paltry benefits came on top of a 40% cut in real terms in two-thirds of the states between 1970 and 1996 when AFDC had existed before it was replaced by TANF in 1996 (Floyd, 2017).

Nor did the future look bright for low-income mothers. Under President Trump's proposed budget in early 2018, TANF would provide even fewer families with cash assistance and work opportunities (Floyd, Schott, & Pavetti, 2018). Trump and Republican leaders hoped to impose work requirements on users of SNAP and Medicaid. Kansas had embarked on a harsh revision of their TANF program in 2011, leading to "unsteady work and earnings below half the poverty line" as well as deep cuts in enrollments (Mitchell, Pavetti, & Huang, 2018). The lack of a robust welfare program for single mothers has had harsh implications for many as illustrated by their plight in the Great Recession. Although many of them received SNAP and/or Medicaid, as well as some help from relatives, charity, and food banks, they had no regular source of cash. It is not surprising that women often resorted to desperate, and sometimes illegal, measures. Some sold SNAP coupons, sold blood, skipped meals, doubled up with friends, scavenged trash cans, and shoplifted. Some returned to violent partners who had abused them and their children (DeParle, 2012). Researchers discovered that 4% of households with children lived on less than $2 per day—or twice the rate in 1996 (DeParle, 2012). The Bureau of the Census discovered that 10% of households headed by women lived in "deep poverty" with less than $9,000 per year—or the highest level in 18 years even as Paul Ryan declared TANF to be an "unprecedented success" (DeParle, 2012).

TANF has failed to reduce poverty among families with children. If 68% of families with children in poverty received TANF in 1996, only 23% received TANF in 2018 (Floyd, 2018; Pavetti, 2018c). TANF cash benefits had fallen by more than 20% in most states in 2017—and continued to erode (Floyd, 2018). These data about TANF are distressing on many counts. The vast majority of poverty-stricken families receive no assistance from TANF even when many studies show that children age two to five are more likely to improve later achievement in school when parents' incomes are raised (Sherman & Mitchell, 2018). Its mandatory work programs are costly and have limited long-term impact (Pavetti, 2018a). Its job preparedness programs are often so underfunded that they are ineffective. The proposed budget of President Trump would provide even fewer families with cash assistance and work opportunities (Floyd, Schott, & Pavetti, 2018). Cuts in TANF and harsh work penalties have mostly led to unsteady work and earnings below half of the poverty line (Mitchell, Pavetti, & Huang, 2018).

Deep bias against low-income mothers in the United States causes public officials to seek simplistic solutions. We now know that significant investments are needed to promote opportunity and upward mobility for low-income mothers. It costs between $10,501 and $13,750 per participant to help significant numbers of low-income mothers with limited education, skills, or work experience to move upward. Take the case of Project QUEST (Quality Employment Through Skills Training) in San Antonio, Texas, where single mothers received training to work in multiple sectors including health services and information technology. Participants earned an average of $28,204 after six years in QUEST—or $5,080 more than persons not selected to participate. By contrast, the "2019 Trump budget failed to invest in core job training programs" and focused on expanding work requirements "despite clear evidence that such requirements do not change individuals' employment trajectories over the long term" (Pavetti, 2018a).

General Relief (GR)

Although providing welfare assistance to children in destitute families (ADC) and later to one or both of their parents (AFDC), to elderly destitute persons (OAA), and to blind and then deaf persons (AB), the Social Security Act failed to enact welfare provisions for destitute persons not attached to a family unit, such as single nonelderly men or women, in 1935. They left welfare for these persons to local units of government or, in some cases, to those states that provided this assistance. This decision reflected widespread animus toward these persons because they were not seen as "deserving" as children, mothers, and elderly persons.

Nor have local units of governments and states been benevolent toward persons who receive GR. It was inevitable that these governments would fund GR benefits at lower levels than TANF or SSI because they received no matching federal funds. The offices that serve GR clients are often harsh, such as those with bulletproof windows and security staff. Many recipients of GR grants are single men and women who have been recently released from prison—persons who often find it difficult to find work because of their prison records. Others include single homeless persons. Still others are single veterans. Still others include married couples with no children. It makes no sense not to contribute federal funding to GR because it is inhumane to fund its benefits at such low levels. These low levels of benefits contribute to re-incarceration of persons released from prisons because it provides them with low benefits at a time when its recipients need resources to survive in the community.

Supplemental Security Income (SSI)

President Nixon developed the idea of joining means-tested programs for elderly and disabled persons funded by the Social Security Act into a single means-tested program in 1969 that soon was enacted as SSI. The federal government funded its benefits and administered it through the Federal Social Security Administration.

Applicants to SSI must meet specific qualifications. They gain access due to blindness if vision in their best eye is no better than 20/200 with glasses or with tunnel vision of 20% or less. Persons with other disabilities can qualify for SSI only if their medical records and/or a physician chosen by the state documents that they have been unable to work, or can be expected to be unable to work, for 12 continuous months or if they possess a disability that is likely to lead to death. Children under 18 can qualify for SSI if they have a medically validated "marked or severe" physical or mental disability that would disallow work if they were adults or that significantly interferes with their daily activities. Persons cannot obtain SSI for alcohol or drug dependence. Persons cannot get SSI any month when they are in prison, in violation of parole or probation, or fugitive from a felony (Los Angeles Coalition to End Hunger and Homelessness, 2010). Some states fund their own cash assistance programs for persons with short-term disabilities, such as injuries caused by automobile accidents or other physical trauma and mental distress caused by traumatic events. Five million adults received SSI in in 2011 with payments of $33 billion and with Medicaid services of $110 billion (Porter, 2012).

Some states offer State Disability Insurance (SDI) for persons with a temporary disability. In California, for example, persons may be eligible for SDI if they cannot work for eight consecutive days and if they have lost wages due to their disability. They must be looking for work (Los Angeles Coalition to End Hunger and Homelessness, 2010).

Social Security and Social Security Disability Insurance (SSDI)

Virtually all American citizens receive Social Security benefits when they retire. It is the most successful anti-poverty program in the United States because every low- and moderate-income senior receives its benefits save only civil servants who receive retirement payments from the government and a small number of persons who were unemployed most of their working years. Sixty-two percent of elderly recipients get at least half of their monthly income from Social Security—and roughly one-third get at least 90% of their monthly income (Williams, 2017). The senior poverty rate of 8.8% would rise to more than 40% if seniors did not receive income from Social Security (Center on Budget and Policy Priorities, 2016, as cited in Williams, 2017). Moreover, benefits of Social Security increase each year at the rate of inflation. Seniors can currently receive benefits just over age 66. Applicants must have 40 lifetime work credits to receive benefits, with a maximum of four earned each year, or usually at least working part time for 10 years. Social Security income contributions to Social Security by employees and employers vary according to the level of employees' contributions that are calculated as a percentage of their earnings.

Affluent people have much higher retirement pensions than low- and moderate-income persons for several reasons. During their working careers, employees and employers each pay a Social Security tax of 6.2% on employee's wages for a total tax of 12.4%. This tax is levied only on employees' annual wages that are less than a cap determined by the federal government. (This cap was $132,900 in 2019). When they retire, employees obtain their Social Security pension that is calculated as a percentage of their total lifetime earnings beneath the annual caps. Employees that have relatively low wages during their life-time employment, or who have worked only episodically or for short periods, have lower Social Security pensions than ones whose earnings are near or above the cap throughout their work career. Affluent persons are also far more likely than low- or moderate-income workers to receive pensions from their employers, whether in the private sector or the government. Most low- and moderate-income persons have pensions near or below $20,000 per year, often exclusively from Social Security, while many affluent persons have pensions far above $100,000—and into millions of dollars for high-level corporate officials. Affluent persons are also far more likely than low- and moderate-income persons to have stocks, property, and other assets that they can sell to fund their retirements.

A simple solution is worth considering: Lift the cap on wages that are taxed from the current $132,900 to $1 million—and use the added revenues to uplift Social Security pensions of low- and moderate-income persons.

Social Security gives many other kinds of benefits. Surviving spouses can receive pensions of their deceased marital partners. Social Security often contributes to college tuitions of children of deceased parents.

Unlike the means-tested SSI Program, SSDI is a universal entitlement for persons with medical certification of lasting and permanent disabilities of the mind or body. They must be under age 65 and have disabling conditions of the body or mind that prevent them from "substantial gainful activity" for at least 12 months or that will result in death. They need specific numbers of social security credits to qualify—credits achieved by paying payroll taxes into Social Security—although this work requirement is waived if they became disabled at or before age 22. They need medical evidence to determine their eligibility. Many persons use third-party disability representatives to make their applications and to appeal adverse eligibility decisions, whether from companies with trained specialists in filing and appealing claims or from law firms from their communities. These representatives screen applicants and may decline to represent them if they believe they will not meet SSDI eligibility requirements. (They are paid from 25% of retroactive awards made to applicants, not to exceed $6,000.) The Social Security Administration sometimes requires that persons with mental disabilities assign someone to disburse their benefits to them or to landlords and others—usually a relative or friend at no fee (Office of the Inspector General, 2010).

SSDI is a huge program, paying $128.9 billion in insurance payments to 10.6 million disabled workers and their family members in 2011 as well as $90 billion in Medicare benefits. Disability payments now constitute almost 20% of total Social Security benefits as compared with 10% in 1990 (Porter, 2012). Persons must often wait eight months to complete the application process due to backlogs—and sometimes more than a year to get an appeals hearing. Roughly 39% of SSDI applications were approved at the state level in 2005.

SSDI has grown rapidly due to the growth in numbers of elderly persons as well as their inability to find jobs due to their low skill and education levels and because many jobs have moved abroad or been supplanted with technology such as robots. Persons with back pain or depression often opt to work when they can find well-paying jobs but otherwise seek SSDI—and they find it easier to obtain SSDI after Congress softened the eligibility criteria in the 1980s, when more weight was given to pain and mental problems like anxiety (Porter, 2012). People can appeal adverse decisions, moreover, before administrative judges without testimony from personnel from Social Security who had rejected them (Porter, 2012). Once persons receive SSDI, they rarely return to work—and they receive no assistance in helping them receive work accommodations that might enable them to rejoin the labor force.

Beneficiaries receive only roughly $1,100 per month as well as Medicare coverage. They can earn $1,000 extra per month, but only 10% of them make any additional money.

The Earned Income Tax Credit (EITC) and Other Tax Benefits

The United States did not have a broad-based federal income tax until the nation was forced to enact one to finance the huge costs of World War II. President Franklin Roosevelt enacted a progressive tax system based on multiple levels of income. Everyone paid the same rate of 19% for the first level of income (up to $2,000) in 1943, for example, but rates were increased for each succeeding level up to the top marginal tax rate of 88% for taxable incomes greater than $200,000. This progressive system of taxation remained mostly intact through the 1950s and early 1960s, with the top marginal rate of 70% even as late as 1980. It promoted a relatively equal distribution of income by exempting very poor persons from any income taxes while levying successively higher marginal rates on persons as they moved up the economic ladder.

The progressivity of the tax system diminished over the ensuing six decades by regressive reforms. President Ronald Reagan cut the top marginal rates substantially to 28% in 1988. (Top marginal rates are taxes imposed on the highest portion of affluent persons' wages.) After rising to 39.6% in the Clinton presidency, it descended to 35% during the presidency of George W. Bush by 2003—and then was modestly increased in the

Obama presidency and reduced to 37% in President Trump's tax legislation in 2018. By contrast, most other industrialized nations have rates from 40% to 55% (Jansson, 2019a). Large numbers of tax loopholes were enacted over many decades that disproportionately favored affluent Americans, such as lower capital gains taxes, lower taxes on dividends, and lower taxes on income kept in foreign nations. The estate tax on affluent Americans was also drastically reduced over many decades. It was markedly reduced to the $5 million base in 2011 to a new $10 million base in 2018 through 2025. These tax cuts mean that many multimillionaires and some billionaires pay federal income taxes of 15% or lower as compared with federal tax rates of 30% or more for many persons who earn less than $100,000.

The federal tax code helps low- and moderate-income persons in several ways. Roughly half of American taxpayers are exempted from federal income taxes for four major reasons. They earn less than $26,400 for a couple with two children with their standard deduction of $11,600 and four exemptions of $3,700. (Sixty percent of households who owed no federal income tax in 2011 had incomes under $20,000.) There are elderly persons who pay no federal income taxes because their Social Security benefits are not taxed. There are persons who receive tax benefits back from the Internal Revenue Service from the EITC, the CTC, and the childcare credit account.

Some conservatives suggest that persons who do not pay federal income taxes are freeloaders. Virtually all employed persons do pay payroll taxes for Social Security and Medicare, however, so that even the poorest fifth of households paid an average of 4% of their incomes in federal taxes in 2007 even when they had average income of only $18,400 (Marr & Huang, 2012). Households with incomes between $20,500 and $34,000 paid 10.6% of their income in federal taxes. Even these figures are misleading, however, because these households pay state and local taxes—so the poor fifth of households paid 12.3% of their incomes in state and local taxes in 2011 (Marr & Huang, 2012). State and local taxes are not progressive, moreover, because the sales tax lands hardest on low-income persons who spend a high proportion of their income on food and other purchases—as do excise taxes on gasoline.

Moderate-income persons benefit from provisions in the federal tax code. They can deduct the interest on their mortgage payments, costs of seeking work, cost of obtaining job training, and work-related costs. They pay lower taxes than some persons with higher incomes. They received some additional benefits from President Trump' tax legislation in 2018.

The big debate in 2017 and 2018 was whether Trump's tax cuts helped the middle class even as they mostly favored affluent Americans. According to the Joint (Congressional) Tax Committee and the Tax Policy Center, his tax legislation provided a benefit for the middle class, but this benefit "pales in comparison to the tax benefits for the wealthy" (Kessler, 2018). It is true: Wealthy people pay more income taxes than non-wealthy people because they have far more money than non-wealthy people. The top 10% paid, for example, 80% of all taxes levied on individuals. Yet non-wealthy people received far less cash benefit from the Trump taxes—often in the range of $1,000 to $2,000 compared with millions of dollars for many affluent persons for the many tax cuts they received. Moreover, the tax cuts were scheduled to expire in 2027 unlike tax cuts for affluent Americans that have no expiration date.

Many states exempt low-income persons, seniors, renters, disabled persons, and veterans from taxes such as state income taxes and property taxes. Some give low-income persons tax benefits like the federal EITC.

The enactment of the federal EITC helped low- and moderate-income working persons in 1975, soon becoming the nation's largest anti-poverty program, with successive expansions in succeeding decades. Individuals and couples who care for qualified children receive a tax refund when they qualify for a tax credit that exceeds the amount of taxes that they owe. A family with three or more qualified children with incomes less than $43,279, if the couple is married, can receive a maximum tax refund, for example, of as much as $5,657 when they fill out a W-5 tax form during the year. Qualified children include those who are 18 or younger, 23 years of age or younger who are full-time students for five calendar months, and of any age who are found by physicians to be permanently and totally disabled. (Benefit tables show maximum amounts available for persons with fewer children as well as for married couples age 25 to 64 with no children.) Sixteen states supplement the federal EITC with their own tax-refund programs for low-wage workers. It lifted 6.5 million persons from poverty, including 3.3 million children, in 2009 (Williams & Johnson, 2010). The EITC provides an incentive to work and offsets Social Security payroll taxes that are otherwise onerous for many low-income persons. Persons can obtain help in getting the EITC by calling its hotline at 1-800-601-5552.

Were the EITC to be repealed, as some conservatives favor, the number of poverty-stricken working people would increase. Use of other welfare programs would increase if the EITC were cut, such as SNAP (Marr & Huang, 2012). A good case can be made to raise eligibility levels of the EITC to decrease income inequality in the United States among low-wage workers (O'Connor, 2011).

Substantial reforms of the federal tax system could not only increase income of moderate-income people but reduce income inequality by decreasing income of affluent Americans (Jansson, 2019a). More people in the lower economic echelons could be exempted from paying federal incomes taxes. Renters could be allowed to deduct some of their rent from their taxable income much as home owners can deduct some of their mortgage payments from their income taxes—or deductions of mortgage interest, mostly received by relatively affluent Americans, could be eliminated. Many parents in low-income families could escape poverty if the CTC was converted to a children's allowance as proposed by Senator Sherrod Brown in 2017 with the American Family Act (https://www.govtrack.us/congress/bills/115/s2018). The federal income tax code is unfair to renters, who cannot deduct all or part of their rents from their taxable income unlike home owners who can deduct their mortgage interest.

Advocacy advocates will push to rescind many of the tax cuts enacted by President Trump in late 2017 and early 2018 when scores of taxes on affluent Americans were cut. Tax increases on affluent Americans would serve three purposes. They would decrease extreme economic inequality in the United States that is higher than 20 other industrialized nations (Jansson, 2019a). They would add revenues that could bolster spending on safety net programs as well as other reforms in the policy sectors discussed in this book. They would provide additional revenues so that spending increases in domestic programs would not increase federal deficits—as took place in the presidencies of Ronald Reagan, George W. Bush, and Donald Trump, when these presidents coupled large increases in spending with large tax cuts. Jansson (2019b) presents various proposals for raising taxes, including the one in Table 9.3 authored by Citizens for Tax Justice in 2014.

Just imagine how safety net and other social programs might benefit from the $3.33 trillion that would be obtained by repealing these tax deductions for affluent Americans

TABLE 9.3 ■ Tax Reforms (Citizens for Tax Justice 2014)

• Repeal "deferral" for multinational corporations	$759 billion
• Repeal accelerated depreciation	714
• Repeal capital gains break	613
• Limit benefits of deductions and exclusions for high incomes	498
• Repeal stock dividends break	231
• Repeal domestic manufacturing deduction	145
• Enact 30% minimum tax for millionaires ("Buffett Rule")	70
• Calculate foreign tax credit on a "pooling" basis	59
• Repeal fossil fuels tax subsidies	51
• Bar interest deductions related to untaxed offshore profits	51
• Restrict excess interest used for earnings stripping	41
• Close payroll tax loophole for S corporation owners	25
• Repeal stock options loophole	23
• Prevent corporate "inversions" for tax purposes	19
• Scale back carried interest loophole	17
• Reform like-kind exchange rules	11
• Limit total savings in tax subsidized retirement plans	4
• TOTAL	$3.33 trillion

Source: Citizens for Tax Justice. (2014). Addressing the need for more federal revenue. Retrieved from http://ctj.org/ctjreports/2014/07/addressing_the_need_for_more_federal_revenue.php.

and corporations—not to mention cutting the $1 trillion increase in the national deficit that will be caused by the tax and spending policies enacted in the Trump presidency.

Unemployment Insurance (UI)

Many persons lose employment even during periods of economic growth, but millions of persons lose it during recessions such as the ones that began in 1991, 2001, and 2007—or in the aftermath of Hurricanes Irma, Harvey, and Maria in 2015 and 2016. Low- and moderate-income persons are particularly harmed by unemployment because loss of even a single paycheck can push them into poverty—or, in the case of impoverished persons, extreme poverty. National unemployment rates exceeded 12% in 2008, remained above 8% in 2012, and receded to 7% in 2014.

Although many persons who lost employment in prior recessions regained it relatively rapidly as the economy improved, many persons did not obtain re-employment within a year during the Great Recession and its aftermath—persons who were disproportionately less than 25 years of age, persons of color, and unskilled workers. Long-term unemployed persons suffer multiple hardships beyond their loss of wages. Their job skills erode. They lose confidence that they can find re-employment. They often develop mental problems such as anxiety and depression. They often develop health problems. Roughly 50% of college graduates under age 25 were unable to find full-time work in 2012, forcing many of them to move back into homes of their parents as they also had an average of $20,000 in college debt.

The United States has relied heavily upon its UI program, a joint federal-state program. UI is financed by payroll taxes of employers with each state determining its tax rates. (The federal government maintains an Unemployment Trust Fund with accounts for each state within it.) Each state runs its own UI programs, such as determining eligibility and benefits. Most states make benefits available for a maximum of 26 weeks, but Congress often extends benefits during economic recessions such as when it gave extended benefits to 2.3 million Americans in 2008—even extending maximum lengths to 99 weeks in 2010 and in 2012. However, persons with relatively low wages receive far lower benefits than persons with relatively high wages.

Unemployed persons must meet specific requirements to receive UI benefits that equal roughly 36% of their average weekly wage. They must be actively seeking work. They must be able to work. They must have lost their work because their employer terminated them, although they can still obtain UI if they left work for good cause. They are not usually eligible if they were discharged for misconduct or a labor dispute. They must usually have worked, full-time, for four of the last five calendar quarters before a claim is filed.

Applicants apply for benefits through their state unemployment agency and must usually wait two weeks to receive their benefits. They can often apply on the Internet or via an automated telephone call. Many states ask individuals to regularly verify that the conditions of benefits are still met. Disqualified or discontinued applicants can appeal these decisions within a specified time.

Unemployed persons are often eligible for other safety net programs simultaneously with UI, including SNAP, Section 8 subsidized housing, Medicaid, and public housing. Persons who exhaust their UI benefits are eligible for other safety net programs and often rely on others, such as parents or relatives, apply for TANF or GR, or obtain one or more part-time jobs. Unemployed persons can appeal denials of UI benefits and receive a hearing before administrative law judges in which roughly one-half of appellants win.

Workers' Compensation

Every state has a Workers' Compensation program that helps workers who have been injured on their jobs. Claims are paid by the state or private insurance company of specific states, whose premiums are paid by employers in those states. Workers' Compensation is a no-fault program so injured employees do not have to prove that their injury was someone else's fault (Los Angeles Coalition to End Hunger and Homelessness, 2010). Persons who have been injured at work file claims at Worker's Compensation offices in their region or hire a private attorney whose fees are set by law. Cases are often heard in a special administrative agency where administrative court judges preside, but appeals can be made to an appeals board and to the state's court system. Successful claimants may receive medical benefits for medical care of their work-related injury that may be treated by the employer's physician of choice or the worker's own physician if treatment extends beyond 30 days. They may receive temporary disability benefits up to two-thirds of wages lost because of the injury. They may receive permanent disability benefits for life. They may receive permanent partial disability. They may receive vocational rehabilitation services if they cannot return to the kind of work they performed prior to their injury. Their dependents may receive death benefits if a worker is fatally injured (Los Angeles Coalition to End Hunger and Homelessness, 2010).

Minimum Wage and Living Wage Policies

Data in 2018 make clear that extreme wage inequality persists in the United States. Wage growth continues to be highest for affluent people as has been the case for four decades. Wage growth for persons with higher education far exceeded other persons. Although women and Latinos made some gains in wages, African Americans continued to lag. The top 1% dramatically increased its share of the nation's income, growing by 148.6% in real dollars since 1980 compared with 21.3% for the lowest 90% (Gould, 2018).

A federal minimum wage was established in 1938 and has been increased by Congress periodically in ensuing years. Critics contend that the federal minimum wage has never been adequate to cover workers' costs—including at its level in 2018 of $7.25. Some employees are exempt from it, including persons reimbursed solely through tips (but their tips must add up to the minimum wage), some seasonal employers like summer camps, youth under age 18 for periods up to 90 days, and some not-for-profit institutions and colleges if they obtain certificates. Many states have their own minimum wages that supersede the federal one if they are higher than it, such as $9.00 in California in 2014—and President Obama signed an executive order in 2014 to raise the minimum wage for federal contractors to $10.10. (See http://www.dol.gov/whd/minwage/america.htm to see the level of your state's minimum wage.) Barack Obama sought an increase in the minimum wage to $15 in early 2014 in hopes it would decrease poverty but was unable to persuade Congress to enact it because many conservatives believed this increase would lead employers to decrease their workforces. A living wage is set at a level needed for workers to meet basic needs sufficient to keep a decent standard of living in workers' communities and to be able to save for future needs and goals. A living wage movement took place that had led 140 state, city, and local governments and universities to enact living wages by 2007 including Boston, Los Angeles, and St. Louis—usually $3 to $7 above the federal minimum wage (Wicks-Lim, 2009). States can establish their own minimum wage laws that can exceed the federal level. California plans to raise its minimum wage to $15 by 2022.

Credit- and Loan-Providing Agencies

The finances of low- and moderate-income persons are often jeopardized by victimization by credit- and loan-providing agencies. Millions of families obtained mortgages before the Great Recession when lending officers often failed to inform them that their interest rates would balloon in coming years beyond their ability to pay. See Table 9.4 for advocacy groups that support consumer-friendly regulations over credit- and loan-providing businesses.

TABLE 9.4 ■ Advocacy Groups Supporting Consumer-Friendly Regulations Over Credit- and Loan-Providing Businesses

1. Consumer Advocacy Group of America
2. Federal Bureau of Financial Protection

Many of them defaulted when they could not make payments. Nor could they sell their houses without sustaining huge losses because the price of their homes plummeted during the Great Recession. Many middle-class families also defaulted when they purchased expensive homes in neighborhoods with highly rated schools. When either parent loses his or her job or when a family member develops a serious health problem, these families must often default on their homes. Predatory lenders and credit card companies devastate the finances of many families as well. Needing cash to keep financially solvent, many families obtain loans with interest rates exceeding 20% only to find collection agencies pursuing them when they cannot repay their loans or their credit card debt.

Congressional Democrats and President Obama enacted the Dodd-Frank Wall Street Reform and Consumer Protection Act of 2010, the most sweeping overhaul of the financial services industry since the 1930s. It established the Federal Bureau of Financial Protection. It outlawed mortgage practices that entice consumers to take loans that they cannot afford or whose rates balloon excessively. It created the Bureau of Financial Protection to regulate the terms and pricing of financial products; eliminated complex fine print in financial contracts that consumers cannot understand; established regulations over payday lenders, check cashers, and other predatory nonbank financial services; and required greater transparency by issuers of credit cards. Consumers are now able to report violations by banks, lenders, and credit card issuers to a single federal agency for the first time in U.S. history at http://www.consumerfinance.gov.

Officials in the Trump administration weakened some of the consumer finance regulations that were enacted by the Obama administration, such as overturning forced arbitration that allows consumers the right to band together in class-action lawsuits over possibly unfair or illegal business practices (Lazarus, 2017). Some Republicans wanted to terminate the agency altogether.

Job-Related Programs

Job-related programs help unemployed persons; train students and welfare recipients; provide job placement and job-seeking skills for unemployed persons or persons who want to change their careers; create jobs; and give workers childcare, transportation, and social-service aids to make it possible for them to work. (See Table 9.5 for advocacy groups seeking expansion of job-related programs and protections.)

Training Students and Displaced Workers for Jobs

Many employment experts worry that large numbers of Americans will suffer from unemployment even when the American economy returns to normalcy because they lack needed skills for a "new economy" in which workers need technical skills. Recall that many unskilled jobs and semi-skilled jobs were lost in the American economy

TABLE 9.5 ■ Advocacy Groups for Improving Job-Related Programs and Protections

1. **Wider Opportunities for Women (WOW).** WOW seeks increased funding for the Workforce Investment Act, the Carl T. Perkins Technical Education Act, Pell Grants, the Women's Bureau, and the Women in Apprenticeship and Nontraditional Occupations Program (WANTO).
2. **Military Family Employment Advocacy Program.** It helps military spouses and dependents find jobs.
3. **Global Policy and Advocacy Jobs.** The Bill and Melinda Gates Foundation funds this program.
4. **Federal Equal Opportunity Employment Commission.** This federal agency works to curtail discrimination in employment.
5. **Economic Policy Institute (EPI).** The EPI advocates for higher minimum wages, workers' rights, and greater economic equality.

as industries moved factories to emerging nations in preceding decades and as factories replaced workers with technology, such as robots on assembly lines.

Many employment experts contend that the United States is not prepared to address these problems. The Carl D. Perkins Career and Technical Education Act was enacted in 1984 and reauthorized in 1998 when it used the term "career and technical education" (CTE) instead of "vocational education." It provided $1.3 billion in grants to states to link academic and technical content in high schools, junior colleges, and colleges. In light of the magnitude of training needs, however, this was a paltry annual sum that had declined to $1.16 billion by 2012 despite the Great Recession's impact on job prospects and was only slightly increased by 2018. Compared with Germany and Japan, American career and technical education often prepared students for relatively unskilled jobs such as offering them internships in fast-food restaurants rather than in industrial positions where they could become certified to perform specific technical tasks. Many secondary students refrained from engaging in vocational education because it had a low reputation.

President Obama proposed sweeping reforms of the Perkins Act in 2012 (U.S. Department of Education, 2012). He wanted to require states to work with workforce and economic development agencies to identify areas of focus for CTE programs embedded in the Elementary and Secondary Education Act (ESSA). He wanted strong collaboration among schools and colleges, employers, and industry partners. He wanted the private sector to contribute funds to CTE programs to strengthen its participation. He wanted to reward local recipients that exceed performance standards in placement rates and earnings of graduates. He wanted to retain CTE funding of $1.1 billion but align it with other administration initiatives "to align classroom teaching and learning with real-world business needs," including $2 billion in Trade Adjustment Assistant grants to strengthen community college programs and workforce partnerships as well as $8 billion for the Community College to Career Fund that seeks to train 2 million workers for high-growth industries and $1 billion to help 500,000 high school students participate in career academies.

Curiously, the Trump administration failed to make progress in reauthorizing the Perkins Act by May 2018 even though Trump promised to create many new jobs during his campaign in 2016 particularly with respect to residents of economically distressed neighborhoods and regions. Implementation of training programs was delayed. Sixty percent of high school graduates have credentials to achieve middle-class incomes of $42,000 or higher, whether through certifications, licenses, associate's, or bachelor's degrees. Another 20% can achieve middle-class incomes through manual labor. Yet 20% of high school students lack credentials and skills to succeed in an economy where most new jobs develop in health care, information technology, or financial services—or to become skilled workers in construction, repair, and machinery operations ("How ESSA and New Federal Funding," 2018). Left to their own devices, most of these students will live in poverty for the rest of their lives. Nor had the Trump administration made progress in enacting legislation to repair the nation's infrastructure by May 2018 other than producing a sketchy proposal that would rely on private money for funding to a remarkable extent.

The Job Corps

The Job Corps provides vocational and academic training for low-income youth ages 16 to 24—and currently serves 60,000 youths at 125 residential centers throughout the United States. Applicants must meet income requirements, be willing to participate in an educational environment, must agree to adhere to a no-violence, no-drug policy, and must agree to dormitory inspection rules. It provides multiple kinds of services, including career planning, on-the-job training, job placement, driver's education, basic health and dental services, and a biweekly living and clothing allowance. Some centers provide childcare for single parents. The program provides vocational training in advanced manufacturing, automotive and machine repair, construction, finance and business, health care and allied health professions, homeland security, hospitality, information technology, renewable resources and energy, retail sales and services, and transportation fields. Youth create a personal career development plan in their first 60 days with help of Job Corps staff before entering a career development phase that links vocational and academic training. They complete their training with a career transition period during which they obtain their first jobs.

Providing Persons With Tools to Find Jobs

The Workforce Investment Act (WIA) of 1998 streamlined federal and state job services by creating a one-stop delivery system to co-locate programs and providers in local workforce investment areas. Workforce Investment Boards (WIBs), chaired by a member of the private sector and with a majority of businesspeople, developed workforce education and career pathway programs for vulnerable populations. They provide youth opportunity grants to youth from high-poverty areas as well as universal access to its one-stop system and its core employment-related services. It allows consumers to select the training program that meets their career needs from certified training providers that bestow certificates upon trainees to help them obtain jobs. WIA programs are funded by federal grants to states, but its funding is inadequate so that only a fraction of persons receive them who need them.

President Obama signed the Workforce Innovation and Opportunity Act of 2014 into law. It streamlined and overhauled the WIA. It created common measures across workforce programs for adults and youth. It mandated a single-state workforce program for all core problems. It eliminated many ineffective programs. It strengthened the links of WIA to other federal training programs and unemployed populations, such as the Job Corps, TANF and GR recipients, returning veterans, and graduates of CTE programs in high schools, junior colleges, and colleges.

WIA training programs often are not effective with TANF recipients because many of them possess mental health and substance abuse problems for which they do not receive specialized help. Many TANF recipients have children with mental health problems or physical disabilities. A survey of job training programs for these TANF recipients suggests that a planned strategy does not exist. Some of them run afoul of TANF time limits and lose their grants. Some receive job training from WIA, but it is greatly underfunded and lacks social services to meet enrollees' needs. Some recipients remain on TANF when TANF administrators waive time limits for them (Zedlewski, Holcomb, & Loprest, 2007). Millions of unemployed Americans do not benefit from training programs. They were often trained for jobs that did not exist. Many were overcharged for training at for-profit colleges. Federal officials distribute the funds and states license the training programs, but neither federal nor state government provide sufficient oversight. Students are insufficiently monitored to see if training programs match their interests and needs (Williams, 2014).

Providing Persons With Childcare, Transportation, and Services

Many persons need childcare, transportation, and services to be able to work. TANF and GR welfare programs require welfare recipients to search for employment. They can use WIB offices to help with job training and placement by visiting one-stop WorkSource Centers that provide computers, fax machines, copiers, and job listings. They are also required to provide persons with career counseling, funds for transportation and childcare, and sometimes quality job training. Recipients can visit community colleges. They can seek high school diplomas for GEDs, learn English as a second language (ESL), and attend job readiness classes and some certificate courses and local public schools.

State departments of rehabilitation provide services for persons with physical or mental disabilities, including substance-abuse problems. These services can include vocational counseling and training, medical treatment, funds for tuition and books, funds for transportation and car modification, and reader and interpreter services (Los Angeles Coalition to End Homelessness and Hunger, 2010).

States' TANF programs usually provide childcare benefits to welfare recipients who have children. Considerable variation exists in the length of the childcare benefit. Many low-income persons cannot afford transportation to places of employment, particularly when places of employment are geographically distant from their residences. They receive transportation subsidies from some TANF programs and from some transit agencies. Some not-for-profit agencies donate or sell used automobiles to low-income persons at reduced prices.

States' abilities to fund childcare and training have been impeded by loose federal regulations that allow them to use TANF funds for other programs. Some of them have "raided" TANF funds to help finance state budgetary deficits. Those TANF recipients who do obtain jobs may receive TANF childcare and transportation subsidies but only for limited periods so that they find it difficult to maintain their employment.

Programs to Protect Workers' Rights

We have already discussed federal and state minimum wages as well as living wage ordinances in some local and state units of government, but these ordinances are sometimes not monitored or enforced (Levine, 2018). Employers may take improper deductions from paychecks, not pay owed wages, pay wages with checks that bounce, fail to give rest breaks, or fail to promptly pay all wages due to workers when they terminate their employment. Workers can obtain advice from their state's labor commissioner or from the U.S. Department of Labor—and use this information to decide whether to file grievances. They can file a wage claim in small claims court for up to $5,000 in California. They can contest discrimination on the basis of race, sex, religion, national origin, citizenship, age, disability, political affiliation, or sexual orientation, as well as sexual harassment, by filing a complaint with the Federal Equal Employment Opportunity Commission (FEEOC). The FEEOC will investigate complaints and give complainants letters authorizing them to file lawsuits if warranted. Workers can also contact their state's fair employment department for any of these problems (Los Angeles Coalition to End Hunger & Homelessness, 2010).

Nutrition-Enhancing Programs

People need sufficient nutrition, yet large numbers of Americans have inadequate diets. The U.S. Department of Agriculture (USDA) identifies four levels of food security:

1. **High food security** in which households have no reported indications of food-access problems or limitations
2. **Marginal food security** in which household members report anxiety over food sufficiency or shortage of food in the house—but with no or little indication of changes in diets or food intake
3. **Low food security without hunger** where reports suggest reduced quality, variety, or desirability of diet—but little or no indication of reduced food intake
4. **Food insecurity with hunger** with reports of multiple indications of disrupted eating patterns and reduced food intake

The USDA discovered that almost one in seven households (or 17.2 million of them) were in Groups 3 and 4 in 2010—and 3.9 million of these households had children in them. Roughly 6.7 million of these families were in Group 4. The median food-secure families in Groups 1 and 2 spent 27% more on food than the median food-insecure family of the same size and household composition (Coleman-Jensen, Nord, Andrews, & Carlson, 2010).

Now fast forward to 2016 by interacting with Figure 9.2. Go to map.feedingamerica.org. Click on your county to obtain food insecurity statistics about it. Compare your county or city with another county or city in your state. Discuss why the rates of food insecurity differ. Discuss how persons and organizations in your county or city could

FIGURE 9.2 ■ Food Insecurity in the United States by County in 2016

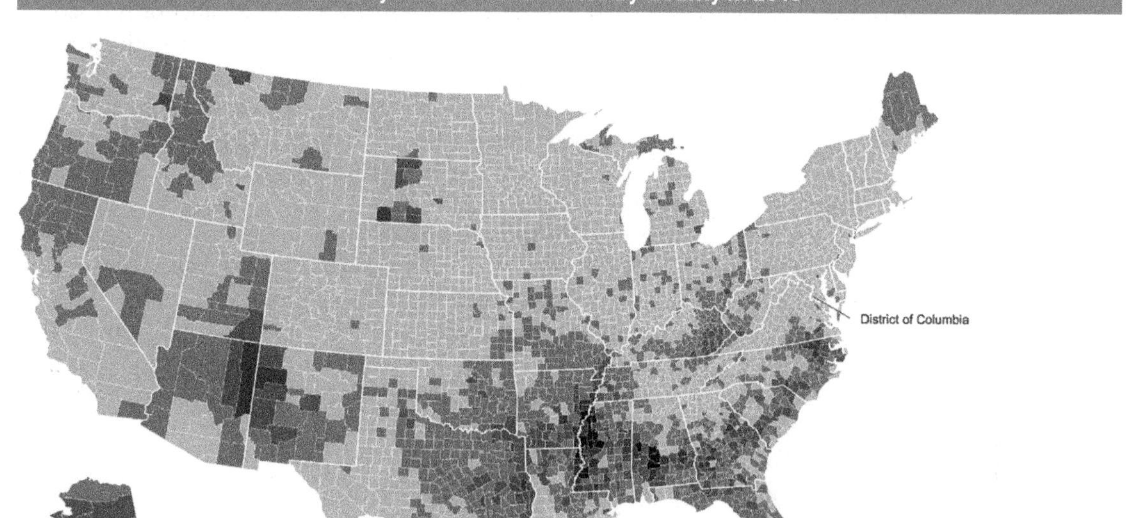

TABLE 9.6 ■ Advocacy Groups That Support Nutrition-Enhancing Programs

1. **Food Research and Action Center (FRAC).** FRAC is the leading not-for-profit organization seeking to eradicate hunger in the United States. It conducts research, monitors programs, coordinates and trains different nutrition programs and advocates, and conducts public information campaigns.
2. **The Feeding America Network.** The Feeding America Network coordinates a nationwide network of 200 member food banks and 60,000 food pantries and meal programs that serve 37 million Americans each year, including 14 million children and 3 million seniors. It also engages in advocacy with respect to public policy concerning hunger among children, rural residents, the working poor, and seniors.

engage in mezzo policy advocacy to increase usage of SNAP. Discuss how you can engage more frequently in micro policy advocacy to help the roughly 50% of eligible residents for SNAP in the United States who do not obtain its benefits. See Table 9.6 for advocacy groups that support nutrition-enhancing programs.

Supplemental Nutrition Assistance Program (SNAP)

SNAP is the largest program in the United States to help low- and moderate-income persons meet their basic food needs, serving 46.5 million persons in January 2012. It is a federal/state program administered by the states but wholly funded by the federal government. It is reauthorized every five years as part of the Farm Bill. Households must meet resource eligibility standards: They cannot have more than $2,000 in countable resources such as bank accounts or $3,520 if at least one person is age 60 or older or is disabled. (Income from most retirement plans is not counted.) A household's net monthly income cannot exceed 100% of poverty or $1,863 for a family of four. Households on TANF or SSI recipients automatically qualify. Able-bodied adults without dependents between 18 and 50 can get SNAP benefits only for three months in a 36-month period if they do not work or participate in a training program. (Go to http://www.fns.usda.gov/snap/ to find current eligibility requirements.)

SNAP benefits are very modest, averaging only $1.44 per person per meal—and the benefit will drop to $1.30 in today's dollars in November 2013, when a temporary increase expires. Enrollees are limited to food items with exclusions including alcoholic beverages, tobacco, soap, cleaning supplies, pet food, and paper products. SNAP benefits increasingly are obtained through electronic benefit transfer using plastic electronic cards obtained from local state or county offices. Almost all food stores are authorized to accept these cards that work like bank debit cards as the cost of groceries are subtracted from a household's account automatically. Congress made major cuts in SNAP in 2014 over the objections of the Obama administration and many congressional Democrats. Its spending levels are uncertain during the Trump presidency, which may impose a work requirement on its beneficiaries.

The Women, Infants, and Children (WIC) Program

The Women, Infants, and Children program (WIC) provides food to pregnant women and up to six weeks after the birth of an infant or the end of pregnancy, postpartum women up to six months after the birth of an infant or the end of the pregnancy, and breast-feeding women up to the infant's first birthday. It serves infants up to their first birthday and children up to their fifth birthday. No residency requirement exists, but some states require applicants to apply to a WIC clinic that serves the area where they live. States' eligibility requirements cannot exceed 185% of the federal poverty level. Recipients of Medicaid and TANF are automatically eligible.

Applicants must be screened by health professionals to determine if they are at nutritional risk, whether their own physician or WIC health professionals. They must have at least one of the medical or dietary conditions on their state's list of WIC nutrition risk criteria, such as anemia, underweight, a history of poor pregnancy outcome, or poor diet (http://www.fns.usda.gov/wic/). More than 9.1 million women, infants, and children used the WIC program each month in 2010. WIC provides food packages designed to meet special nutritional needs of low-income pregnant, breast-feeding, and non-breast-feeding postpartum women, infants, and children who are at nutritional risk. These packages contain a wide variety of foods that include fruits, vegetables, and whole grains—and can accommodate cultural food preferences and allergies. They include infant formula.

Federal School Breakfast, School Lunch, Child and Adult Care, and Summer Nutrition Programs

The Child Nutrition Act of 1966 authorized the School Breakfast Program. All children in participating schools and residential institutions are eligible for these federally subsidized meals, no matter their income, but only children from families with less than 130% of the federal poverty level can obtain free meals, and only children with family incomes between 130% and 185% of the poverty level can obtain reduced price meals. The federal government funds the cost of these meals under the administration of the USDA Food and Nutrition Service, which subsidizes recipient schools with $1.51 per free breakfast, $1.21 for reduced-price breakfasts, and $0.27 for paid breakfasts. More than 87,000 schools and institutions participated in served breakfasts to 11.7 million children from 2010 to 2011, with 83.4% of them receiving free or reduced price breakfast. Yet many children do not receive school breakfast even when they would qualify, such as 1 million low-income children in California when the state ranked 33rd in participation by low-income students in 2009 (Macvean, 2009).

National school lunch programs (NSLPs) served 31.6 million children in almost 100,000 schools and residential institutions, with nearly 20 million children receiving free or reduced-price lunches from 2009 to 2010. The federal government spent $9.7 billion for the NSLP in 2010. Like school breakfasts, school lunches meet nutritional standards and upgrade the nutritional intake of many children and youth. Eighty-eight percent of schools serving NSLPs also serve school breakfast.

Enactment of the Healthy Hunger-Free Kids Act of 2010 allowed the federal government to improve the school lunch and breakfast programs. Nutritional standards were upgraded for school meals. The USDA set nutritional standards for food sold by vending machines at schools. It connected 115,000 new children to the school programs by using Medicaid data to directly certify their eligibility. Twenty-one million at-risk children receive food from not-for-profits that receive federally subsidized food for the Child and Adult Care Program (CACFP) that gives meals to at-risk children after school, whether at dinnertime or at after-school programs.

The federal government also subsidizes summer nutrition programs often at educational enrichment and recreational programs with 2.8 million children participating each weekday. These programs are often administered by a nonprofit organization or a local government agency.

Meals on Wheels

We discuss this program in Chapter 8 because it mostly serves seniors.

Not-for-Profit Food Banks and the USDA

A huge network of food banks, food pantries, and soup kitchens in the United States distribute or provide food to hungry persons. Some are not-for-profit, community-based organizations. Some are faith-based organizations. Some are associated with shelters for homeless persons. They obtain food from grocery stores, restaurants, and private donations. They also obtain food from the USDA's Emergency Food Assistance Program (TEFAP) and the Commodity Supplemental Food Program (CSFP) that distribute food commodities to the states, which supervise their distribution to public or nonprofit organizations. The USDA's Senior Farmers' Market Nutrition Program (SFMNP) selects local government agencies or private not-for-profit organizations to provide low-income elderly persons over age 60 with coupons or checks to purchase fresh fruits and vegetables at authorized farmers' markets, roadside stands, and community supported agriculture programs.

SHELTER-ENHANCING PROGRAMS

No one should have to live on the streets where persons are exposed to the elements, lack privacy, and have poor access to medical care. Exposure to the elements and poor diet cause many illnesses. Employers often won't hire homeless people because they cannot provide a stable address in applications and often cannot groom themselves. Homeless people lack privacy.

The homeless population includes youth, women, women subjected to family violence or sexual harassment, college students unable to afford rent and tuition, families, graduates of foster care, disabled people, men, seniors, veterans, ex-offenders, persons discharged from mental institutions, LGBTQ persons, persons of color, persons evicted from apartments because they cannot pay their rent, persons with substance abuse, runaway youth, prostitutes, and other marginalized populations. About 40% of people in New York City's primary shelter system are children under age 18—or its single largest population.

Despite the development of programs to decrease homelessness, national statistics demonstrate little overall success for a variety of reasons. The cost of housing has markedly increased particularly in large cities with expensive housing. In cities like Seattle, San Francisco, Los Angeles, New York City, Chicago, and Dallas, high-wage employees in banking, technology, and other corporations have placed upward pressure on housing costs. Landlords have seized upon the tight housing market to raise rents such as $3,500 or more per month for a two-bedroom apartment. The wages of persons in the bottom 50% of the economic echelon have stagnated for almost four decades even as rents have moved rapidly upward. Remarkable numbers of persons are evicted when they cannot afford their rent in a humiliating process in which local law enforcement officials place their possessions on the street. Often unable to find new housing on the heels of eviction, individuals and entire families end up on the streets with their possessions. Lacking resources to store their possessions in storage facilities, homeless people often place them in tent-and-tarpaulin encampments.

Homelessness is also caused by untreated or poorly treated mental illness and substance abuse. Roughly 40% of homeless people have one or both of these conditions that are often not addressed by mental health and health facilities. These problems not only disrupt family relationships but limit the ability of homeless people to find and sustain employment that in turn makes them not able to afford rental housing. Most new construction in major cities focuses on upscale housing that is out of the reach of many low- and moderate-income persons. The workforce in major cities, which used to contain jobs in manufacturing industries, has veered toward white-collar jobs for persons with considerable education.

The response to homelessness in the United States lags policies in many other nations. Stockholm, Helsinki, and Copenhagen, for example, have small homeless populations because Sweden, Finland, and Denmark invest far greater resources in housing and services for homeless people than the United States.

Homeless people often draw the ire of local residents and businesses that have led to punitive laws. The Los Angeles City Council, for example, enacted legislation by a 13–1 vote that limits storage on sidewalks, parkways, and alleys to a 60-gallon trash container that is roughly the size of a city trash bin and allowed homeless people to be cited or arrested for failing to take down their tents between the hours of 6 a.m. and 9 p.m. (Holland, 2016).

Local politics often impedes the development of shelters and housing for homeless people. When the Los Angeles City Council and the Los Angeles County Board of Supervisors approved bond issues totaling billions of dollars to develop housing and shelters for homeless people, they encountered the wrath of many local citizens under the mantra of not in my backyard (NIMBY). They flooded hearings of local officials, initiated lawsuits, and threatened to withdraw support from local politicians who failed to block these shelters and housing units. This protest emanates from a combination of prejudice against homeless people, a belief that property values decline in areas where they live, and actual experiences with them in their lives. In Los Angeles, for example, with 58,000 homeless persons in Los Angeles each day, drivers exiting freeways encounter homeless people requesting money; libraries, train stations, and public parks become refuges for them; and people often see their encampments in residential neighborhoods and commercial districts (Times Editorial Board, 2018).

It is likely that homelessness cannot be effectively addressed without rent control in major cities where the cost of renting is out of reach even for many middle-class people. Rent control exists in major cities but in relatively restricted areas. Rent control pits landlords, realtors, and developers against advocates to end homelessness. Some cities, such as Austin, Texas, are experimenting with small houses that are relatively inexpensive to construct—often less than $100,000 rather than $300,000 for a conventional house with two bedrooms—and that do not take up the limited unoccupied space in urban areas.

The American remedies have been patchwork and underfunded, relying on a combination of programs and housing from public, not-for-profit, and for-profit sources. Many non-profit shelters for homeless people exist in many cities that provide residence for various lengths of time. The U.S. Department of Housing and Urban Development (HUD) developed and funds Housing First, an evidence-based program that moves homeless people through a specific sequence that includes a reception stage, a period of shared housing with supportive services, a time-limited occupation agreement based on special conditions such as receiving support services to help with mental health and substance abuse problems, and a regular dwelling unit with a rent contract and supportive services. HUD researchers discovered in multiple cities, Australia, and Canada that placing homeless persons immediately in housing rather than giving supportive services to them as homeless people cut overall homelessness by 72%, reduced incarceration rates by 76%, and cut substance abuse detox by 82% (see https://pathwaystohousingpa.org/). Social workers have become an integral part of support services for homeless people across the United States.

New York City has an elaborate system of housing and street services for homeless people, spending $1.8 billion per year for shelters, apartments, hotel rooms, and programs (Stewart, 2018). Homeless persons proceed through specific stages. They register at the city intake center called Prevention Assistance and Temporary Housing (P.A.T.H.), where homeless people apply and are interviewed. They receive temporary shelter for up to 10 days while the city determines if an applicant is homeless. They proceed to placement in a long-term shelter where they reside for 414 days on average. They then apply for rental assistance and search for an affordable apartment. The city has roughly 77,000 persons who are homeless; that has increased from 57,000 persons in 2012 even as it has spent millions of dollars on rental assistance programs and legal assistance to stop

TABLE 9.7 ■ Advocacy Groups That Seek Adequate Shelter

1. National Coalition for the Homeless
2. National Law Center on Homelessness and Poverty
3. American Civil Liberties Union (ACLU)

evictions. Housing coordinators assume the role of matchmakers who persuade landlords to rent units to homeless people who are traditionally shunned by landlords and who try to persuade them not to evict tenants or to raise rents so high that they cannot afford them (Stewart, 2018).

People need secure and adequate shelter, yet roughly 170,000 families with children lived in temporary shelters in 2010 according to HUD. Roughly four times as many families doubled up or occupied other unstable home situations according to the U.S. Department of Education. Considerable research documents the negative effects of housing instability on children's mental health and school performance as well as their health (Price, 2011). See Table 9.7 for advocacy groups that seek adequate shelter.

Section 8 Rental Housing Subsidies

Section 8 rent subsidies are provided by HUD in privately owned rental units, senior-citizen housing, subsidized housing units owned by the government with sliding rates of monthly rent or mortgage payments, and housing constructed by not-for-profit or religious groups such as Habitat for Humanity. They include shelters for temporary stays for homeless persons that are financed by not-for-profit or religious groups or by local, state, and federal governments. They include housing for homeless veterans.

Shelter for Veterans

Multiple housing programs assist veterans. They include U.S. Department of Housing and urban Development—VA Supportive House (HUD—VASH) that combines housing vouchers with supportive services to help homeless veterans and their families. It includes VA Home Loans for Veterans as well as housing grants. They include VA home loans to take cash out of home equity to help finance college and other expenses.

Continuum of Care Systems (COC)

The McKinney-Vento Homeless Assistance Act provides funds for competitive grants provided by HUD to local jurisdictions for a coordinated, community-based process of identifying needs and building a system to address them. Coordinated programs require the development of shelters, including use of a supportive housing program, rental assistance for homeless persons with disabilities, and single-room occupancy programs (SROs) that give rental assistance to homeless persons. Strategies must be developed to address physical, economic, and social unmet needs of homeless persons. HUD funds, as well, an Emergency Solutions Grant (ESG) program to provide street outreach and emergency shelter services as well as rapid-rehousing assistance that helps homeless individuals and families obtain permanent housing quickly. All of these programs are inadequately funded to address homelessness—and need to be supplemented with funding from localities and states. Information about housing resources in local areas is found on the HUD website at the home page of the Homelessness Resource Exchange (HRE).

Public Housing

Federally funded public housing programs were initiated in the New Deal for low-income persons, including seniors. They developed a poor reputation in subsequent decades because they often contained many units that were located in low-income areas. Some of them have been demolished; others have been gradually converted into units owned by their residents. They remain an important source of housing for low-income persons.

ASSET-CREATING PROGRAMS

People not only need current income to meet their basic needs, but also need assets—that is, backup resources when they face adversities such as loss of work or uninsured medical costs, want education for themselves or family members, want to care for parents and relatives, and want to purchase homes. They can use assets as collateral to obtain loans or mortgages. Assets allow persons to take risks, such as leaving a specific job to find a better one. Low- and moderate-income persons usually have no or few assets as compared with more affluent persons, making them less able to meet these contingencies. (See Table 9.8 for advocacy groups that help members of vulnerable populations obtain assets.)

Social Security and Private Pensions

Many persons need economic assistance after they retire because they no longer receive their work-related income. Moreover, many elderly persons retire with no assets, such as homes, and are in debt.

Social Security is not a generous program. It replaces, on average, roughly 40% of workers' earnings before they retired with an average benefit of $12,530 in 2007. The

TABLE 9.8 ■ Advocacy Groups to Enhance Asset Creation by Vulnerable Populations

1. **American Association of Retired Persons (AARP).** AARP advocates for maintaining Social Security, although some critics view it as accommodating excessively to groups that favor benefit cuts.
2. **National Academy of Social Insurance (NASI).** NASI seeks to retain benefits of Social Security by expanding taxes on employers and employees.
3. **National Committee to Preserve Social Security & Medicare.** This organization has assumed a leading role in fighting efforts to privatize Social Security and Medicare.
4. **Center for Enterprise Development (CFED).** CFED engages in policy advocacy at federal, state, and local levels to enable low-income families to increase their savings, start businesses, attend college, obtain health insurance, and increase economic mobility.

United States ranks low in 30 nations with advanced economies in the generosity of its public pensions. Even with its modest payments, it keeps 20 million Americans out of poverty. (Roughly one-half of retirees have no assets when they retire including houses—and often possess credit card and other kinds of debt.) It helps stabilize the American economy by giving persons spendable resources. It provided benefits to 55 million persons in 2012.

Social security has become even more important because seniors have found that two other sources of income during retirement have become less reliable: employer-funded pensions and personal savings. Many employers have cut back or eliminated their contributions to employees' private pension plans, such as 401(k) or 403(b) plans, without paying taxes on these savings until they receive them in annual payments after they retire. (Relatively affluent workers disproportionately receive private pensions because their employers are more likely to match their contributions than employers of less affluent workers.)

Many conservatives say they want to save Social Security from bankruptcy by cutting its benefits, but many persons argue that its benefits should be expanded (Hiltzik, 2012). Benefits are indexed annually for inflation but not sufficiently to reflect seniors' actual living costs such as medical care and the goods and services they consume. Social Security's formula for computing benefits discriminates against women because they currently spend only an average of 27 years in the labor force due to their caregiving with children, disabled family members, and parents. (Benefits are calculated on the best-paid 35 years of workers' working lives.) Terry O'Neill, president of the National Organization of Women, wants a "caregiver credit" for women in which Social Security would count caregiver years at a value equivalent to half the median wage for full-time work, which is $41,000 (Hiltzik, 2012).

Benefits might be expanded and the solvency of the Trust Fund protected, some persons contend, if the payroll tax was expanded to include all of workers' income rather than only their income to the current cap of $132,900.

Survivor's Benefits

Benefits for survivors of social security beneficiaries were added to Social Security in the 1950s and 1960s. The following persons receive specific percentages of benefit amounts of deceased workers up to a family limit of 150% to 180% of the basic benefit rate:

- Widows or widowers can receive 100% at full retirement age or older.
- Widows or widowers can receive 71.5% to 99% of the deceased worker's basic amount at age 60.
- Disabled widows or widowers, age 50 through 59, receive 71.5%.
- Widows or widowers of any age receive 75% when caring for a child under age 16.
- Children under age 18 or disabled receive 71% (or up to age 19 if still in elementary or secondary school).
- Dependent parent(s) of the deceased worker, age 62 or older, receive 82.5% (for one surviving parent) or 75% to each parent (for two surviving parents).

The survivors' benefits of children should be expanded to cover part or all of their college tuition costs, which had been added to their benefits in 1965 by Congress through age 21 but were removed by President Reagan in a cost-cutting move in 1981. This benefit would pay for itself because college graduates pay far higher payroll taxes than non-graduates because they earn 60% to 70% higher wages than high school graduates (Hiltzik, 2012).

HOME OWNERSHIP PROGRAMS

HUD has a variety of programs geared toward helping persons own their own homes. It provides subsidies to not-for-profit agencies to construct homes to be purchased by low- and moderate-income persons. The Federal Housing Agency (FHA), Fannie Mae, and Freddie Mac issue many of the nation's mortgages along with private banks. The federal government has several programs geared toward helping persons who are "underwater" (whose mortgages exceed the market value of their homes) to avoid foreclosure even if they have been relatively ineffective.

INDIVIDUAL DEVELOPMENTACCOUNTS (IDAS)

We have discussed how low-income persons often lack assets, such as houses and savings accounts, that they could invest or use as economic backups. IDAs are savings accounts in which persons receive an additional deposit, called a match, every time they place their own funds in them. These plans are usually sponsored by not-for-profit agencies that recruit members and offer them financial counseling. They target low-income populations to help them obtain resources to purchase houses, send children to college, and make other personal investments. The Personal Responsibility and Work Opportunity Reconciliation Act of 1996 authorized states to fund IDAs with use of federal grants. Professor Michael Sherraden at the George Warren Brown School of Social Work at Washington University has championed IDAs and directs their Center for Social Development (Sherraden, 1991).

UNDERSTANDING HOW SEVEN CORE PROBLEMS EXIST IN THE SAFETY NET SECTOR

Core Problem 1: Engaging in Advocacy to Protect Persons' Ethical Rights, Human Rights, and Economic Justice—With Some Red Flag Alerts

- Red Flag Alert 9.1. Help persons apply to specific safety net programs. Many people do not apply to specific safety net programs that might help them meet their basic needs. They may be unaware of the programs. They may falsely believe they do not qualify. They may believe they will be stigmatized as others learn about their enrollment—not realizing that many safety net programs require them to maintain strict confidentiality. They may not know where to go. They may fear they cannot fill out forms because they have limited literacy or because they speak another language. They may fear they cannot locate needed pay stubs and the many documents that need to be assembled. They may fear they will have to wait for many weeks to receive benefits or services due to bureaucratic delays. These non-enrolled persons and families sacrifice billions of dollars of income and nutrition in aggregate.

Social workers should sometimes consider macro policy interventions. Perhaps application procedures for specific programs can be simplified to make them more consumer friendly. Perhaps long waits can be shortened. Perhaps brief videos can be developed to educate persons about how to apply for a program.

- **Red Flag Alert 9.2.** Help persons file appeals when decisions by local, state, or federal officials appear incorrect or discriminatory. Specific appeal procedures are defined for each safety net program. Clients should be encouraged to file appeals whenever they believe that they have not received benefits or services to which they have been entitled. They should be encouraged to file appeals when they believe that staff has discriminated against them on the basis of their gender, race, disability, age, or any other personal characteristic. Social workers engage in micro advocacy when they inform applicants that they have a right to request a hearing to appeal a specific decision. They may get a favorable decision by an agency just by asking for a hearing. Applicants should insist on a hearing even when told it will be fruitless. They should retain all documents and obtain copies of documents they receive or bring. They have a right to see relevant statutes and regulations. They should keep good records. They have a right to see their case file and to copy any documents from it. Persons who believe they are treated unfairly because they are disabled or have health problems can send a letter to the Civil Rights Division of the Department of Justice, P.O. Box 66118, Washington, DC 20035-6118. In the case of TANF, applicants should complain to their worker's supervisor when they are not satisfied with any decision or call HELPline for their welfare office. If this fails, they should ask to speak to the deputy director and then the director of their welfare office. If they are still dissatisfied, they should file for a fair hearing by writing to the appeals office in their region or by calling a designated toll-free number, possibly seeking assistance from an advocate, legal aid, a local government official, or a local legislator. (The request for a fair hearing should take place before the date that their benefits will be reduced or stopped as explained in a special notice that the public agency is required to send them notification that explains why the action was taken and when it will occur.) The states send a notice to the applicant that tells him or her when the hearing will occur and where, usually in three or four weeks—but the applicant can request a delay to prepare for it. An appeals hearing specialist is assigned to the case to represent the public agency's position. Hearings for disabled or homebound persons can be held by phone if requested. An applicant can file a lawsuit to contest rulings at these hearings. Applicants can get free legal help from federally funded legal aid programs if their incomes roughly qualify them for TANF.

- **Red Flag Alert 9.3.** Help persons appeal decisions or obtain information from supervisors. Applicants to safety net programs should ask to talk with supervisors when they believe that intake or service staff have inadequate information or are making erroneous decisions.
- **Red Flag Alert 9.4.** Help persons obtain legal and related assistance. Legal aid programs provide invaluable assistance to social workers and clients who want to know if they have been wrongly denied benefits or services from safety net programs such as the state programs that implement TANF. An inquiry from an attorney may lead eligibility and other public personnel to reconsider their decisions.
- **Red Flag Alert 9.5.** Refer persons to specialized advocacy groups. In many cases, too, social workers can refer specific clients to not-for-profit advocacy organizations that specialize in income, housing, nutrition, and other human needs in their communities. Their staff not only will be familiar with existing eligibility and service policies, but may act as intermediaries with specific public organizations. They include in Los Angeles County, for example, the Bus Riders Union, California Food Policy Advocates, California Partnership, the Coalition for Humane Immigrant Rights of Los Angeles, the Community Coalition, Hunger Action LA, L.A. Alliance for a New Economy, L.A. Community Action Network, Legal Aid Foundation of Los Angeles, Maternal & Child Health Access Project, and Neighborhood Legal Services of Los Angeles.
- **Red Flag Alert 9.6.** Seek higher eligibility and benefit levels in a polarized context. Social workers engage in policy advocacy in a polarized context as they seek to raise eligibility and benefit levels for many safety net programs. Does social justice require expansion of these levels to enhance social justice in a society with marked economic inequality?

Barriers

Widespread animus in the United States toward persons who need or use many safety net programs makes it difficult to obtain sufficient resources, shelter, and benefits for low- and moderate-income persons. Current policies are sometimes punitive in nature, such as federal law that precludes persons with prison records from using TANF—even women exiting prison for minor drug offenses who must house, feed, and cloth their children. Some staff persons in local welfare offices are themselves hostile to their clients. It is difficult for many persons to understand the complex rules and accounting procedures of many safety net programs such as TANF.

POLICY ADVOCACY LEARNING CHALLENGE 9.3

CONNECTING MICRO, MEZZO, AND MACRO POLICY ADVOCACY INTERVENTIONS TO ADDRESS ETHICAL ISSUES IN THE SAFETY NET SECTOR

Do the Math to Decide What a Family Needs to Survive

Some persons contend that eligibility levels for many safety net programs are excessively high, allowing persons to gain benefits from them when they do not truly need them. Find the eligibility level of the food stamps program (now called SNAP) for a single mother with two dependent children on the Internet. The eligibility level of TANF varies by state. Find the eligibility level for your state.

Determine likely expenditures for the single mother and her two dependent children, including rent, living expenses like utilities and heat, food, transportation, clothing, necessary personal items, entertainment, medical and dental expenses, and miscellaneous expenditures. Use your personal experience to make your estimates.

Compare the eligibility level of SNAP and TANF with your calculations. Are they excessively high? Compare the actual benefits this woman and her children would receive under SNAP and TANF. Are they excessively high?

Now assume that the mother finds work at the minimum wage in your state. (It is the higher of the federal minimum wage or the minimum wage established by your state.) Her working hours are from 8 a.m. to 5 p.m. with a one-hour release time for lunch. Assume she has to travel 15 miles to get to work or 30 miles round trip. Assume she has to use childcare for both children from 3 p.m. (when preschool ends) to 5:45 p.m. (when she returns from work). Select a probable rent level for this woman and her family in a relatively safe neighborhood. Would SNAP and TANF suffice to meet the family's expenses?

Where do you position yourself in the debate about whether safety net programs are excessively generous or punitive? Would you support or oppose political candidates who want to markedly cut or markedly increase eligibility levels and benefits of SNAP?

Many Americans believe that poor persons turn to safety net programs unnecessarily, sometimes using words like "lazy," "immoral," "welfare queens," and "parasites" to describe them. Develop your own position with respect to Policy Advocacy Learning Challenge 9.2.

Core Problem 2: Engaging in Advocacy to Promote the Quality of Safety Net Programs—With Some Red Flag Alerts

- Red Flag Alert 9.7. Conduct a safety net needs assessment. Many members of low-income and other vulnerable populations need benefits from one or more safety net programs, yet they often lack personal knowledge of them. Social workers should routinely conduct assessments of their clients' needs for benefits and services of safety net programs to which they may be entitled. They can hand them, for example, a list of safety net programs with their rough income or other eligibility requirements clearly stated—and ask them to identify which ones might help them meet their basic needs. They can provide them with handouts about relevant programs or give them (if they are computer savvy)

POLICY ADVOCACY LEARNING CHALLENGE 9.4

CONNECTING MICRO, MEZZO, AND MACRO POLICY ADVOCACY INTERVENTIONS IN THE SAFETY NET SECTOR

Initiating a Community Outreach Campaign

A social worker was worried that many of her clients from low- and moderate-income families were not using SNAP to the detriment of their food intake as well as their family finances. She discovered a tendency of many of them to curtail food expenses, such as by organizing their meals around low-cost items like macaroni and cheese to the detriment of also eating vegetables, fish, fruit, chicken, and low-fat meat. She discovered on a website established by the Food and Nutrition Service of the U.S. Department of Agriculture (FNS) that federal officials welcomed collaboration with state and local agencies, advocates, employers, and community and faith-based organizations in reaching out to eligible low-income persons not using SNAP. Go to their website at http://www.fns.usda.gov/snap/outreach/, and develop some ideas for an outreach strategy in your local area or region.

websites to locate relevant information. When specific clients urgently need specific benefits and services, they can monitor their progress in seeking and obtaining them.

- Red Flag Alert 9.8. Broaden community awareness. Social workers should participate in local publicity campaigns in liaison with the mass media so that persons learn about specific safety net programs—possibly in liaison with state or federal campaigns such as ones to publicize WIC and SNAP. (See Policy Advocacy Learning Challenge 9.4.)
- Red Flag Alert 9.9. Bring applications to completion. Many people fail to complete the application process for specific safety net programs. They may disagree with a staff person about some conditions. They may lack important documents. They may not understand the staff worker. They may believe that they have an emergency that will not be addressed by bureaucratic procedures that may not allow them to obtain SNAP, TANF, Medicaid, or other benefits for weeks or even months. Or they may become frustrated with the laborious application process and leave. Social workers should coach clients prior to their visits, such as highlighting situations where they could ask to see a supervisor, ask for a translator, and ask permission to bring certain documents at a later date but without delaying the likely date they will receive benefits.
- Red Flag Alert 9.10. Applicants to some safety net programs are unaware that specific exceptions exist in their regulations depending on the circumstances of specific persons. In TANF, for example, persons may be able to obtain cash immediately when they are being evicted from an apartment, when they are subject to intimate partner violence, and when they have specific medical needs. Agency staff may not understand these exceptions or may sometimes direct an applicant elsewhere. An applicant may have to talk with a supervisor or call legal aid if they persist in failing to honor exceptions.

- **Red Flag Alert 9.11.** Many persons often fail to renew their eligibility. Perhaps they have changed locations or jobs and wrongly believe that it will be difficult to renew their services or benefits. Perhaps they had bad experiences when they applied on a prior occasion, such as long waits in an office, hostile treatment from staff, or long waits to obtain benefits or services. Micro advocates need to help these persons understand that they may cause themselves or their families harm by not renewing their eligibility such as poorer nutrition, lack of needed resources, and evictions that could have been forestalled if they had additional funds to pay rent. Advocates can work with public agencies to make it less difficult to reapply for specific benefits or services. Persons who have already qualified on one or more prior occasions might, for example, receive expedited service, simpler application forms, and mail-in options.
- **Red Flag Alert 9.12.** Many persons do not apply for complementary programs. Someone who receives TANF may be eligible for Medicaid, SNAP, housing assistance, temporary shelter, rent subsidies, help with the cost of moving, special payments for evictions, extra money for higher food costs due to a medical condition or breast-feeding, welfare-to-work services like counseling and job training, additional funds for pregnancy, a monetary bonus for a woman who receives child support, and a higher payment for those children who qualify for foster care payments. Persons who use combinations of programs improve their economic condition more dramatically than ones who use only single programs. Social workers should ask agencies to inform their clients of complementary programs for which they are eligible and give them fact sheets, referrals, and other assistance to use them. A macro advocate could attempt to promote formal linkages among these programs at higher levels of administration of each of them by asking them to form partnerships of outreach in communities with large numbers of poor persons.

Core Problem 3: Engaging in Advocacy to Promote Culturally Competent Safety Net Programs—With Some Red Flag Alerts

- **Red Flag Alert 9.13.** Clients are denied benefits or services based on discriminatory actions of service personnel. Civil rights statutes and regulations at federal, state, and local levels of government protect persons who use safety net programs from discrimination.
- **Red Flag Alert 9.14.** Persons with limited English proficiency are not given translation services under Title VI of the 1964 federal Civil Rights Act, which declares that "no persons in the United States shall, on the ground of race, color, or national origin, be excluded from participation in, be denied the benefits of, or be subjected to discrimination under any program or activity receiving financial assistance."
- **Red Flag Alert 9.15.** Persons with disabilities are not given special assistance in filling out forms, gaining access to offices, obtaining forms, or using benefits or services. The Americans with Disabilities Act (ADA), as well as other federal, local, and state legislation, require that they be given this assistance.

- Red Flag Alert 9.16. Cultural preferences of persons are not heeded when providing services and benefits in specific safety net programs. The federally funded school breakfast and lunch programs, as well as WIC, have increasingly honored food preferences of persons from specific ethnic groups.

Core Problem 4: Engaging in Advocacy to Promote Preventive Safety Net Programs—With Some Red Flag Alerts

- Red Flag Alert 9.17. Help persons avoid adverse cascading events. Social workers need skills in predicting when persons and families are likely to experience adverse cascading events. Low-income families and single heads of households are particularly vulnerable to adverse cascading events that precipitate a downward cycle. Assume, for example, that a single head of household manages to hold three part-time jobs to support her two children, including funding their childcare, medical bills, dentist bills, and survival needs. When she loses one of the jobs and her income plummets, she no longer can fund an adequate diet for her children and has to default on her mortgage. Even though SNAP helps her, her remaining income disqualifies her from TANF—and she cannot receive unemployment insurance because she still holds two part-time jobs. When the bank moves aggressively to evict her from her home, she loses her shelter—and the family resorts to living in their car.

Advocates may arrest adverse cascading events through job training and placement services, gaining eligibility for one or more safety net programs, finding (and keeping) affordable housing, avoiding foreclosure, helping homeless persons find shelter, improving management of personal finances, and finding social and mental health services for depression and anxiety.

- Red Flag Alert 9.18. Reform policies that encourage dependency. Some safety net programs create dependency excessively and need to be reformed. TANF creates excessive dependency when it fails to provide effective job training and childcare services to recipients—benefits that have been cut by many states due to their deficits during the Great Recession and beyond. Social Security encourages persons to leave the workforce prematurely and at great cost to the broader society (see Policy Advocacy Learning Challenge 9.5).

Core Problem 5: Engaging in Advocacy to Promote Affordable and Accessible Safety Net Programs—With Some Red Flag Alerts

- Red Flag Alert 9.19. Many safety net programs impose excessive non-monetary costs on persons who use them, however, such as long waits for Section 8 subsidized rental housing, public housing, SSI, SSDI, and TANF and excessively complex applications. Policy advocates should seek to diminish these non-monetary costs. They should also reduce fees and deductibles associated with many other public programs.

POLICY ADVOCACY LEARNING CHALLENGE 9.5

CONNECTING MICRO, MEZZO, AND MACRO POLICY ADVOCACY INTERVENTIONS IN THE SAFETY NET SECTOR

Reforming SSDI to Encourage Work

Even as the health of the American population has markedly improved since 1990, annual SSDI disability payments to Social Security recipients have increased from roughly $200 billion to more than $700 billion—or from 10% to 18% of total Social Security benefits (Porter, 2012). This change partly occurred because Congress relaxed eligibility criteria by giving more weight to subjective factors like pain, anxiety, back pain, and other muscular problems as compared with traditional medical diagnoses. Rejected applicants were allowed to appeal before administrative judges—and 50% to 60% of them were successful.

Many SSDI recipients are relatively young persons who are harmed by leaving the workforce prematurely. Policy advocates should seek reforms that would help workers stay in the workforce. They could avoid the current either/or situation by allowing workers to apply for SSDI benefits while still working to encourage them to take less demanding jobs at lower wages. They could penalize employers who send many workers into disability programs by asking them to pay more into the system. They could increase EITC payments to adults with no dependents to encourage disabled persons to work. They could seek job training to help disabled persons obtain positions that require less physical effort.

Core Problem 6: Engaging in Advocacy for Services to Help Reduce Mental Distress of Consumers Using Safety Net Programs—With Some Red Flag Alerts

- Red Flag Alert 9.20. Provide mental health and health services to users of specific safety net programs. Persons who use many safety net programs experience disproportionate levels of health and mental health problems, whether because they are often poverty stricken or because they experience cascading adverse events (Barr, 2008). The 2005/2006 National Health Interview Survey discovered that "anywhere from 10 to 40 percent of TANF recipients have a disability and almost one-fifth have a family member with a disability"—and SNAP recipients have similar disability rates (Loprest & Maag, 2009). Residents of public housing often possess unaddressed mental health, substance abuse, and other problems. Safety net programs should hire staff or contract with mental health agencies who can screen enrollees and provide mental health assistance to ones who need help.

Core Problem 7: Engaging in Advocacy to Link Safety Net Programs to Communities—With Some Red Flag Alerts

- Red Flag Alert 9.21. Address economic and social factors that promote use of safety net programs. Safety net programs distribute benefits and services to applicants without analyzing why persons make disproportionate use of

POLICY ADVOCACY LEARNING CHALLENGE 9.6

CONNECTING MICRO, MEZZO, AND MACRO POLICY ADVOCACY INTERVENTIONS TO MENTAL HEALTH FOR TANF RECIPIENTS

Discuss how a social worker might engage in micro policy advocacy to obtain mental health/substance abuse services for a specific TANF recipient. How might he or she use mezzo policy advocacy to place mental health services within TANF offices or offices that take applications for SNAP? How might he or she use macro policy advocacy to change local or state TANF regulations related to specialized services for recipients with mental health or substance abuse challenges.

their benefits and services in specific communities. Nor do staff in safety net programs work in tandem to improve economic and social conditions in these communities.

THINKING BIG AS POLICY ADVOCATES IN THE SAFETY NET SECTOR

With renewed interest in the sheer level of economic inequality in the United States in the aftermath of the Great Recession, new impetus for reforms of safety net programs has developed—and could come to fruition if congressional gridlock abates. For the first time in decades, public interest in reducing economic inequality has increased. Discuss the following reforms:

1. Expanding eligibility and benefits for SNAP and many other safety net programs markedly
2. Greatly increasing advertising and outreach for SNAP, the EITC, and Medicaid so that far more eligible persons actually receive benefits from these programs
3. In tandem replacing TANF with a national guaranteed income that places an economic floor under all citizens
4. Removing work requirements for Medicaid and SNAP if they are put in place by the Trump administration
5. Markedly increasing taxes on affluent Americans—and using these funds to buttress safety net programs and to fund a national guaranteed income
6. Establishing a public jobs program that guarantees work for long-term unemployed persons with particular priority on younger persons
7. Guaranteeing work for everyone including during recessions, as has been proposed by Senator Bernie Sanders in 2018
8. Phasing in a national minimum wage that exceeds $20 per hour along with work and training programs for persons who lose their jobs when employers

dismiss employees when they cannot afford these costs, as was proposed by Senator Bernie Sanders in 2016

9. Increasing Social Security pensions by raising the ceiling on income that is subject to payroll taxes so wealthy people pay far more taxes into the Social Security Trust Fund that funds them

Learning Outcomes

You are now equipped to:

- Identify an array of safety net programs that are needed by vulnerable populations
- Analyze how these safety net programs began and evolved in the United States
- Place these safety net programs in the American cultural and political context
- State the strengths and weaknesses of specific safety net programs after critically analyzing them
- Describe possible micro, mezzo, and macro policy advocacy initiatives in the safety net sector that might help consumers of service
- Identify and discuss some policy reforms in the safety net system

References

Alvarez, L. (2012, April 18). No savings are found from welfare drug tests. *New York Times*, p. A1.

Barr, D. (2008). *Disparities in the United States: Social class, race and ethnicity*. Baltimore, MD: Johns Hopkins University Press.

Ben-Shalom, Y., Moffitt, R. A., & Scholz, J. K. (2011). An assessment of the effectiveness of anti-poverty programs in the United States [No. w17042]. Retrieved from https://www.nber.org/papers/w17042.pdf

Cohn, J. & Delaney, A. (2018). Paul Ryan's whole career was about sticking it to the poor and elderly. *Huffington Post*. Retrieved from https://www.huffingtonpost.com/entry/paul-ryan-social-safety-net-assault_us_5ace41e7e4b0648767761edf

Coleman-Jensen, A., Nord, M., Andrews, M., & Carlson, S. (2010). Household food security in the United States in 2010. [ERR-125]. U.S. Department of Agriculture. Retrieved from http://www.ers.usda.gov/Publications/err125/

Corn, D. (2013, March 4). Mitt Romney's "twisted" defense of his 47 percent rant. *Mother Jones*. Retrieved from http://www.motherjones.com/mojo/2013/03/mitt-romneys-twisted-defense-his-47-percent-rant

Delaney, A. (2018). GOP tax bill mostly benefits the wealthy, Tax Policy Center finds. *Huffington Post*. Retrieved from https://www.huffingtonpost.com/entry/jct-tax-reform-bill_us_5a00b5a3e4b0baea2633fbf5

DeParle, J. (2012, April 7). Welfare limits left poor adrift as recession hit. *New York Times*, p. 1.

Floyd, I. (2017, October 23). TANF cash benefits have fallen by more than 20 percent in most states and continue to erode. Center on Budget and Policy Priorities. Retrieved from https://www.cbpp.org/research/family-income-support/tanf-cash-benefits-have-fallen-by-more-than-20-percent-in-most-states

Floyd, I., Pavetta, L., & Schott, L. (2017, December 13). Center on Budget and Policy Priorities. Retrieved from https://www.cbpp.org/research/family.../policy-brief-tanf-reaching-few-poor-families

Floyd, I., Schott, L. L., & Pavetti, L. (2018, April 2). Under president's budget, TANF would likely provide fewer families with cash assistance and work opportunities, new details show. Center on Budget and Policy Priorities. Retrieved from https://www.cbpp.org/research/family-income-support/under-presidents-budget-tanf-would-likely-provide-fewer-families-with

Gould, E. (2018, March 18). The state of American wages, 2017. Economic Policy Institute. Retrieved from https://www.epi.org/.../the-state-of-american-wages-2017-wages-have-finally-recovere...

Greenstein, R. (2012, April 17). Testimony before the House Budget Committee hearing on strengthening the safety net. Center on Budget and Policy Priorities. Retrieved from http://www.cbpp.org/cms/index.cfm?fa=view&id=3745

Grogger, J. (2003). The effects of time limits, the EITC, and other policy changes on welfare use, work, and income among female-headed families. *Review of Economics and Statistics, 85*(2), 394–408.

Hiltzik, M. (2012, April 25). Let's *expand* social security benefits. *Los Angeles Times*, pp. B1, B5.

Holland, G. (2016, March 30). L.A. council oks law limiting homeless people's belongings to what can fit in a trash bin. *Los Angeles Times*, LA Now Section.

How ESSA and New Federal Funding Rules Will Impact Career and Technical Education. (2018, April 23). [White paper.] Retrieved from ctepolicywatch.acteonline.org/hea/

Jansson, B. (2019a). *Reducing inequality.* San Diego, CA: Cognella Academic Publishing.

Jansson, B. (2019b). *The reluctant welfare state.* Boston: Cengage.

Katch, H., Wagner, J., & Aron-Dine, A. (2018, August 13). Taking Medicaid coverage away from people not meeting work requirements will reduce low-income families' access to care and worsen health outcomes. Center on Budget and Policy Priorities. Retrieved from https://www.cbpp.org/.../taking-medicaid-coverage-away-from-people-not-meeting-w...

Kessler, G. (2018, January 12). Is the Trump tax cut good or bad for the middle class? *Washington Post*. Retrieved from https://www.washingtonpost.com/.../is-the-trump-tax-cut-good-or-bad-for-the-middle-cl...

Krugman, P. (2018, April 12). The Paul Ryan story: From flimflam to fascism. *New York Times,* Opinion Section.

Lazarus, D. (2017, October 27). By a single vote, Republicans throw out years of work on consumer protection. *Los Angeles Times*, Business Section.

Leonhardt, D. (2017, November 3). The tax bill deserves to die. *New York Times.* Opinion Section.

Levine, M. (2018, February 2). Behind the minimum wage fight, a sweeping failure to enforce the law. Politico. Retrieved from https://www.politico.com/story/2018/02/18/minimum-wage-not-enforced-investigation-409644

Loprest, P. & Maag, E. (2009, May). *Disabilities among TANF recipients: Evidence from NHIS.* Urban Institute.

Los Angeles Coalition to End Hunger and Homelessness. (2010). *The people's guide to welfare, health, and other services* (33rd ed.). Retrieved from http://cfpa.net/LosAngeles/ExternalPublications/LACEHH-PeoplesGuideEnglish-2010.pdf

Macvean, M. (2009, December 12). California students underutilize School Breakfast Program. *Los Angeles Times*. Retrieved from http://www.latimes.com

Marr, C. & Huang, C. (2012). Misconceptions and realities about who pays taxes. Center on Budget and Policy Priorities. Retrieved from http://www.cbpp.org/cms/index.cfm? fa=view&id=3505

Mitchell, T., Pavetti, M., & Huang, V. (2018, February 20). Life after TANF in Kansas: For most, unsteady work and earnings below half the poverty line. Center on Budget and Policy Priorities. Retrieved from https://www.cbpp.org/research/family-income-support/life-after-tanf-in-kansas-for-most-unsteady-work-and-earnings-below

O'Connor, K. (2011, March 3). Sen. John Kerry proposes tax relief to low-income families. *Herald News*. Retrieved from http://www.heraldnews.com/news/x1174965957/Sen-John-Kerry-proposes-tax-relief-to-low-income-families

Office of the Inspector General. (2010). Social Security Administration: Disability impairments on cases most frequently denied by disability determination services and subsequently allowed by administrative law judges. Social Security Administration. Retrieved from http://oig.ssa.gov/disability-impairments-cases-most-frequently-denied-disability-determination-services-and

Pavetti, L. (2018a, April 12). Mandatory work programs are costly, have limited long-term impact. Center on Budget and Policy Priorities. Retrieved from https://www.cbpp.org/blog/mandatory-work-programs-are-costly-have-limited-long-term-impact

Pavetti, L. (2018b, April 12). Opportunity-boosting job preparedness takes significant investment, evidence shows. Center on Budget and Policy Priorities. Retrieved from https://www.google.com/search?source=hp&ei=rGHXW9HRCMO9gged-KiIDQ&q=Pavetti%2C+L.+%282018c%3B+April+12%29.+Opportunity-boosting+job+&btnK=Google+Search&oq=Pavetti%2C+L.+%282018c%3B+April+12%29.+Opportunity-boosting+job+&gs_l=psy-ab.3..33i299.1713.1713..2666...0.0..0.206.304.1j0j1......0....1j2..gws-wiz.....0.0-UaySq4KHM

Pavetti, L. (2018c, January 10). Work requirements don't work. Center on Budget and Policy Priorities. Retrieved from https://www.cbpp.org/blog/work-requirements-dont-work

Porter, E. (2012, April 25). Disability insurance causes pain. *New York Times*, pp. B1, B10.

Price, D. (2011). Off the charts: Hardship in America, Part 3. Homelessness growing among families with children. Center on Budget and Policy Priorities. Retrieved from https://www.cbpp.org/blog/hardship-in-america-part-3-homelessness-growing-among-families-with-children

Sachs, J. (2005). *The end of poverty*. New York, NY: Penguin Books.

Sherman, A. & Mitchell, T. (2017, July 17). Economic security programs help low-income children succeed over long term, many studies find. Center on Budget and Policy Priorities. Retrieved from https://www.cbpp.org/research/poverty-and-inequality/economic-security-programs-help-low-income-children-succeed-over

Sherraden, M. (1991). *Assets and the poor: A new American welfare policy.* New York: M. E. Sharpe.

Stein, J. (2018, April 23). Bernie Sanders to announce plan to guarantee every American a job. *Washington Post.* Retrieved from https://www.washingtonpost.com/news/wonk/wp/2018/04/23/bernie-sanders-to-unveil-plan-to-guarantee-every-american-a-job/

Stewart, N. (2018, February 18). Homelessness, step by step. *New York Times.* Retrieved from https://nyti.ms/2C6smnZ

Times Editorial Board. (2018, March 1). How can a place with 58,000 homeless people continue to function. Retrieved from http://www.latimes.com/opinion/editorials/la-ed-homelessness-impact-on-others-20180301-htmlstory.html

U.S. Department of Education. (2012). Investing in America's future: A blueprint for transforming career and technical education. Retrieved from http://www2.ed.gov/about/offices/list/ovae/pi/cte/transforming-career-technical-education.pdf

Wicks-Lim, J. (2009). Should we be talking about living wages now? *Monthly Review Press.* Retrieved from http://www.peri.umass.edu/fileadmin/pdf/other_publication_types/magazine___journal_articles/0309wicks-lim.pdf

Williams, S. (2017, October 1). 5 expected changes in Social Security in 2018. *Motley Fools.* Retrieved from https://www.fool.com/retirement/2017/10/01/5-expected-social-security-changes-in-2018.aspx

Williams, T. (2014, August 17). Seeking new start, finding steep cost: Workforce Investment Act leaves many jobless and in debt. *New York Times*, Home Section.

Zedlewski, S., Holcomb, P., & Loprest, P. (2007). *Hard-to-employ parents: A review of their characteristics and the programs designed to serve their needs* [Paper 9]. Urban Institute.

PRACTICING POLICY ADVOCACY IN THE MENTAL HEALTH AND SUBSTANCE ABUSE SECTOR

This chapter was coauthored by Judith A. DeBonis and Eri Nakagami

LEARNING OBJECTIVES

In this chapter you will learn to:

1. Develop an empowering perspective about mental health and substance abuse
2. Analyze the evolution of the mental health and substance abuse sector in the United States
3. Understand the sheer magnitude of mental and substance abuse disorders in the United States
4. Analyze the effects of income inequality on mental health and substance abuse
5. Analyze the political economy of mental health services and substance abuse
6. Recognize the magnitude of other addictions such as smoking and poor dietary habits
7. Identify ways to engage in micro, mezzo, and macro policy advocacy to redress seven core recurring problems in the mental health and substance abuse sector
8. Apply the eight challenges in the multilevel policy advocacy framework

Note: Judith A. DeBonis and Eri Nakagami made invaluable contributions to this chapter in the first edition of this book. Bruce Jansson was the sole author of the second edition, editing and updating contributions from each original contributor.

DEVELOPING AN EMPOWERING PERSPECTIVE ABOUT MENTAL HEALTH AND SUBSTANCE ABUSE

Mental illness was a scourge in the United States until relatively recently. Persons with mental problems were often placed in poorhouses and institutions through the 19th century. They encountered stigma and often did not work. Beneficial medications had not been invented. Mental health professionals lacked evidence-based clinical tools. Nor did the nation have evidence-based medications or clinical tools to address alcoholism or drug addictions. Persons with serious mental illness were warehoused in asylums in rural locations.

Considerable progress has been made in the intervening decades. A large number of mental illnesses and drug addictions have been identified, such as those listed in the *Diagnostic and Statistical Manual of Mental Disorders (DSM-5)* of the American Psychiatric Association (2013). Many medications have been developed that help persons with mental problems cope with their conditions, allowing them to live in their communities, have healthy networks, and hold jobs. Evidence-based clinical tools have been developed that support the effectiveness of cognitive therapy, mindfulness, and motivational therapy with specific kinds of persons.

Moreover, social policies have been developed in past decades that have greatly helped persons with mental health problems and addictions. Civil rights laws forbid discrimination against persons with mental health problems and addictions in places of work. Medicare, Medicaid, and private health insurance plans have greatly expanded coverage of services and medications for mental health problems and addictions. Increasing numbers of employers give persons with mental disabilities and substance abuse issues work accommodations, such as reduced hours or different job assignments. The population of mental health practitioners has been greatly expanded to include social workers, psychologists, psychiatrists, and marriage and family counselors.

Even with these advances, the United States has remarkable numbers of persons with mental health and substance abuse problems—and many of them are poorly addressed. Nearly one in five U.S. adults lived with mental illness in 2016, or 44.7 million persons. Their mental illnesses ranged in severity from mild to moderate to severe. Two broad categories exist: any mental illness (AMI) and serious mental illness (SMI). AMI includes all mental, behavioral, or emotional disorders and can vary from no impairment to mild, moderate, and severe impairment. SMI is a "mental, behavioral, or emotional disorder resulting in serious functional impairment, which substantially interferes with or limits one or more major life activities" (National Institute of Mental Health, data from the 2016 National Survey on Drug Use and Health collected by the Substance Abuse and Mental Health Services Administration). The distribution of AMI, overall, was 18.3% (44.7 million adults) with 21.7% for women and 14.5% for men. Roughly 21.5% of persons from 18 to 49 years of age had AMI as compared with only 14.5% of persons over age 50. Across race and ethnicity, 22.8% of American Indians and Alaskan Natives, 19.9% of whites, 15.7% of Hispanics, 16.7% of Native Hawaiians and other Pacific Islanders, and 12.1% of Asian persons had AMI. The prevalence of persons with SMI overall was 4.2% of all U.S. adults with young adults from 18 to 25 having the highest prevalence of SMI (5.9%) compared with 5.3% for persons age 26 to 49 and 2.7% for persons age 50 and older.

The prevalence of mental illness includes 1.1% of American adults living with schizophrenia (2.4 million), 2.6% with bipolar disorder, 6.9% with major depression, and 18.1% with anxiety disorders that include panic disorder, obsessive-compulsive disorder, posttraumatic stress disorder, generalized anxiety disorder, and phobias—or 18.7% of adults (42 million).

About 26% of homeless adults in shelters have serious mental illness. About 24% of state prisoners have a recent history of mental illness (National Alliance on Mental Illness, 2018). Mental illness costs the United States $132.9 billion in lost wages every year. It disrupts families. Roughly 90% of those who commit suicide—the tenth-leading cause of death—have an underlying mental illness.

Roughly 60% of adults with mental illness did not receive mental health services in the previous year—and nearly 50% of youth age 8 to 15 did not receive them in the previous year (National Alliance on Mental Illness, 2018).

One in eight American adults is alcoholic, rising from roughly 8% of adults over age 18 in 1991–1992 to 12.7% in 2012–2013 (Ingraham, 2017)—or a rise of 49% in the first decade of the 21st century. This includes 10.2 million adults with co-occurring mental health and addiction disorders.

Many problems prevent the United States from empowering tens of millions of its residents who possess mental illness and substance abuse. The Centers for Disease Control and Prevention (CDC, 2018a) says that 50% of Americans experience mental health issues over their lifetimes. About 25% of adults experience mental health issues in a given year, including anxiety and mood disorders such as depression—but less than half receive assistance despite the fact that roughly 85% are treatable (McClain, 2015). It is unacceptable that 9.2 million persons have mostly untreated substance abuse.

Nor has the United States eliminated the stigma that many persons with mental illness and addictions encounter. Persons with mental health and substance abuse problems (MHSAPs) encounter many kinds of stigma. When labeled with terms like "schizophrenia," "suicidal," "manic-depressive syndrome," "neurotic," "anti-social," "alcoholic," or "drug abuser," people often choose not to receive help for their MHSAPs. People contemplating suicide may not discuss this problem with anyone else or seek help to address it because they are ashamed they have clinical depression as illustrated by suicides of celebrities Anthony Bourdain and Kate Spade in 2018. People who disclose mental health problems fear they will be shunned, not be hired, and even be fired. Persons who admit substance abuse problems may fear they will not be hired or even fired even when they have stellar work records. Many people wrongly believe that persons with psychoses are dangerous even when evidence demonstrates that they commit fewer violent crimes than other persons (Brekke, Prindle, Bae, & Long, 2001).

Stigma leads American policy makers to fund programs that address cancer and many other health programs rather than also funding mental health and substance abuse programs sufficiently. The federal budget devoted only $1.167 billion to mental health and only $2.195 billion to substance abuse in 2016. The Trump administration sought cuts in federal mental health spending in 2018 of $252 million while increasing spending on substance abuse only by roughly $500 million at a time when 72,000 persons had died in 2017 from poisoning from opioid drugs (U.S. Department of Health and Human Services, 2018).

Many persons with mental illness and addictions cannot afford health services for their problems despite augmentation of insurance coverage in recent years by Medicare, Medicaid, and private insurance plans, as we discuss subsequently.

Many Americans have dual diagnoses of mental illness and chronic diseases because individuals with serious mental illness are more likely to have a chronic medical condition—yet many of them receive care for only one condition (Colton & Manderscheid, 2006). They are likely to have life expectancies that are 25 years shorter than persons without these dual conditions largely because the American medical system fails to address their chronic medical conditions (Manderscheid, Druss, & Freeman, 2008). Males with schizophrenia have life expectancies of roughly 63 years primarily because they receive poor medical care for heart disease and other ailments.

Why Social Workers Work to Improve Poor Outcomes in This Sector

Social workers can assume a pivotal role in reforming the mental health and substance abuse sector because they have advocacy skills at micro, mezzo, and macro levels. They have a code of ethics that asks them to contest policies and practices that discriminate against vulnerable populations that include persons of color, LGBTQ persons, low-income persons, children, seniors, and many other populations, as we discussed in Chapter 1. All social workers are versed in social justice that mandates them to hew to the ethos of Jane Addams and other social work leaders. They have all taken at least one social policy course.

Let's start at the macro level and then move to mezzo and micro levels. Politicians love to cut funding for mental health institutions while not funding community-based services for people with MHSAPs who have exited these institutions. The Congress and president failed to simplify the maze of insurance, Medicare, and Medicaid programs for persons with MHSAPs—not to mention the complex network of public, not-for-profit, and for-profit institutions that help people with MHSAPs. It is a challenge for consumers of service to find out where to go for which problem, to understand fees, and to determine which professionals have the requisite knowledge about evidence-based treatments. Waits for physicians to obtain medicine for opioids were as much as one year in 2016—and the demand for such medicines has grown exponentially in the succeeding years as deaths from overdoses soared to 72,000 persons in 2017 (Vestal, 2016).

Social workers seek greater resources for programs that help persons with mental health and substance abuse problems. They can ask to view their state's application for federal resources from the Community Mental Health (MHBG) and Substance Abuse Block Grant (SABG), which were enacted in 1981. Each state applies for federal funds for these block grants, including in fiscal years 2018 and 2019, when states were asked for the first time to develop joint applications for mental health and substance abuse. States have to identify in remarkable detail the populations their block grants serve, the kinds of services they provide, and assessments of mental health and substance abuse clients who receive help from these block grants. They discuss the extent they use Medicare, Medicaid, and other resources to fund these services. They describe services provided by public, not-for-profit, and for-profit agencies. (Go to http://www.samhsa.gov to view application forms under the title Block Grants, "SABG and MHBG" and examine the application form, including one that asks them to identify "resources." Then contact your state's mental health and substance abuse agencies to view the actual plans they submitted to the federal government for fiscal years 2018 and 2019.)

Because many persons have to wait months to get treatment for their substance abuse problems, macro policy advocates can demand states and the federal government increase access for them. They can demand that the federal government fund and distribute lifesaving medications to persons with opioid poisoning, which we discuss subsequently. They can mandate the use of evidence-based medications and cognitive therapy for persons with substance abuse issues as compared with 12-step programs, like Alcoholics Anonymous (AA), which are not evidence based. They can seek more funding for primary health clinics in low-income areas that have insufficient services for persons with mental health and substance abuse problems. They can diminish stigma in many ways. They can link clients to advocacy groups that emphasize the strengths of persons with MHSAPs. They can educate employers about the strengths of persons with MHSAPs including how to help them make work adjustments that increase their productivity. They can seek social policies that ban discrimination against persons with MHSAPs in hiring. They can seek better enforcement of the Americans with Disabilities Act (ADA) that prohibits discrimination by employers against persons with MHSAPs during hiring processes.

Social workers can engage in mezzo policy advocacy to improve services for persons with MHSAPs. They can work to increase residential mental health services for persons who languish in emergency rooms and hospitals because professionals are prohibited from discharging them before they are medically stabilized and before they can find licensed residential facilities. They can work to increase staff and clinics in hospitals with expertise in persons with MHSAPs at a time when relatively few hospitals have sufficient numbers of staff with expertise in addictions. They can initiate training sessions with community residents and leaders about specific mental health and substance abuse. They can seek cross-disciplinary teams in primary clinics and hospitals that integrate contributions of physicians, psychologists, and social workers.

Social workers can help persons with MHSAPs through micro policy advocacy in many ways. They can help them navigate the complex funding systems that the United States has developed, including coverage offered by private insurance, Medicare, and Medicaid. They can help them navigate public, not-for-profit, and for-profit providers. They can help them find providers that use evidence-based programs (EBPs). They can help persons with MHSAPs seek and retain employment. They can help them find temporary residential centers where they can recover sufficiently to reenter their communities after they have been discharged from emergency rooms and hospitals. They can help persons with MHSAPs who are often impoverished due to their lack of employment or low wages to find assistance from safety net programs including the Supplemental Nutrition Assistance Program (SNAP), food banks, rent subsidy programs, and other programs discussed in Chapter 9. They can help them find and fund caregivers when they need them, such as tapping into Medicaid funds targeted to persons with MHSAPs. They can help them obtain legal assistance from federal administrative judges when their rights are violated.

Indicators of Poor Services in Mental Health and Substance Abuse Areas

Several indicators suggest that many Americans receive inadequate care for mental health and substance abuse problems.

1. Primary care physicians deliver most mental health services in the United States, yet most have no or only elementary knowledge of mental health problems. They usually treat patients in a single session during which they prescribe medications rather than referring them to mental health specialists who can diagnose,

give medications, and provide follow-up services. There is an overlap between physical and mental health conditions. Up to 70% of all primary care visits involve a mental health concern—and nearly 68% of people with mental illness have chronic health conditions such as diabetes, hypertension, or heart disease. Integrated care models that coordinate physical and mental health services can improve care, coordinate physical and mental health care, and reduce costs (National Alliance on Mental Health, 2015).

2. Clients often do not have access to an array of interventions because primary care physicians, as well as other helpers, lack knowledge of an array of interventions that include the following (Jansson, 2011):

    - Diagnose biopsychosocial factors that contribute to specific mental and substance abuse problems (*differential assessment*).
    - Identify specific medications that are useful in treating specific mental health and substance abuse problems (*psychiatric or medical interventions*).
    - Help consumers resolve their internal conflicts and uncertainties (*counseling*).
    - Help consumers cope with adverse news (*supportive counseling*).
    - Facilitate the diagnosis of specific mental health conditions that might otherwise not be detected (*screening*).
    - Help consumers resolve familial tensions and conflicts (*family counseling*).
    - Help consumers resolve tensions and conflicts within their social networks (*social network counseling*).
    - Help consumers find meaning and dignity when they confront disabilities, chronic diseases, and terminal illness (*existential counseling*).
    - Help consumers deal with grief, bereavement, and trauma (*grief, bereavement, and trauma counseling*).
    - Help consumers deal with persisting psychological and social effects of trauma (*posttraumatic stress counseling*).
    - Gauge the cognitive competence of specific persons to determine if they can make decisions about their health care—and help find proxies when necessary (*assessing competence*).
    - Diagnose why specific consumers do not adhere to treatment recommendations and develop interventions to improve adherence (*adherence counseling*).
    - Help stabilize persons with psychotic and other disorders who are disruptive in health and other settings (*stabilizing services*).
    - Help consumers deal with specific crises (*crisis intervention services*).

- Help consumers with specific mental health interventions that have proven effectiveness with specific problems (cognitive, motivational, behavioral, psychosocial, and other therapeutic strategies).
- Provide social service and mental health interventions as part of a health care team to achieve specific health objectives like weight loss, including psychosocial and behavioral counseling (*adjunctive counseling in a health team*).
- Educate physicians and other health staff about mental health problems (*mental health education for professionals*).
- Use evidence-based yoga, meditation, biofeedback, and other interventions (*nontraditional interventions, often called mindfulness programs*).
- Help consumers avoid learned helplessness (*empowerment strategies*).
- Help consumers surmount environmental stressors, such as those involving economic problems, housing, and the community (*coping with environmental stressors*).
- Help consumers obtain knowledge about psychological conditions that often accompany physical illness, mental health problems, and substance abuse problems—and how to address them (*psychoeducation*).

Mental health researchers and administrators have to decide which of these interventions are relevant to which clients, for what duration, and at what cost. Acute mental illnesses often can be addressed in less than 10 visits with some follow-up services. Some mental illnesses can be managed with medications and periodic follow-up services. Many substance abuse problems, including addictions that include smoking and drug addictions, may require extended treatment including use of medications, monitoring, and cognitive therapy. Chronic mental health and substance abuse problems may require episodes of treatment coupled with ongoing monitoring as clients contend with these problems. Some clients may need inpatient care in hospitals or in residential facilities for short or longer periods. Many mental health and substance abuse problems recur, requiring clients to resume treatments on some or many occasions.

3. Clients that need access to inpatient care, whether in hospitals or residential facilities, often cannot be placed in them due to a paucity of beds or residential facilities. Some patients remain in emergency rooms or hospital beds for days or weeks because they cannot be discharged when they need residential care (Jansson, 2011).
4. Many patients do not receive evidence-based care for alcoholism or addictions to drugs. They are often referred to 12-step programs developed by AA even though evidence-based data fails to uphold their service. Considerable evidence supports, instead, the coupling of medications with cognitive therapy for varying lengths of time. Relapses are common, so medications and cognitive therapy may need to be used on several or more occasions. Many clients do not receive evidence-based care for substance abuse, such as for opioids or other drugs. They do not receive medications with cognitive therapy. They do not receive medications when

they have overdosed sufficiently that they may lose their lives. They do not have access to emergency medical care due to lack of sufficient emergency room staff and emergency vehicles. Physicians and dentists often perpetuate drug epidemics by over-prescribing addictive drugs and not monitoring patients, even though considerable progress has been made in curtailing prescriptions in some locations. Drug companies often distribute addictive drugs to areas, such as rural towns, where drug epidemics exist (Macy, 2018).

5. The United States lacks universal health insurance so that the nation is divided into different subpopulations with varying levels of insurance coverage for mental health and substance abuse services and medications. Despite federal legislation that requires private insurance companies to cover mental health conditions and substance abuse on equal terms with physical health problems, coverage of mental health and substance abuse services varies. Despite augmentation of coverage for mental health and substance abuse in Medicare and Medicaid programs, certain gaps still exist.

6. Roughly 37% of adults with clinical depression fail to receive medications, as well as therapy, that help patients control this disease (National Institute of Mental Health, 2017). Absent therapy and medications, many depressed persons cannot hold jobs, sustain relationships, or engage in other activities. Some of them commit suicide.

7. Insufficient numbers of halfway houses exist for persons recovering from psychoses or other chronic health conditions—nor are their costs sufficiently covered by some health insurance policies (Jansson, 2019).

8. Large gaps exist in the professions that provide mental health and substance abuse services. Few physicians have expertise in substance abuse treatment out of a workforce of roughly 1 million physicians. Most people who provide help to persons with substance abuse are themselves recovering addicts who lack professional degrees. The nation has a substantial number of social workers, but hospitals vary widely in the number that they hire—and relatively few social workers specialize in addictions.

9. Mental health and substance abuse problems are insufficiently addressed by professionals and other staff in the eight policy sectors discussed in this book including the following:

    - In the education sector many students in middle schools and many students in high schools have mental health problems of varying seriousness. Considerable numbers of students commit suicide. Many students have substance abuse issues with alcohol, smoking, cocaine, heroin, and marijuana. Some students have sexual identity issues. Bullying often takes place. Some students engage in mass shootings of fellow students, teachers, and administrators (see further discussion in Chapter 12 of this book). Many students do not graduate from high school, junior colleges, and colleges because they possess substance abuse and mental health problems. The National College Health Assessment revealed that the number of students suffering from depression rose from 32.6% in 2013 to 40.2% in 2013—and those attempting suicide from 1.3% to 1.7% (Brody, 2018). Many schools have augmented their mental health and substance abuse staff, but most cannot address students' problems sufficiently. Preschools and schools are ill equipped to provide them with any mental health professional on their premises. The vast majority of students who attend junior colleges do not graduate—and those students on the spectrum or with serious mental problems are often included in their ranks due to the lack of sufficient support services. Nor are most institutions of higher education equipped to provide mental health services to students, leading to high dropout rates of students (Brody, 2018). Youth with other chronic mental illnesses receive inadequate services from preschools or primary and secondary schools. It is not surprising that many of these youth receive inadequate services. Once they graduate from high school, many youth with chronic and serious mental illnesses have no place to live because their parents are unwilling or unable to care for them in their homes. Left to their own devices, these youth often end up on the streets or in jail.

- In the criminal justice sector, roughly 40% of inmates have mental health and substance abuse problems, as we discuss in Chapter 14. Although prison populations have recently declined in some states that have released nonviolent offenders, probation departments contend with high rates of mental illness and substance abuse with insufficient staff.
- The child and family sector possesses populations with high levels of mental illness and substance abuse, yet lacks sufficient staff to give them adequate care. Child welfare departments have insufficient links to health, mental health, and substance abuse services. When youth are emancipated at age 18 in some states, 21 in other states, and even later in other states, they often end up on the streets or in jail because their mental health and substance abuse problems have not been addressed, as we discuss in Chapter 11.
- In the immigration sector adults and youth who are undocumented suffer from anxiety and depression related to the uncertainties of their legal status and from separations from family members. They rarely use existing medical and mental health services because they fear deportation, as we discuss in Chapter 13.
- People in the safety net sector often have serious food, housing, and employment needs. We have already discussed the substance abuse and mental health problems of many homeless people that are often exacerbated by the stress of lacking secure housing, as is discussed in Chapter 9. We can hypothesize that many persons who use food banks and SNAP have mental health and substance abuse issues caused by their poverty and economic uncertainty.

10. Mental health and substance abuse programs often face an ethical dilemma when clients are imminent threats to other people—but who refuse treatment. Mental health personnel are compelled to inform the police when a client threatens to harm or kill another person by a ruling by the U.S Supreme Court. These persons include people who commit violent acts in families, schools, and places of work against other persons or who attempt suicide. They include some homeless persons who refuse to enter government-supplied or financed housing at considerable cost to the broader society. Although this policy alerts police and potential victims of danger, police are often reluctant to provide indefinite protective services to persons in danger. Some states, such as California and New York, allow mental health and substance abuse personnel and/or police officers to seek court orders to require specific persons to take medications and counseling that have been proven to decrease violent actions or even to induce homeless persons to accept housing or residential treatment centers or to accept other forms of protective custody. Such court orders are viewed by some critics as violations of clients' rights to informed consent, but judges have to balance clients' rights with the rights of community members and the community at large. Protective custody might, for example, have averted the massacre at Marjorie Stoneman Douglas High School in Parkland, Florida, in 2018.

11. Funding for mental health services by state governments has decreased from 1980 to the present despite substance abuse epidemics, high levels of suicide, burgeoning homeless populations, high levels of anxiety, high rates of mental illness among secondary and college populations, and many other indicators of unmet mental health and substance abuse diseases (National Alliance for Mental Illness, 2011). However, states that chose to participate in marked increases in the eligibility of Medicaid in the wake of enactment of the Affordable Care Act (ACA) in 2010 gave millions of Americans mental health and substance abuse coverage that they would not otherwise have had. By late 2018, 37 states and the District of Columbia had made this expansion, leaving only 14 states that did not increase Medicaid's eligibility. The federal government paid for this expansion, including for mental health and substance abuse services funded by Medicaid (FamiliesUSA, 2018).

12. The federal MHBG and SABG, enacted in 1981, supplanted many so-called categorical federal grant programs. Governors hoped that the block grant funds would give them more latitude to develop their state mental health and substance abuse programs—but block grants were funded at levels lower than the funds of their prior categorical programs (Jansson, 2019). Nor has funding of the MHBG and the SABG kept up with inflation. For example, the MHBG has received funding of roughly $1 billion for many years.

13. An unfulfilled promise of parity took place in the wake of federal legislation that required private insurance companies to equalize access to mental health and substance abuse services as compared with their health benefits. Prior to this legislation, many persons found that their private insurance policies failed to cover any mental health services or only meager ones. In a survey conducted in 2015, many persons could not locate mental health providers in their networks, including inpatient care in hospitals and residential services (including social workers and

POLICY ADVOCACY LEARNING CHALLENGE 10.1

Select one of the 15 indicators of major, unsolved shortcomings in American mental health and substance abuse services. Identify a policy or policies that might address the shortcoming. Identify whom you might contact at federal, state, or local levels to start the ball rolling.

psychologists), as compared with general or specialty medical care. Only 75% of respondents could find an in-network mental health therapist, for example, as compared with 91% who could find an in-network medical specialist. This meant they had to pay higher costs for out-of-network mental health and substance abuse services. They had to pay higher out-of-pocket costs for these mental health and substance abuse services than for services for their physical health problems. These findings suggest that the federal government has to monitor health systems to upgrade their numbers of mental health professionals and so that consumers of mental health services do not have to pay a higher percentage of these costs than patients using traditional medical care (National Alliance on Mental Illness, 2016). Their out-of-network and out-of-pocket obstacles have proven to be greater for mental health services than when they sought general or specialty medical care from their networks in violation of federal legislation that required parity.

14. Half of all mental illness emerges by the age of 14 and three-quarters by age 24—and early identification has been shown to prevent crises and allow children to avoid serious mental problems. In the United States, however, an average lag of eight to 10 years exists between the onset of a mental health condition and the start of treatment. Although one in five American youth live with a mental health problem, less than half receive treatment. The federal Medicaid law requires mental health screening as part of the Early and Periodic Screening, Diagnostic and Treatment Program, but many states do not comply with this mandate (National Alliance on Mental Health, 2015).

15. Early interventions for persons with first episode psychoses (FEP) dramatically improve outcomes for persons with them—but many of these persons do not receive timely care. Congress has required states to use 5% of the 2015 and 2016 MHBG to encourage FEP programs (National Alliance on Mental Health, 2015).

ANALYZING THE IMPACT OF INCOME INEQUALITY ON MENTAL HEALTH AND SUBSTANCE ABUSE

All persons, regardless of age, race, religion, or income, are vulnerable to mental illness. The causes of specific mental illness and addiction often include a combination of brain disorders affected by biological and genetic vulnerabilities as well as environmental stressors (National Institute of Mental Health, 2017). Low-income people have markedly higher rates of anxiety, depression, and many other forms of mental illness than relatively affluent persons. They commit suicide at higher rates than affluent persons. They have far poorer access to mental health services than other persons, whether medications or counseling. They have higher rates of mental illness that stem from multiple causes. Some researchers contend that stress induced by poverty exposes persons to ongoing economic uncertainty as well as a quest for survival (Kolbert, 2018). Psychologist Keith Payne contends that the subjective experience of feeling poor—not limited to persons in the lowest economic quintile—causes substantial stress that can lead to depression, risky behaviors like gambling, feeling inferior, and subscribing to conspiracy theories. A study of British civil servants found that people ranking themselves in terms of status was a better predictor of their health than their education level or their actual income (Payne, 2017). Yet other researchers implicate poverty itself in causing mental health and substance abuse problems. Poor people often have poor nutrition even with assistance from SNAP and food banks. They experience high rates of eviction when they cannot pay their rents—a demeaning process during which police place their possessions on the street (Barr, 2008). They experience uncertainties in their neighborhoods such as gangs, drug dealers, and random shootings. Their mental illness is less likely to be effectively addressed by mental health services than affluent persons because they have poorer access to counseling and to medications due to a paucity of emergency rooms and outpatient mental health and substance abuse clinics. Incomes of low-income persons with mental health and addiction disorders are further diminished by their disorders, making it difficult for them to find and hold employment (Barr, 2008). Access to mental health and substance abuse services is particularly poor in those 17 states that chose not to increase their Medicaid eligibility levels as allowed by the ACA. Recall that the ACA gave states the option of expanding Medicaid eligibility to nearly all low-income persons with incomes at or below 138% of the poverty line and many moderate-income families—or $28,676 for a family of three in 2018 (Garfield, Damicao, & Orgera, 2018). In states that did not expand coverage, the median income limit for persons seeking care from the ACA's insurance policies is just 43% of poverty. Low-income whites, African Americans, and Latinos in states that did not increase Medicaid eligibility levels, such as Florida, Georgia, and Texas, disproportionately lack access to mental health and substance abuse services that are reimbursed by Medicaid.

Roughly 32% of adults beneath the poverty level smoke, but only 17% of persons twice the level of poverty do. Roughly 28% of persons with less than a high school education smoke cigarettes, but less than 10% of college graduates do. Low-income persons are just as likely as affluent persons to attempt to quit smoking cigarettes but are far less likely to quit. These disparities in smoking levels partly stem from the targeting of low-income areas by tobacco companies and outlets (CDC, 2018b). Opioid consumption is highest in rural areas partly because pharmaceutical companies flooded them with drugs, but the relationship between social class and use of opioids is complex and varying (Macy, 2018). Disparities in alcoholism between low-income and affluent persons is not as discernible as smoking.

ANALYZING THE POLITICAL ECONOMY OF MENTAL HEALTH AND SUBSTANCE ABUSE

Mental health and substance abuse services have traditionally been viewed as stepchildren of the medical system. Until relatively recently, mental health was given poorer coverage than physical health conditions—and it still lags behind, as we discuss subsequently. Psychiatrists have lower status in the medical profession than surgeons. Primary care physicians have the lowest status in the medical hierarchy. Moreover, primary care physicians often have little training in counseling, mental health, and substance abuse disorders. They often rely excessively on medications to provide most mental health services within the medical system—with little or no follow-up.

Discussion of the opioid epidemic as well as the epidemic of alcoholism reveal the political economy of the mental health and substance abuse sector.

The opioid tragedy from roughly the mid-1990s to the present is a case study of how political and economic factors impede the development of evidence-based and fully funded interventions in the mental health and substance abuse sector. Large pharmaceutical companies realized that they could flood specific areas of the nation with drugs that are prescribed by physicians and dentists to allay the pain of persons with dental problems, arthritis, injuries from car accidents, workplace injuries, sciatica, and many other illnesses (Macy, 2018). The U.S. Food and Drug Administration (FDA) approved OcyContin's new time-release mechanism on the grounds that it reduced addiction—and sales representatives from drug companies convinced physicians to prescribe many other pain-reducing opioids, including Vicodin, Percocet, Lortab, and immediate-release opioids. Unfortunately, these pain-reducing drugs are highly addictive. Drug companies hired 5,000 doctors, pharmacists, and nurses to give seminars to convince doctors that OxyContin was not addictive. They argued that opioids could be used not just for intense pain but for moderate pain. They gave bonuses to physicians who prescribed large amounts of these drugs, even as high as $100,000 every three months (Macy, 2018). Dentists joined the fray by prescribing addictive drugs for pain. Opioid distributors spent a billion dollars on political lobbying and campaigns to be certain that Medicaid and physicians would fund and prescribe opioids. The drug company Perdue reformulated opioids in 2010—and inundated small towns with the drug, including 9 million pills to just one town in West Virginia. Drug companies eventually manufactured and widely distributed Fentanyl, another highly addictive painkilling drug because of its small size, its potency at 50 times heroin, and the ease with which it can be mixed with heroin (mostly imported from India and China). They convinced dentists and physicians to prescribe it. When addicts could not obtain Fentanyl, they moved to heroin that they obtained from drug dealers (Macy, 2018). Meanwhile, rehabilitation centers continued to use the ineffective 12-step model, rather than medication-assisted therapies, even though it is effective in preventing deaths only for 11% of patients as compared with roughly 50% for medication-assisted therapies (Macy, 2018).

Other family members and acquaintances soon obtained the drugs, whether by raiding places where parents or relatives stored their drugs or by purchasing drugs from dealers. Drug dealers found ways to find the drugs through underground markets or from smugglers. With a dearth of physicians or other professionals who were versed in addiction science, few addiction clinics, and overtaxed emergency rooms, an addiction epidemic grew from relatively few deaths in 1996 to 72,000 deaths from opioids in 2017 (CDC, 2018a).

Stigma assumed a major role in fueling the drug epidemics. Many persons traveled to other states to secure drugs to hide their addictions from friends and relatives—only often to die in these states. Even well-connected and wealthy people refrained from getting help for substance abuse and mental health problems because of stigma. Congressman Patrick Kennedy, for example, chose not to inform the Mayo Clinic that he had a substance abuse problem because he feared his problem would become public. He finally surmounted his fears by championing congressional legislation in 2006 and later years that required private insurance companies to cover substance abuse and mental health at levels equivalent to Medicare and Medicaid.

The drug epidemic was so lethal because the health system and residential centers lacked evidence-based treatments. They referred drug addicts to the 12-step program initiated by AA decades earlier. They referred them to residential centers that also used this strategy or other counseling techniques. These strategies were not effective for the vast majority of addicts (Glaser, 2015).

Because the nation's attention was diverted to the epidemic of addictive painkilling drugs, few experts realized that alcoholism rose by 49% from 2000 to 2017. Roughly 12.7% of the American population met the criteria for "alcohol use disorder" in 2017, with annual deaths from it rising to 88,000 per year in 2017 (Ingraham, 2017). These deaths were caused by hypertension, cardiovascular disease, cirrhosis of the liver, and other diseases caused by alcoholism, as well as automobile accidents. Alcoholism rates where 16.7% for men, 16.6% for Native Americans, 14.3% for persons below the poverty line, and 14.8% for people living in the Midwest. Remarkably, 23.4% of adults under age 30 suffer from alcoholism (Ingraham, 2017).

Most physicians and other health professionals in outpatient settings, as well as staff in residential facilities, had long assumed that the 12-step model used by AA was the only effective cure (Coy, 2010; Glaser, 2015). Remarkably, only 582 physicians in the United States out of almost 1 million identified themselves as addiction specialists (Glaser, 2015). Most treatment providers call themselves "addiction counselors" or "substance abuse counselors," which requires only a high school degree or a GED, with many of them also recovering from substance abuse (Glaser, 2015).

Curiously, many physicians in the United States fixated on a single approach to addressing alcoholism for decades rather than exploring alternative approaches that are effective with specific persons (Glaser, 2015; Macy, 2018). That single approach has been the faith-based 12-step program (five of the steps mention God) provided by AA, in which alcoholics and former alcoholics meet in groups to support one another to achieve and retain abstinence. Some people are helped by AA, but others often feel defeated if they relapse because AA's "Big Book" states "those who do not recover are people who cannot or will not completely give themselves to this simple program, usually men and women who are constitutionally incapable of being honest with themselves. These are such unfortunates; they seem to have been born that way" (Glaser, 2015). Critics note that AA's approach has not been verified by scientific studies. The Cochrane Collaboration, a leading health research group, maintained in 2006 that no scientific studies demonstrate that the 12-step method reduces alcohol dependence (Glaser, 2015). A retired psychiatry professor from Harvard estimated from available data that AA's success rates falls between 5% and 8% (Dodes, 2015). Some experts contend that abstinence often causes binging, citing animal and human studies (Sinclair, cited by Glaser, 2015).

Toward Evidence-Based Treatments for Both Drug Addictions and Alcoholism

Experts in the fields of drug addiction and alcoholism are moving toward evidence-based treatments. They prefer medications to remove or decrease the urge for addictive substances. They prefer medications to help addicts avert death when near death after taking life-threatening doses or amounts of alcohol and drugs.

Scientists have used disulfiram (Antabuse), which causes persons to have bad reactions when they drink alcohol by blocking its processing in the body. Other medications have appeared more recently, such as naltrexone, that greatly reduces the urge to drink alcohol. John David Sinclair, a clinical psychologist, discovered that patients who coupled naltrexone with cognitive therapy achieved a 75% success rate among 5,000 Finnish alcoholics over 18 years—a treatment costing only $2,500 compared with tens of thousands of dollars in the United States for each patient placed in rehabilitation units for roughly 30 days (Sinclair, cited by Glaser, 2015). High rates of success were obtained in a dozen clinical trials including one funded by the National Institute on Alcohol Abuse and published by the *Journal of the Medical Association* in 2006 (Glaser, 2015). Other medications became available to patients but were not funded by Medicare or Medicaid.

Although many health professionals seek to terminate intake of alcohol altogether, some experts contend that treatments that help alcoholics control the quantity of alcohol they consume have been shown to be effective (Glaser, 2015). They contend that alcoholism is often not a progressive disease that inevitably worsens. The CDC found that nine out of 10 heavy drinkers are not dependent on alcohol but can change their intake with a brief intervention from a medical professional (Glaser, 2015). The 2015 diagnostic manual of the American Psychiatric Association asserts that only 15% of persons with "alcoholic-use disorder" have a severe problem—although the rest are often ignored by researchers and clinicians and do not receive individualized treatment options (Glaser, 2015).

Yet other research was published in late 2018 that drew upon data of tens of thousands of persons in many nations. It concluded that any use of alcohol has negative health effects on persons who use it (CNN, 2018). Our discussion of alcohol usage suggests that considerable controversy exists even as we move beyond the 12-step model to strategies to control its usage. Although some experts aim for total cessation of drinking, others want merely to control its usage. Still others want to combine cognitive therapy with medications to help persons not to drink to levels that constitute clinical alcoholism.

Similar progress was made with respect to opioid addictions as naltrexone and other medications decreased the urge to take them. Scientists also discovered that naloxone (a drug with the brand name Narcan) reverses drug overdoses from heroin and other painkillers by quickly restoring normal breathing.

Persons addicted to alcohol and opioid drugs often relapse even when they appear to have recovered. Many of them remember the highs they obtained by their addictions and are tempted to return to them. Or they fail to return for prescriptions or cannot afford them. Or they do not engage in cognitive therapy that allows them to realize that addictions have ongoing negative impacts on their lives, including disrupting their families, employment, and education.

Many addicts also recover from their addictions without assistance, even if the process may take decades and despite negative impacts on their lives. These evidence-based approaches provide a superior alternative for most of them because they may save their lives and decrease negative impacts.

Addicts often cannot afford some effective medications that have not been approved by the FDA and that are not funded by Medicaid. To help addicts buy them, some clinics gave them twice the amount they need so they can sell the half they don't need and use the remaining funds to buy their share of the drug (Macy, 2018).

Many physicians have not yet adopted these new approaches. Many physicians shun drug addicts whether because they don't want them in their waiting rooms or because they don't have waivers that allow them to use the medications (Macy, 2018).

Drug courts and law enforcement officers often do not like medication-assisted treatment. They wrongly view the lifesaving medications as addictive. They wrongly subscribe to the 12-step model. They wrongly believe residential centers are effective. They began changing their views only in 2018 (Macy, 2018).

Three Additional Epidemics: Smoking, Poor Diets, and Unaddressed Mental Health Problems

The United States also has to address three additional major epidemics: smoking, poor diets that lead to obesity, and unaddressed mental health problems. Roughly 480,000 Americans die each year from smoking due to lung cancer and other cancers, heart disease, and many other ailments it causes (CDC, 2018a). Although health and longevity problems from smoking are often viewed as health problems, they also can be viewed as mental health problems. Smoking is an addiction to a drug—nicotine and other chemicals in tobacco products. The United States has, at best, an inadequate method for dealing with this problem. Roughly half of primary physicians do not ask patients if they smoke. Yet smoking is an addiction. Here too, some medications help people curtail smoking when combined with cognitive therapy and regular monitoring by health professionals.

Food is another addiction for tens of millions of Americans. Roughly 35.7% of American adults age 20 to 39 years are obese as well as 42.8% of adults age 40 to 59 years and 41% of adults age 60 and older (CDC, 2018c). Here, too, an epidemic exists with rates of obesity rising from 33.7% in 2007–2008 to 39.6% in 2015–2016—and with rates of severe obesity rising from 5.7% to 7.7% (CDC, 2018c). With respect to obesity, considerable scientific evidence refutes the notion that obese persons need to undergo drastic weight reduction because researchers have discovered that obese persons almost always regain lost weight (Mayo Clinic, 2015).

Persons often have better outcomes when they combine modest weight reduction with regular exercise. Many scientists urge interventions in families that have obese children to modify diets, modify meals at schools, reduce eating sugar and fat content, especially at fast-food outlets, and educate parents about healthy nutrition. Some people with extreme obesity undergo surgery to diminish the size of their stomachs because they cannot otherwise control their weight. They realize that they are otherwise likely to have shortened life expectancy from heart disease and other diseases that are linked to extreme obesity.

The challenge of changing eating habits to include less fat, meat, dairy products, sugar, and salt is formidable, but other nations have succeeded. The Finnish government launched an attack on early deaths in the North Karelia province of Finland in 1972 by attacking high blood pressure, cholesterol, and smoking. They used "media campaigns, community meetings, chats in people's kitchens, carrots and sticks for farmers and food producers—up to and including village-versus-village competitions over cutting back on smoking or reducing cholesterol counts" (Willingham, 2018). As of 2012, deaths from heart disease plummeted by 82% among middle-age men. When the project was converted into a national one, the death rate of these men plummeted by 80% by 2012. According to Professor Glorian Sorensen at the Harvard T.H. Chan School of Public Health, the project demonstrated "how social factors may really shape health behaviors that are associated with chronic disease" (Willingham, 2018).

Mental health problems can also be viewed as an epidemic. We have already discussed the prevalence of mood disorders such as anxiety and depression as well as the roughly 1% of the American population who have schizophrenia. We have discussed the sheer numbers of persons with serious mental illness who receive no help for their conditions. We view with dismay the sheer number of mass shootings often conducted by persons with serious mental illness. We see many homeless persons on the streets for mental health problems. Many persons fail to graduate from high school, junior colleges, colleges, and universities due to unaddressed mental health problems, as I discuss in Chapter 12. I discuss in prior and ensuing chapters the sheer number of unaddressed or poorly addressed mental health problems in the health, gerontology, safety net, child and family, education, immigration, and criminal justice sectors.

Many persons are members of two or more of these epidemics. A person may be alcoholic and obese. Another person may have a severe mental condition and smoke. Odds of early death likely increase with dual or greater membership in these epidemics.

It is now certain that these epidemics require collaboration among the health, mental health, and substance abuse sectors. Combinations of medications, counseling, and social support systems need to work in tandem to address them, as we have discussed with respect to each of the five epidemics.

A prominent law professor, Elyn Saks, reveals that she could not have had her highly productive career without using medications for her schizophrenia as well as periodic visits with her psychiatrist (Saks, 2007). Persons with these problems need to work with health professionals to determine the dosage levels that relieve symptoms but minimize side effects.

Many difficult issues need to be addressed moving forward. Many experts also contend that children and adolescents are overmedicated in the United States for conditions such as attention deficit hyperactivity disorder. They argue that excessive medications used by children and youth may harm their brains because they are not fully developed. Many adults may be overmedicated for a variety of disorders including sleep problems and anxiety. Unlike many other nations, the United States allows mass media to advertise medications widely to the public rather than relying on health professionals to inform patients about medications that might be useful to them. Nor does the United States allow Medicare and Medicaid officials to bargain with pharmaceutical companies about the cost of medications—or encourage American consumers to shop for medications in Canada, Mexico, and elsewhere.

Considerable prejudice against persons with problems in the five epidemics will slow the progress in addressing them. Many people believe that persons with schizophrenia or autism are inherently dangerous, when in fact, they commit fewer crimes than persons in the broader population (Brekke et al., 2001). Some biologists and geneticists contend that current descriptions of mental illnesses and substance abuse in the *DSM-5*, published by the American Psychiatric Association, lack scientific rigor, even though many clinicians find it to be useful. Its authors, they contend, relied on descriptions of patients' symptoms and behaviors to identify mental health and substance abuse problems rather than seeking their underlying genetic and physiological causes. The National Institute of Mental Health seeks genetic and biological factors that cause mental health and substance abuse problems so that medications and other physiological interventions can be developed to treat them, but its leaders admit that this approach may not yield substantial results for a decade (National Institute of Mental Health, 2017).

What Will It Take to Curtail the Substance Abuse, Alcohol, Smoking, Obesity, and Mental Health Epidemics?

The nation spends about $35 billion per year on alcohol and substance abuse treatment (Glaser, 2015). Absent other estimates, assume that at least $50 billion in additional resources is needed over the next 10 years just to get medication-assisted treatments into the medical system for substance abuse as well as cognitive therapy. Physicians have to be retrained. Thousands of physicians have to be recruited to treat persons with substance abuse. Residential treatment centers have to be banned or not reimbursed if they fail to provide medication-assisted treatment. Primary care physicians need to be informed about evidence-based care so that they can make appropriate referrals. Emergency rooms and ambulances have to be able to administer lifesaving medications that revive persons who are near death from overdosing (CNN, 2018). Support and monitoring systems have to be developed, such as when persons with substance abuse relapse. Everyone needs to view different addictions as diseases rather than as a moral fault. Unfortunately, House Republicans and the Trump administration proposed minuscule resources in 2018, even seeking deep cuts in Medicare and Medicaid, not to mention repealing the ACA (Macy, 2018).

Absent other estimates, let's assume that $50 billion is needed for each of the other epidemics in the next 10 years because the nation needs similarly to retool its existing non-systems for addressing alcoholism, smoking cessation, and obesity, not to mention mental health problems. That would require at least $200 billion over 10 years in addition to the new funds allocated to the substance abuse sector—and possibly far more. The nation can afford these expenditures as illustrated by the nearly $2 trillion in tax cuts in 2018 that mostly benefited corporations and wealthy persons. Otherwise, the nation will needlessly lose hundreds of thousands of lives each year, not to mention diseases, loss of employment, premature deaths, and other problems that compromise the well-being of persons encumbered by any of these five epidemics. These investments would be partially offset by fewer emergency room visits, lower surgical costs, higher tax revenues from increased employment, lower medical costs, and greater productivity of workers.

Primary care doctors need to refer persons in the five epidemics to teams of professionals that develop and monitor treatment rather than assuming the primary role. Community clinics need to be developed to augment access and monitoring. Interdisciplinary teams are needed, as are currently common with respect to chronic diseases like diabetes. Social workers need to be important members of these teams. Sufficient numbers of staff need to be funded to eliminate long waits, such as six-month waits that commonly occur with respect to persons seeking help with substance abuse. Telemedicine should be widely used to allow frequent interactions with patients and to allow fast responses when they relapse. Policy planning at federal and state levels needs to be developed by policy and budget teams vested with each of these epidemics. The teams need to cut across federal agencies, including the FDA for medications, the Department of Health and Human Services for financing of the four initiatives, and the Office of Management and Budget for funding of the initiatives. Funds need to be planned for 10-year periods so that Congress does not frequently cut them. Unlike President Trump, who failed to invest sufficient resources to stem the opioid epidemic, the nation needs to invest huge resources for a sufficient period to stanch these epidemics. Trump did sign a package of bills to address the opioid crisis but failed to fund a "wide and sustained expansion of addiction treatment" (Lopez, 2018).

ANALYZING THE EVOLUTION OF THE MENTAL HEALTH AND SUBSTANCE ABUSE SECTORS

Grob (1994, 2008) points out that throughout history, mental health policy goals have been shaped and transformed by the shifting of funding responsibility among the local, state, and federal sectors of government. The following timeline depicts the evolution of the American mental health care system from the colonial era to the present:

- **Mental Illness in Colonial Times.** Low incidence of mental illness reported; not defined as a medical problem or a matter of social concern. There were no effective treatments; families were responsible for providing care to members unable to work (Grob, 1973, 1994; Deutsch, 1937).
- **The Late 17th Century.** As numbers grew, the mentally ill were housed with other dependent populations (the aged, infirm, or criminals) in local workhouses or almshouses (Grob, 1983; Jansson, 2019) or in the basements of early private hospitals (Grob, 1973).
- **The Moral Age.** Progressive communities applied Christian principles and established well-ordered asylums to cure the effects of the outer world (Grob, 1983; Jansson, 2019).
- **The Public Mental Hospital Movement.** In 1848, social reformer Dorothea Dix advocated for more humane treatment and state financing for building 95 asylums (Jansson, 2019).
- **The Late 1800s.** Optimism diminished as state hospitals offered custodial care but little hope of cure; and costs for care became the largest piece of the budget; state regulations pressed for scientific basis for provisions (Grob,1983; Deutsch,1937; Mechanic, 1989).
- **The Mental Hygiene Movement to the Progressive Era.** Following institutional care failures, many medical treatments shifted to an outpatient services focus (Grob, 1994). Explorations in neurology and psychiatry, including Freudian insights into the mind, created hope that understanding and cures were possible (Grob, 1983). Treatment innovations became supplements to institutional care rather than a substitute for it; by 1940 admissions to state hospitals increased to 455,000 (Grob, 1973, 1983).
- **Impact of the Great Depression, World War I, and World War II on Psychiatric Care.** There developed increased public awareness that given sufficient stress, anyone could develop psychiatric symptoms (Grob, 1983);

America's involvement in wars exposed the need for crisis-oriented and group treatments (Grob, 1973). New laws and the creation of the National Institute of Mental Health (NIMH) provided the impetus for research to develop preventive and therapeutic services (Grob, 1994).

- **The Deinstitutionalization Movement.** In the early 1940s, public reports of state hospital overcrowding and mistreatment of patients combined with evidence of the debilitating effects of institutionalization (Grob, 1991) made deinstitutionalization an appealing opportunity to shift costs to the federal government (LaFond & Durham, 1992).
- **Development of Antipsychotic Drugs.** The introduction and use of antipsychotic drugs in the treatment of the mentally ill during the late 1950s was credited with the successful release of large numbers of patients from the state hospitals (Grob, 1994; Deutsch, 1937).
- **The Liberal Era: Great Society and Community Mental Health.** In 1963, the federal budget allocation to create community mental health treatment centers (Kennedy, 1990; Koyanagi & Goldman, 1991; Jansson, 2019) resulted in fewer hospitalized patients but failed to produce positive treatment outcomes. By the early 1970s, many former hospital patients were abandoned to psychiatric ghettos, jails, or the ranks of the homeless (Dear & Welch, 1987; LaFond & Durham, 1993).
- **The Carters.** In 1978 the Carter administration increased funding for community-based services and Rosalynn Carter made mental health her major domestic issue (Grob, 2005; Jansson, 2019) in response to sharp criticism of care provided for serious mental illness (Koyanagi & Goldman, 1991). Unfortunately, funding was inadequate to implement real reform.
- **The Neoconservative Era.** Between 1980 and 1993 the Reagan administration relinquished federal responsibility for the mentally ill and slashed state block funding and support of the national institutes (National Institute on Alcohol Abuse and Alcoholism [NIAAA], National Institute on Drug Abuse [NIDA], and NIMH). States were allowed to ignore mental health issues (Mechanic, 1989), and the numbers of mentally ill who became homeless, incarcerated, or confined to nursing homes increased (Jimenez, 2009). Public outcry, grassroots advocacy, and public education efforts were organized by the newly formed National Alliance for the Mentally Ill (NAMI).
- **Social Reform in a Polarized Context: President Bill Clinton and Vice President Al Gore.** By the 1990s Medicaid and Medicare became the source of most mental health funding, helping restore state authority for mental health services policy, whereas the federal government covered more of the costs (Grob, 2008).
- **The Mental Health Parity Act (1996).** This act aimed to improve mental health reimbursement for private insurance recipients (Mental Health America, 2012) but failed to require employers to offer mental health coverage, and many employers dropped coverage completely (Jimenez, 2009).

- **George W. Bush Era.** The New Freedom Commission on Mental Health (2003) committed to early treatment of mental disabilities similar to treatable physical illness and the need to address barriers to reform (stigma, treatment limitations, and delivery system fragmentation).
- **The Paul Wellstone and Pete Domenici Mental Health Parity and Addiction Equity Act** (2008; Substance Abuse and Mental Health Administration, 2012a). This act required that large insurance plans provide parity of coverage for behavioral health. Plans could be exempt from the law if they excluded mental health and substance health coverage and were allowed to retain higher premiums/co-pays for use of those services (Substance Abuse and Mental Health Services Administration, 2012a).
- **Barack Obama and the Audacity to Hope.** The Affordable Care Act (ACA) was signed into law in 2010, expanding coverage to 32 million uninsured Americans. The ACA reformed the parity law to integrate the care and managing of physical health and mental health conditions and disorders as well as requiring them to be equally funded in the hope of lessening or removing the historic barriers (U.S. Department of Health and Human Services, 2011).
- **The Opioid Crisis During the Obama and Trump Presidencies.** A massive epidemic of addiction to opioid drugs grew rapidly during the presidency of Barack Obama, growing to roughly 72,000 deaths in 2017. President Trump and his administration dealt with the crisis haltingly even though many of these deaths took place in the white blue-collar population in rural and semi-rural areas. Congress enacted the Substance Use-Disorder Prevention Act that was signed by President Trump on October 24, 2018. Critics viewed it as underfunded. See Table 10.1 for advocacy groups serving persons with mental illness and substance abuse.

TABLE 10.1 ■ Some Advocacy Groups for Mentally Ill Persons and Persons With Substance Abuse
American Psychiatric Association (APA) is a leading advocate for mental health services: http://www.psych.org.
American Psychological Association (APA) adopts a variety of policy statements: http://www.apa.org.
National Alliance on Mental Illness (NAMI) mobilizes persons with mental illness to improve mental health services: www.nami.org.
National Council for Behavioral Healthcare (NCBH) is a leading advocate for improved mental health services: http://www.thenationalcouncil.org.
National Association of Social Workers (NASW) injects social workers' policy preferences into public arenas: http://www.socialworkers.org.

ADDRESSING SEVEN CORE PROBLEMS IN THE MENTAL HEALTH SECTOR WITH ADVOCACY

Core Problem 1: Engaging in Advocacy to Promote Consumers' Ethical Rights, Human Rights, and Economic Justice—With Some Red Flag Alerts

- **Red Flag Alert 10.1.** Ethical features associated with a patient's care are violated because health care staff may not be aware of the individual's right to confidentiality, self-determination, informed consent to treatments and medication, accurate information, and safety related to being a danger to self or others.
- **Red Flag Alert 10.2.** Patients may experience a lack of empathy from providers because of differences in life experiences, attitudes, personal values, and culture.
- **Red Flag Alert 10.3.** Patients experience provider negative stereotyping, non-therapeutic attitudes, and behavior.
- **Red Flag Alert 10.4.** Patients may be misdiagnosed or experience diminished expectations for improvements due to clinician negative bias or inaccurate stereotypes, such as that individuals with mental illness are not intelligent and cannot change or that older adults are all frail, ill, or inflexible.
- **Red Flag Alert 10.5.** Patients may work with health care providers who are not transparent about their lack of clinical expertise about mental health conditions.
- **Red Flag Alert 10.6.** Some patients do not have conservators appointed through special court proceedings even though they cannot provide for their basic needs, engage in treatment voluntarily, or engage in hostile or disruptive behaviors at a high level.
- **Red Flag Alert 10.7.** Do insufficient numbers of mental health staff exist in specific settings, neighborhoods, regions, or states to help persons with mental problems or substance abuse?

Background

According to the National Association of Social Workers (NASW, 2012) Code of Ethics, mental health consumers have the right to have access to services, resources, and other needed information; to be treated with dignity, respect, and worth; as well as to self-determination to refuse treatment. Mental health consumers also have the right to receive care from providers who are competent; are committed to consumers; act respectfully, responsibly, and honestly; acknowledge the importance of human relationships; and act respectfully with regard to lifestyle choices.

Power disparities during treatment can create difficulties for consumers. Mental health providers tend to be placed in the expert role, which creates a power and role imbalance (McCubbin, 1994; Sullivan, 1992; Ware, Tugenberg, & Dickey, 2004). Tarrier and Barrowclough (2003) affirm that "without an equality and collaboration between those who provide and those who use professional mental health services there is always a risk of paternalism, stigmatization, and coercion" (p. 240).

In contrast to the goal of effective patient–provider partnerships, consumers may be undervalued and underestimated in their capacity to think and speak for themselves. These commonly held negative beliefs, perceptions, and attitudes about persons with mental illness can lead providers to a type of "self-fulfilling prophesy" about the potential for the person to participate fully in treatment or to recover (Rosenhan, 1973).

Resources for Advocates

Mental health advocates can draw upon many existing policies to protect patients' ethical rights at the micro level. These include professional ethical values and standards of accreditation agencies that license health professionals.

The Patient Self-Determination Act (PSDA) of 1991 mandated that health care facilities receiving Medicare or Medicaid reimbursement inform individuals of their right to engage in treatment decisions (Bradley, Wetle, & Horwitz, 1998). The ability for individuals with mental illness to participate in shared decision-making about treatment choices and goals or to complete advance directives are important matters to clients, families, and their providers.

Each state has a Board of Behavioral Sciences (BBS) that serves as a consumer protection agency with the goal of protecting consumers by determining and upholding standards for competent and ethical behavior by the professionals under its jurisdiction. For example, the BBS in California has adopted guidelines that identify the types of violations and range of penalties or disciplinary actions that may be imposed on practitioners. Detailed information about the violations of statutes and regulations under the jurisdiction of the BBS and the appropriate scope of penalties for each violation can be found at http://www.bbs.ca.gov/pdf/publications/dispguid.pdf.

Historically, the lack of mental health awareness and supportive services to the mentally ill has contributed to unnecessary deaths and injuries in hospitals, prisons, and the community (Jansson, 2019). Further, community agencies, services, and other professionals who have interactions with persons who have mental health symptoms or disorders may require additional training and protocol to act in an ethical and safe way.

The *Washington Post* calculated that roughly 500 people with mental illness were fatally shot by police in the United States—or one in four police shootings (Roth, 2018). Although law enforcement agencies typically have provided training to their officers and staff so that they can safely and appropriately communicate with persons who are homeless or mentally ill, training is not as extensive as the training offered to mental health professionals. In addition, staff may not be able to learn and implement specific techniques and approaches for responding to certain populations in particular settings, such as with persons who have psychotic symptoms and are un-medicated and homeless. Many police departments have instituted go-slow and back-off policies to counteract the tendency of many officers to take actions that escalate violence. Many police departments employ social workers and women to take prominent roles in reducing tension in confrontations between law enforcement and persons with mental problems. Police have increasingly been taught how to differentiate mentally ill persons and persons inebriated or on drug highs from other people in confrontations so that they refrain from shooting at them.

Public officials have historically responded to problems or disruption in the community related to mental health by proposing additional laws to enforce mandatory treatment of individuals with mental disorders, particularly in the wake of publicized violent incidents in which either the victim or perpetrator is mentally ill. For example, Andrew Goldstein, a 29-year-old man with schizophrenia, pushed Kendra Webdale into a subway train, leading to her death. Reports of Mr. Goldstein's refusal of treatment and medicine non-compliance resulted in Kendra's Law in New York State (1999), which gives judges the authority to mandate individuals with severe mental illness to receive outpatient psychiatric treatment or be subjected to involuntary inpatient state hospitalization. Opponents of involuntary treatment argue that the notion that individuals with mental and substance use disorders are "dangerous" is an exaggerated bias and unsubstantiated fact that has influenced public policy (Stuart & Arboleda-Flórez, 2001). Statistics show that only 3% of violent, incarcerated offenders have committed crimes that are attributable to a primary non-substance-use-related disorder (Stuart & Arboleda-Flórez, 2001) and that individuals diagnosed with a serious mental illness are 14 times more likely to be a victim of a violent crime than to be arrested as the perpetrator of one (Brekke et al., 2001).

Conservators are often appointed through special court proceedings for persons with mental illness that lead them to be able to care for their basic needs and not to engage in disruptive behaviors (Jansson, 2011).

POLICY ADVOCACY LEARNING CHALLENGE 10.2

CONNECTING ADVOCACY TO PROTECT A PATIENT'S ETHICAL RIGHTS AT MICRO, MEZZO, AND MACRO LEVELS

Josie Mora, a 35-year-old woman, was diagnosed with schizophrenia (disorganized type) at the age of 22. Despite large doses of psychotropic medication, Josie exhibited disruptive and bizarre behavior, disorganized speech, inappropriate affect, and daily auditory hallucinations. For example, she called out to people using derogatory language, often falsely accusing them of trying to harm her. At home, she required constant limit setting as she would drink continuous pots of coffee and large amounts of soft drinks unless stopped. During the night she would get up several times to pace and clang pots and pans in the kitchen. As a result, she required ongoing supervision and was unable to participate in any psychiatric rehabilitation program activities offered at the mental health center.

The Mora family lived in the same community since they emigrated from Mexico over thirty years ago. The parents preferred to speak in Spanish, although they understood and spoke limited English. Josie preferred English, although Spanish was her first language. Josie lived at home and her parents were her primary caretakers. Several adult siblings and other extended family members lived in the same city and participated in Josie's care. Mrs. Mora had not worked outside the home since the onset of Josie's illness. During the day while Mr. Mora was at work, Mrs. Mora relied on a network of family and friends to help her with Josie, particularly whenever Mrs. Mora had her own medical appointments or needed to run certain errands. Eventually, Mr. Mora opted for early retirement from his factory job because he wanted to help his wife with Josie's care.

Both parents always accompanied Josie to her medication management appointments and monthly meetings with a social worker. The Moras took turns sharing the highlights of the month regarding Josie's behavioral outbursts involving family members, friends, and neighbors. Often they shared certain successful outcomes, like going to the park for a family gathering where Josie was able to tolerate being around many people without causing much disruption. Because of her severe impairment and her extensive need for supervision, on several occasions the social worker gently raised the option of board and care or other supervised residential care for Josie. The Moras considered the various options presented, but always expressed their willingness to accept their parental responsibility for Josie's caregiving, and their great hope that Josie would get better. The Mora family wanted to raise enough money for a trip to Mexico City. They hoped to take Josie to a special church to receive a holy blessing that would lead to healing and possibly a miracle. Several years later Josie was prescribed a new atypical anti-psychotic medication and with additional rehabilitation she made a substantial improvement in her social functioning. The Moras expressed that this progress was more than they had hoped Josie would achieve.

Source: Barrio, C. & Yamada, A. M. (2005). Serious mental illness among Mexican immigrants: Implications for culturally relevant practice. *Journal of Immigrant & Refugee Services*, 3(1/2), 87–106.

Learning Exercise

1. How does the social work value of self-determination apply to this case example?
2. Does the social worker respect the values of interdependence and the family's sense of hope?
3. What other advocacy actions could a social worker take to protect the patient's ethical rights on micro, mezzo, and macro levels? For example, do the laws of your state allow judges, under specific circumstances and safeguards, to require specific patients to take anti-psychotic medications—and would you agree that this is ethical in these specified circumstances?

Core Problem 2: Engaging in Advocacy to Promote Quality Care—With Some Red Flag Alerts

- **Red Flag Alert 10.8.** Clients receive one eclectic or generic treatment from an agency or practitioner that is used for all clients.
- **Red Flag Alert 10.9.** Clients are treated by practitioners who are not fully trained to implement EBPs.
- **Red Flag Alert 10.10.** Clients are offered treatments that are provided by clinicians who do not have access to ongoing supervision or training and are not up-to-date on the empirically supported treatments.

- **Red Flag Alert 10.11.** Clients may be treated by clinical staff or at agencies with negative attitudes about EBPs or new interventions.
- **Red Flag Alert 10.12.** Clients are recommended for treatment interventions without attention to their individual preferences.
- **Red Flag Alert 10.13.** Clients are treated with EBPs that have not been adequately researched, such as with specific ethnic/racial populations.
- **Red Flag Alert 10.14.** Social workers' distinctive focus on diagnoses and treatments based on a biopsychosocial model are not sufficiently recognized in those settings that rely on EBPs. (See Policy Advocacy Learning Challenge 10.3).

Background

As the emphasis on implementing EBPs and effective empirically supported treatments (ESTs) has gained stronger literature support, mental health providers are expected to choose and implement the best possible services and EBPs that target the client's specific needs and desired health care outcomes. Treatments need to be implemented with fidelity (sometimes called adherence or integrity) to preserve the components that made the original practice effective.

Despite extensive evidence of effectiveness, mental health practitioners and programs often underutilize EBPs with the majority of clients with mental health disorders and in other human services (Kirk, 1990; Brooks, 2016). Professionals often lack training in the EBP process, possess insufficient provider education to implement specific ESTs, have limited resources, and lack ESP in community practice settings (Bellamy et al., 2008; Bledsoe et al., 2007; Brekke, Ell, & Palinkas, 2007). They lack reimbursement from Medicaid and other health insurance for EBP (Ganju, 2003). They face competing organizational demands, insufficient length of client visits or number of sessions, and excessive reliance on existing treatments (Cohen et al., 2008), as well as inadequate preparation to interpret research findings (Bellamy et al., 2008; Murray, 2009).

It must be acknowledged that considerable controversy exists about the science of mental health. Some mental health experts questioned the validity of some diagnostic information provided in the *Diagnostic and Statistical Manual of Mental Disorders (DSM-4)* of the American Psychiatric Association (2003). Bryan King, director of the Seattle Children's Autism Center and who served on the task force of the American Psychiatric Association, influenced the decision in the fifth edition of this manual to eliminate many autism spectrum diagnoses such as Asperger's syndrome, pervasive developmental disorder, and childhood disintegrative disorder and to consolidate them in a single diagnosis of autism spectrum disorder. It must also be acknowledged that some medications produce side effects that offset their positive effects for some patients, including recently developed anti-psychotic medications. It is often difficult to diagnose patients with borderline or multiple conditions. See Table 10.2 for sources that provide EBPs for mental health and substance abuse.

TABLE 10.2 ■ Sources That Provide EBPs for Mental Health and Substance Abuse

1. SAMHSA: Evidence-Based Practices (EBPs) Resource Center on the web at https://www.samhsa.gov/ebp-resource-center
2. National Guideline Clearinghouse™ (NGC), http://www.guideline.gov
3. California Evidence-Based Clearinghouse for Child Welfare (CEBC), http://www.cachildwelfareclearinghouse.org/
4. Suicide Prevention Research Center: Best Practice Registry, http://www.sprc.org/featured_resources/bpr/index.asp—information about best practices that address the specific objectives of the National Strategy for Suicide Prevention
5. *Diagnostic and Statistical Manual of Mental Disorders*, fifth edition (*DSM-5*), American Psychiatric Association

Resources for Advocates

Current Council on Social Work Education (CSWE) Educational Policy and Accreditation Standards (EPAS) (CSWE, 2008) have established 9 professional social work competencies that specify that social workers in accredited MSW programs learn to use research to inform practice and their practice to inform research (e.g., CSWE, 2015). Ones particularly relevant to policy practice include Demonstrate Ethical and Professional Behavior (Competency 1), Engage in Diversity and Difference in Practice (Competency 2), Advance Human Rights and Social, Economic, and Environmental Justice (Competency 3), Engage in Policy Practice (Competency 4), and Engage in Practice-Informed Research and Research-Informed Practice (Competency 5).

POLICY ADVOCACY LEARNING CHALLENGE 10.3

CONNECTING MICRO, MEZZO, AND MACRO LEVEL POLICY INTERVENTIONS TO ADVANCE QUALITY CARE

Adams, LeCroy, and Matto (2009) believe that social workers do not share the medical model philosophy of treatment (focusing on symptoms and diseases) that typically underlies an EBP model because it may not adequately reflect sufficient focus on the individual and environmental factors that social workers view as essential to quality care. Consider an example of a discouraged and caring family who presents in treatment with a family member who has previously been diagnosed and treated for major depressive disorder and, despite treatment motivation and adherence, is currently so seriously depressed that he or she is unable to work or care for personal daily needs. The potential complexity in the case may not fit neatly into an EBP model.

(Continued)

(Continued)

Learning Exercise

Discuss the following questions posed by Adams, LeCroy, and Matto (2009):

1. How would a social work frame of reference, which includes the biopsychosocial model; the developmental stage of the person; the cultural, spiritual, and community factors; as well as ecosystems context assess and choose an EBP?
2. What other aspects of the case would the social worker frame suggest as important to the assessment and treatment planning for this case?
3. How does the social work emphasis on alliance or therapeutic relationship get incorporated into the treatment response?
4. What mezzo and macro policy advocacy interventions might social workers consider in these circumstances?

Core Problem 3: Engaging in Advocacy to Promote Culturally Competent Care—With Some Red Flag Alerts

- **Red Flag Alert 10.15.** A diagnosis is assigned of a mental health disorder based on the presentation of one or more symptoms taken out of context of the person's situation.
- **Red Flag Alert 10.16.** A person is judged as incompetent based on his or her failure to respond in the language and rules of the predominate culture.
- **Red Flag Alert 10.17.** Assumptions are made that a person has certain mental health problems or a diagnosis based on demographic characteristics (education level, age, and gender).
- **Red Flag Alert 10.18.** There is a failure to consider the importance of cultural values and beliefs or include them in the assessment or treatment of the patient's mental health.
- **Red Flag Alert 10.19.** Insufficient numbers of personnel, members of boards of directors, and public officials exist to engender multicultural services in specific programs or agencies.

Background

Access to health care is a leading health indicator and barriers to access include cost, language differences, lack of information about mental health services, scarce presence of mental health services in their native country, stigma toward mental health services, and competing cultural practices (Betancourt, Green, Carrillo, & Ananeh-Firempong, 2003). Lack of interpreter services or culturally or linguistically appropriate health education materials is associated with patient dissatisfaction, poor comprehension and compliance, and ineffective or lower quality care. Bureaucratic intake processes and long waiting times for appointments have also been particularly cited by ethnic minorities as major obstacles to obtain health care (Phillips, Mayer, & Aday, 2000).

Resources for Advocates

In response to the surgeon general's report, the President's New Freedom Commission on Mental Health (2003) was established in 2002, and a call for the elimination of disparities in health services due to cultural or geographic factors was identified.

The National Center for Cultural Competence (2007) reported that California, New Jersey, and Washington had passed laws mandating the integration of cultural and/or linguistic competence into curricula, continuing education, and licensure requirements for health and mental health care professionals.

POLICY ADVOCACY LEARNING CHALLENGE 10.4

CONNECTING MICRO, MEZZO, AND MACRO LEVEL POLICY INTERVENTIONS TO PROMOTE CULTURALLY COMPETENT CARE

Ensuring Culturally Competent Care

An 80-year-old Italian immigrant woman named Maria, who had lived in the United States for 60 years, arrived at an emergency room of a local hospital with her Italian friend and neighbor. She had fallen from her chair as she tried to pick up something that she had dropped and as a result cut her eyebrow on the edge of an end table. Her family was called, and when her daughter-in-law and adult granddaughter arrived, the daughter-in-law went with a nurse to fill out some paperwork, and the granddaughter stood in the waiting room anxious to find out where her grandmother was and how she was doing. A resident emerged and told the granddaughter in a matter-of-fact tone that he did not have good news. "Your grandmother has blood collecting in the sinus cavity above her eye, and it will eventually go to her brain and kill her."

"Isn't there anything that can be done?"

"Well, she's very old and likely has a type of dementia so I'm sure your family will not want to put her through surgery. Her quality of life is already poor. What medical conditions does she have? How old is she?"

The granddaughter knew that her grandmother had no medical conditions, had planted her garden the day before, and had dinner with the family as was usual for a Sunday, and she had seemed fine. "Demented? How did you assess that?"

The resident went on to say that when the patient was first seen in the hospital, she was unable to answer any questions they asked and wasn't making sense.

"Did they have someone speak to her in Italian?" At that moment the women's son arrived, and the resident repeated to him that he was sorry that there was nothing to be done for his mother. Hearing what had happened, the son requested that before the resident decide that it was his mother's time to die, he tell the family what he would do for this type of problem if his mother was 50 years old instead of 80.

"I'd do a CAT scan to see the full extent of the trouble and then likely surgery to remove the blood and repair the bleed." The son suggested he do the CAT scan. He then went to talk to his mother in her dialect and to ask how she felt. She spoke to him normally, apologizing for taking him away from his workday. Clearly, as the son suggested later to the resident, his assessment was that his mother's mental faculties were working as usual. The resident was surprised to see the

(Continued)

(Continued)

patient was able to respond to her son when he translated questions for her and to laugh when he joked with her; she looked like she had been in a brawl (her eye was swollen from the fall, but this was a very petite and mild-mannered woman who hardly ever raised her voice above a whisper). After the CAT scan, the resident returned in astonishment and reported that the woman had no sinus cavity on the right side of her head (where she had sustained the injury) and therefore the black area that had shown up on the X-ray that was thought to be blood was really only the absence of the sinus cavity (a condition that the patient was born with but had not caused any problem). The women received five stitches over her right eye and left the hospital with her family. She lived happily and independently on her own for another 17 years. This was her first and only trip to a hospital in her entire life. A week after her visit, she removed her own stitches, reporting to her son that she was feeling fine and didn't need them anymore.

Learning Exercise

1. Which of the problems encountered by this patient could benefit from micro, mezzo, and/or macro policy interventions?
2. What assumptions were made by the practitioner? How did those shape the assessment?
3. What advocacy skills do families need when a member of the family is determined to be "incompetent"?

Core Problem 4: Engaging in Advocacy to Promote Prevention of Mental Distress—With Some Red Flag Alerts

- **Red Flag Alert 10.20.** Prevention screenings are not universal but only offered to persons based on selection criteria in hospitals, schools, places of employment, and elsewhere.
- **Red Flag Alert 10.21.** Clients are not screened for behavioral health risks or problems that were not included as the presenting problem due to practitioner discomfort in medical settings, schools, places of employment, and elsewhere.
- **Red Flag Alert 10.22.** Clients have insufficient insurance coverage for behavioral health screenings.

Background

The lack of knowledge about the etiology and causes of mental illness and absence of formalized tests (such as a blood test) to detect or confirm the presence of the illness are serious barriers to preventive services. Although it is currently believed that many genetic and environmental factors contribute to mental health, there is no certainty or understanding as to the exact mechanisms of mental illnesses, and therefore scientific support for the creation of prevention programs is lacking. There has been a lack of funding from private insurance companies, Medicare, and Medicaid for services intended to prevent an illness. There also has been less money for research to show the efficacy of preventive interventions in part because of the expense of conducting the longitudinal studies necessary to show positive results. Because of this fact, which is inherent in the definition of prevention, research and interventions related to prevention may appear to be more costly in that they do not provide cost savings immediately. Finally, stigma associated with mental illness continues to stand as a barrier to prevention.

Resources for Advocates

Primary care physicians are often the first point of contact for many people as they enter the health care system. This fact offers primary care physicians an opportunity for the early identification of mental health concerns, to educate individuals and families about behavioral health, and to help facilitate referrals where necessary. There is a recognition of the need for primary care physicians to participate more fully in early screening and intervention aimed at new models of integrated health that focus on the impact of all health conditions on the person. These efforts encourage health care practitioners to view the person as a whole and to consider mental health, physical health, substance use, and behavioral health as essential components of health. Screenings to treat all known risks and manage existing conditions can improve the person's health status, reduce the need for additional health care services, and reduce financial and human costs. It is essential for mental health providers to collaborate and coordinate with primary care providers to screen for physical illness and substance abuse.

As part of the ACA, a new Prevention and Public Health Fund was created to expand the infrastructure needed to prevent disease through early detection and support the management of existing health conditions at the lowest severity level possible. Inclusion of additional prevention measures (the Prevention and Public Health Fund and the National Prevention Strategy) signals a significant shift in the focus of the ACA and movement in our country from a focus on sick care toward a system that advances health and health equity, saving money and lives (Jansson, 2011).

Core Problem 5: Engaging in Advocacy to Promote Affordable and Accessible Mental Health Services—With Some Red Flag Alerts

- **Red Flag Alert 10.23.** Persons who do not have health insurance cannot get coverage for mental health treatments, including immigrants.
- **Red Flag Alert 10.24.** Persons have health insurance, but their coverage offers inadequate coverage of mental health services, including its amount, length, and coverage of certain types of service.
- **Red Flag Alert 10.25.** Insurance deductibles or co-insurance for mental health services and medications place financial burdens on persons seeking treatment.
- **Red Flag Alert 10.26.** Persons with limited resources (lower socioeconomic status or fixed income) give priority to expenditures for health conditions (such as a life-threatening condition or a chronic medical condition) over treating mental health concerns.

POLICY ADVOCACY LEARNING CHALLENGE 10.5

CONNECTING MICRO, MEZZO, AND MACRO LEVEL POLICY INTERVENTIONS FOR PREVENTION IN MENTAL HEALTH

Benefits and Risks of Prevention

The *Diagnostic and Statistical Manual* (*DSM-5*) was currently revised and considered the inclusion of a category or disorder that would reflect a pre-diagnosis risk group, specifically for individuals who present with psychotic-like or attenuated symptoms that may convert and eventually meet the criteria for a DSM disorder. Although supporters of the risk category highlight that "early intervention may help delay or prevent exacerbation into psychosis," others feel that the evidence to distinguish between "ill and non-ill persons is difficult," making the likelihood of false positive diagnosis higher (Carpenter, 2009, p. 841).

Given the controversy of screening for psychotic risk and treating individuals before they have a diagnosis, Carpenter (2009) suggests that we consider whether the benefits of early intervention intended to help prevent a psychotic exacerbation outweigh the negative damage of labeling individuals with a mental health disorder when it is not certain that it will develop. Further, Carpenter (2009) asks practitioners to consider whether placing a person in a diagnostic risk category could "do more harm than good because of stigma or the unwarranted administration of treatment with a poor benefit/risk ratio?" (p. 841). Yet screening for mental health problems is widely practiced in health settings, such as by asking patients to respond to several questions that have proven effectiveness in diagnosing depression.

Recent research suggests, however, that early interventions are effective with persons who have signs and symptoms that place them at high clinical risk of psychosis but who have not yet developed full symptoms. These interventions delay the onset of first-episode psychosis and improve the outcomes of first-episode psychosis. NIMH has required states to allocate a substantial share of their MHBGs to funding early interventions for persons who are at high risk of approaching first-episode psychosis (Fusar-Poli, McGorry, & Kane, 2017). Further evidence may suggest that early intervention may improve subjects' future health and response to treatment. By using a team approach that gives families supportive assistance, this intervention may reduce family trauma associated with the onset of psychosis of family members.

Learning Exercise

1. What are your thoughts about this controversy?
2. As an advocate, what direction would you recommend? Why?
3. Do other efforts to prevent problems associated with mental health serve to educate the public or to reinforce the stigma?
4. If we do not seek early detection in schools, hospitals, and other settings, do we risk increasing unnecessary suicides or outbreaks of violence?
5. Do we need to be attentive to research findings to identify evidence-based findings that do, or do not, support early interventions for specific mental health and substance abuse problems?

- **Red Flag Alert 10.27.** Patients are not compliant with mental health treatment recommendations due to costs rather than personal motivation or health beliefs.

Background

The financial structure of the health care system in the United States has presented as one of the primary barriers to addressing the health care needs of individuals with mental illness:

- Health care systems lack reimbursement for coordinated care across service systems, health education, and supportive services; inadequate case management services to promote self-management and linkage to services; poor coordination between physical and mental health care systems; and lack of integrated treatment for dual diagnosis disorders.
- Many individuals with mental health problems are uninsured or underinsured, putting them at a disadvantage for receiving integrated health care.
- The majority of individuals who seek services in community mental health centers have health care insurance through Medicaid. To be effective serving patients in the public mental health system, Medicaid and Medicare must become partners in designing and implementing new strategies to improve access to care.
- Community Mental Health Centers (CMHCs) are not nationally required to serve the uninsured. Currently, a state policy mandate is being considered to require CMHCs to treat those who are uninsured. A decrease in non-Medicaid funding to CMHCs at the state level has led to more patients choosing Community Health Centers (CHCs), thereby giving them a lower quality of mental health care.
- Medicare has a large gap between mental health and physical health care coverage. Much larger co-pays are required for mental health visits than physical health visits. These cost differences force many individuals to pay high prices for mental health care or to seek mental health diagnosis and treatment through their physical health providers.

Resources for Advocates

The 1996 Mental Health Parity Act did not offer coverage for substance use/abuse disorders and increased insurance premiums for some, so California voters enacted Prop 63 (SB1136) to increase resources and access to mental health services to underserved or uninsured persons. Prop 63, or the Mental Health Services Act of 2004, was funded by a 1% tax on those individuals with incomes greater than $1 million; the associated funding has established full service partnerships across California and allowed for service expansion, innovation, and outreach programs addressing the concern for the rising numbers of those with mental illness who also are homeless. Similar legislation was not enacted in most other states, however.

Although the federal parity law of 2008 provided necessary coverage to persons with private insurance, millions of Americans remained uninsured and did not qualify for federal Medicaid or Medicare programs. The ACA attempts to close the service gap or the loopholes created under the federal parity law with new rules to the parity law that require that mental health treatment must be treated equally with standard medical and surgical coverage in terms of out-of-pocket costs, benefit limits, and practices such as prior authorization and utilization review. These practices must be based on the same level of scientific evidence used by the insurer for medical and surgical benefits. The ACA's provisions include the following:

- The ACA creates additional incentives to coordinate primary care, mental health, and addiction services.
- Grants and Medicaid reimbursement will be available for the creation of health homes for individuals with chronic health conditions, including mental illness and substance use disorders. Studies demonstrate that integrated and coordinated care is ultimately beneficial as it can help detect health problems before they become more serious concerns. It can also ensure that if a person gets a life-threatening diagnosis, he or she is also seen by a psychiatric professional for emotional health needs.
- The ACA provides improved coverage of mental health and substance abuse conditions.
- Mental health parity law prohibits insurers and/or health care service plans from discriminating between coverage offered for mental illness, serious mental illness, substance abuse, and other physical disorders and diseases. The ACA carries forward and considerably expands the requirements set forth in the parity law.
- The ACA prioritizes services in the home and community instead of within institutions.
- To promote the coordination of care, the ACA provides Medicaid payments for medical homes or health homes that coordinate care for people with chronic physical and/or behavioral health conditions.

Core Problem 6: Engaging in Advocacy to Promote Mental Health Services for Underserved Populations—With Some Red Flag Alerts

We discuss in this section two specific underserved populations: persons in places of employment and current and retired military personnel and their families.

- **Red Flag Alert 10.28.** Employees report that they fear that their mental health problems, if known by colleagues, will lead to adverse repercussions such as termination.

POLICY ADVOCACY LEARNING CHALLENGE 10.6

CONNECTING MICRO, MEZZO, AND MACRO LEVEL POLICY INTERVENTIONS TO INCREASE ACCESS

The ACA Cuts Consumers' Mental Health Costs

The ACA has greatly decreased disparities between the cost of care to patients of health care and mental health care. For example, a patient who was treated for diabetes and bipolar disorder prior to its implementation paid a 20% co-pay for diabetes care but a 50% co-pay for bipolar disorder. The new health care law will change this so that there is parity in co-pays. Think about some policy implications of this major shift in policy.

Learning Exercise

1. Are patients aware that mental health treatment is now much cheaper—and, if not, what kinds of community outreach and education might be considered?
2. Have health providers added sufficient mental health personnel, including social workers, to their rosters to provide the increased levels of mental health services needed by patients who will now receive them?

See how one patient advocate sees this issue at the following Internet site: http://www.npr.org/templates/story/story.php?storyId=92751635.

- **Red Flag Alert 10.29.** High absenteeism, dropouts, resignations, or discharges stem from untreated mental conditions as well as poor performance.
- **Red Flag Alert 10.30.** Persons currently in military service, as well as veterans, often fail to receive services for posttraumatic stress disorder, traumatic brain injuries, substance abuse, and family violence.

Background

Stigma, as well as lack of mental health personnel, interferes with mental health services in many institutional settings, such as schools and universities, military units, prison systems, nursing homes, and workplaces (Carter, Golant, & Kade, 2010). The U.S. National Comorbidity Survey of Americans ages 15 to 54 found that 18% of those who were employed reported that they experienced symptoms of a mental health disorder in the previous month. In 1990, mental health disorders cost the U.S. economy almost $79 billion in lost productivity (Rice & Miller, 1996, as cited in U.S. Department of Health and Human Services, 1999). Mood disorders cost more than an estimated $50 billion per year in lost productivity and resulted in 321.2 million lost workdays (Kessler et al., 2006). In addition, serious mental illnesses, which afflict about 6% of American adults, cost society $193.2 billion in lost earnings per year (Insel, 2008).

The failure of the Veterans Administration to provide timely and effective mental health services to military personnel and veterans has received extraordinary publicity as veterans have returned from the wars in Afghanistan and Iraq during the past 10 years.

The toll on these veterans has been particularly linked to the sheer prevalence of traumatic brain injury from improvised explosive devices in these wars. Veterans from the Korean and Vietnam wars also have complained about lack of treatment for long-standing cases of posttraumatic stress disorder. A new head of the Veterans Administration, Robert McDonald, was appointed in the summer of 2014 with the mandate to cut delays in mental health and medical services that have been experienced by tens of thousands of veterans in recent years.

Resources for Advocates

We have already discussed augmented health insurance for mental health care by the ACA, but the ACA also benefits veterans' mental health care because the ACA counts Veterans Administration health care as insurance coverage along with TRICARE, other military health plans, Medicaid, and Medicare. President Obama signed an executive order in 2012 authorizing additional funding to improve access to mental health services for veterans, service members, and military families (The White House, 2012), including expansion of crisis services, additional staffing for the Veterans Administration health care system, improved research on posttraumatic stress disorder, and care for traumatic brain injuries (O'Gorman, 2012). These services remained grossly inadequate, however, in 2014 and into 2015. Military personnel and veterans were also underserved in areas of substance abuse and family violence. It remained to be seen if services would be dramatically increased in coming years. Automatic access to medical care is provided to all veterans.

Mental health services are inadequately provided in many places of employment. Some corporations possess relatively large human resource departments with Employee Assistance Programs (EAPs) that offer free or low-cost mental health services. They provide short-term counseling either with their own hired staff or through contracts with external providers. Some employees are reluctant to seek help for mental problems because of stigma or because they fear that their diagnoses will be disclosed to their employers even when EAPs pledge confidentiality.

Core Problem 7: Engaging in Advocacy to Promote Care Linked to Consumers' Communities—With Some Red Flag Alerts

- **Red Flag Alert 10.31.** A national network of mental health agencies based in communities that help persons with substance abuse does not currently exist. Many patients who are discharged from hospitals, particularly in low-income areas, cannot obtain mental health or substance abuse services because of this problem.

Background

Historically, the care for mental illnesses has been offered by separate facilities, programs, and systems of care. Even as there is increasing evidence that a person-centered approach, which integrates services for health, mental health, and substance use disorders, improves health outcomes, the integration and need for more effective and collaborative community-based services, including outreach and transitional

POLICY ADVOCACY LEARNING CHALLENGE 10.7

CONNECTING MICRO, MEZZO, AND MACRO LEVEL POLICY ADVOCACY INTERVENTIONS FOR UNDERSERVED POPULATIONS

The Impact of Support From Fellow Employees

Without the insistence of her colleagues who knew that this episode of psychotic symptoms was somehow different than previous episodes that they had observed, Elyn Saks, an accomplished law professor at the University of Southern California and recipient of the MacArthur Foundation genius award, might have died of encephalitis. Brought to the emergency room by her colleagues due to worsening psychosis, she was immediately assessed to be psychotic and might have been dismissed by the hospital with additional antipsychotic medications except that her work colleagues advocated strongly that doctors check her for something other than psychosis. They knew how Dr. Saks looked when her mental illness worsened, and they knew that she was behaving differently. Dr. Saks, whose symptoms and mental illness and treatment resulted in hospitalization, forced treatment, isolation, and restraints (Carter, Golant, & Cade, 2010), has reported that one of the pillars that offered her support was her workplace and doing work that she loved. Saks and other tenured college professors with mental illness, including depression, have spoken openly about their initial fear of disclosing information about their diagnoses due to fear that they might be looked upon as unstable or less competent or be ostracized by peers.

Learning Exercise

1. What micro, mezzo, and macro policy level actions would you promote in specific employment settings to address employees' mental health problems, such as help from EAPs? Should social workers be central to EAPs in light of their biopsychosocial orientation?
2. What mezzo and micro policy advocacy initiatives might social workers lead to obtain more responsive care by the Veterans Administration and other health systems for military personnel and veterans?

programs, exists. Different philosophies of care, missions, and mandates that are not complementary and also separate funding sources act as barriers to collaboration.

Resources for Advocates

Many common mental illnesses are similar to the common physical chronic illnesses (such as diabetes, asthma, and cardiovascular disease) in that they are less likely to be "cured" but with proper care can effectively be managed over a person's lifetime. To successfully manage any chronic illness, patients will require treatment that is integrated and collaborative so that all aspects of their health are considered (they are treated as a "whole" person). Members of their health care team would also be able to communicate about ongoing treatment planning and decisions effectively. Access to community care that is integrated and collaborative will allow for smooth transitions from one type of treatment to another, when and if necessary, to manage the person's condition. For example, hospitalization to stabilize a person on a new medication may be necessary in a crisis, but having outreach services and an emergency on-call staff available to the person may help him or her return to home sooner with family support. The link between health care institutions and community agencies is essential to provide a continuum of care that can offer the patient the best and safest care in the least restrictive environment, many of which are also cost-effective and more desirable for the patient.

The ACA has begun to implement changes that will promote collaborative care. The ACA provides new flexibility in existing Medicaid state plan options for covering home- and community-based services. Outreach services, which began in 2014, will help enroll vulnerable and underserved populations in Medicaid and will target individuals living with mental illness (Mental Health America, 2012). Unfortunately, many neighborhoods currently lack mental health services that are easily available to many Americans in medically underserved areas. We discuss subsequently a policy initiative (The Excellence in Mental Health Act) that would fund neighborhood centers.

POLICY ADVOCACY LEARNING CHALLENGE 10.8

CONNECTING MICRO, MEZZO, AND MACRO LEVEL POLICY INTERVENTIONS TO DECREASE FRAGMENTED CARE IN THE MENTAL HEALTH SECTOR

Fragmented Care

The *Los Angeles Times* (Morocco, 2007) reported on a story that highlights the negative impact created by a fragmented system of care. The story tells of the struggle to find a way to respond to the complex needs of a woman who is homeless, 22 weeks pregnant, and mentally ill. Brought to the hospital emergency room by two good Samaritans who found her mumbling and wandering the street naked, her physical exam and lab tests were normal except for her mental status. After hours of trying to locate a place for this unnamed person ("Jane Doe"), she was unacceptable to the psychiatric unit in the hospital (they don't treat pregnant women who are more than eight weeks pregnant), had no medical reason to be admitted to the hospital, and was refused by a county psychiatric center, which never called back after reading her faxed record. Even after almost a day of effort on the part of the emergency room staff, and locating the name of a brother (who would not take responsibility for his sister), the hospital failed to find any possible support for the woman and started the process of recalling every call they had previously made in hopes of eventually being successful.

Learning Exercise

1. What failures in the current health care system are identified in this case?
2. How would you advocate and at what level to assist this woman in getting the help she needs?
3. What would you prioritize as her needs? Are they medical, psychiatric, or situational?

THINKING BIG AS POLICY ADVOCATES IN THE MENTAL HEALTH AND SUBSTANCE ABUSE SECTOR

We've identified five epidemics that include unaddressed or poorly addressed problems of alcoholism, substance abuse, obesity, smoking, and mental health. Assume the United States enacted Medicare for All that would replace the current Medicare and Medicaid programs. Assume Medicare for All would fund services for each of these epidemics that include cross-disciplinary teams that provide medical, psychological, and social support, outreach services, regional service facilities, and a network of community agencies. Take one of these epidemics. Discuss how social workers might develop and improve services by using micro policy advocacy. Discuss how social workers would build services at the community level (HINT: Reread the Finnish mezzo policy initiatives that led citizens to have better diets and exercise). Discuss how social workers would use macro policy advocacy to convince members of Congress to develop policies to address one of these five epidemics within the Medicare for All program.

Learning Outcomes

You are now equipped to:

- Describe the numbers of Americans with specific mental health and substance abuse problems
- Analyze the impact of poverty and income inequality on mental health and substance abuse
- Analyze flawed policies in the United States with respect to substance abuse and mental health problems as well as promising initiatives
- Identify how mental health and substance abuse policies evolved in the United States
- Identify and analyze seven core problems in the mental health and substance abuse sector
- Think big in the mental health and substance abuse sector by developing initiatives with respect to mental health, alcoholism, substance abuse, obesity, and smoking

References

Adams, K. B., LeCroy, C. W., & Matto, H. C. (2009). Limitations of evidence-based practice for social work education: Unpacking the complexity. *Journal of Social Work Education, 45*, 165–186.

American Psychiatric Association. (2013). *Diagnostic and statistical manual of mental disorders* (5th ed.). Arlington, VA.

American Psychiatric Association (2003). *Diagnostic and statistical manual of mental disorders* (4th ed.). Washington, D.C.

Barr, D. (2008). *Health disparities in the United States: Social class, race, ethnicity, and health.* Baltimore, MD: Johns Hopkins University Press.

Barrio, C. & Yamada, A. M. (2005). Serious mental illness among Mexican immigrants: Implications for culturally relevant practice. *Journal of Immigrant & Refugee Services, 3*(1/2), 87–106.

Bellamy, J., Bledsoe, S. E., Mullen, E. J., Fang, L., & Manuel, J. (2008). Agency-university partnership for evidence-based practice in social work. *Journal of Social Work Education, 44*(3), 55–75.

Betancourt, J. R., Green, A. R., Carrillo, J. E., & Ananeh-Firempong, O. (2003). Defining cultural competence: A practical framework for addressing racial/ethnic disparities in health and health care. *Public Health Reports, 118,* 293–302.

Bledsoe, S. E., Weissman, M. M., Mullen, E. J., Ponniah, K., Gameroff, M., Verdeli, H., . . . Wickramaratne, P. (2007). Empirically supported psychotherapy in social work training programs: Does the definition of evidence matter? *Research on Social Work Practice, 17,* 449–455. doi: 10.1177/1049731506299014

Bradley, E. H., Wetle, T., & Horwitz, S.M. (1998). The Patient Self-Determination Act and advance directive completion in nursing homes. *Archives of Family Medicine, 7,* 417–423.

Brekke, J. S., Ell, K., & Palinkas, L. A. (2007). Translational science at the National Institute of Mental Health: Can social work take its rightful place? *Research on Social Work Practice, 17,* 123–133.

Brekke, J. S., Prindle, C., Bae, S. W., & Long, J. D. (2001). Risk for individuals with schizophrenia who are living in the community. *Psychiatric Services, 53*(4), 485.

Brody, J. (2018, July 2). Preventing suicide among college students. *New York Times.* Retrieved from https://www.nytimes.com/2018/07/02/.../preventing-suicide-among-college-students.htm...

Brooks, J. (2016, December 6). Stanford social innovation review. Retrieved from https://ssir.org/articles/entry/making_the_case_for_evidence_based_decision_making

Burt, M. R. (2001). What will it take to end homelessness? Retrieved from http://www.urban.org/UploadedPDF/end_homelessness.pdf

Carey, B. & Hartocollis, A. (2013, January 16). Warning signs of violent acts often unclear. *New York Times,* pp. A1, A15.

Carpenter, W. T. (2009). Anticipating DSM-V: Should psychosis risk become a diagnostic class? *Schizophrenia Bulletin, 35*(5), 841–843.

Carter, R., Golant, S. K., & Cade, K. E. (2010). *Within our reach: Ending the mental health crisis.* Rodale Books.

Centers for Disease Control and Prevention (CDC). (2018a). CDC fact sheet. Retrieved from https://www.cdc.gov/tobacco/data_statistics/fact_sheets/fast_facts/index.htm

Centers for Disease Control and Prevention (CDC). (2018b). Cigarette and tobacco use among people of low socioeconomic status. Retrieved from https://www.cdc.gov/tobacco/disparities/low-ses/index.htm

Centers for Disease Control and Prevention (CDC). (2018c). Rates of obesity. Retrieved from https://www.cdc.gov/obesity/data/adult.html

Centers for Disease Control and Prevention. (n.d.). National Center for Injury Prevention and Control. Web-based Injury Statistics Query and Reporting System (WISQARS). Retrieved from http://www.cdc.gov/ncipc/wisqars

Cohen, D. J., Crabtree, B. F., Etz, R. S., Balasubramanian, B. A., Donahue, K. E., Leviton, L.C., & Green, L. W. (2008). Fidelity versus flexibility translating evidence-based research into practice. *American Journal of Preventive*

Medicine, 35, S381–S389. doi:10.1016/j.amepre.2008.08.005

Colton, C. W., & Manderscheid, R. W. (2006). Congruencies in increased mortality rates, years of potential life lost, and causes of death among public mental health clients in eight states. *Preventing Chronic Disease, 3*(2). Retrieved from http://www.pubmedcentral.nih.gov/articlerender.fcgi?tool=pubmed&pubmedid=16539783

Council on Social Work Education (CSWE). (2015). Educational policy and accreditation standards. Retrieved from https://www.cswe.org/Accreditation/Standards-and-Policies/2015-EPAS

Coy, A. L. (2010). *From death do I part: How I freed myself from addiction.* Three in the Morning Press.

Dear, M. J. & Wolch, J. R. (1987). *Landscapes of despair: From deinstitutionalization to homelessness.* Cambridge: Polity Press.

Deutsch, A. (1937). *The mentally ill in America.* New York: Columbia University Press.

Dodes, L. (2015). *The sober truth: Debunking the bad science behind 12-step programs and the rehabilitation industry.* Beacon Books: Boston.

Edwards, H. (2015, February 20). Should mentally ill people be forced into treatment? *Time Magazine.* Retrieved from http://time.com › health ›healthcare

FamiliesUSA. (2018). A 50-state look at Medicaid expansion. Retrieved from https://familiesusa.org/product/50-state-look-medicaid-expansion

Fox, C. (2018, March 9). How US gun culture compares with the world in five charts. *CNN.* Retrieved from https://www.cnn.com/2017/10/03/americas/us-gun-statistics/index.html

Fusar-Poli, P., McGorry, P., & Kane, J. (2017). Improving outcomes of first-episode psychosis: An overview. *World Psychiatry, 16*, 251–265.

Ganju, V. (2003). Implementation of evidence-based practices in state mental health systems: Implications for research and effectiveness studies. *Schizophrenia Bulletin, 29*(1), 125–131.

Garfield, R., Damico, A., & Orgera, K. (2018, June 12). The coverage gap: Uninsured poor adults in states that do not expand Medicaid. *KFF.* Retrieved from https://www.kff.org/medicaid/issue-brief/the-coverage-gap-uninsured-poor-adults-in-states-that-do-not-expand-medicaid/

Glaser, G. (2015, April). The irrationality of Alcoholics Anonymous. *The Atlantic.* Retrieved from https://www.theatlantic.com/magazine/archive/2015/04/the...of-alcoholics.../386255/

Goldman, W. (2001). Is there a shortage of psychiatrists? *Psychiatric Services, 52*(12), 1587–1589.

Grob, G. N. (1973). *Mental institutions in America: Social policy to 1875.* New York: The Free Press.

Grob, G. N. (1983). *Mental illness and American society, 1875–1940.* Princeton, New Jersey: Princeton University Press.

Grob, G. N. (1991). *From asylum to community: Mental health policy in modern America.* Princeton, New Jersey: Princeton University Press.

Grob, G. N. (1994). Government and mental health policy: A structural analysis. *The Milbank Quarterly, 72*(3), 471–500.

Grob, G. N. (2005). Public policy and mental illness: Jimmy Carter's presidential commission on mental health. *The Milbank Quarterly, 83*(3), 425–456.

Grob, G. N. (2008). Mental health policy in the liberal state: The example of the United States. *International Journal of Law and Psychiatry, 31*, 89–100. doi: 10.1016/j.ijlp.2008.02.003

Ingraham, C. (2017, August 11). One in eight American adults is an alcoholic, study says. *Washington Post*. Retrieved from https://www.sciencealert.com/one-in-eight-american-adults-is-an-alcoholic-study-says

Insel, T. R. (2008). Assessing the economic costs of serious mental illness. *American Journal of Psychiatry, 165*(6), 663–665. doi: 10.1176/appi.ajp.2008.08030366

Jansson, B. S. (2011). *Improving healthcare through advocacy: A guide for the health and helping professions*. Wiley.

Jansson, B. S. (2019). *The reluctant welfare state: Engaging history to advance social work practice in contemporary society*. Belmont, CA: Brooks/Cole Cengage Learning.

Jimenez, J. (2009). *Social policy and social change: Toward the creation of social and economic justice*. SAGE.

Kennedy, E. M. (1990). Community-based care for the mentally ill: Simple justice. *American Psychologist, 45*(11), 1238–1240.

Kessler, R. C., Akiskal, H. S., Ames, M., Birnbaum, H., Greenberg, P. A., Robert, M., . . . Wang, P. S. (2006). Prevalence and effects of mood disorders on work performance in a nationally representative sample of U.S. workers. *American Journal of Psychiatry, 163*, 1561–1568. Retrieved from http://ajp.psychiatryonline.org/cgi/reprint/163/9/1561

Kirk, S. A. (1990). Research utilization: The substructure of belief. In L. Videka-Sherman & W. J. Reid (Eds.), *Advances in clinical social work research* (pp. 233–250). Washington, DC: National Association of Social Workers.

Kolbert, E. (2018, January 15). The psychology of inequality. *The New Yorker.* Retrieved from https://www.newyorker.com/magazine/2018/01/15/the-psychology-of-inequality

Koyanagi, C. & Goldman, H. H. (1991). The quiet success of the national plan for the chronically mentally ill. *Hospital and Community Psychiatry, 42*(9), 899–905.

LaFond, J. Q. & Durham, M. (1992). *Back to asylum*. New York, NY: Oxford University Press.

Lopez, G. (2018, October 24). Trump just signed a bipartisan bill to confront the opioid epidemic. *Vox*. Retrieved from https://www.vox.com/policy.../trump-opioid-epidemic-congress-support-act-bill-law

Lopez, S. R. (2011, September 21). Kelly Thomas death should be a call to action. *Los Angeles*. Retrieved from https://www.nhpf.org/.../background-papers/BP66_MedicaidMentalHealth_10-23-08....

Macy, B. (2018). *Dopesick: Dealers, doctors and the drug company that addicted America*. Goodreads.

Manderscheid, R., Druss, B., & Freeman, E. (2008). Data to manage the mortality crisis. *International Journal of Mental Health, 37*(2), 49–68.

Mayo Clinic. (2015, June 10). Obesity: diagnosis and treatment. Retrieved from https://www.mayoclinic.org/diseases-conditions/obesity/diagnosis.../drc-20375749

McClain, M. (2015, October 29). Mental health in the spotlight Thursday on Capitol Hill. *Washington Post.* Retrieved from https://www.washingtonpost.com/...health/.../2015/.../mental-health-in-the-spotlight-toda...

McCubbin, M. (1994). Deinstitutionalization: The illusion of disillusion. *Journal of Mind and Behavior, 15*, 35–53.

Mechanic, D. (1989). *Mental health and social policy.* Englewood Cliffs, New Jersey: Prentice Hall.

Mental Health America. (2012, 28 June). Mental Health America hails ruling on ACA as a tremendous victory. Retrieved from http://www.mentalhealthamerica.net/go/about-us/pressroom

Morocco, M. (2007). He's willing to be her safety net. *Los Angeles Times.* Retrieved from http://articles.latimes.com/2007/jul/09/health/he-inpractice9

Murray, C. E. (2009). Diffusion of innovation theory: A bridge for the research-practice gap in counseling. *Journal of Counseling and Development, 87,* 108–116.

National Alliance on Mental Illness. (2011, November). State mental health cuts: The continuing crisis. Retrieved from https://www.nami.org/getattachment/About-NAMI/.../NAMIStateBudgetCrisis2011.pd...

National Alliance on Mental Illness. (2015, December). *State mental health legislation 2015: Trends, themes & effective practices.* Altoona, PA: Author.

National Alliance on Mental Illness. (2016). *Out-of-network, out-of-pocket, out-of-options: The unfulfilled promise of parity.* Altoona, PA: Author.

National Alliance on Mental Illness. (2018, September 5). Retrieved from https://www.nami.org/learn-more/mental-health-by-the-numbers

National Association of Social Workers. (2012). *Code of ethics.* Washington, DC: Author. Retrieved from http://www.socialworkers.org/nasw/ethics/ProceduresManual.pdf

National Center for Cultural Competence. (2007). *A guide for advancing family-centered and culturally and linguistically competent care.* Retrieved from http://nccc.georgetown.edu/documents/fcclcguide.pdf

National Institute of Mental Health (2017, November 1). Major depression. Retrieved from https://www.nimh.nih.gov/health/statistics/major-depression.shtml

National Institute of Mental Health. (n.d.). Suicide in the U.S.: Statistics and prevention. Retrieved from http://www.nimh.nih.gov/publicat/harmsway.cfm

New Freedom Commission on Mental Health. (2003). Achieving the promise: Transforming mental health care in America. Final report [DHHS Pub. No. SMA-03-3832]. Rockville MD: Department of Health and Human Services. Retrieved from http://govinfo.library.unt.edu/mentalhealthcommission/reports/FinalReport/downloads/Execsummary.pdf

Nuechterlein, K. H. & Dawson, M. E. (1984). A heuristic vulnerability/stress model of schizophrenic episodes. *Schizophrenia Bulletin, 10*(2), 300–312.

O'Gorman, K. (2012, August 18). The U.S. army reports record high suicide rates for July. *The Huffington Post.* Retrieved from http://www.huffingtonpost.com/kate-ogorman/army-suicides-veterans_b_1797199.html

Payne, K. (2017). *The broken ladder: How inequality affects the way we live.* Penguin Books. New York.

Phillips, K. A., Mayer, M. L., & Aday, L. A. (2000). Barriers to care among racial/ethnic groups under managed care. *Health Affairs, 19*(4), 65–75.

Rice, D. P. & Miller, L. S. (1996). The economic burden of schizophrenia: Conceptual and methodological issues, and cost estimates. In

M. Moscarelli, A. Rupp, & N. Sartorious (Eds.), *Handbook of mental health economics and health policy: Vol.1. Schizophrenia* (pp. 321–324). New York: John Wiley and Sons.

Rosenhan, D. L. (1973). On being sane in insane places. *Clinical Social Work Journal, 2*(4), 237–256.

Roth, A. (2018, May 30). Worried mom wanted the police to take her mentally ill son to the hospital, they shot him. *Vox*. Retrieved from https://www.vox.com/.../police-shootings-mental-illness-book-vidal-vassey-mental-he...

Saks, E. (2007). *The center cannot hold: My journey through madness*. New York: Hyperion.

Shirk, C. (2008). Medicaid and mental health services. Background Paper, 66. Retrieved from https://www.nhpf.org/.../background-papers/BP66_MedicaidMentalHealth_10-23-08...

Stuart, H. & Arboleda-Florez, J. (2001). Community attitudes toward people with schizophrenia. *Canadian Journal of Psychiatry, 46*(3), 245–252.

Sullivan, W. P. (1992). Reclaiming the community: A strengths perspective and deinstitutionalization. *Social Work, 37*(3), 204–209.

Tarrier, N. & Barrowclough, C. (2003). Professional attitudes to psychiatric patients: A time for change and an end to medical paternalism. *Epidemiologia e Psichiatria Sociale, 12*(4), 238–241.

Tavernise, S. (2016, February 12). Disparity in the life spans of the rich and the poor is growing. *New York Times*. Retrieved from https://www.nytimes.com/2016/02/13/health/disparity-in-life-spans-of-the-rich-and-the-poor-is-growing.html

U.S. Department of Health and Human Services. (2011, January 3). Obama administration issues rules requiring parity in treatment of mental, substance use disorders. Retrieved from http://www.hhs.gov/news/press/2010pres/01/20100129a.html

U.S. Department of Health and Human Services (2018). FY 2018 budget in brief. Retrieved from https://www.hhs.gov/about/budget/fy2018/budget-in-brief/samhsa/index.html

Vestal, C. (2016, February 11). Waiting lists grow for medicine to fight addiction. Pew Research Center. Retrieved from https://www.pewtrusts.org/.../waiting-lists-grow-for-medicine-to-fight-opioid-addictio...

Ware, N. C., Tugenberg, T., & Dickey, B. (2004). Practitioner relationships and quality of care for low-income persons with serious mental illness. *Psychiatric Services, 55*(5), 555–559.

Weissman, M. M. & Sanderson, W. C. (2001). *Promises and problems in modern psychotherapy: The need for increased training in evidence based treatments*. Prepared for the Josiah Macy, Jr. Foundation Conference, Modern Psychiatry: Challenges in Educating Health Professionals to Meet New Needs, Toronto, Canada.

The White House. (2012, August 31). Executive order—improving access to mental health services for veterans, service members, and military families. Office of the Press Secretary. Retrieved from http://www.whitehouse.gov/the-press-office/2012/08/31/executive-order-improving-access-mental-health-services-veterans-service

Willingham, E. (2018, March 7). Finland's bold push to change the heart health of a nation. *Knowable Magazine*. Retrieved from https://www.knowablemagazine.org/.../health-disease/.../finlands-bold-push-change-he...

BECOMING POLICY ADVOCATES IN THE CHILD AND FAMILY SECTOR

This chapter was coauthored by James David Simon, Ph.D., LCSW, and Anamika Barman-Adhikari, Ph.D.

LEARNING OBJECTIVES

In this chapter you will learn to:

1. Develop an empowering perspective with respect to the child and family sector
2. Understand the current state of children
3. Strengthen families by meeting their basic needs
4. Understand the journey of children through the child welfare system
5. Analyze the quest for algorithms to predict the odds that specific children will be harmed
6. Analyze the challenges of emancipation
7. Identify advocacy groups in the child welfare sector
8. Analyze the evolution of the child welfare sector and child welfare policies
9. Understand the scope of child abuse and neglect and where to find relevant statistics
10. Understand the political economy of the child welfare sector
11. Recognize the impact of an inegalitarian nation on the child welfare sector
12. Understand how seven core problems are often experienced by children and families
13. Think big in the child and family sector

Note: James David Simon, Ph.D., LCSW, and assistant professor at Cal-State San Bernardino, and Anamika Barman-Adhikari, Ph.D. and assistant professor at the University of Denver, made invaluable contributions to this chapter in this book's first edition. Bruce Jansson was the sole author of the second edition, editing and updating contributions from each original contributor.

The nation's child welfare system serves as the most important safety net for children who have been abused or neglected. This chapter argues, however, that child welfare should be revised to include far greater health prevention strategies with the use of micro, mezzo, and macro policy advocacy.

DEVELOPING AN EMPOWERING PERSPECTIVE WITH RESPECT TO THE CHILD AND FAMILY SECTOR

Every nation depends upon families to have children and to care for them in a humane way. Much is at stake: Children who have supportive and constructive relations with their parents, relatives, and siblings are more likely to mature into adults who will have supportive and constructive relations with their children. This cycle of birth, marriage, children, and youth has many positive by-products, such as a productive workforce, low levels of mental distress, low levels of crime, and high levels of happiness. Yet many families do not provide children with supportive and constructive relations whether due to economic and other hardships, lack of parenting skills, substance abuse, mental illness, or other reasons. Nor do they connect them with persons who can support their economic and employment success.

Relatively affluent families have high amounts of social capital, including high levels of education, income investments in real estate, financial investments, safe neighborhoods, excellent schools, nutritious meals, access to health and mental health professionals, and positive social networks (Bourdieu, 1986). Their children are highly likely to remain in the middle, upper-middle, and upper classes. Their children are likely to complete high school, obtain higher degrees, and join professions.

Many families lack all or most of these amenities. They live below, at, or near poverty levels. Their parents may not have completed high school or proceeded beyond high school. They may have intermittent unemployment. They do not own property, may not have access to health and mental health services, and are in economic jeopardy if they lose a single paycheck or if someone becomes ill. They may only have a single parent in the household, leading to lower family income that forces many single mothers to live in poverty, work several jobs, and scramble to fund childcare and other family expenses. They often lack the social connections that help them gain upward mobility. Families with relatively little social capital include disproportionate numbers of African Americans and Latinos as well as white families in rural and semi-rural areas.

Social workers and other professionals should engage in primary prevention with families with relatively little social capital. They should help the parent or parents obtain skills and education to avoid unemployment and low-paying work. They should help children succeed in school and to move into higher education. They should link children, youth, and families to health and mental health services. They should engage in secondary prevention by identifying families with at-risk characteristics such as substance abuse, poor parenting skills, chronic unemployment, and incarceration. They should provide an array of services to at-risk families. They should identify evidence-based interventions that decrease the likelihood of dysfunctional parenting.

In the actual world, social workers in child welfare departments devote most of their time to working with families with reported neglect or abuse rather than seeking to avoid these negative outcomes. Child welfare programs primarily respond to reports that a child has been neglected (lacks needed essentials like food, clothing, and shelter) or abused (has been physically or mentally assaulted by a parent or other family member). If they do not dismiss the case, they launch an investigation to see if the reports suggest likely neglect or abuse. They write reports that are forwarded to court when they believe neglect or abuse has occurred. Judges determine whether neglect or abuse has occurred or whether a high risk of neglect or abuse exists. If the court answers these questions in the affirmative, a judge decides whether to seek immediate removal of children from specific households, whether to foster parents or a relative, or whether to return a child to natural parents, who receive counseling and other assistance. If they are placed in foster care, a judge decides at a disposition hearing, often held within a year, to determine whether to return a child to natural parents or whether to seek an adoptive parent for the child. Due to inadequate numbers of foster homes and adoptive parents, children often live in multiple foster homes. Child welfare is a critical but last-resort service to identify and help children who are neglected or abused. It needs to be expanded to assume a larger role in helping at-risk families obtain greater social capital, gain access to health and mental health services, help children gain high school and higher-level degrees, and improve parenting skills with the goal of reducing the number of cases of child neglect and abuse.

Children from relatively affluent families also have serious problems, including substance abuse, school violence, suicide, venereal disease, family violence, and mental health problems. They may engage in cyber-bullying and in-person bullying—or be affected by it. Child welfare agencies seldom engage these children because reports of possible neglect or abuse are not given to child welfare agencies. Affluent families and their children are more likely to receive help from psychiatrists and psychologists, from counseling professionals in schools, or from child guidance clinics.

UNDERSTANDING THE CURRENT STATE OF CHILDREN

Differences exist among children from different racial and ethnic backgrounds when measured on the Race for Results Index developed by the Annie E. Casey Foundation (2017a). The foundation compared children of African American, American Indian, Latino, White, and Asian and Pacific Islander groups with respect to 12 indicators. The children in these groups did not differ significantly with respect to whether babies were born at normal birthweight, children ages three to five were enrolled in school, and the extent females delayed childbearing until adulthood. Significant differences existed, however, with respect to the other nine indicators. Whites and Asian and Pacific-Islanders scored far higher than African Americans, American Indians, and Latino groups with regard to school performance and graduation, including whether fourth graders scored at or above proficiency in reading, eighth graders scored at or above proficiency in math, high school students graduated on time, and young adults ages 25 to 29 had completed associate's degrees or higher. Whites and Asian and Pacific Islanders were more likely to live in two-parent families and with a householder who has at least a high school diploma. White and Asian and Pacific

Islanders were more likely as well to be in school or working between the ages 19 and 26. Whites and Asian and Pacific Islanders were far more likely to live above 200% of poverty and far less likely to live in low-poverty areas. This data suggests that children from White and Asian and Pacific Islander communities are more likely to be raised in families with considerable social capital. It suggests they are more likely to be accumulating social capital themselves by relatively high levels of school performance, completing one or more higher degrees, and either working or studying from 19 to 26.

The data shows, as well, that many children and youth in African American, Latino, and American Indian groups defy the odds and score relatively high on indicators. Also, many white and Asian and Pacific Islander youth score lower in specific indicators than their peers. But the pattern is clear: Children and youth in poverty households and who live in low-poverty areas disproportionately fall behind whites and Asian and Pacific Islanders in education and post-education success.

We should not conclude that African American, Latino, and American Indian children and youth score lower on many of these indicators due to character and family defects. As the foundation staff suggest, "many children of color are growing up in communities where unemployment and crime are higher; schools are poorer; access to capital, fresh produce, transit, and health care is more limited; exposure to environmental toxins is greater; and family supports and services are fewer" (Annie E. Casey Foundation, 2017b). Children and youth with relatively low scores on the indicators disproportionately live in deep southern states, Nevada, Arizona, New Mexico, and West Virginia—states that invest relatively low levels of resources in schools, safety net programs, and other social investments. Levels of child poverty differ markedly among states such as from only 11% in New Hampshire to 31% in Mississippi. As compared with the child poverty rates for whites of 12% in 2015, the child poverty rates for African American, American Indians, and Hispanics, respectively, were 36%, 34%, and 31%.

These data confirm the argument that social workers should seek greater funding to engage in primary and secondary prevention to help children, youth, and parents who are developing or possess low social capital to increase it to higher levels. Some progress has been made in the past five years, such as rising high school graduation rates. But 21% (15 million) of American children lived in families below the poverty line in 2015.

Those parents or relatives who are reported for child neglect or abuse overwhelmingly come from African American, Latino, and American Indian populations and poor families (Bartholet, Wulczyn, Barth, & Lederman, 2011). This outcome is not surprising. Impoverished family heads encounter extraordinary stress as they seek to feed and clothe themselves and their children, obtain and retain suitable housing, obtain medical and mental health services, and find quality schools. Higher levels of substance abuse exist in low-income families—abuse that can precipitate child abuse and neglect. Many male heads of households in African American and Latino families have been incarcerated for varying lengths of time—often for nonviolent offenses. It is more difficult for parents to keep track of their kids in neighborhoods with gang and gun violence. Parents who have not graduated, or only completed high school, are less able than other parents to help their children navigate high schools, particularly when their children attend low-quality inner-city schools. Low-income parents with substance abuse or other mental health problems have difficulty obtaining help when public hospitals are underfunded and when they lack insurance to seek help elsewhere.

STRENGTHENING FAMILIES BY MEETING BASIC NEEDS

Every social worker, no matter in which sector they work, should engage in micro, mezzo, and macro policy advocacy to enhance primary and secondary prevention. They can use micro policy advocacy to link families to the scores of safety net programs that we discuss in Chapter 9 because roughly 50% of families below or near federal poverty lines do not use the Supplemental Nutrition Assistance Program (SNAP), Medicaid, Supplemental Security Income (SSI), the Earned Income Tax Credit, Section 8 Housing vouchers, school breakfast and lunch programs, and many other programs. These programs are essential to a family's economic well-being because poverty is an early warning sign of an array of social problems, including disability, incarceration, family violence, homelessness, infant mortality, poor school performance, and short life span. They may not know about substance abuse programs funded by mental health departments and medical clinics. They may not know about vocational programs in specific school districts. They may not realize that an aging parent, living with his or her family, is eligible for Meals on Wheels and other services.

Social workers should strive to break down so-called silos that make it difficult for family members to use resources that cross health, mental health, education, safety net, gerontology, and other sectors. Perhaps tutoring programs are available to children who cannot read at grade level. Perhaps early intervention programs exist in the Department of Mental Health to avert substance abuse problems.

Families headed by single mothers particularly need robust public programs because many of them receive no or little child support. Absent fathers often do not make alimony payments, partly because they lack cash reserves themselves. Imagine parenting two children while working two, or even three, jobs. Policy makers should augment incomes for these families so that mothers have to work only a single job to fund childcare, food, and rent. They should increase housing subsidies. They should provide mentors who help them navigate working and parenting as well as job training and vocational education options. They should fund high school completion programs, enrollments in community colleges, and enrollments in four-year colleges that allow the mothers to survive with extended education and training benefits.

UNDERSTANDING THE JOURNEY OF CHILDREN THROUGH THE CHILD WELFARE SYSTEM

We now turn to a substantial subset of children in the United States who become the clientele of departments of child welfare each year. We have already discussed the sequence of events when a child welfare department receives a report from a community member, a family member, a teacher, a medical professional, or anyone else who believes a child has been neglected or abused. We will discuss some common mistakes that are made by screeners, higher-level staff who oversee the screeners, and judges. These mistakes sometimes lead to the injuries and deaths of children.

Social workers, lawyers, and judges may do the following:

- Underestimate the current danger of neglect and abuse by deciding at the outset that neglect or abuse has not taken place when, in fact, it has. Parents or others may cleverly conceal neglect or abuse—and children may fear to disclose these actions. Or relatively modest infringements of a child's well-being may become more serious after the initial investigation, such as when a parent engages in substance abuse or develops serious mental illness. The persons who made the initial report may have left the scene so that no one observes neglect or abuse subsequently. Many instances of injuries and deaths of children occur when they are mistakenly returned to their natural parents. Perhaps a natural parent convinces social workers, other investigators, and the judge that reports of neglect or abuse were erroneous. They may contend that neglect and abuse did not rise to a level sufficient to warrant removal of a child from their custody. They may argue that counseling and medical treatment will allow them to improve their parenting when it does not.
- Underestimate the risk of future neglect and abuse even when investigators find no current danger. A mother may not disclose substance abuse, for example, even though she is addicted to multiple substances. Investigators may not realize that a stepfather or boyfriend committed abuse against another child in a different family at a prior point and also had and still has substance abuse problems.
- Overestimate the existence of neglect or abuse or the likelihood of future neglect or abuse, leading to the unnecessary placing of a child in foster care or with a relative—or recommending the severing of parental ties through adoption. This mistake deprives a family of its parental rights. Persons of color contend, for example, that social workers, judges, and others overstate dangers of parents of color to their child as compared with white parents. If a child is wrongly placed in foster care or adopted, the child may suffer trauma from these unnecessary actions. Time and resources are wasted when social workers and others invest considerable time in investigating children and families who do not have serious problems.

These errors can lead to bad consequences. Overestimating danger can lead to the violation of parents' rights to their children. It can add additional costs to child welfare systems such as unnecessary placing of children in foster care. It can bring trauma to children taken from their natural parents. It can disrupt families who lose a family member. Underestimating the danger can cause unnecessary injuries or death to a child. Unnecessary injuries and deaths demoralize the staff of child welfare departments—and give the departments adverse publicity. The underestimating of danger can lead to litigation against a child welfare department and its employees.

Other systemic errors can take place once the child is removed from his or her natural parents. Foster parents need sufficient remuneration for their work—work that is often difficult because they become quasi-parents to children who have been traumatized by neglect and abuse. Many of them have children of their own already in their households.

Foster children not only need necessities like food and clothing but to be transported to dentists, doctors, psychologists, and other professionals. What's needed is an elite cadre of foster parents who are carefully selected and monitored.

Partly because foster parents are underpaid and because their work is difficult, they turn over rapidly. They often fund clothing, food, and other amenities from their own resources. Child welfare departments often cannot attract sufficient high-quality foster parents because they pay them insufficiently. Because many child welfare departments lack sufficient numbers of social workers, they often do not monitor foster care staff sufficiently or give them backup support and training. Some foster parents themselves engage in neglect and abuse.

Child welfare social workers themselves turn over at high rates in child welfare. Their work is made even more difficult by large caseloads that can include as many as 100 foster families. Some of them do not like the quasi-police role of investigating and overseeing families and foster children to detect neglect or abuse. Some are not used to working in concert with attorneys and judges. Some dislike personnel policies of child welfare agencies, such as excessive bureaucracy, lack of in-house training, and inaccessibility of supervisors.

High turnover rates of foster parents in many jurisdictions require many foster children to migrate between different foster homes rather than having stable placements. This migration increases trauma of foster children who already have experienced the trauma of leaving their natural parents.

It is optimal for foster children who cannot return to their natural homes to be adopted by new parents as soon as possible after a judge has determined that they cannot return to their natural parents. They often have to remain in foster care, however, for years because inadequate numbers of persons volunteer to become adoptive parents. Although adoptive parents receive some inducements, such as funding of their adopted children's medical care by public authorities, they, too, self-fund some of the costs of their adopted child. Like many foster parents, they often care for their own children as they adopt foster children. It is particularly difficult to find persons who are willing to adopt teenagers. It is often difficult to find people to adopt foster children from minority backgrounds, although considerable progress has been made in finding African American, Latino, and Native American adoptive parents.

ANALYZING THE QUEST FOR ALGORITHMS TO PREDICT THE ODDS THAT SPECIFIC CHILDREN WILL BE HARMED

Recall that under- and overestimates of the risk of neglect or abuse have often led to the deaths and injuries of children or the unnecessary removal of them from their natural parents. These problems will be lessened by the advent of algorithms that use large amounts of data to estimate the odds that a particular child in a particular family will be neglected or abused. Recall that screeners and investigators know relatively little about the backgrounds and life events of natural parents, stepparents, and other adult figures such as boyfriends who assume major roles in households with children. Social workers and other investigators, as well as judges, are forced to rely on their best guesses.

Let's discuss the dilemmas of Thomas Byrne, a social worker in the Department of Children, Youth and Families in Allegheny County in Pennsylvania (Hurley, 2018). He received a call from a hotline in 2016 from a preschool teacher who told him that a three-year-old girl had told him that a man, a friend of her mother, "was bleeding and shaking on the floor and the bathtub." Byrne checked agency records. He discovered prior reports of substance abuse, domestic violence, and inadequate food, as well as medical neglect and sexual abuse by an uncle with respect to an older sibling. None of these prior reports had been substantiated, and none met the minimal legal requirement to send a caseworker to the home to open an investigation. When required to estimate the risk to the child's future well-being on a form, Byrne, wrote "low risk" and "no safety threat" on his computer. Roughly 42% of the 4 million allegations received nationally are screened out whether because they did not contain charges of serious physical harm to a child or because of judgment calls, opinions, biases, and beliefs by screeners. Yet 1,670 children died in the United States of neglect or abuse in 2015 (Hurley, 2018).

Fortunately, Byrne worked in child welfare in Allegheny County in Pennsylvania, which had become the first jurisdiction in the United States that had compiled a predictive-analytic algorithm consisting of data about members of thousands of households that had made contact with his agency during at least the four prior years. The data had been painstakingly collected about jails, drug clinics, public welfare, police, psychiatric clinics, drug and alcohol treatment centers, and many other organizations. Using data from these sources, researchers Emily Putnam-Hornstein and Rhema Vaithianathan identified 100 criteria that accurately predicted whether prior clients of the child welfare agency in Allegheny County actually engaged in, or experienced, neglect and child abuse. They were able to calculate risk scores for former clients based on this prior data.

The two researchers who developed the algorithm were astonished to find that screeners in the recent past had screened in 48% of the lowest-risk children and had screened out 27% of the highest-risk families. Moreover, they identified eight cases in which a child had been screened out who was subsequently killed or seriously injured during the prior four years.

Once the researchers were certain that their algorithm yielded risk scores for prior clients, they could move to apply these findings to children and families named in reports that screeners were currently receiving. Armed with this methodology, they hoped they could help screeners not waste their time on very low-risk children and devote more of their time to children with high risk. Recall that Thomas Byrne had written "low risk" and "no safety threat" on his computer about the report from a preschooler to her preschool teacher. Byrne then decided to recheck the level of risk in this case by applying the algorithm to this child's family. He discovered that the score for this child's family was 19 out of a possible 20—with 20 indicating maximum risk. Armed with this data, Byrne and his supervisor ordered an investigation of members of this household to be conducted within 24 hours. The social worker investigator found no immediate threat: Sufficient food was in the refrigerator, and no signs of neglect or abuse existed. But she remained vigilant in light of the high-risk score. She visited the home on successive occasions, but the child's mother was not there. The mother refused to reveal test results at an alcohol/drug clinic to see if she was still using drugs, but the social worker soon discovered that she had failed three tests with cocaine and opiates in her urine. The child welfare team decided to act due to red flags that included the mother's substance abuse and her resistance to visits by a social worker. They received an emergency custody authorization from a judge that approved taking the child from the mother, who, while angry, divulged names of family members to whom to take the child. At a hearing, the presiding judge told the mother to get clean before the child could be returned to her. We don't know that the child would have been neglected or abused, but the data from the algorithm suggested a high likelihood of these outcomes.

Some reservations exist against use of algorithms in child welfare (Hurley, 2018). Critics of algorithms fear they are biased against African American families because low-income members of these families show up more frequently in the caseloads of welfare departments, prisons, and substance abuse clinics than members of white families. An independent evaluation of this algorithm concluded that African American family members do use many public agencies more frequently than white persons, not because of discrimination but because they qualify for the eligibility criteria of the agencies. Because African Americans have higher rates of poverty than whites in Allegheny County, their names appeared more frequently than whites on caseloads of these public agencies that were tapped by the researchers who developed the algorithm. The independent evaluation also revealed that screeners and investigators were spending far less time on low-risk clients and far more time on high-risk ones when they used the algorithm. The independent evaluation also found that black and white families were treated more consistently on their risk scores than on their race. The independent evaluation discovered that high-risk children were more consistently screened in by social workers after they used the algorithm as compared with their practice prior to using it.

It was too soon to tell if the number of injuries and fatalities of children will decrease post-algorithm, but the independent evaluator said, "They appear to be screening in the kids who are at real risk" (Hurley, 2018).

ANALYZING THE CHALLENGES OF EMANCIPATION

The emancipation of foster children at age 18 or age 21 poses its own risks. Foster children traditionally were left to fend for themselves when they left the jurisdiction of child welfare departments. It is not surprising that many of them encountered difficulties due to the traumas they had experienced from neglect, abuse, and severance from their natural parents during years of foster care. Many of them had not received adequate health and mental health services during foster care, attended relatively low-quality, inner-city, or rural schools and had lived in multiple foster homes prior to emancipation. Many of them had problems as foster children, such as poor school performance, brushes with the law, use of drugs, mental health issues, and (rarely) prostitution and drug dealing.

Foster children urgently need better services prior to emancipation. Some progress has been made for youth after their emancipation. A structured innovation was developed for emancipated youth, the "Village," aka, YVLifeSet. It provides these youth with intensive and customized one-to-one support to develop the skills set for youth to succeed independently in a central location without housing the youth. Youth received these services over a nine-month period. However, only modest improvements were discovered among youth in the experimental versus the control group after two years of initiating the program (Skemer & Valentine, 2016). Further research is needed to find strategies with larger effects, such as combining augmented services before the youth emancipate with extended services post-emancipation.

Identifying Advocacy Groups in the Child and Family Sector

Sites that are helpful to policy advocates in child welfare include the following:

Child Welfare League of America—http://www.cwla.org/

Child Welfare Information Gateway—http://www.childwelfare.gov/

First Five Los Angeles—http://www.first5la.org/

First Five California—http://www.ccfc.ca.gov/

Annie E. Casey Foundation—http://www.aecf.org/

Children's Defense Fund—http://www.childrensdefense.org/

Alliance for Children and Families—http://www.alliance1.org/

American Humane Society—http://www.americanhumane.org/

Court Appointed Special Advocates for Children—http://www.casaforchildren.org

National Conference of State Legislatures—http://www.ncsl.org (see the child welfare enacted legislation database to locate legislation in your state by year from 2012 to the present on adoption, child fatality/near fatality, child protection, child sex trafficking, courts and legal representation, disproportionality, education of children in foster care, foster care, funding of child welfare services, health and mental/behavioral health of children involved in the child welfare system, infant abandonment and safe surrender, kinship care, oversight, administration and interagency collaboration, prevention of child abuse and neglect, report of child abuse and neglect, services for older youth in foster care, siblings, termination of parental rights, tribal issues, and workforce)

ANALYZING THE EVOLUTION OF THE CHILD AND FAMILY SECTOR

We provide a brief chronology of the evolution of child welfare policies in the United States.

- The concept of *parens patriae*, which today has been expanded to justify the intervention of the court to protect minors, evolved in late 17th-century England—a power that allowed magistrates to remove children from their parents and allowed authorities to place children with other families (Areen, 1975; Murray & Gesiriech, 2005; Myers, 2008).
- Many children in the 1800s were either abandoned or sent to live in almshouses with other poor families, sick adults, and people declared insane (as cited in Schene, 1998).

- Criticism of the horrible conditions of children living in almshouses gave rise to orphanages and children's asylums, where many children were subsequently moved in the 19th century. The Children's Aid Society sent upwards of 150,000 children on orphan trains to live in rural homes throughout the Midwest (Schene, 1998).
- From the colonial period through the 19th century, childhood was not regarded as a special period for their development, and children were regarded as almost "miniature adults" (Trattner, 1999). They were often thrust into work and family responsibilities early in their lives. However, as time passed, due to the work of a number of scholars and religious ministers, experts believed that children needed to be nurtured for them to grow into healthy and well-adjusted individuals later in their lives. They needed extended periods in high school and post-high school education (Trattner, 1999).
- The beginning of organized child protection started in 1874 with the formation of the New York Society for the Prevention of Cruelty to Children in response to the physical abuse and neglect of a child with no child protective services agency or juvenile court to intervene on his or her behalf (Myers, 2008).
- Anti-cruelty societies grew in numbers across the United States, which gradually led to the establishment of child protection legislation and juvenile courts toward the end of the 19th century (Fogarty, 2008).
- The federal government held the White House Conference on the Care of Dependent Children in 1909 and established the Children's Bureau in 1912 (Costin, Karger, & Stoesz, 1996).
- Although there were hundreds of anti-cruelty societies assisting needy families and intervening legally to protect abused children in the 1920s, government agencies slowly took over these responsibilities (Brittain & Hunt, 2004). In January 1921, the Child Welfare League of America formed from the Bureau for the Exchange of Information among Child-Helping Agencies (Anderson, 1989).
- The federal government became a major funder of children's programs with the enactment of the Social Security Act through which programs under Title IV funded government child welfare agencies in the various states, survivors' benefits funding dependents of deceased workers, and the Aid to Dependent Children funded welfare for low-income families—which later became the Aid to Families with Dependent Children (AFDC) in the 1950s (Schott, 2009).
- Child protection became recognized as a profession that encompassed both social casework and child welfare services in the 1950s (Anderson, 1989).
- Many safety net programs were enacted during the Great Society that provide major assistance to relatively poor families including food stamps (now called SNAP), Medicaid, and Head Start, initiated as part of the Community Action Program of the War on Poverty.

- Many safety net programs were enacted during the presidency of Richard Nixon including the Earned Income Tax Credit, Section 8 rent vouchers, SSI, and mainstreaming of disabled children into public schools under the Education for All Handicapped Children Act.
- The Child Abuse Prevention and Treatment Act of 1974 was passed in response to the public outcry over battered child syndrome to provide fiscal aid to programs for the prevention, detection, and treatment of child abuse and neglect and to establish a National Center on Child Abuse and Neglect (Pecora, 2000).
- Title XX Amendment to the Social Security Act was enacted in 1975, which provided states with flexibility to fund social service programs for children and their families and provided funding for children in foster care who had special needs and for foster parents to receive extra services that were not provided earlier (Twiname, 1975).
- The Indian Child Welfare Act (ICWA), another piece of landmark legislation, was passed in 1978 to protect the integrity and support the continued existence of Native American children. ICWA established a minimum federal standard for removal of a Native American child from his or her home and provided guidelines for placement in foster or adoptive homes. If out-of-home placement became necessary, then the preference for placement was with another Native American family (Pecora, 2000; Wilkins, 2004). Tribal courts were given the right to determine what is in the best interest of the child.
- The Adoption Assistance and Child Welfare Act (AACWA) amended Title IV-A and Title IV-B of the Social Security Act in 1980 and created Title IV-E (Pecora, 2000). AACWA provided federal subsidies for adoption of children from foster care and was instrumental in reducing the financial hurdles that deterred a lot of people from adopting children (Child Welfare Information Gateway [CWIG], 2011). It provided funds to help children with special needs. The act was the first to provide both state and federal assistance for children with special needs, which included help with medical care (CWIG, 2011).
- The Child Care and Development Block Grant was enacted in 1981.
- President Clinton enacted the Multi-Ethnic Placement Act, in 1994, which eliminated policies that favored same-race placements and prohibited agencies that receive federal dollars from delaying or denying placements based on race or ethnicity.
- The Adoption and Safe Families Act (ASFA) was enacted in 1997. It established timelines to move foster children into permanency, provided bonuses to states for adoptions, and expanded funds available for time-limited reunification services, adoption promotion, and support services (Child Welfare League of America, n.d.).
- The ACA has provisions that have significant and far-reaching implications for children and youth who are a part of the child welfare system (Lehmann & Guyer, 2012). It allows former foster youth who have aged out of the system to continue to receive Medicaid coverage until the age of 26 with the federal government providing matching Medicaid funds to the states to fund this

extended benefit (Lehmann & Guyer, 2012). It has established community-based health homes to provide integrated care for children and youth who have a chronic condition (Lehmann & Guyer, 2012).

- The Child Tax Credit was enacted in the Tax Cuts and Jobs Act of 2018. It provides up to $2,000 for each qualifying child up to age 17 for persons earning between $2,500 and $200,000.

UNDERSTANDING THE SCOPE OF CHILD ABUSE AND NEGLECT

The federal government first developed policies to deal with child abuse and neglect in 1935. Beginning in the 1970s, mandated reporting laws were passed. As soon as they were passed, experts addressed this question: "How do we define child abuse and neglect for this and other purposes?" It was decided that the federal government would establish minimum standards, whereas each state would define maltreatment within civil and criminal contexts. Federal legislation provides a foundation for states by identifying a minimum set of acts or behaviors that define child abuse and neglect. The Federal Child Abuse Prevention and Treatment Act, as amended by the Keeping Children and Families Safe Act of 2003, defines child abuse and neglect as "any recent act or failure to act on the part of a parent or caretaker which results in death, serious physical or emotional harm, sexual abuse or exploitation; or an act or failure to act which presents an imminent risk of serious harm." This definition of child abuse and neglect refers specifically to parents and other caregivers. A "child" under this definition generally means a person who is under the age of 18 or who is not an emancipated minor (CWIG, 2009).

For policy advocates to do their jobs effectively, they need access to accurate and reliable data. Data are essential to establishing the scope, magnitude, and urgency of the problem. Main data sources of child abuse statistics in the United States include the National Child Abuse and Neglect Data System (NCANDS), the National Incidence Study (NIS), the National Survey of Child and Adolescent Well-Being (NSCAW), and the Administration for Children and Families. Both the NCANDS and NIS include suspected child abuse reports investigated by child protective services (CPS) agencies. The Centers for Disease Control and Prevention (CDC) provides annual statistics through its National Violent Death Reporting System and its web-based Injury Statistics Query and Reporting System. NSCAW includes only reports of child abuse and neglect investigated by CPS but adds clinical measures measuring child and family well-being (Child Protective Services, n.d.; Wulczyn, 2009).

Using data from NCANDS, 3.4 million children were the subjects of an investigation in 2015—with 683,000 determined to be victims of neglect and physical abuse and 63.4% of these children suffering neglect only (National Child Abuse and Neglect Data System, 2016). Wang and Holton (2007) estimated that the cost to society in 2007 for child victims was an estimated $103.8 billion, including direct costs of hospitalization, mental health care, child welfare services, and law enforcement and indirect costs of special education, juvenile delinquency, mental health and health care, adult criminal justice system, and lost productivity to society.

UNDERSTANDING THE POLITICAL ECONOMY OF THE CHILD AND FAMILY SECTOR

A pendulum exists in child welfare that fluctuates between a cautious approach that emphasizes child protection through the removal of children and a more family-oriented approach that emphasizes family preservation (Gelles, 2001). When agencies focus on family preservation, specific instances of child abuse in families lead to protests that child welfare workers mistakenly keep children in unsafe situations. CPS agency directors and their staff are often held culpable and fired when a child under CPS supervision dies. When they focus upon removing children, however, other critics contend that they are usurping the roles of families as they remove children from their natural parents.

The child welfare system is unique because of its intricate relationship with the legal system. Specifically, child abuse and neglect are legal terms that only courts with appropriate jurisdiction can designate after complying with due process and equal protection of law (Brittain & Hunt, 2004). There are three types of courts that handle legal matters dealing with child abuse and neglect: juvenile court, civil court (domestic relations courts), and criminal court. Juvenile courts, which encompass family courts with juvenile jurisdiction, typically use state statutes to guide court procedures, and they can utilize state and federal case law when considering definitions of child abuse and neglect. State statutes that define child abuse and neglect guide civil courts, or domestic relations courts, and these courts often utilize case law when interpreting criminal and juvenile statutes. Criminal courts are the only courts in which state criminal codes are applicable, and these courts use statutes that can apply to a person or to juveniles. If a perpetrator of child abuse and neglect is a relative or caregiver, the matter can be heard in criminal court as well as juvenile court and domestic relations court. If a perpetrator of child abuse and neglect is a not a relative or caregiver, the matter can be heard only in criminal court (Brittain & Hunt, 2004).

Child welfare is shaped by federal, state, and local policies. Federal legislation and regulations provide a national framework for the child welfare system. Within this framework, states have discretion to establish legal and administrative structures and programs within their jurisdictions. States differ in the manner they have interpreted and implemented federal policy, resulting in a range and quality of services available to at-risk children and families (Lind, 2004).

The federal government frames and implements national policy by passing rules, monitoring state performance, and conducting compliance reviews. The Department of Health and Human Services (DHHS) is the principal federal agency that regulates and partially funds services to maltreated children and their families. The Administration for Children and Families and the Centers for Medicaid and Medicare Services within DHHS oversee services provided to children and families involved with the child welfare system (Reed & Karpilow, 2002).

Most state governments, as well as the District of Columbia, directly implement child welfare programs. They establish child welfare departments throughout the state. They hire their staff. They oversee their operations. These states include Alabama,

Alaska, Arizona, Arkansas, Connecticut, Delaware, Florida, Georgia, Hawaii, Idaho, Illinois, Indiana, Iowa, Kansas, Kentucky, Louisiana, Maine, Massachusetts, Michigan, Mississippi, Missouri, Montana, Nebraska, New Hampshire, New Jersey, New Mexico, Oklahoma, Oregon, Rhode Island, South Carolina, South Dakota, Texas, Tennessee, Utah, Vermont, Washington, West Virginia, and Wyoming. Three hybrid states—Maryland, Nevada, and Wisconsin—are partially administered by the state and by counties. Nine states have county-administered child welfare programs, including California, Colorado, Minnesota, New York, North Carolina, North Dakota, Ohio, Pennsylvania, and Virginia.

New York State is an example of a county-administered system. Mayor Bill de Blasio oversees New York City's Administration for Children's Services. He appointed Gladys Carrion to be commissioner of the agency in 2014, who promised to tighten "the system so that no child fall through the cracks" (Stewart, 2016). A series of child deaths ensued. Some were lapses in investigations, whereas others reflected the difficulties in predicting neglect or abuse among families in extreme poverty. Although many associates and the president of Social Service Employees Union Local 371 wanted to retain her, Ms. Carrion acknowledged some blame. While Mayor de Blasio was overseeing the agency, New York State ordered an investigation into a six-year-old boy who died in September 2017 after social workers failed to remove him from the care of his mother and boyfriend, who were charged with his murder. Mayor de Blasio appointed a new commissioner in early 2017 "who was expected to rely more on data to assess performance and determine priorities" (Stewart & Fortin, 2017).

Had similar problems occurred in states where child welfare is administered by the states, such as Texas, Massachusetts, Michigan, and Florida, the governor would have taken the role of Mayor de Blasio when addressing malfunctions in their child welfare agency.

Child welfare agencies differ considerably. The effective functioning of service delivery systems requires what are known as "organizational components" (Pecora, 2000). These components include a clear organizational mission, program capacity, careful personnel recruitment and training, reasonable caseloads, adequate clerical supports, supervision, performance data, and fiscal support that is consistent with the outcome requirements. Researchers have discovered that organizational context greatly influences the effectiveness of the services of specific child welfare agencies. The organizational climate, broadly defined as employees' shared perceptions, for example, can affect the quality of service and child welfare outcomes (Glisson, 2009). In a study that looked at the emotional and behavioral outcomes of 1,640 children in 88 child welfare agencies, maltreated children served by investigative caseworkers from positive organizational climates (high in personal accomplishment and low in depersonalization) made significant improvements in their psychosocial functioning 36 months later as opposed to children served by investigative caseworkers from negative organizational climates (Glisson, 2009).

The federal government assumes a major role in funding child welfare departments through Title IV and Title XIX of the Social Security Act (Reed & Karpilow, 2002). These funds are passed through to the states, and in states like New York and California, they are further distributed to the counties. More than 80% of California's foster children

TABLE 11.1 ■ Child Welfare Financing Structure

Major Federal Child Welfare Funding Streams				
Funding Source	**Type of Funding**	**Authorized Services**	**Eligibility**	**Funding Level (in millions) Fiscal Year 2017**
Title IV-B of the Social Security Act				
Stephanie Tubbs Jones Child Welfare Program	Discretionary	Broad array of prevention, family reunification, and permanency services	Defined by the state	$269
Promoting Safe and Stable Families Program and Child and Family Services	Part capped mandatory and discretionary	Family support, family preservation, time-limited family reunification, and adoption promotion and support services	Defined by the state	$651
Title IV-E of the Social Security Act				
Foster Care	Open-ended entitlement	Maintenance payments to foster families, administration, and training	Based on old AFDC need standards	$5,302
Adoption Assistance	Open-ended entitlement	Maintenance payments to adoptive families, administration, and training	Based on old AFDC need standards	$2,658
Adoption Incentive Payments	Discretionary	Any allowable IV-B or IV-E service	Generally defined by the state	$38
Chafee Foster Care Independence Program	Part capped state entitlement, part discretionary	Broad array of independent living support services	Adolescent foster youth and former foster youth up to age 21	$140

Source: Stoltzfus, E. (2017, June 1). Child welfare funding in brief, FY2017 final funding and the president's FY2018 request. Congressional Research Service.

are eligible for and receive partial funding from the federal government for board, care, and medical costs, with the balance covered by state and county funds. Foster children who are not eligible for federal funds are supported by state, county, and private funds. Federal funding should be sharply increased in light of the complex issues encountered by foster children as well as the many social problems that they present at age 18 and during the first years of their emancipation. The flow of federal money to states in 2017 is illustrated in Table 11.1.

The mass media plays a pivotal role in not only shaping public opinion about child welfare services but also exposing cases of child abuse. The influence of the media was illustrated by an infamous sexual abuse scandal involving Jerry Sandusky, the former assistant football coach at Pennsylvania State University and founder of a charity for abused children. After news broke in 2011 that he had sexually abused eight young boys at the university and that university officials allegedly covered up or ignored the abuse, an investigation resulted in his conviction on 45 counts of child abuse and a life sentence in prison (Penn State, n.d.). Two top officials of Penn State, including its president, were convicted of a cover-up in 2018. Another example of the role of the media comes from Los Angeles County. A Blue Ribbon Commission was formed to examine the child protection system in Los Angeles County after the widely publicized death of an eight-year-old boy, Gabriel Fernandez, in 2013. Despite six investigations in which social workers were told that he had numerous bruises and despite a suicide note, Gabriel remained in his home. In May 2013, paramedics took Gabriel to the hospital, where he died of injuries related to severe physical abuse (Therolf, 2013). This tragedy sparked the formation of a Blue Ribbon Commission that declared a state of emergency and developed scores of reforms, such as establishing a single entity with sole responsibility for child protection that would collaborate with other public agencies including the Department of Mental Health, Department of Public Social Services, law enforcement agencies, the Office of Education, and Dependency Court (Bluecommissionla.com, n.d.).

RECOGNIZING CHILD AND FAMILY PROBLEMS THAT ARE CREATED BY AN INEGALITARIAN NATION

Fifteen million children live in poverty—or 21% of all children. Research shows, however, that families need an income roughly twice the federal poverty level to cover their basic expenses. So roughly 43% of children live in families that lack resources to cover their basic needs (National Center for Children, 2017).

Childhood poverty cost the United States $1.03 trillion in 2015, or 5.4% of gross domestic product, or 28% of the federal budget (Rank, 2018). It exacts huge costs on the nation in many ways including costs of welfare, health care, and social programs; reduced economic productivity; child homelessness; and engagement in crime. Each dollar spent on reducing childhood poverty would save the nation at least $7 (Rank, 2018).

UNDERSTANDING HOW SEVEN CORE PROBLEMS EXIST IN THE CHILD WELFARE SECTOR

Core Problem 1: Engaging in Advocacy to Protect Children's Ethical Rights, Human Rights, and Economic Justice—With Some Red Flag Alerts

- Red Flag Alert 11.1. Foster children are not made aware of their rights or these rights are not respected. This places social workers in a unique position to advocate for these children and ensure that their rights are being protected.
- Red Flag Alert 11.2. Children do not receive the services they are entitled to. They are often not consulted on their case plans and are actually sidelined when it comes to making pivotal decisions that affect their lives.
- Red Flag Alert 11.3. Children in foster care are often lost in the system. They get to see their social workers rarely and often do not even know the lawyers who are representing them.

Background

The child welfare system is deluged with huge caseloads and high rates of worker burnout. Many serious problems continue to plague a service delivery system that is high on stress and short on resources, has high staff turnover, and deals with very difficult family situations (Arkansas Advocates, 2005). Therefore, ethical rights are sometimes not addressed and are sometimes compromised given all of these structural problems. In addition, many clients in child welfare are often unaware of their ethical rights, especially clients in vulnerable populations such as foster youth, low-socioeconomic status families, and undocumented families. Because social workers work with such vulnerable populations, extra care should be taken to ensure that these children and families are made aware of their rights (CWIG, n.d.).

Many minority youth experience a violation of their basic civic and human rights simply because of the color of their skin. For example, youngsters in the inner city, city, and suburban lower and middle class who are of African American, Hispanic, or Asian descent are oftentimes suspected of being gang members and are subject to racial profiling (Morales, Sheafor, & Scott, 2011). The shooting of unarmed African American 17-year-old teenager Trayvon Martin by neighborhood watchman George Zimmerman grabbed the nation's attention as a prime example of the ethically dangerous practice of racial profiling. On February 26, 2012, George Zimmerman shot Trayvon Martin fatally after a brief altercation in a gated condominium complex in Sanford, Florida, where Trayvon was staying temporarily. It has been alleged that Trayvon was profiled primarily because of his race, which ultimately led to his death. Unfortunately, many unarmed youth from minority groups have been slain in the ensuing years with only marginal improvements in training police how not to use violence, disciplining police who use needless violence, and recruiting officers with orientations that reduce the odds they will use needless violence.

Ethical barriers also exist in the context of educating the country's children and youth. Education is a known indicator for the financial success persons experience as they develop into adults; the higher the educational level attained, the smaller the difficulty experienced in finding employment and supporting one's family. However, because black and Hispanic children are more likely to live in poverty than their white and Asian counterparts, they are included in a snowball effect of circumstances that lead to negative outcomes. For example, ethnic minority children are more likely to attend worse schools, have poorer health and nutrition, experience violent crime, perform worse on standardized tests, and drop out of high school at higher rates (Lee, 2002; Sampson & Wilson, 1995; Fryer & Levitt, 2004; Harding, 2003; Menchik, 1993).

Resources for Advocates

Children in foster care are entitled to a bill of rights. The rights are legal, and it is illegal to violate these rights. A number of these rights are delineated in the Adoption and Safe Families Act of 1997. According to this act, the child's safety and well-being is of paramount concern when planning his or her case plan. In 2001, California enacted the following bill of rights, along with a provision to its Health and Safety Code, requiring that foster care providers must give every school-age child and his or her authorized representative an age-appropriate orientation and an explanation of the child's rights (Doherty, 2005). Some of the most important rights are as follows:

- To live in a safe, healthy, and comfortable home where he or she is treated with respect.
- To receive medical, dental, vision, and mental health services.
- To be free of the administration of medication or chemical substances, unless authorized by a physician.
- To contact family members, unless prohibited by court order, and social workers, attorneys, foster youth advocates and supporters, court-appointed special advocates, and probation officers.
- To visit and contact brothers and sisters unless prohibited by court order.
- To make and receive confidential telephone calls and send and receive unopened mail unless prohibited by court order.
- To attend religious services and activities of his or her choice.
- To attend school and participate in extracurricular, cultural, and personal enrichment activities consistent with the child's age and developmental level.
- To attend court hearings and speak to the judge.
- At 16 years of age or older, to have access to existing information regarding the educational options available, including, but not limited to the course work necessary for vocational and postsecondary educational programs and information regarding financial aid for postsecondary education.

In 1997, the ASFA was enacted and modified many of the goals and policies set forth by the AACWA. Some important modifications include the following policy requirements:

Enhancing the safety of the child—in home and in foster care. The law clarifies the meaning of the "reasonable efforts" condition, which appears in the Family Preservation and Support Services Act (P.L. 96-272). The 1984 law stated that the state must take reasonable measures to prevent or eliminate the need for removing a child from his or her home, or if the child has been removed, then the state must make reasonable efforts to reunify the child with his or her family within a suitable time frame (Welte, 1997). The new law emphasized that the child's well-being, safety, and security shall be of paramount concern in arriving at a decision about a logical and rational plan of action. This law also allows for dual planning, which is the process of looking at suitable prospective adoptive families while reunification services are still being provided (Welte, 1997).

Shortening the time frame for permanent placement. Services to reunify families funded under Title IV-B should not extend beyond 15 months (Welte, 1997). These services include counseling, substance abuse treatment services, domestic violence services, and temporary childcare and related services. A petition to terminate parental rights shall be filed for parents whose child has been in foster care for 15 of the last 22 months, if a court has determined a child is an abandoned infant, or in the circumstances described under "reasonable efforts" above.

Providing incentives for adoptions or other permanency placements. The law provides states with cash incentives to find permanent homes for children in foster care. A state will receive $4,000 in federal funds for each foster child adoption, which exceeds a base number of foster care adoptions in a fiscal year and an additional $2,000 for special needs adoptions (Welte, 1997).

Decreasing geographic barriers to adoptions. To facilitate timely adoptions for waiting children across state and county jurisdictions, states are required to develop plans to utilize cross-jurisdictional resources. Title IV-E Foster Care and Adoption Assistance payments to the state are also predicated upon a state's cooperation in processing a child's adoptive placement if an approved family is available outside of the jurisdiction (Welte, 1997).

Establishing outcome measures to assess state performance. The Secretary of Health and Human Services (HHS), in discussion with public officials and child advocates, will develop outcome measures, as well as a rating system, to assess states' performances in child protection and child welfare programs. Measures would include length of stay in foster care, the number of placements, and the number of adoptions and, to the extent possible, use data available from the established Adoption and Foster Care Analysis and Reporting System. States will report their performance on each outcome measure, and the secretary of HHS will provide an annual report to Congress (Welte, 1997).

Expanding health coverage for special needs children and independent living services. States must provide health insurance coverage for any child with special needs for whom it would be hard to find placement without providing for these services. States

POLICY ADVOCACY LEARNING CHALLENGE 11.1

CONNECTING MICRO, MEZZO, AND MACRO POLICY INTERVENTIONS TO PROTECT CHILDREN'S ETHICAL RIGHTS

Who Will Uphold the Rights of a Native American Child?

A. J. is a four-year-old Native American child in foster care within the purview of the ICWA. The newborn was considered a medically fragile child, experienced drug withdrawal symptoms, and exhibited signs of fetal alcohol syndrome. The mother denied serious drug use and did not want to participate in services or raise A. J., and his father wanted custody but was deemed unfit to raise a medically fragile child. It was under these circumstances and because no other family member offered to care for A. J. that he was detained in foster care, and dependency proceedings were initiated.

Based on the mother's membership in her tribe, the child welfare agency in that county notified the tribe of the dependency proceedings pursuant to ICWA. Meanwhile, A. J. was moved to a foster home at the age of one. Although A. J. presented with flat affect and did not engage with his parents, A. J. engaged with his foster mother and used her as a reference base for most of his activities. According to the child's therapist, A. J. needed to be either reunited with his parents or placed in a permanent placement as soon as possible to reduce attachment difficulties and avoid future mental health problems. The tribe, however, first wanted to explore options within the tribe, and A. J.'s grandmother was put forward as a possible option. The tribe's council approved a resolution authorizing placement of A. J. with the grandmother for the purposes of legal guardianship.

In the meantime, the child protective services agency modified its recommended permanent plan for A. J. and alternatively recommended legal guardianship with the foster mother or adoption by the foster mother because A. J. had no relationship with the grandmother. In addition, A. J. had been in his current placement for two years and had developed a significant bond with his foster mother. In conflict with ICWA, ASFA (1997) indicates that "the child's well-being, safety and security shall be of paramount concern in arriving at a decision regarding the child's permanency options." However, the social worker attached to this case discovered that ICWA has a provision that allows ASFA to take precedence if she could prove that the child's best interest was to remain with his foster mom. The social worker was able to get expert testimony from bonding experts and ICWA experts who testified that whereas the law prefers placement with a Native American family, they could see how this could prove to be counterproductive to A. J.'s developmental goals and is, therefore, not in the right spirit of the ASFA.

The courts agreed with the testimony of the expert witnesses and the social worker and gave the following ruling:

> It is obvious that the grandmother has had very limited contact with A.J. and her failure to engage him in any meaningful way. Her lack of effort to visit, interact and develop a relationship with A.J. was striking. Although there was evidence that a child who has an attachment can attach to another care provider, it was undisputed a child's ability to attach also depended on the new care provider's skill. Under the circumstances of this case, the juvenile court reasonably could infer it was unlikely A.J. could develop a healthy attachment to the grandmother. (Fearnotlaw.com, n.d.)

(Continued)

(Continued)

Learning Exercise

1. How does this vignette illustrate conflicting provisions provided by the two laws and the reasoning that the safety of the child needs to be of paramount concern?
2. Which skills came in handy for the social worker in this situation to advocate for this child?
3. Could the social worker have taken any other steps to resolve the situation?

can provide health coverage, including mental health, through the Medicaid option or another program that is at least equivalent to Medicaid. Independent Living Services are extended to young people whose assets do not exceed $5,000 rather than the current $1,000 cap. Services are designed to assist young people in preparing for living independently when they leave foster care (Welte, 1997).

Defining legal and standby guardianship. The phrase ***legal guardian*** is defined in the statute as a permanent relationship between child and caretaker and transfers parental rights to the caretaker for the child's protection, education, care, custody, and decision making. The new law urges states to adopt laws and procedures to allow parents who are chronically ill or near death to designate a standby guardian for their children without surrendering their parental rights (Welte, 1997).

Core Problem 2: Engaging in Advocacy to Improve Quality of Care—With Some Red Flag Alerts

- **Red Flag Alert 11.4.** High turnover of children and youth often exists in the child welfare system leading, for example, to multiple foster homes for specific children and youth.
- **Red Flag Alert 11.5.** Lack of respect often exists between social workers and clients.
- **Red Flag Alert 11.6.** Many standard services are provided that do not meet the individual needs of clients. Clients' unique needs are often not sufficiently assessed when determining what type of services to provide.
- **Red Flag Alert 11.7.** Some services are provided without any consideration of empirical evidence to back up their effectiveness.
- **Red Flag Alert 11.8.** Insufficient numbers of social work staff, as well as their excessive turnover, impede the quality of services.
- **Red Flag Alert 11.9.** Insufficient numbers and turnover of foster parents, as well as insufficient training and monitoring of them, impede the quality of services.
- **Red Flag Alert 11.10.** Insufficient numbers of adoptive families impede the quality of services.

Background

Funding for social welfare services is never enough because the number of people who require those services expands enormously each year, yet the level of funding often remains constant. Because the resources are sparse, agencies should strive to use these resources in the most cost-efficient and effective manner to increase the likelihood of the accessibility of their services. Patton and Sawicki (1993) state, "Efficiency is measured in dollars per unit of output (benefit)," which can be operationally defined by examining all program costs (i.e., direct, indirect, operation, maintenance, etc.), number of participants, the delivery of service, and what services will be provided. Effectiveness is based on whether the program or policy achieves the desired goals and objectives (Bardach, 1996). Accessibility can best be described as the availability of programs and services to the population being served.

At-risk children and families are typically poor and face multiple challenges, such as substance abuse, mental illness, domestic violence, and inadequate housing. Supports and services geared toward addressing these needs can help mitigate sources of stress and instability that may contribute to child abuse and neglect. By making a continuum of care available to at-risk children and families, states and counties can not only intervene but also successfully prevent child abuse and neglect (Lind, 2004). Unfortunately, this is not the situation in many cases. The reality is that services are scattered, and there is little systematization or consolidation of these resources, so people are often left scrambling to find resources.

Adding to this complexity, many of the programs and practices in child welfare lack specific evidence about their effectiveness. To catch up with fields such as medicine and public health, child welfare has recently begun implementing more evidence-based practices and evidence-based policies. As the volume of research evidence grows, social workers will be able to complement their practice-based knowledge and clinical judgment to include practices that have been shown to be effective (CWIG, n.d.).

For more information about evidence-based practice in child welfare, explore the following links:

Child Welfare Information Gateway: http://www.childwelfare.gov/management/practice_improvement/evidence/ebp.cfm

California Evidence-Based Clearinghouse: http://www.cebc4cw.org/

Annie E. Casey Foundation: http://www.aecf.org/work/evidence-based-practice/

Resources for Advocates

In accordance with the Child and Family Services Improvement Act of 2006 (P.L. 109-288), which came into effect on September 28, 2006, and amended the Social Security Act Title IV-B, a new purpose for child welfare services programs was established that allowed for an expansion of services and flexibility. With this expansion, different states could change their programs as needed to increase the breadth of services (CWIG, n.d.). The act also allowed for competitive grants to regional partnerships that were able to provide collaborative, integrated services and programs to improve children's safety in out-of-home care and to children at risk of entering out-of-home care due to parent or caregiver substance use.

POLICY ADVOCACY LEARNING CHALLENGE 11.2

CONNECTING MICRO, MEZZO, AND MACRO POLICY ADVOCACY TO IMPROVE QUALITY OF CARE

Imagine that you are an emergency response social worker for the Los Angeles Department of Children and Family Services (DCFS). Your client needs a parenting class that caters to parents with children with developmental disabilities. After completing your child abuse investigation and finding no evidence of abuse, you have to close the case but need to find appropriate services to help the family, which is difficult because you are only able to find generic parenting classes. After reviewing the policy on Community Response and Alternative Response Services (Department of Children and Family Services [DCFS], 2014) and noting that the investigation should be closed once a family links to services, you decide to engage in micro advocacy. As a micro policy advocate, you should be detailed and specific when locating resources for the family so that the receiving agency can easily understand the identified problem. You contact the agency beforehand and explain the extent of the problem and see if they could cater their services to the client's needs. It helps to ask the parenting class instructor to address particular issues. As part of mezzo policy advocacy, you discuss the lack of specialized classes with a supervisor and with an administrator, once given approval by a supervisor. If the problem appears to be common and affects many clients, you should inquire about the process of changing a policy within the agency. As a macro policy advocate, you might check into statewide regulations regarding parenting classes and advocate for new legislation that permits such parenting groups to customize their services to the needs of particular clients. This legislation should also provide additional funding to design, implement, and evaluate those services.

Learning Exercise

1. How does this vignette illustrate the difficulties of finding resources that meet a client's need?
2. Why does the social worker call the agency beforehand instead of just providing the client with the telephone number for the agency?
3. When would it be appropriate for a social worker at DCFS to engage in mezzo or macro policy advocacy to address a limitation of services?

Core Problem 3: Engaging in Advocacy to Promote Culturally Competent Care—With Some Red Flag Alerts

- **Red Flag Alert 11.11.** Culture is not taken into consideration in case planning.
- **Red Flag Alert 11.12.** Cultural practices are misinterpreted to form the basis of neglect or abuse charges.
- **Red Flag Alert 11.13.** Lack of staff training on cultural sensitivity exists.
- **Red Flag Alert 11.14.** Clients express that their cultural needs are not being met.

- **Red Flag Alert 11.15.** Lack of consideration of cultural needs exists when staff tries to engage clients.
- **Red Flag Alert 11.16.** There is lack of staff diversity to match a diverse client population.

Background

Children and families from ethnic minority groups are at considerable risk for developing several problems, such as substance abuse, low grades, delinquency, and poor self-esteem, and there is an increasing need for effective culturally sensitive interventions designed for this population (Jackson, 2009). Children and families who are from minority racial and ethnic backgrounds are also more likely to be poor. Family income has a substantial effect on child and adolescent well-being. Poor children experience emotional and behavioral problems more often than their more privileged counterparts, which include emotional issues such as aggression, fighting, acting out, anxiety, social withdrawal, and depression, and the incidence of out-of-wedlock births among poor teens is nearly three times more frequent than the rate among non-poor families (Brooks-Gunn & Duncan, 1997). In addition, poor children face higher rates of adverse health and developmental outcomes than their non-poor counterparts, such as low birth weight and infant mortality, which are both significant indicators of health in children. For example, low birth weight is associated with an increased likelihood of persistent physical health, cognitive and emotional problems, grade retention, learning disabilities, lower levels of intelligence, and decreased math and reading achievement. Furthermore, adverse birth outcomes are more likely in unmarried women, persons with low educational levels, and black mothers, three populations that have high poverty rates (Brooks-Gunn & Duncan, 1997).

Early childhood interventions can be crucial in reducing the undesirable effects of poverty on the lives of children (Brooks-Gunn & Duncan, 1997). Other measures to mitigate the adverse effects of poverty in children include nutrition programs, especially those targeted toward the most undernourished, lead abatement, enhanced parental education, and involvement in a home-learning environment, income policy reforms, and in-kind support programs.

Resources for Advocates

According to the Government Accounting Office (1998), approximately 60% of children in foster care belong to minority groups, and they wait twice as long to be placed in foster homes or be adopted. The Multi-Ethnic Placement Act (1994) eliminated delaying or denying placements based on race or ethnicity, although children under the ICWA are exempt from the Multi-Ethnic Placement Act provisions because same-race placements for Native American children should be the preferred course of action when it concerns placement decisions (Pecora, 2000). However, the Multi-Ethnic Placement Act could be counterproductive, as well, because minority children might not be a good fit in households or families that are not sensitive to their culture or their linguistic needs.

POLICY ADVOCACY LEARNING CHALLENGE 11.3

CONNECTING MICRO AND MACRO POLICY INTERVENTIONS TO IMPROVE CULTURALLY COMPETENT CARE FOR CHILDREN

Ignoring Cultural Factors

Anastacia is a 13-year-old Romanian girl who identifies herself as a "gypsy." She was removed from her parents on allegations that they had intentionally allowed her to be molested by an influential older man (who was a family friend) and videotaped the entire incident to financially blackmail that person. Anastacia is bilingual and conversational in both English and Romanian. The parents are, however, not fluent in English. In addition, because they strongly identify with the gypsy culture, they believe that a girl should be homeschooled and refused to send their daughter to a regular school. The social worker does not take the necessary steps of understanding the culture of the family and accuses the family of educational neglect along with the abuse charges. In addition, the social worker does not offer interpreter services to the parents and uses Anastacia to interpret for her parents.

Learning Exercise

1. How does this vignette illustrate a tendency to ignore culture in defining parental responsibilities and providing relevant services?
2. What laws could have protected this family and the child's best interests?
3. Could a social worker in this clinic have engaged in macro policy advocacy to resolve this situation?

The DCFS in Los Angeles implemented the Permanency Partners Program (P-3) in October 2004 to find permanent homes and long-term connections for youth ages 12 to 18 placed in long-term foster care. The staff consists of retired caseworkers who review case records to find relatives or other adults who were connected to the youth. By July 2011, P-3 had serviced 4,635 youth, of which 37% (1,696) have a legal permanent plan. Go to the following link to read about P-3 and other programs offered by DCFS in Los Angeles (DCFS, n.d.): http://dcfs.co.la.ca.us/community/Pomona/index.html.

Core Problem 4: Engaging in Advocacy to Promote Prevention—With Some Red Flag Alerts

- **Red Flag Alert 11.17.** Your agency's services are reactive and not preventive.
- **Red Flag Alert 11.18.** There is little support recourse for families struggling with poverty.
- **Red Flag Alert 11.19.** Your agency does not collaborate with other agencies to provide preventive services.

- **Red Flag Alert 11.20.** No resources are provided to families who do not meet the criteria to receive services from your agency.

Background

The lack of prevention is felt across all social sectors, and it is especially true for the child welfare sector. Child welfare is a multilevel systemic problem that might have its origins in other root problems embedded in the social environment or the personal circumstances of the individual or the family.

Social workers can engage in primary and secondary prevention with families that lack economic resources, have family members with mental and physical problems including substance abuse, exhibit early signs of serious family conflict, lack suitable housing or are homeless, have one or more children who have left the home for brief or longer periods, have children who are truant or have behavioral problems in school, have children who are failing classes or not moving ahead with classmates, or who exhibit other early signs of family problems, and have children that have been arrested. They can engage in "family rescue" no matter in which sector they are employed.

Social workers who are employed by child welfare departments can expand their roles to include primary and secondary prevention. Assume that their department received reports of child neglect and child abuse that did not rise to a level that required a court hearing—and that the department routinely closed these cases. Also assume that some of these reports *did* suggest the emergence of serious problems in specific families. Rather than dropping these cases, families might be referred for counseling developed by child welfare staff to improve family communications. Or families might be referred to a mental health program to help a parent who routinely yelled at the children even if he or she did not engage in physical abuse.

Child welfare staff can engage in secondary prevention with respect to youth under their care. Assume, for example, that many of their foster children attend a specific inner-city high school. Also assume that these staff learned that a high percentage of these foster children lagged in one or more of three areas: attendance, behavioral problems, and educational performance. Also assume that they could document that a high percentage of these foster children were not graduating from high school or were not graduating on time. They could engage in mezzo policy advocacy to seek a reduction in their caseloads from their department of child welfare to allow them to work with school staff to address these problems and to increase graduation rates. They could seek funding from the child welfare department and the high school to support the program.

CPS is a reactive agency and appears after child abuse occurs. In many cases, because of this post-incident reaction, the problem has become acute. For example, there are natural linkages between the welfare and child welfare systems. More than half the children who enter the child welfare system come from families eligible for welfare, and poverty is strongly associated with an increased risk of child maltreatment. Children in families with annual incomes less than $15,000 are 22 times more likely than children in families with annual incomes more than $30,000 to be abused or neglected (Lind, 2004).

It is essential to prevent child abuse and neglect because it has detrimental effects throughout a child's life span. Abuse and neglect have been found to negatively affect attachment style and aggression in children (Weizman, Har-Even, Shnit, Finzi, & Ram, 2001) and have negative effects on the developing brain (De Bellis et al., 1999). In a longitudinal, national study of more than 15,000 adolescents that examined the effects of childhood maltreatment on adolescent health behavior, each type of childhood maltreatment, that is, physical abuse, emotional abuse, sexual abuse, or neglect, was associated with increased smoking, alcohol use, drug use, and violent behavior during adolescence (Kotch, Chang, & Hussey, 2006).

A well-known study discovered linkages between adverse childhood experiences (ACEs) and health risk behaviors and diseases (Felitti et al., 1998). ACEs include early child abuse and neglect. Adults with four or more ACEs relative to adults with none experienced a four- to 12-fold increase in substance use, depression, and suicide attempts; a two- to fourfold increase in smoking, higher self-rated poor health, a larger number of sexual partners, more sexually transmitted diseases, and about a 1.5-fold increase in severe obesity and physical inactivity. In addition, adults with more ACEs were more likely to develop cancer, liver disease, and chronic lung disease.

Resources for Advocates

The passage of AACWA in 1980 was a direct response to the discontent over the inadequate functioning of the child welfare system at that time. It amended Title IV-A and Title IV-B of the Social Security Act and created Title IV-E, thus shifting the focus from child abuse detection to that of prevention and permanence planning for children in out-of-home placements (Pecora, 2000). Some of the requirements that states and counties had to implement to receive federal funding consisted of the following:

1. Establish a state inventory of children in out-of-home placements to include all children who have been in foster care during a period of six months or more.
2. Establish a statewide information system with a database whereby the sociodemographic characteristics, legal status, and goals set for a child receiving foster care services for a period of more than 12 months could be recorded and available for review if needed.
3. Implement pre-placement preventive services whereby agencies document that they had offered services to a family that could have facilitated the retention of the child within the confines of his or her home.
4. Establish reunification or permanent planning services to reunify the child with his or her birth family or to find an adoptive home for the child.
5. Implement periodic case reviews with a judicial or administrative review every six months and a dispositional review by a court after 18 months of the child being placed in the system.
6. Establish standards for care that prescribe best efforts to place children in the least restrictive settings, preferably close to their parents and their relatives.
7. Encourage participation of parents and children in the formulation of their case plans and their permanency goals as much as possible.

8. Provide adoption subsidies for children with special needs with additional subsidies for families who adopt special needs children.

POLICY ADVOCACY LEARNING CHALLENGE 11.4

CONNECTING MICRO, MEZZO, AND MACRO POLICY INTERVENTIONS TO EXPAND PREVENTIVE SERVICES FOR CHILDREN

Imagine that you are a social worker for the Los Angeles DCFS, and you screen telephone calls from people calling to report suspected abuse and neglect. A grandmother calls the child protection hotline concerned that her daughter, a 19-year-old first-time mother, is not properly caring for her one-year-old child. When asked for more details, the grandmother states that she parties a lot, and they argue about how to raise the child. After ascertaining that the young mother does not abuse her child, does not leave the child alone, and is meeting the child's medical needs, you decide that the call does not meet the criteria for abuse or neglect and determine that the family would benefit from assistance. As you speak with the grandmother further, she indicates that she and her daughter have communication problems, and she wishes her daughter would enroll in a parenting class.

As part of micro and mezzo advocacy, you consult your resource guides and provide the family with several telephone numbers to local agencies within their vicinity including 211, a Los Angeles County-operated 24/7 telephone system for resources. Before providing the family with a list of numbers, you call them to ensure that the family is eligible for services and that there is no waitlist. You also make a follow-up telephone call to speak with the young mother and inform her of the services in her area. After listening to her discuss how she and her mother argue about how to raise her daughter, you empathize with her and discuss the many services of the close-by agencies. You recommend that she contact the agencies for assistance with child care, baby food, baby clothing, parenting classes, and education. She agrees to call the agency, and you follow up with her in a week to ask whether she had any trouble obtaining assistance.

Regarding mezzo and macro policy advocacy, the social worker speaks to the supervisor and administrators regarding the possibility of creating a new policy or protocol that would provide services to families who need assistance but do not meet the criteria for abuse and neglect. You also discuss the possibility of forming partnerships with other agencies for the creation of primary preventive services for families that come to the attention of community agencies early who are at risk for child maltreatment.

Learning Exercise

1. How does this vignette illustrate the difficulty of providing preventive services?
2. Would you describe these services as preventive or reactive services?
3. What are some disadvantages to providing services only if a family meets the criteria for child abuse and neglect?
4. Can you think of additional micro, mezzo, or macro policy advocacy that could be done in this situation?

Core Problem 5: Engaging in Advocacy to Promote Affordable and Accessible Care—With Some Red Flag Alerts

- Red Flag Alert 11.21. A client's recovery is impeded by lack of funding for services.
- Red Flag Alert 11.22. The political atmosphere does not support increased funding.
- Red Flag Alert 11.23. Families are reluctant to engage in services due to their immigration status.
- Red Flag Alert 11.24. Vulnerable families, such as immigrants, are excluded from financial benefits.

Background

Many families are loosely affiliated with child welfare. They include families that were subjects of reports but for whom no investigations were initiated. They include families whose child or children were temporarily removed but returned when a judge decided that the family could be reunited due to behavioral changes of the natural parents. They include foster parents with children who have experienced neglect and abuse. They include emancipated youth who venture into the world alone or with others.

Most of these families and individuals need resources, housing, medical and mental health services, education, preschool services, childcare, financial planning, and (in some cases) legal aid. Yet they soon discover that the American welfare state is divided into silos, with different regulations, eligibility requirements, missions, and professions. Silos have different funding and policy streams.

Some progress has been made in developing services across sectors. Case managers orchestrate cross-sector coordination. Experienced and skilled social workers develop informal working relationships with staff and supervisors in different sectors. Some agencies develop formal partnerships. Some agencies develop a cafeteria of services.

Even with these cross-sector relationships, many families use only a fraction of the resources, services, and programs that could be useful to them. They do not know about them. They fear a hostile response from some service personnel. They do not know the regulations or eligibility requirements. Social workers use micro, mezzo, and macro policy advocacy to facilitate the flow of resources and services across silos to these families and individuals.

Consider this book, then, to be a map that cuts across eight policy sectors and that introduces you to micro, mezzo, and macro policy advocacy.

Core Problem 6: Engaging in Advocacy to Promote Care for Mental Distress—With Some Red Flag Alerts

- Red Flag Alert 11.25. There is no communication between the child protective services agency and the agency providing therapy.
- Red Flag Alert 11.26. There are unaddressed mental health concerns.

POLICY ADVOCACY LEARNING CHALLENGE 11.5

CONNECTING MICRO, MEZZO, AND MACRO ADVOCACY

Funding Substance Abuse Services for Jose's Father

Jose is a 10-year-old Hispanic male who has been in the foster care system for a year; he was removed from his father's care because of his father's substance abuse problems. His mother's whereabouts are unknown, and Jose is receiving family reunification services. His father is court ordered to attend a residential substance abuse program; Jose's father is motivated and completes his program quickly. The father relapses soon after, and Jose is placed back in foster care. Upon investigation, it is found that Jose's father could no longer access continued substance abuse and mental health services because he did not qualify for services under Medi-Cal. With no support, Jose's father returned to his drug abuse.

As a social worker assigned to this case, you try to find the resources through which Jose's father can access drug counseling or support groups. You find that Medicaid is no longer an option because of their regulations. Therefore, you look for other nonprofits in the community and find one that offers counseling and support groups for free if the client offers to volunteer for one of their services. However, you are still frustrated that there is no broad-based policy at the department level that might have been able to help Jose's father. The social worker found that Delaware had used their Title IV-E funds to hire substance abuse specialists in each of their offices to take care of cases like this case (U.S. DHHS, 2010). The social worker presented a plan to the director of the agency detailing a similar program with data supporting the programmatic benefits and cost-effectiveness of such programs.

Learning Exercise

1. How does this vignette highlight the risks of not supporting families in recovery?
2. What are some disadvantages of having short-term case planning?
3. What type of policy advocacy needs to be done here?

- **Red Flag Alert 11.27.** The family's functioning is severely affected by mental health problems.
- **Red Flag Alert 11.28.** A client is receiving mental health services but they are either too little or too much.
- **Red Flag Alert 11.29.** The psychological services are not targeting the person who needs to be receiving help. For example, a family brings their child in for behavioral problems, but the child is acting out as a result of the fighting at home between the parents.

Background

The incidence of youth mental health problems in foster care is overwhelming. Roughly 80% of foster children have mental health disorders compared with 18% to 22% of the general population (Dore, 2005). A Foster Care Alumni Study of the Casey Family Programs in 2003 measured mental health disparities between foster care alumni and the general population by comparing 1,087 former foster youth and 3,547 adults from the general population during the past 12 months (Pecora et al., 2003; Pecora et al., 2005). The alumni of foster care had experienced foster care for a year or longer—and matched the general population in age, gender, and race/ethnicity. They experienced high levels of posttraumatic stress disorder, generalized anxiety, depression, social phobia, and panic disorder. The 12-month rate of panic disorder among alumni was more than three times that of the general population, seven times the rate of drug dependence, two times the rate of alcohol dependence, seven times the rate bulimia, and five times for posttraumatic stress disorder. Yet the alumni who received mental health care proved resilient, with recovery rates for all of the mental health problems similar to the general population, except for posttraumatic stress disorder. The data strongly suggest the need for mental health services not only for current foster children and youth but for ex-foster children and youth. The data provides clues as to why foster children and youth have relatively low educational performance and subsequent difficulty in maintaining employment. Children in out-of-home placements not only struggle to cope with the tremendous loss of their family but also recurrently blame themselves for being removed.

Mental health services should be provided to foster youth according to their Bill of Rights, but research shows that less than one-third of foster children received mental health services in 2004 (Austin, 2004). Findings from a national study that examined specialty mental health service use for children involved in child welfare across 97 U.S. counties indicated that younger children had significantly lower rates of specialty mental health service use, although they had higher levels of clinical need (Hurlburt et al., 2004). Another study, which measured emotional and behavioral outcomes for children placed in foster care, comparing children who reunified with their families with those who did not reunify six years later, found that children who reunified with their families had worse emotional and behavioral outcomes than their non-reunified peers (Taussig, Clyman, & Landsverk, 2001). Both of these studies point to areas of need within CPS: The former highlights the unmet needs of children left in their home, and the latter indicates that some of these same problems remain even if children are detained and later reunified with their families. The provision of mental health services is further complicated by the growing needs of seriously emotionally disturbed children and the inconsistent availability of foster parent training and support, which can result in high foster parent turnover, placement disruption, inconsistent treatment, and increased trauma for the child (Austin, 2004). "Wrap-around services are community-based care that literally wraps individualized services around a specific child to maintain that child in a community setting" (Dore, 2005). All necessary services are provided that are needed to stabilize the child in home, school, and community to avoid placement in residential treatment centers and psychiatric hospitals. A team includes the child and his or her parents, a care manager, other community partners, and mental health providers and provides services. Foster parents provide the primary mental health intervention in their homes with help from mental health training, consultation, and clinical support (Dore, 2005). Cross-training among systems providing care should be provided, thorough mental health assessments and screening should be provided on a yearly basis for all children and youth in care, coordination should be improved across systems of care, and accessibility to and continuity of mental health care should be increased (Polihronakis, 2008; Vulin-Reynolds et al., 2008).

POLICY ADVOCACY LEARNING CHALLENGE 11.6

CONNECTIVE MICRO, MEZZO, AND MACRO POLICY INTERVENTIONS TO IMPROVE FAMILIES' MENTAL HEALTH SERVICES

The Rodriguez Family Needs Mental Health Services

The Rodriguez family consists of 47-year-old Jennifer Rodriguez, 15-year-old Julie Rodriguez, and 12-year-old Bobby Rodriguez. The family came to the attention of the Los Angeles DCFS because of an allegation of general neglect after Julie was admitted to a psychiatric hospital for attempting to overdose on sleeping pills. As part of the allegation, it was reported that the mother was not meeting Julie's unique psychiatric needs and constantly yelled at her children.

After reading the relevant policy and identifying the mental health concerns to be addressed by the Department of Mental Health (DMH), the social worker immediately called the psychiatric hospital to find out when Julie would be released. Due to the mental health concerns and the three previous referrals for general neglect, the social worker and supervisor decided that this referral needed a team decision meeting (TDM) to best coordinate services with the family's participation. In preparation for the TDM, the social worker summarized the DCFS history and the psychiatric treatment recommendations. At the TDM, the social worker presented the information and discussed all of the possible options with the family, the DMH worker, and DCFS supervisor to come to a mutual agreement. The mother felt she could not control her daughter's behavior and was worried that her daughter would run away as soon as she got home. Because of the seriousness of Julie's mental health problems and because her mother felt as if she could not handle Julie on her own, it was agreed to open up a voluntary case and to voluntarily place Julie with a foster family while her mother received support to deal with Julie's unique mental health needs.

The social worker attempted to make Julie's transition into foster care as easy as possible by explaining the process and by having her family visit on a regular basis. Her mother was enrolled in parenting classes, and Julie began receiving therapy immediately because of the coordination with DMH. After the situation stabilized, Julie was returned to her family, and they received family preservation services to assist with the transition back home.

Learning Exercise

1. How does this vignette illustrate the importance of coordination and collaboration among service agencies?
2. What examples of micro policy advocacy did you notice in this vignette? What other type of micro policy advocacy could the social worker have done for the family?
3. Could a social worker in this clinic have engaged in mezzo or macro policy advocacy in this scenario? Give some examples.

Resources for Advocates

Eligibility for Temporary Assistance for Needy Families (TANF) funding for alcohol and drug treatment was expanded to include individuals at risk of involvement. In 1999, TANF significantly expanded services for family safety program clients. Fifteen million dollars (non-recurring funds) were allocated to fund substance abuse treatment services, substance abuse prevention services, children's mental health services, and residential treatment for substance abuse clients, with priority for family safety clients. More recently, DHHS approved up to 10 states to waive certain requirements of Title IV-E to conduct demonstration projects to deliver and finance child welfare services through greater flexibility of the funds according to Section 1130 of the Social Security Act. The projects allow for Title IV-E foster care funds to be used to test different approaches for implementing child welfare services to improve safety, permanency, and well-being for children.

Core Problem 7: Engaging in Advocacy to Promote Care Linked to Communities—With Some Red Flag Alerts

Examples of Red Flag Alerts at the Micro, Mezzo, and Macro Level

- **Red Flag Alert 11.30.** Your agency cannot provide the services your client desperately needs.
- **Red Flag Alert 11.31.** Your agency competes with other agencies instead of collaborating.
- **Red Flag Alert 11.32.** There are no additional agencies that can provide the services that your client needs.
- **Red Flag Alert 11.33.** Your client fears receiving services because of his or her national origin.
- **Red Flag Alert 11.34.** Information among agencies involved with the same client is not shared, and there is no coordination of services.

Background

Social workers need to engage in mezzo and macro policy advocacy extensively if rates of neglect and abuse are to be reduced, if effectiveness of foster care is to be increased, or emancipated youth are to reduce rates of incarceration, homelessness, and other problems. Child welfare programs that only focus on identifying children at risk of neglect and abuse and removing them from their natural homes fail to see the larger picture: millions of children and youth who need greater resources, services, mental and medical care, and education lest they, too, encounter not only neglect and abuse but myriad social problems.

We have already discussed the challenges faced by foster youth after they have been emancipated by child welfare agencies. Social workers need to engage extensively in mezzo policy advocacy as they work to improve the effectiveness of foster care so that its graduates do not have high rates of incarceration, arrests, unemployment, and homelessness in the first place.

- Develop partnerships between different agencies to serve foster children and emancipated youth so that fully funded wraparound services exist that are customized to each child and to each teenager.
- Give child welfare departments sufficient resources to fund services from collaborating agencies so that these services actually materialize. All too often,

"collaboration" fails to take place because the collaborating agencies already are deluged with clients that they already are not funded to service. Remember that foster children and teenagers have to be prioritized because of the sheer number of mental health, educational, legal, substance abuse, and other problems that they possess as we have discussed with respect to Core Problems 1 through 6.

- Have monitors regularly check to see if customized packages of services are developed for each child and teenager in foster care.
- Develop community groups that pressure local, state, and federal officials to augment funding for foster care programs and programs that help emancipated youth.
- Conduct needs assessment studies to identify unmet needs of foster children and emancipated youth.
- Link emancipated youth to jobs and to higher education. If child welfare agencies do not adequately prepare and support young people to enter, pay for, and complete college, the chances are good that their children, in turn, will struggle, thus resulting in a cycle of poverty (Promising Practices, 2009). Child welfare graduates need scholarships in each state that are reserved for them, whether funded by local governments, states, or the federal government.
- Cut the rates of turnover of foster parents so that they can participate in efforts to expand wraparound services for the children and youth that they parent. They are the cornerstone of the entire child welfare system, yet they are poorly reimbursed, not given sufficient training, and not helped to participate in developing a broader set of services geared around wraparound services.

Resources for Advocates

The ASFA of 1997 introduced legislation that extended independent living services to young people whose assets did not exceed $5,000, which differed from the former $1,000 cap. According to the provisions in the act, these services should be designed to assist young people in preparing to live independently when they leave foster care (Welte, 1997). However, based on expert testimony and the realization that separate legislation is needed to allocate services and resources separately for this group of foster youth, the Foster Care Independence Act of 1999 (also named the Chafee Act after the late John H. Chafee, the senator who sponsored the bill) was passed by Congress (Graf, 2002). The main thrust of the Foster Care Independence Act was to expand the provisions for independent-living programs by doubling the allotment provided for these programs under Title IV-E, and it allowed for more flexibility in terms of providing independence-oriented services. The Chafee Act also allowed for extensions of Medicaid to youth until the age of 21 and strengthened the focus on accountability of states by allocating 1.5% of the total allotment (or $2.1 million) for the development and implementation of a national evaluation and provision of technical assistance to states in assisting youths (Graf, 2002).

In California, the Higher Education Outreach and Assistance Act for Emancipated Foster Youth (California Education Code 89340) offers more needed assistance for former foster youth by requiring California state colleges and universities to provide them with outreach services and technical assistance. It also requires California state colleges and universities to evaluate programs, improve delivery of services, track the progress of former foster youth, and engage in other activities designed to advance services and outreach for former foster youth. In addition, the McKinney-Vento Act clarified some of the educational responsibilities of child welfare agencies. According to this act, the educational goals of a foster child or youth are the shared responsibility of the child welfare agency and the schools. Thus, the McKinney-Vento Act made it clear that child welfare agencies had their share of responsibilities in supporting the educational attainment of their wards. In addition, this act mandated that child welfare agencies implement and design their own policies and practices to maximize placement stability, consider the educational consequences of any action in the child welfare case, and make sure the act is implemented efficiently and youths derive maximum benefit out of these programs (Julianelle, 2009).

There are other federal programs and resources that also can assist these youth in reaching their vocational training goals. A program called Youthbuild operated by Housing and Urban Development (HUD) provides grants on a competitive basis to assist high-risk youth between the ages of 16 and 24 to learn housing construction job skills and to complete their high school educations. Program participants are able to augment their skills as they construct and/or restore affordable housing for very low-income and homeless persons or families. During the past seven years, HUD has released more than $300 million in grants to Youthbuild programs around the nation (NACO, 2008). Also, there are other resources for obtaining educational and vocational grants such as the Federal Pell Grants (AECF, 2001). The Pell Grants authorize up to $3,000 per student per year to support postsecondary training and education. Also, Welfare to Work and TANF programs have funding available for educational and vocational training. In addition to that, the Workforce Investment Act (WIA) has a specifically allocated a part of its funding to support work experience and postsecondary vocational training for disadvantaged youth. These nontraditional funding sources can supplement the funding provided by the Foster Care Independence Act (Promising Practices, 2009).

Assembly Bill 12 (AB12) was passed in California in 2010, which allows youth to remain in foster care voluntarily until they are 21 as long as they meet one of the following conditions: completing high school or an equivalent program, enrolling in college or a vocational program, working at least 80 hours a week, participating in an employment program, or living with a medical condition that renders them unable to do any of the above (Aspiranet, n.d.). AB12 took effect in 2012 and uses Title IV-E funds to provide assistance to eligible youth involved with child welfare and probation (California Department of Social Services [CDSS], n.d.a.). The effects of AB12 are being felt in many areas of child welfare including kinship care, the Kinship Guardian Assistance Program (Kin-Gap), transitional housing, and adoptions (CDSS, n.d.b).

Child welfare programs should consider developing a jobs program similar to one created by Father Greg Boyle, a Jesuit priest, who developed a job program for gang members called Home Boys. It gave jobs to gang members for 18 months in which they operated eating establishments, with positive results. See legislation developed in 21 states by selecting "workforce" under the 2017 child support enactments at the Child Welfare Enacted Legislation Database on the website of the National Conference of State Legislators: http://www.ncsl.org/research/human-services/workforce-and-foster-care.aspx.

Children in the child welfare sector often find themselves simultaneously involved in other public sectors such as public social services, mental health, criminal justice, immigration, education, and public social services, which are sometimes referred to as cross-sector services (Drake, Jonson-Reid, & Sapokaite, 2006). The literature reveals that from 35% to 85% of children entering foster care have significant mental health problems (Reed & Karpilow, 2002). Parental substance abuse is a factor in an estimated two-thirds of cases with children in foster care. Child abuse and domestic violence most often go hand in hand, occur in the same household, and have a common perpetrator. Foster children are more likely than other children to perform poorly in school and develop conduct problems. Juvenile justice departments have also been consistently reporting a rising tide of children from foster care landing on their doorsteps. Therefore, policies should be geared toward streamlining services in agencies so that people can transition from one agency to another without any bureaucratic red tape.

Changes in TANF laws have prevented many individuals and families from receiving the financial assistance they require (Stoesz, 1999). This prevention has pushed many families beyond their limits and contributed to the growing number of child neglect cases owing to a lack of financial resources (Stoesz, 1999).

Another sector that frequently crosses over with child welfare is juvenile justice. In juvenile justice, children who are maltreated are at risk for delinquency and often end up in the juvenile justice system (Bender, 2010). It is estimated that a quarter of foster youth will serve time in jail in the first two years after high school, and nearly 70% of inmates in state penitentiaries have been in foster care (Jitfosteryouth.org, n.d.) These youth are often referred to as crossover youth or dually involved youth, and the majority of them have problems involving school, mental health, and/or drug use (Herz et al., 2012). Furthermore, many crossover youth have witnessed domestic violence and have parents with mental health problems, substance abuse issues, and criminal justice involvement (Herz et al., 2012). National estimates indicate that children whose caregivers have been arrested have higher rates of drug use, domestic violence, and poverty (Phillips & Dettlaff, 2009).

Almost one-fourth of all children in the United States have at least one parent who is an immigrant, as I discuss in more detail in Chapter 13. Although the numbers of children living with immigrant families are not certain because these statistics are usually not collected by CPS agencies, national estimates indicate that this population makes up approximately 8.6% of children reported to CPS (Dettlaff, Earner, & Phillips, 2016). Related to immigration, human trafficking has become a growing problem. With the passage of the Victims of Trafficking and Violence Protection Act in 2000, identification and treatment of victims of human trafficking has considerably improved (Fong & Berger Cardoso, 2012)—but its resources have mostly gone to adult victims rather than children. According to the Department of Justice, there were 2,515 human trafficking incidents opened for investigation between January 2008 and June 2010 (Banks & Kyckelhahn, 2011). Of these, 1,016 children were investigated for prostitution or child exploitation, accounting for more than 40% of all the investigated incidents.

The following website is relevant to the crossover between immigration and child welfare:

Immigration and Child Welfare—Child Welfare Information Gateway: https://www.childwelfare.gov/pubs/issue-briefs/immigration/

The following websites provide services and advocacy to victims of human trafficking:

http://www.acf.hhs.gov/programs/orr/programs/anti-trafficking

http://www.gems-girls.org/

http://www.acf.hhs.gov/programs/orr/resource/fact-sheet-child-victims-of-human-trafficking

http://www.state.gov/documents/organization/192587.pdf

POLICY ADVOCACY LEARNING CHALLENGE 11.7

CONNECTING MICRO, MEZZO, AND MACRO POLICY INTERVENTIONS THAT LINK CHILDREN TO THEIR COMMUNITIES

Finding Community Resources

Imagine that you are a social worker in the Department of Children and Family Services (DCFS) who has been assigned to 17-year-old Tommy Jones. Tommy is African American, and he walked into your office today after running away for one month. He will not say where he was and states that he returned to the office because he was tired of living on the streets. Tommy has not participated in therapy and has not taken his psychotropic medication since he went AWOL from his last group home. You look in his file and note that he was last prescribed Seroquel for psychotic disorder NOS. He also has not been in school in more than six months and was on probation, although it is unclear whether it is still active.

While sitting at your desk, you notice that Tommy is staring at people and laughing for no reason at all. You ask him about drug use, and he states that he smokes tobacco and marijuana every once in a while. You ask him whether he has ever been in a psychiatric hospital, and he states that he has been hospitalized twice in the last six months (he reports that he does not remember why). You ask Tommy about whether he'll go talk to a psychiatrist to get his medication filled again, and he states that he does not want to take medication because he is Christian and God will take care of him. You further read in his file that his aunt dropped him off at a DCFS office when he was 16 because she could no longer handle his strange behavior. He still visits her once a month, but she states that she can't have him at her house for more than a few hours at a time because of his strange behavior. Aside from this aunt, his other relatives' whereabouts are unknown. When you ask him what he wants, Tommy says that he just wants a place to stay.

Learning Exercise

1. How does this vignette illustrate the multiple sectors that a client can be involved with?
2. Before any type of policy advocacy can occur, what is the first step that should be taken when working with Tommy?
3. What kind of policy advocacy needs to be done with Tommy considering he is a crossover youth?

THINKING BIG AS POLICY ADVOCATES IN THE CHILD AND FAMILY SECTOR

The foster youth caucus of the U.S. House of Representatives, chaired by Democratic Congresswomen Karen Bass of Los Angeles, often develops legislation to assist foster children. Go to its website at https://fosteryouthcaucus-karenbass.house.gov/ and click on "Resources for Foster Youth," "Supports for Foster Families," and "Resources for Families." Using concepts developed in Chapter 6 on policy analysis, develop some policy options that the House legislators might consider when drafting a legislative proposal with regard to one of their prioritized legislative or issue areas. Develop some criteria to be used to assess the options, such as cost and effectiveness. Decide which option appears to be meritorious.

You may also tackle another common problem in foster care: the inability of social workers to obtain education records of foster children from their schools on the grounds that schools do not want to breech student confidentiality. This policy, while well intended, often has bad consequences for foster children because they must often switch schools when they are shifted from one foster home to another. Consequently, they often repeat courses unnecessarily, severely delaying their schooling and jeopardizing their graduation rates.

Another option is to propose and seek enactment of a "children's allowance" that other industrialized nations provide to parents of children. Such a resource would be invaluable to stressed-out, single parents who often work two to three jobs just to make ends meet. Congress enacted a children's tax credit in 2018 that gives only negligible benefits, such as $70, to low-income families. Advocates should seek to replace it with a children's allowance that could cut in half the number of children who live in poverty (Marr et al., 2017). You can examine this topic in your own state with an interactive map that identifies the extent of child poverty and the low benefits the children's tax credit will give to millions of low-income children by going to https://www.cbpp.org/research/federal-tax/house-tax-bills-child-tax-credit-increase-excludes-thousands-of-children-in-low. Be sure to click on "Interactive." Also read about the benefits of a children's allowance sponsored by Senators Sharrod Brown (D, Ohio) and Michael Bennet (D, Colorado) at https://www.vox.com/policy-and-politics/2017/10/26/16552200/child-allowance-tax-credit-bill-michael-bennet-sherrod-brown. Discuss the positive effects on children and their families if child poverty was cut in half.

Learning Outcomes

You are now equipped to:

- Develop an empowering perspective with respect to the child and family sector
- Understand the current state of children
- Understand the journey of children through the child welfare system
- Analyze the quest for algorithms to predict the odds that specific children will be harmed
- Analyze the challenges of emancipation
- Identify advocacy groups in the child welfare sector

- Analyze the evolution of the child welfare sector and child welfare policies
- Understand the scope of child abuse and neglect and where to find relevant statistics
- Understand the political economy of the child welfare sector
- Recognize the impact of an inegalitarian nation on the child welfare sector
- Understand how seven core problems are often experienced by children and families
- Think big in the child and family sector

References

AECF. (2001). Promising practices: School to career and postsecondary education for foster care youth, a guide for policymakers and practitioners. Retrieved from http://collegeforamerica.org/reports/promisingpractices2.pdf

American Humane Association. (n.d.). *Disparities in child welfare.* Retrieved from http://www.americanhumane.org/children/programs/disparities-in-child-welfare.html

Anderson, P. G. (1989). The origin, emergence, and professional recognition of child protection. *Social Service Review, 63*(2), 222–244. doi:10.1086/603695

Annie E. Casey Foundation. (2017a). *Race for results: Building a path to opportunity for all children.* Retrieved from http://www.aecf.org

Annie E. Casey Foundation. (2017b). State trends in well-being. Retrieved from http://www.aecf.org

Areen, J. (1975). Intervention between parent and child: A reappraisal of the state's role in child neglect and abuse cases. *Georgetown Law Journal, 63*(4), 887–937.

Arkansas Advocates. (2005). *The Arkansas child welfare system.* Retrieved from http://www.aradvocates.org/_images/pdfs/ArkChildWelfareSystem.pdf

Aspiranet: Transition Age Youth Services. (n.d.). *Building the bridge between foster care and independence.* Retrieved from http://www.aspiranetthpplus.org/ab12-benefits-aging-out-of-foster-care/

Austin, L. (2004). *Mental health needs of youth in foster care: Challenges and strategies.* Retrieved from http://www.casanet.org/library/foster-care/mental-health-%5Bconnection-04%5D.pdf

Banks, D. & Kyckelhahn, T. (2011). *Characteristics of suspected human trafficking incidents, 2008–2010.* US Department of Justice, Office of Justice Programs, Bureau of Justice Statistics. Retrieved from http://bjs.ojp.usdoj.gov/index.cfm?ty=tp&tid=40#data_collections

Bardach, E. (1996). The eight-step path of policy analysis. In *Best practices.* Berkeley, CA: Berkeley Academic Press. Retrieved from http://www.newwaystowork.org/initiatives/ytat/practices.html

Bartholet, E., Wulczyn, F., Barth, R., & Lerderman, C. (2011). Race and child welfare. Chapin Hall Issue Brief. Retrieved from http://www.chapinhall.org

Bender, K. (2010). Why do some maltreated youth become juvenile offenders? A call for further investigation and adaptation of youth services. *Children and Youth Services Review, 32*(3), 466–473. doi:10.1016/j.childyouth.2009.10.022

Bluecommissionla.com. (n.d.). *The road to safety for our children*. Retrieved from http://ceo.lacounty.gov/pdf/brc/BRCCP_Final_Report_April_18_2014.pdf

Bourdieu, P. (1986). The forms of capital. In J. Richardson (Ed.), *Handbook of theory and research for the sociology of education* (pp. 242–258). New York: Greenwood.

Brittain, C. & Hunt, D. E. (2004). *Helping in child protective services: A competency-based casework handbook*. Cambridge Oxford University Press.

Brooks-Gunn, J. & Duncan, G. J. (1997). The effects of poverty on children. *Future of Children*, *7*(2), 55–71.

California Department of Social Services (CDSS). (n.d.a). *Assembly Bill 12*. Retrieved from http://www.childsworld.ca.gov/PG2902.htm

California Department of Social Services (CDSS). (n.d.b). *Title IV-E Child Welfare Waiver Demonstration Capped Allocation Project (CAP)*. Retrieved from http://www.dss.cahwnet.gov/cfsweb/PG1333.htm

Child Protective Services (CPS). (n.d.). In Wikipedia. Retrieved from http://en.wikipedia.org/wiki/Child_Protective_Services#Child_Protective_Services_Recidivism_in_the_United_States

Child Welfare Information Gateway. (2009). *Adoption assistance for children adopted from foster care*. Retrieved from http://www.childwelfare.gov/can/defining/federal.cfm

Child Welfare Information Gateway. (2011). *Adoption assistance for children adopted from foster care*. Washington, DC: U.S. Departments of Health and Human Services, Children's Bureau. Retrieved from https://www.childwelfare.gov/pubs/f_subsid.pdf#page=1&view=Introduction

Child Welfare League of America. (n.d.). *Summary of the Adoption and Safe Families Act of 1997 (P.L. 105-89)*. Retrieved from http://www.cwla.org/advocacy/asfapl105-89summary.htm

Childrenunitingnations.org. (n.d.). Retrieved from http://www.childrenunitingnations.org/who-we-are/foster-care-statistics/

Costin, L. B., Karger, H. J., & Stoesz, D. (1996). *The politics of child abuse in America*. New York: Oxford University Press.

De Bellis, M. D., Keshavan, M. S., Clark, D. B., Casey, B. J., Giedd, J. N., Boring, A. M., & Ryan, N. D. (1999). Developmental traumatology part II: Brain development. *Biological Psychiatry, 45*, 1271–1284.

Department of Children and Family Services. (2014). *Community response services alternative response services and up-front assessments 0070-548.00 | Revision Date: 07/01/14*. Retrieved from http://policy.dcfs.lacounty.gov/default.htm#POE_ARS_Communit.htm#Topic1

Doherty, S. (2005). *Rights of children in foster care*. Retrieved from http://www.hunter.cuny.edu/socwork/nrcfcpp/downloads/rights-children-foster-care.pdf

Dore, M. (2005). Child and adolescent mental health. In G. Mallon and P. Hess (Eds.), *Child welfare for the twenty-first century: A handbook of practices, policies and programs* (pp. 148–172). New York: Columbia University Press.

Drake, B., Jonson-Reid, M., & Sapokaite, L. (2006). Rereporting of child maltreatment: Does participation in other public sector services moderate the likelihood of a second maltreatment report? *Child Abuse and Neglect*, *30*(11), 1201–1226. doi:10.1016/j.chiabu.2006.05.008

Fearnotlaw.com. (n.d.). *In re A.J.* Retrieved from http://www.fearnotlaw.com/articles/article28570.html

Felitti, V. J., Anda, R. F., Nordenberg, D., Williamson, D. F., Spitz, A. M., Edwards, V., & Marks, J. S. (1998). Relationship of childhood abuse and household dysfunction to many of the leading causes of death in adults. *American Journal of Preventive Medicine*, *14*(4), 245–258. doi:10.1016/S0749-3797(98)00017-8

Fogarty, J. (2008). Some aspects of the early history of child protection in Australia. *Family Matters*, *78*, 52–59.

Fong, R. & Berger Cardoso, J. (2010). Child human trafficking victims: Challenges for the child welfare system. *Evaluation and Program Planning*, *33*(3), 311–316.

Fryer, R. G., Jr., & Levitt, S. D. (2004). Understanding the black-white test score gap in the first two years of school. *Review of Economics and Statistics*, *86*(2), 447–464.

Gelles, R. J. (2001). Family preservation and reunification: How effective a social policy? In *Handbook of youth and justice* (pp. 367–376). New York: Kluwer Academic/Plenum Publishers.

Glisson, C. (2009). Organizational climate and culture and performance in the human services. In R. Patti (Ed.), *The handbook of social welfare management* (pp. 119–141). Thousand Oaks, CA: SAGE.

Government Accounting Office. (1998). *Foster care implementation of the multiethnic placement act poses difficult challenges* [Report to the chairman, Subcommittee on Human Resources, Committee on Ways and Means, House of Representatives]. Retrieved from http://www.gao.gov/archive/1998/he98204.pdf

Graf, B. (2002). *Foster Care Independence Act*. Retrieved from http://www.hunter.cuny.edu/socwork/nrcfcpp/downloads/information_packets/foster_care_independence_act-pkt.pdf

Harding, D. J. (2003). Counterfactual models of neighborhood effects: The effect of neighborhood poverty on dropping out and teenage pregnancy. *American Journal of Sociology*, *109*(3), 676–719.

Herz, D., Lee, P., Lutz, L., Stewart, M., Tuell, J., & Wiig, J. (2012). *Addressing the needs of multi-system youth: Strengthening the connection between child welfare and juvenile justice*. Washington, DC: Georgetown University Center for Juvenile Justice Reform, and Boston: RFK Children's Action Corps. Retrieved from http://cjjr.georgetown.edu/pdfs/msy/AddressingtheNeedsofMultiSystemYouth.pdf

Hurlburt, M., Leslie, L. K., Landsverk, J., Barth, R. P., Burns, B. J., Gibbons, R. D., . . . Zhang, J. (2004). Contextual predictors of mental health service use among children open to child welfare. *Archives of General Psychiatry*, *61*(12), 1217–1224. doi:10.1001/archpsyc.61.12.1217

Hurley, D. (2018, January 2). Can an algorithm tell when kinds are in danger? *New York Times*. Retrieved from https://www.nytimes.com/2018/.../magazine/can-an-algorithm-tell-when-kids-are-in-dan...

Jackson, K. (2009). Building cultural competence: A systematic evaluation of the effectiveness of culturally sensitive interventions with ethnic minority youth. *Children and Youth Services Review*, *31*(11), 1192–1198.

Jitfosteryouth.org. (n.d.). Retrieved from http://jitfosteryouth.org/outcomes/

Julianelle, P. (2009). *The McKinney-Vento Act and children and youth awaiting foster care placement*. National Association for the Education of Homeless Children and Youth. Retrieved from http://www.naehcy.org/dl/mv_afcp.pdf

Kotch, J. B., Chang, J. J., & Hussey, J. M. (2006). Child maltreatment in the United States: Prevalence, risk factors, and adolescent health consequences. *Pediatrics, 118*(3), 933–942. doi:10.1542/peds.2005-2452

Lee, J. (2002). Racial and ethnic achievement gap trends: Reversing the progress toward equity. *Educational Researcher, 31*(1), 3–12.

Lehmann, B. & Guyer, J. (2012). *Child welfare and the Affordable Care Act: Key provisions for foster care children and youth.* Georgetown, Washington, DC: Center for Children and Families. Retrieved from http://ccf.georgetown.edu/wp-content/uploads/2012/07/Child-Welfare-and-the-ACA.pdf

Lind, C. (2004). *Developing and supporting a continuum of child welfare services.* Retrieved from http://www.financeproject.org/publications/developingandsupportingIN.pdf

Marr, C., et al. (2017, November 17). Senate tax bill's child tax credit increase provides only token help to millions of children in low-income working families. Center on Budget and Policy Priorities. Retrieved from https://www.cbpp.org/...tax/interactive-how-does-the-senate-republican-tax-bill-affect-...

Menchik, P. L. (1993). Economic status as a determinant of mortality among black and white older men: Does poverty kill? *Population Studies, 47*(3), 427–436.

Morales, A., Sheafor, B. W., & Scott, M. E. (2011). *Social work: A profession of many faces.* Boston: Allyn and Bacon.

Murray, K. & Gesiriech, S. (2005). *A brief legislative history of the child welfare system.* Retrieved from http://pewfostercare.org/research/docs/Legislative.pdf

Myers, J. E. B. (2008). Short history of child protection in America. *Family Law Quarterly, 449*, 449–463.

NACO. (2008). *Youth aging out of foster care: Strategies and best practices.* Retrieved from http://www.naco.org/Content/ContentGroups/Issue_Briefs/IB-YouthAgingoutofFoster-2008.pdf

National Center for Children in Poverty. (2017). Child poverty. Retrieved from http://www.nccp.org/topics/childpoverty.html

National Child Abuse and Neglect Data System. (2016). Child maltreatment. Retrieved from https://www.ndacan.cornell.edu/datasets/dataset-details.cfm? ID=210

Patton, C. V. & Sawicki, D. S. (1993). *Basic methods of policy analysis & planning* (2nd ed.). Englewood Cliffs, NJ: PrenticeHall.

Pecora, P. J. (2000). *The child welfare challenge: Policy, practice and research.* New York: Aldine de Gruyter.

Pecora, P. J., Kessler, R. C., Williams, J., O'Brien, K., Downs, A. C., English, D., . . . Holmes, K. E. (2005). *Improving family foster care: Findings from the Northwest Foster Care Alumni Study.* Seattle, WA: Casey Family Programs. Retrieved from http://www.casey.org

Pecora, P. J., Williams, J., Kessler, R. C., Downs, A. C., O'Brien, K., Hiripi, E., & Morello, S. (2003). *Assessing the effects of foster care: Early results from the Casey National Alumni Study.* Seattle, WA: Casey Family Programs. Retrieved from http://www.casey.org

Penn State. (n.d.). In Wikipedia. Retrieved from http://en.wikipedia.org/wiki/Penn_State_child_sex_abuse_scandal

Phillips, S. D. & Dettlaff, A. J. (2009). More than parents in prison: The broader overlap between the criminal justice and child welfare systems. *Journal of Public Child Welfare, 3*(1), 3–22.

Polihronakis, T. (2008, April). *Mental health issues of children and youth in foster care.* National Resource Center for Family-Centered Practice and Permanency Planning at the Hunter College School of Social Work.

Promising Practices. (2009). School to career and post-secondary education for Foster youth. Retrieved from http://www.workforcestrategy.org/publications/promisingpractices2.pdf

Rank, R. (2018, April 17). Childhood poverty costs US more than $1 trillion a year. *New York Times.* Retrieved from https://nonprofitquarterly.org/2018/.../childhood-poverty-costs-us-1-trillion-year-rese...

Reed, F. D. & Karpilow, K. (2002). Understanding the child welfare system in California: A primer for service providers and policymakers. Retrieved from http://www.ccrwf.org/pdf/ChildWelfarePrimer.pdf

Sampson, R. J. & Wilson, W. J. (1995). Toward a theory of race, crime, and urban inequality. In *Race, crime, and justice: A reader* (pp. 177–190). Retrieved from http://faculty.washington.edu/matsueda/courses/371/Readings/Sampson%20Wilson.pdf

Schene, P. A. (1998). Past, present, and future roles of child protective services. *Future of Children, 8*(1), 23–38.

Schott, L. (2009). *Policy basics: An introduction to TANF.* Retrieved from http://www.cbpp.org/cms/? fa=view&id=93

Skemer, M. & Valentine, E. (2016, November). Striving for independence: Two-year impact findings from the Youth Villages transitional living evaluation. MDRC. Retrieved from https://www.mdrc.org/sites/default/files/YV_2016_FR.pdf

Stewart, N. (2016, March 29). How New York's child welfare chief is trying to fix his agency's image. *New York Times.* Retrieved from https://www.nytimes.com/.../29/.../how-new-yorks-child-welfare-chief-is-trying-to-fix-hi...

Stewart, N. & Fortin, J. (2017, February 21). De Blasio picks new commissioner for troubled child welfare agency. *New York Times.* Retrieved from https://www.nytimes.com/.../david-hansell-new-york-administration-for-childrens-serv...

Stoesz, D. (1999). Unraveling welfare reform. *Society, 36*(4), 53–61.

Stoltzfus, E. (2017, June 1). Child welfare funding in brief, FY2017 final funding and the president's FY2018 request. Congressional Research Service, Retrieved from http://greenbook.waysandmeans.house.gov/sites/greenbook.waysandmeans.house.gov/files/2011/images/R42027_gb.pdf

Taussig, H. N., Clyman, R. B., & Landsverk, J. (2001). Children who return home from foster care: A 6-year prospective study of behavioral health outcomes in adolescence. *Pediatrics, 208*, E10. Retrieved from http://pediatrics.aappublications.org/cgi/content/full/108/1/e10#otherarticle

Therolf, G. (2013, May 31). Boy's death prompts outrage, red flags of child abuse "ignored." *Los Angeles Times.* Retrieved from http://articles.latimes.com/2013/may/31/local/la-me-ln-boy-killed-20130531

Trattner, W. I. (1999). *From poor law to welfare state.* New York: The Free Press.

Twiname, J. (1975). *Using Title XX to serve children and youth*. Retrieved from http://www.eric.ed.gov/ERICDocs/data/ericdocs2sql/content_storage_01/0000019b/80/31/5b/a7.pdf

U.S. Department of Health and Human Services. (2010). *Substance abuse specialists in child welfare agencies and dependency courts considerations for program designers and evaluators*. Substance Abuse and Mental Health Services Administration, Administration for Children and Families. Retrieved from https://www.ncsacw.samhsa.gov/files/SubstanceAbuseSpecialists.pdf

U.S. Department of State. (2012). *Trafficking in persons report, June 2012*. Washington, DC.

Vulin-Reynolds, M., et al. (2008). School mental health and foster care: A logical partnership. Retrieved from https://www.tandfonline.com/doi/pdf/10.1080/1754730X.2008.9715726

Wang, C. T. & Holton, J. (2007). *Total estimated cost of child abuse and neglect in the United States*. Chicago, IL: Prevent Child Abuse America.

Weizman, A., Har-Even, D., Shnit, D., Finzi, R., & Ram, A. (2001). Attachment styles and aggression in physically abused and neglected children. *Journal of Youth and Adolescence, 30*(6), 769–786. doi:10.1023/A:1012237813771

Welte, C. (1997). *Detailed summary of the Adoption and Safe Families Act*. Retrieved from http://www.casanet.org/reference/asfa-summary.htm

Wilkins, A. (2004). *The Indian Child Welfare Act and the states*. Retrieved from http://www.ncsl.org/IssuesResearch/StateTribal/TheIndianChildWelfareActandtheStates/tabid/13275/Default.aspx

Wulczyn, F. (2009). Epidemiological perspectives on maltreatment prevention. *Future of Children, 19*(2), 39–66.

BECOMING POLICY ADVOCATES IN THE EDUCATION SECTOR

This chapter was coauthored by Elain Sanchez Wilson, MPP, and Vivian Villaverde, PPSC, LCSW

LEARNING OBJECTIVES

In this chapter you will learn to:

1. Develop an empowering perspective with respect to education
2. Analyze the evolution of the American education sector
3. Recognize problems in education that are created by an unequal nation
4. Analyze the political economy of the education sector
5. Analyze seven core issues in the American education sector
6. Think big as policy advocates in the education sector

DEVELOPING AN EMPOWERING PERSPECTIVE WITH RESPECT TO EDUCATION

The education system lies at the heart of American democracy. Students who get multiple degrees from high school, junior colleges, colleges, graduate schools, and professional programs are likely to have relatively high salaries and wealth as compared with

Note: Elain Sanchez Wilson, MPP, and Vivian Villaverde, PPSC, LCSW, and clinical associate professor at the Suzanne Dworak-Peck School of Social Work, made invaluable contributions to this chapter in this book's first edition. Bruce Jansson was the sole author of the second edition, editing and updating contributions from each original contributor.

students who do not receive high school diplomas who are likely to experience poverty and an array of social problems. Unfortunately, the American educational system often fails its students. Its public schools are often poorly funded as witnessed by teacher strikes in West Virginia, Colorado, and Oklahoma in 2018. It lacks sufficient mentors, social workers, and psychologists to help students with social problems that include bullying, substance abuse, autism, attention deficit disorder, anxiety, and depression. Some students go to their beds hungry, while others have no beds at all. The schools that serve low-income youth of color in inner cities, as well as ones in rural areas, are particularly underfunded and understaffed. Although graduation rates from high school have improved, students of marginalized populations often have higher drop-out rates than white students—a phenomenon that harms them and that harms the broader society by increasing joblessness and rates of incarceration (Sum, Khatiwada, & McLaughlin, 2009). The nation has an archaic financing system for schools that is linked to property taxes in thousands of local jurisdictions—a system that yields far higher revenues for schools attended by affluent students in suburbs and far lower revenues for students in inner-city schools. Students in junior colleges and colleges often drop out from them due to mental, substance abuse, and other problems—such as completion rates of only 20% for many junior colleges. The United States not only should pay its teachers better but should make a national effort to diversify its population of teachers, not just to decrease prejudice among white students but to provide models of success for students from marginalized populations (Ingersoll & May, 2011; Donlevey, Meierkord, & Rajania, 2016).

Social workers can assume a critical role in helping students in these institutions. They can provide micro, mezzo, and macro policy advocacy that draws upon their unique empowerment perspectives.

ANALYZING THE EVOLUTION OF THE AMERICAN EDUCATION SYSTEM

This timeline represents the historical forces that influenced the current processes and policies of the U.S. educational system particularly at the level of primary and secondary education:

- The Massachusetts Act of 1642 mandated colonial-era parents and masters to teach their children and apprentices about moral character as well as their new homeland's laws.
- The U.S. Constitution mandated separation of church and state that led to separate systems of public schools and schools funded by churches.
- The United States implemented its first national system of public schools, beginning with primary and elementary grades in the middle part of the 19th century and enacting compulsory high school education in its latter part and in the early part of the 20th century.

- Despite making civil rights gains in the aftermath of the Civil War, African Americans endured a major blow at the hands of the Supreme Court, which validated the "separate but equal" doctrine in its 1896 *Plessy v. Ferguson* decision. This ruling led to the legitimization of separate educational institutions that were largely inferior in the case of minorities.
- The United States funded "land grant institutions of higher education" by federal land grants or funds by the Morrill Acts of 1862 and 1890, including Texas A&M University, Pennsylvania State University, Purdue University, Ohio State University, Cornell University, Rutgers University, and Colorado State University. The second Morrill Act in 1890 targeted former confederate states—and required them to establish separate land grant colleges for black people if they were excluded from existing land grant colleges. Many historically black colleges and universities were formed prior to or after the Civil War such as Cheyney University of Pennsylvania and Shaw University.
- Compulsory education laws, first seen in Massachusetts during the colonial period, were adopted nationwide by states by 1918.
- Enacted in 1944, the G. I. Bill provided massive educational subsidies to hundreds of thousands of veterans, whether in higher education or vocational training.
- In the famous court case in 1954, *Brown v. Board of Education*, the U.S. Supreme Court overturned *Plessy v. Ferguson* by ruling that separate educational institutions were inherently unequal when they separated white from black students. Fast forward to 2018, and most African American and Latino students nonetheless attend segregated schools.
- The Elementary and Secondary Education Act (ESEA) of 1965 provides federal assistance to public schools with relatively high concentrations of low-income students. It also allows private schools to share books and other materials that had been purchased by public schools.
- School busing was used on a large scale in the late 1960s and early 1970s to desegregate public schools but disappeared in the wake of political discontent and court rulings.
- Head Start was initiated in 1964 by the so-called War on Poverty. It provides quality preschool education to low-income children.
- The Education for All Handicapped Children Act of 1975 required secondary schools to mainstream children with disabilities, including deafness, mental retardation, and mental illness, into public schools.
- The Bilingual Education Act of 1968, later amended in 1974 and 1988, granted funding to school districts to set up their own bilingual educational programs, teacher training, and materials.

- The federal Pell Grant program was created in 1972 when it was originally named the Educational Opportunity Grant. These scholarships are funded by the federal government with average grants totaling $3,740 in 2017.
- The National Association of Social Workers (NASW) Council on Social Work in Schools held its first meeting in 1973, when it discussed the issues that field workers encountered, including troubled parent-child relationships, emotionally disturbed students from low-income areas from all races, overwhelming caseloads, excessive referrals, and unclear job descriptions.
- The first major mass killing of students and teachers in a secondary school took place at Columbine High School in Littleton, Colorado, on April 30, 1999. Many other school shootings took place in secondary schools and colleges in ensuing years, such as the slaying of 17 students and teachers at a Parkland, Florida, high school on February 14, 2018.
- President Bill Clinton obtained enactment of the Charter School Expansion Act of 1998. It allowed charter schools to be established in states as not-for-profit or for-profit institutions. They could compete for local and state funds used to fund public schools. They could legally prohibit unions from organizing their teachers. They could develop innovative curricula.
- Passing Congress in 2001 with widespread bipartisan support, No Child Left Behind (NCLB) is the latest version of the omnibus 1965 Elementary and Secondary Education Act. The legislation, proposed and signed into law by President George W. Bush, aimed to improve student achievement, especially among disadvantaged students, through an expanded federal role in education.
- The Post-9/11 G. I. Bill was enacted in 2008 and continued a tradition of providing access to college and training by former military personnel. It was expanded in 2017 with enactment of the Harry W. Comery Veterans Educational Assistance Act—often called the Forever G. I. Bill.
- President Barack Obama developed a revised version of Bush's legislation in 2009 called Race to the Top as part of the American Recovery and Reinvestment Act. It gave states greater latitude to develop their educational systems while still requiring states to develop their own methods of evaluating the quality of students' education as long as they included students' test scores in their evaluations of teachers and principals.
- President Trump named Betsy DeVos to be his secretary of education. Critics contend that DeVos favors charter schools over regular public schools that most American students attend, fails to seek gun control after numerous mass shootings during her tenure, fails to address disproportionate discipline given students of color in schools, and does not require private schools that receive federal funding to follow federal civil rights laws that prohibit sexual, racial, and religious discrimination.

RECOGNIZING EDUCATIONAL PROBLEMS THAT ARE CREATED BY AND CAUSE AN INEGALITARIAN NATION

The extreme economic inequality of the United States—higher than 20 other industrialized nations—is partly caused by its educational system. Children and youth from low-income families of all ethnicities as well as white backgrounds, whether from inner cities or rural areas, attend schools that receive far lower funding than schools used by children and youth from relatively affluent families. They have more pupils in their classes, teachers who receive lower pay, poorer facilities, larger classes, and fewer advanced classes than their more affluent counterparts. They are more likely to encounter teachers who have lower expectations that they will attend colleges and join professions. They are more likely to use outdated textbooks. They are less likely to have advanced classes in math, English, history, and other topics. They are less likely to see nurses, psychologists, and social workers in their schools. They are less likely to have enriching experiences, such as summer internships.

Poorer educational outcomes are caused, as well, by the life experiences of many low-income youth and their parents. They are exposed to fewer words than their more affluent counterparts. They are less likely to see and read books at home. They are less likely to have parents who have received higher education. Often focusing on survival needs, it is not surprising that low-income youth have lower scores than more affluent students on standardized tests, lower grade point averages, and lower graduation rates from high schools and institutions of higher education (Rumberger, 2013). Extreme poverty extends, as well, to each of the 16 marginalized populations we discussed in Chapter 1. There are many exceptions to these statistics. Graduation rates of African Americans and Latinos from high schools markedly increased from 2010 to 2018. More students of color attend institutions of higher education, whether community colleges, four-year colleges, or post-college institutions. These "exceptions" would be far greater, however, if schools with many low-income students were funded at equal levels as those with relatively affluent students.

Research suggests, as well, that educational attainment of students of color would increase if they attended integrated schools partly because segregated schools are inferior to integrated ones with lower turnover of teachers, more experienced teachers, and greater resources for high-quality instruction (Catalan, 2014). When the Supreme Court ruled that *de jure* (planned and officially sanctioned) segregated schools are unconstitutional in *Brown v. Board of Education* in 1954, many reformers hoped that the nation would integrate all of its schools. These reformers hoped that integration would place students of color in better-funded schools than in segregated systems that often receive less funding than schools attended primarily by white students. But the Supreme Court did not rule that *de facto* (unplanned, naturally occurring) segregated schools were unconstitutional. Although segregated schools decreased somewhat in the North and the South in the 1950s and 1960s, they increased in number in succeeding decades so that roughly 39% of students of color attended schools in 2017 with less than 10% whites in the United States in 2017—and students of color are far more likely than whites to attend schools where more than 60% of students live in poverty (Tatum, 2017). When civil rights advocates attempted to overcome this problem by instituting school busing in the 1960s and 1970s, white parents rallied against it—and adverse court rulings prohibited it. Segregation of schools is linked to housing that is segregated by race and social class partly due to discriminatory practices by banks and realtors who "steer" persons of color to segregated neighborhoods and fail to finance mortgages for persons of color. Segregation by social class takes place because low- and moderate-income persons cannot afford rentals and cannot afford to purchase houses in many white areas.

Persons of color and low-income persons of all ethnic and racial groups, including low-income white students, often find it more difficult to be admitted to colleges than affluent whites—and to graduate from colleges when they are admitted. They often have lower scores on college admission tests such as SATs and have completed fewer advanced courses. Elite universities like Harvard and other Ivy League universities remain bastions of privilege by race and social class. But many other colleges and universities, including many public ones, are not affordable for low-income students even when they are in the top 50% of their high school classes and even when they receive federal Pell Grants. Forty-seven percent (500,000) of these students do not complete even a postsecondary certificate, much less a degree from an institution of higher education. Three-fourths of 551 public four-year institutions with on-campus housing are not affordable to average low-income students who receive federal, state, and institutional grant aid, take out loans, have federal work-study jobs or similar part-time work, and work full-time over the summer (National College Access Network, 2018).

When they do gain entry to four-year colleges, they are far more likely not to graduate. Some of them have to earn money for themselves or family members during their educations. Yet many low-income students of all races and ethnic groups have graduated from colleges and professional schools including Supreme Court Justice Sonia Sotomayor, Michelle and Barack Obama, and J. D. Vance (2016), the author of *Hillbilly Elegy* about white communities in coal country.

Community colleges give millions of low- and moderate-income persons a stepping stone to careers and (in some cases) to four-year colleges. Some of them take all or most applicants. Yet graduation rates can be as low as 20%—and persons often take six years to graduate. Low graduation rates are caused by many factors. Community colleges often lack sufficient support staff. They often cannot provide scholarships. Their students often have to work full or part time as they attend them, not just to meet their own needs but to help their parents and siblings. They have small support staffs to help students with personal problems and financing of their education.

Curiously, no reformer has called for drastic increases in federal funding education from preschool through college. Why vest the cost of public education primarily in local governments when the future of the nation will be powerfully shaped by its educational system? Why not invest more resources in a school system in which average black or Hispanic students at 17 years old performed worse on standardized tests than at least 80% of their white classmates in 2003—with better but similar results in 2018. Why accept a situation in 2003 that still persists in 2019 that "for many students . . . the die is cast by eighth grade . . . students without the appropriate math and reading skills by that grade are unlikely to acquire them by the end of high school" (Thernstrom & Thernstrom, 2003). School achievement profoundly impacts life earnings, longevity, and social well-being. Why not invest in summer- and after-school tutoring programs for these students as well as mentors and support staff in their schools? Why not equalize funding of schools so that no school falls beneath a national floor of funding? Isn't it time to catch up with funding of public schools in many European nations, Canada, and Japan, where teachers receive higher pay than American teachers who sometimes work two or three jobs to survive, as was revealed in strikes of public school teachers in West Virginia, Oklahoma, Seattle, Kentucky, Arizona, North Carolina, Colorado, and elsewhere in 2018 (Karp & Sanchez, 2018)?

ANALYZING THE POLITICAL ECONOMY OF THE EDUCATION SECTOR

The relatively poor funding of secondary schools, particularly in low- and moderate-income neighborhoods, is partly linked to the decision by Americans to fund them by local governments that rely on sales and property taxes rather than by the federal government that obtains revenues twice as large as the combined revenues of state governments. (Federal revenues come substantially from income and corporate taxes as well as payroll deductions for Medicare and Social Security.) Although 41 states have income taxes, they established them at far lower rates than the federal government. States expend, moreover, huge amounts of their revenues on Medicaid, prisons and related institutions, and infrastructure—curtailing resources for schools. The federal government funds only 10% of costs of secondary and postsecondary schools—much of it through the ESEA that was enacted in 1965 to provide supplemental funds to schools with large numbers of low-income students. Title I of ESEA provides local schools roughly $500 to $600 per low-income student—and no research has shown that these funds produce positive results overall for evidence-based services. Much of these funds are used for teacher professional development, which has not been shown to be effective and which teachers do not find valuable. Performance gaps between affluent and disadvantaged students can be addressed only by identifying evidence-based programs and/or by spending five to eight times as much resources as currently spent by Title I to generate large reductions in class size, which has been shown to be effective (Dynarski & Kainz, 2015).

Education is therefore underfunded in all states—and particularly in communities mostly occupied by low- and moderate-income persons that raise meager funds from property taxes as compared with affluent communities. Fierce competition for scarce resources often takes place between educational institutions. Relatively affluent suburban schools lure experienced teachers from inner-city schools by offering them higher salaries and smaller classes. Elite state universities compete with community colleges and less prestigious state colleges for resources. Private colleges and universities obtain revenues from students' tuition and income from their endowments as well as from Pell Grants and other subsidies from their states that decrease resources for community colleges and state colleges and universities.

Economic inequality extends, as well, to preschool programs. Children from affluent families obtain preschool education by self-funding it, whether through private nursery or preschool programs. Less-affluent families rely on Head Start funded by the federal government. Rigorous evaluation of Head Start reveals that it provides both short- and long-term positive benefits to children, including improving school achievement in elementary school and increasing high school graduation rates. Head Start served only 42% of children who were eligible for it in 2013.

President Bill Clinton secured enactment of the Charter School Expansion Act of 1998 that allowed charter schools to be organized in states whether as for-profit or not-for-profit institutions. They were given prerogatives that included the power to prohibit unions from organizing their teachers, the ability to develop their own curriculums rather than curriculums approved by public schools, and the right to receive the same per-capita funding from public sources that are given to public schools in their jurisdictions. Their defenders argue that charter schools give parents alternatives to ineffective public schools. They contend that charter schools are more effective than public ones, that they hire higher-quality teachers than public schools, and that they are more innovative than public schools. It is ethically correct, they argue, to give parents choices, including low-income black and Latino parents. They contend that they can dismiss low-quality teachers more easily than at public schools whose teachers are protected by unions (Chait, 2016). Some charter schools are highly successful, but others produce poor educational results.

President Donald Trump chose Betsy DeVos, a militant supporter of charter schools, as his secretary of education, a choice that opponents of charter schools opposed (Terkel, 2017). Her opponents also accused her of not fining and closing for-profit colleges with low graduation rates and inadequate curriculum, not vigorously supporting policies and regulations to decrease sexual harrassment of women in colleges, not seeking major increases in federal funding of schools with low-income students, and not seeking ways to address disparities in discipline given to students of color as compared with white students.

Opponents of charter schools argue that many charter schools refuse admission to youth with social, educational and mental problems—forcing public schools to serve all of them. Teachers unions contend that blame is too quickly placed on its members for not improving test scores of nonaffluent students and students of color. Opponents of charter schools contend they drain resources from public schools that, together, serve far more students than charter schools taken together. Many teachers work well past school hours, crafting lessons and grading assignments into the night. Salaries are modest, at best, and classrooms are overcrowded, and with today's tight budgets, many teachers face job insecurity. Teacher strikes in West Virginia, Colorado, Oklahoma, and many other states in 2018 drew attention to the low pay of public school teachers and poor supplies, forcing many of them to buy supplies themselves and take two or three jobs just to meet their survival needs. Many public schools in low-income areas lack air-conditioning and have poor facilities. Rural school districts often cannot recruit and retain quality teachers because of their poor pay and poor facilities. By contrast, teachers in Finland, the top-rated school system in the world, earn much higher salaries, are selected only from high-performing graduate students, and have far lower turnover. Defenders of public schools also cite meritorious "magnet schools" that recruit high-achieving students in the performing arts, science, medicine, and other subjects.

American students and their parents foot the cost of much of their postsecondary education unlike students and parents in many other industrialized nations that fund their schools more generously than the United States. School debt burdens many students for years. Millennial students had an average of $27,000 in college debt that often is not repaid for decades (Jansson, 2019). Millennial students applauded President Barack Obama's proposal to fund tuition of students attending community colleges and Senator Bernie Sanders's proposal in the presidential campaign of 2016 to fund tuition of

TABLE 12.1 ■ Selected Advocacy Groups for Educational Reform

Education Reform Now seeks educational reforms for low-income students in 14 states: http://www.edreform.org

The Education Trust engages in advocacy for low-income students at all levels of education: http://www.edtrust.org

Stand For Children advocates for public education: http://www.stand.org

Democrats for Education Reform advocates for accountable public schools: http://www.dfer.org

National Indian Education Association seeks to improve schools for American Indians: http://www.niea.org

National School Boards Association represents 90,000 local school boards and advocates for equity in funding: https://www.nsba.org

The Mexican American Legal Defense and Educational Fund (MALDEF) seeks to improve schools for Latino children: http://www.maldef.org

students attending public colleges and universities. The federal government has funded Pell Grants for many low- and moderate-income students in postsecondary education since the mid-1970s, but maximum grants were only $5,820 in 2016 and 2017. States need to greatly increase spending on schools to be coupled with enhanced federal spending.

Many advocacy groups seek to improve funding and policies of public education through lobbying and other kinds of advocacy (see Table 12.1).

ANALYZING SEVEN CORE PROBLEMS IN THE AMERICAN EDUCATION SYSTEM

Core Problem 1: Engaging in Advocacy to Promote Ethical Rights, Human Rights, and Economic Justice—With Some Red Flag Alerts

Many ethical issues arise in secondary and postsecondary institutions.

- **Red Flag Alert 12.1.** Schools refuse to honor a parent's request for their child's academic record.
- **Red Flag Alert 12.2.** Students are suspended for personal views or political views.
- **Red Flag Alert 12.3.** A student is forced to sit in a time-out for refusing to recite the Pledge of Allegiance.
- **Red Flag Alert 12.4.** A student receives a slap on the wrist for classroom disruption but is not given counseling.
- **Red Flag Alert 12.5.** A student is expelled without a hearing.

- **Red Flag Alert 12.6.** School personnel search student lockers without providing a reason.
- **Red Flag Alert 12.7.** School administrators prevent a school newspaper from publishing an editorial about teen pregnancy.
- **Red Flag Alert 12.8.** A group of students are prohibited from assembling a gay-straight alliance.
- **Red Flag Alert 12.9.** Juvenile bullying is a frequent occurrence.
- **Red Flag Alert 12.10.** Schools discriminate against lesbian and gay students and members of other marginalized groups.
- **Red Flag Alert 12.11.** Educational institutions fail to discipline students and faculty that engage in sexual harassment.

Background

Most elementary, middle, and high school students are not aware of their personal rights. Although some pupils may be equipped with this knowledge, the teacher-student dynamic makes it challenging for youngsters to assert and defend these rights. They may feel intimidated or fear punishment from school administration or parents.

Members of marginalized groups are at risk of experiencing violations of their rights, including LGBTQ students, students of color, students on the autism spectrum, and students with mental conditions. Non-English-speaking parents and students may not understand teachers and fellow students. Some Asian American parents and students are reluctant to question persons in authority.

Resources for Advocates

Education advocates can use the different education policies, statutes and constitutional rights, state and federal regulations, and court rulings to guide their advocacy.

Statutes and Constitution. Article 26 of the Universal Declaration of Human Rights states that every person is entitled to an education. It endorses a free public elementary and secondary education. "Education shall be directed to the full development of the human personality and to the strengthening of respect for human rights and fundamental freedoms," it proclaims, adding that parents have the right to select the type of education their children receive. Nevertheless, school choice is not always available to families. Some low-income students are not offered an alternative to their dilapidated classrooms, non-credentialed faculty, and lack of educational resources.

A strong ethical case can be made to increase school integration. Marginalized persons of color often receive inferior education because their schools receive less funding than white schools, have less-experienced teachers, and have teachers who receive less pay than teachers in suburban schools (Kahlenberg, 2016; Villegas & Irvine, 2010).

The Family Educational Rights and Privacy Act, enacted in 1974, protects the privacy of student records (20 U.S.C. 1232g; 34 CFR Part 99). Schools that receive any Department of Education funds are subject to this policy; therefore, private and parochial schools are not legally bound to comply. Parents, or eligible students 18 years of age or older, have the right to review students' academic records and request changes if they believe the document is inaccurate or misleading. If a school does not grant the change, parents can ask for a formal hearing. If the review panel also denies the request, parents have the right to include a written statement about the contested information in their child's record. Schools are prohibited from sharing information from a student's education record except in a few cases, such as when students are transferred to another academic institution or to the juvenile justice system.

The Equal Access Act, ratified in 1984, provides that secondary schools that utilize federal funds must grant their students equal access to extracurricular activities. If the institution already possesses a "limited open forum," or in other words already permits one student-led club to meet outside class time, it must allow additional clubs to organize and be given equal access to meeting spaces and publications.

Court Rulings. A number of court cases have also shaped the ethical standards that are upheld in schools. For example, in the 1943 *West Virginia State Board of Ed. v. Barnette* trial, the Supreme Court ruled that students could not be forced to salute the American flag or recite the Pledge of Allegiance in school. The 1969 *Tinker v. Des Moines Independent Community School District* decision produced what is today known as the Tinker test, which courts apply in assessing whether academic disciplinary decisions violate students' First Amendment rights for free speech. In this particular case, students Mary Beth Tinker, John Tinker, and Christopher Eckhardt wore black armbands to school in protest of the Vietnam War and were subsequently suspended. In its majority opinion, the Court concluded, "It can hardly be argued that either teachers or students shed their constitutional rights to free speech or expression at the schoolhouse gate" (393 U.S. 503). Exceptions to this policy include speech that infringes on the rights of others; interferes with school operations; is obscene, lewd, or offensive; and is school-sponsored. Students' freedom of expression did not fare so well in the 1988 ruling in *Hazelwood School District et al. v. Kuhlmeier et al.*, when it was ruled for the first time that school administration can censor content in school-sponsored student publications.

Secondary and postsecondary schools have an ethical duty to protect marginalized populations, including women, LGBTQ students, disabled students, and students of color. Yet many lapses exist as revealed by an epidemic of sexual harassment of women in secondary and postsecondary institutions.

Core Problem 2: Engaging in Advocacy to Promote Quality Education—With Some Red Flag Alerts

Although more than six decades have passed since the Supreme Court's *Brown v. Board* decision declared that separate schools are inherently unequal, children across the country still receive very separate and very unequal educations. An achievement gap exists as divergent test scores and college attendance rates between black and white Americans suggest (Fox, Connolly, & Snyder, 2005).

- **Red Flag Alert 12.12.** A student who is failing in math lives in a one-parent household in a distressed neighborhood.
- **Red Flag Alert 12.13.** A student is habitually absent from science class.

POLICY ADVOCACY LEARNING CHALLENGE 12.1

CONNECTING ADVOCACY AT MICRO, MEZZO, AND MACRO LEVELS TO PROTECT STUDENTS' ETHICAL RIGHTS

Alison came from a loving household with parents who supported her. As a result of the trust she developed with them, she came out as a lesbian and introduced them to her girlfriend Caroline. Both seniors in high school, the two girls were excited to attend the upcoming homecoming ball, and they naturally wanted to go together. Although Alison's parents were proud that their daughter had found happiness and was brave enough to share it with the world, they worried about the fallout she might experience in her school. Despite her parents' warnings, Alison decided to list Caroline as her date when she RSVP'd for the event.

A week later, the student homecoming coordinator, also a fellow classmate of Alison's and Caroline's, approached the girls in the cafeteria. She proclaimed that she did not agree with their "chosen lifestyle" and that students were allowed only to bring dates of the other sex. The pair refused to comply and said they would not change their plans to arrive at the dance together. The next day, Alison and Caroline wore matching gay pride T-shirts to school. After the homecoming coordinator noticed this act of defiance, she informed the school's principal about the girls' intentions. He immediately called Alison and Caroline into his office. Not only were they prohibited from going together to homecoming, but they were also suspended from school for violating the dress code. During periods of suspension, students are not allowed to participate in any school-sponsored events or activities. As a result, Alison and Caroline missed their homecoming ball.

Learning Exercise

1. How could a case advocate have assisted with Alison and Caroline's situation?
2. What are some of the challenges an advocate might encounter in addressing this controversial issue?
3. What were some of the federal policies that school administration neglected to follow?
4. How can a school social worker initiate policy advocacy at the district level or beyond?

- **Red Flag Alert 12.14.** A school uses decades-old textbooks in poor condition.
- **Red Flag Alert 12.15.** A school is not equipped with more advanced technology, such as computer stations.
- **Red Flag Alert 12.16.** A student is reading at a grade level well below that of others of her age.
- **Red Flag Alert 12.17.** A student is prevented from enrolling because he or she is without documentation.
- **Red Flag Alert 12.18.** A high school dropout wishes to enroll back in school to earn his diploma.

- **Red Flag Alert 12.19.** A parent has expressed a desire for his or her child to attend a local charter school but is not assisted in exploring this option by local school officials.
- **Red Flag Alert 12.20.** A teacher frequently arrives late to class and is known to call in sick for work.
- **Red Flag Alert 12.21.** Secondary and postsecondary educational institutions fail to provide sufficient support staff for students who fail to pass courses or have attendance problems.

Background

Although government must ensure that children have access to an education, it does not enforce access to a quality education. Neighborhood schools may be equipped with inept personnel, outdated learning materials, and inadequate funds. Schools may have cultures that encourage bullying. Young students face risks on their journeys to adulthood, including but not limited to poverty, racial discrimination and injustice, limited opportunities for education and employment, child abuse, parental conflict, and biomedical problems (Kirby & Fraser, 2004). Furthermore, affluent persons are more likely to exercise school choice as compared with low-income students and students from marginalized groups (Orfield & Eaton, 1996; Smrekar & Goldring, 1999). Secondary and postsecondary schools fail to give sufficient assistance to poor-performing students.

Resources for Advocates

With the adoption of the Fourteenth Amendment to the U.S. Constitution in 1868, all persons in every state were granted equal protection under the law. After the *Brown v. Board* decision, Title VI of the Civil Rights Act of 1964 made its way through Congress and forbade discrimination based on race, color, creed, or national origin in federally sponsored programs. These pieces of legislation paved the way for the Equal Educational Opportunities Act of 2010 that declares that all public school attendees are entitled to "equal educational opportunity" without regard to race, color, sex, or national origin. It directed local educational agencies to restructure dual school systems, which had historically separated students. Furthermore, it acknowledged that "excessive transportation" disrupted the educational process, noting that the assignment of a student to a school "nearest his place of residence which provides the appropriate grade level and type of education for such student is not a denial of equal educational opportunity" unless the intention is to segregate students deliberately (20 U.S.C. Sec. 1705). This legislation admits that government cannot help the fact that students are bound to their local schools, even if those academic institutions are not of the highest quality, as long as there is no proof of racial motivation. Yet the truth remains that these schools are often subpar even if they are easy to access—and considerable research suggests that educational outcomes of minority students improve when they attend integrated schools (McDaniels, 2017). Considerable research suggests, as well, that students of color perform better if they have teachers and counselors of color (McDaniels, 2017).

Undocumented youth were historically precluded from attending schools—a policy overridden in 1982 when the U.S. Supreme Court ruled in *Plyler v. Doe* that a Texas law that denied enrollment to illegal immigrant children was unconstitutional. The Court's ruling maintained that access to a quality education for undocumented youth would likely lead to "the creation and perpetuation of a subclass of illiterates within our boundaries, surely adding to the problems and costs of unemployment, welfare, and crime." This policy was restricted to primary and secondary schools, however, but many states also did not allow undocumented students to enter state colleges and universities or to receive in-state tuition and financial aid. President Obama was unable to enact the Development, Relief, and Education for Alien Minors (DREAM) Act to provide federal subsidies and a path to citizenship for undocumented youth brought to the United States by undocumented parents when they were children. Unable to enact the DREAM Act due to Republican opposition in the Senate in 2011, he signed an executive order in 2012 to grant them work permits for three years and to allow them to stay in the nation if they registered with the federal government and provided fingerprints and addresses. Many of them, now under the Deferred Action for Childhood Arrivals (DACA), accepted this offer and many continued their education in American high schools and colleges. Unfortunately, the Congress during the presidency of Donald Trump failed to guarantee a long-term path to citizenship in 2018, so DACA youth remained in limbo.

Considerable data demonstrates that poor-performing students in secondary education benefit from support services from interdisciplinary teams that include social workers, psychologists, and educators. Balfanz and Fox (2011) and DePaoli, Balfanz, Atwell, and Bridgeland (2018) demonstrated, for example, that timely assistance to sixth-grade students with poor attendance, behavioral problems, and failing grades in English and/or math greatly improved their odds of graduating from high school. Greater support is needed, as well, for students in junior colleges, where as many as 80% of students fail to graduate for economic and psychological reasons as well as lack of needed skills in English and math.

Core Problem 3: Engaging in Advocacy to Promote Culturally Competent Education—With Some Red Flag Alerts

- **Red Flag Alert 12.22.** Faculty are mostly white in secondary and postsecondary schools.
- **Red Flag Alert 12.23.** Parents with limited English skills attend their child's disciplinary hearing without the presence of a translator.
- **Red Flag Alert 12.24.** Schools send home announcements that are only written in English.
- **Red Flag Alert 12.25.** A Hispanic child acts as translator for his or her parents during parent-teacher conferences.
- **Red Flag Alert 12.26.** Asian American children or girls are hesitant to participate in class.
- **Red Flag Alert 12.27.** Black students are frequently referred to the principal's office.
- **Red Flag Alert 12.28.** An immigrant student consistently shows up to school without completed homework.

- **Red Flag Alert 12.29.** A bilingual student is recommended for a general track program.
- **Red Flag Alert 12.30.** School curriculum lacks sufficient content about marginalized populations.

POLICY ADVOCACY LEARNING CHALLENGE 12.2

CONNECTING MICRO, MEZZO, AND MACRO POLICY ADVOCACY TO PROMOTE QUALITY EDUCATION

Ms. Lopez is a school social worker working with a 10th-grade male student named John. His mother came to school asking for help transferring her son to a new charter school in the neighborhood. Juan's mother expressed that she is concerned that Juan is falling through the cracks and not getting the attention he needs in school to perform at grade level. John is very shy and tends to disappear in a big class. John has had minor behavior problems, and he is a C average student. The mother had expressed that when they lived in another district with a smaller class size, his grades were better. The mother is scared that John is starting to regress and isolate. He is now at risk of failing.

The mother attempted to enroll her son at the new charter school and was told that her son does not meet the qualification to be accepted. The mother was confused because it was said that there are still openings at the school and they are accepting new students. The school social worker agreed to contact the charter school and follow up. Ms. Lopez received the same information. John does not meet the grade requirement, and his behavior and attendance history do not make him a good candidate. The school social worker tried to advocate for the student but to no avail. The school social worker was concerned about this, so she did a little bit of research. The charter school has the right to set the requirements and criteria for acceptance in her school district and state. The California Department of Education (2011) states that in general, charter schools are not mandated to follow most district, state, or federal guidelines. Further research showed that this particular charter school has a reputation for targeting and accepting students who are high performers and college bound. In fact, there was a recent transfer of the school's high-performing students to the charter school when it first opened. Ms. Lopez informed the mother of this. The mother was outraged, but she had no choice but to keep John at the same school. The mother has limited income and cannot afford to send John to a school outside of the neighborhood.

Learning Exercise

1. What could the school do to prevent losing their high-performing students to a charter school?
2. How could John's mother advocate for him? How can Ms. Lopez and the school help advocate for John?
3. What impact would the student or family's ethnicity have on the advocacy efforts?
4. Is it legal for the charter school to exclude certain students from their school?
5. What can local schools and districts do to prevent losing their high-performing students to charter schools from a policy perspective? How can they increase inclusion?
6. What is the responsibility of the state or federal government in this process? How can it be more equitable?

Background

Not every child learns the same way, and when one throws race and ethnicity into the mix, the notion of what effective and appropriate teaching methods are becomes even more complicated. Take Asian Americans, for example. As a group, they are seen as a model minority due to their academic achievement and scores on standardized testing. Yet, a closer look into the effects of their cultural experience reveals a much grimmer depiction of their educational paths.

Immigration stress for children manifests as behavioral disorders in the classroom (Aronowitz, 1984). Immigrant students indicate school and communication as two of the biggest areas of cultural adjustment difficulties (Yeh & Inose, 2003). Parental involvement, or rather the lack thereof, is a reality for many Asian American students. In one study, Chinese American children commonly felt like they lack the necessary support from parents, who routinely work long shifts (Chan & Leong, 1994). Some immigrant parents do not give their children adequate attention, especially during early resettlement, because they themselves are dealing with demands that necessitate time and emotional energy (Moon & Lee, 2009). The long work hours of working-class parents, who also have limited educational attainment, prevent them from assisting children with homework. On the other hand, for Asian American parents who exhibit a desire to engage in their child's education, their inability to navigate American educational institutions greatly restricts their level of participation (Ngo, 2006). Because of their parents' lack of knowledge, students commonly manage the daunting college admissions process on their own or with the help of friends and older siblings (Ngo, 2006).

Persons from many other ethnic and racial groups also experience problems in schools where teachers and other schools are insensitive to their cultures, including Latino, African American, Native American, and LGBTQ students. School personnel can also be inattentive to the needs of girls by allowing boys to dominate classroom discussions. Considerable research suggests that students from marginalized populations benefit from having teachers and mentors from marginalized populations (Mazama & Lundy, 2012; Wells, Fox, & Cordova-Cobo, 2016).

Resources for Advocates

Teacher Recruitment. Many college-educated African Americans worked as teachers before the passage of civil rights legislation (Cole, 1986). However, after the Civil Rights Act of 1964, 38,000 members of the black teaching force of 82,000 lost their jobs. Schools hired white teachers to handle the influx of desegregated school populations, whereas blacks were either demoted or dismissed. According to the Urban Institute, reasons for the underrepresentation of minority teachers today include inadequate academic preparation, the attraction of other careers, unsupportive working conditions, lack of cultural and support groups, increased standards and competency testing, and financial considerations, among others. The absence of teachers of color may have a negative effect on minority students, as research has suggested that having a teacher of the same race may result in positive gains in student performance (Dee, 2016).

Bilingual Education. Ohio was the first state to pass a law permitting bilingual education, which allowed instruction in German upon parental request in 1868. Louisiana soon followed, giving the green light to French-English instruction, and in the following decade New Mexico said yes to Spanish-English instruction. Nevertheless, a number of laws passed in American history took on a more nationalistic zeal. Some laws required Native American children to be taught in English rather than their own languages—a policy made more onerous when they were taken off reservations and received their education in boarding schools. The influx of immigrants at the turn of the 19th century alarmed the federal government to issue the 1906 Nationality Act, mandating that all immigrants speak English as a requirement for citizenship. Thirty-four states had English-only instruction laws in place by the 1920s.

The federal government has gradually recognized the needs of immigrant students. The 1965 ESEA set aside federal funds for bilingual education, paving the way for the Bilingual Education Act of 1968. The act specifically aimed to promote experiential educational programs for non-English native speakers entering into the American school system. The grant-in-aid program provided school districts with funds to foster native-language instructional models and, later, English-only programs. It must be recognized, however, that as many as 5 million students have limited English skills that impact their learning (Payan & Nettles, n.d.).

Bilingual education advocates suffered a setback in 1998, when California passed Proposition 227, and again in 2000, when Arizona enacted Proposition 203. As part of the English for the Children campaign, the initiatives sought to eliminate bilingual education in the United States. The California proposition is written in language that allows for flexibility, and the state has affirmed parents' rights to choose bilingual education for their children rather than vesting it in school boards' and administrators' decision-making authority. The Arizona proposition contains much stricter language, by contrast, that essentially mandates English-only instruction by severely constraining parental choice (Wright, 2005).

References to bilingualism as an education goal were eliminated in the NCLB program, and the Bilingual Education Act was renamed the English Language Acquisition, Language Enhancement, and Academic Achievement Act. States are required to demonstrate marked progress in the academic achievement of limited English proficiency (LEP) students. The legislation also set up the Office of English Language Acquisition, charged with identifying the needs of English language learners (ELLs), students who study English as a second language, and assisting school districts with issues such as academic standards, accountability, professional training, and parental involvement. Specifically, the office distributes funding for language programs designed for LEP students. The National Clearinghouse for English Language Acquisition collects and analyzes data about LEP instructional programs.

Translation Services. Per the requirements of Title IV of the 1964 Civil Rights Act and the Equal Educational Opportunity Act of 1974, school districts must adequately notify LEP parents about school-related activities that it reports to other parents. In other words, administrations must provide written or oral translation of school announcements and notices. If the U.S. Department of Justice discovers noncompliance, it may call on the offending district to implement specific practices such as the following:

- Securing adequate interpreter and translation resources within the district and outside of it, when needed, to meet the language needs of LEP parents.
- Ensuring district staff has access to these resources in a timely manner.
- Developing district procedures for the timely and competent provision of translation and interpreter services, and training for district staff regarding these procedures.
- Providing notice to parents about the availability of translation and interpreter services and how to request them.
- Requiring translations of documents containing essential information.
- Prior to conducting an individualized education program meeting, notifying parents of the availability of interpreters, and providing interpreters on request with reasonable notice.
- Prohibiting the use of students to provide translation or interpretation services except in the event of an emergency.

President Bill Clinton issued an executive order in 2000 for federal agencies to "develop and implement a system by which limited-English-proficient (LEP) persons can meaningfully access the agency's services" (Executive Order No. 13166). Following the announcement, many state departments of education issued guidelines that directed public schools in how to provide adequate translation and interpretation services. In California, for example, schools are instructed not to use children as intermediaries in parent-teacher communications. The California Department of Education declared in 2012 that "some discussions with families involve discipline, medical or mental issues, or academic performance and may make the child uncomfortable and produce biased results" (California Department of Education, 2012). Critics contend that individuals and groups who work on language-access issues say districts are still not adhering to protocols (Zehr, 2011).

Core Problem 4: Engaging in Advocacy to Prevent Students' Personal and Educational Problems—With Some Red Flag Alerts

- **Red Flag Alert 12.31.** A parent asks for support for her son who she thinks is being bullied. She was asked to report it every time it happens, but no services were provided.
- **Red Flag Alert 12.32.** A parent is not responding to the school's effort to link a student to services in school.
- **Red Flag Alert 12.33.** A parent refuses to sign the consent form for an anger management group that school officials direct his son to attend.

POLICY ADVOCACY LEARNING CHALLENGE 12.3

CONNECTING MICRO, MEZZO, AND MACRO POLICY ADVOCACY TO PROVIDE CULTURALLY SENSITIVE EDUCATION AND SERVICES

Linda was a social worker assigned to a public high school in the heart of Chinatown. This was no ordinary academic institution; rather, it was designed as a transfer program for non-native English speakers who either were recent immigrants or had failed out of traditional schools. Because of its location, the majority of students were of Chinese descent. Many felt a special connection to Linda as she was the only social worker who shared their nationality. When parents were not showing up to parent-teacher conferences, Linda prodded the school to schedule these meetings later in the evenings to accommodate busy schedules of mostly restaurant workers. When a teacher reported that a student had disrespected him in class, Linda explained that the student's evasion of eye contact was actually a sign of respect in China. When a student brought up her interest in nursing, Linda linked her with a volunteer candy-striping program at a nearby hospital and paired her with a supervising mentor at the facility.

Lately, Linda has been feeling overworked and underappreciated. She is frequently called upon to serve as a translator during meetings, although her fluency is in Mandarin, not Cantonese and other dialects. Her office is always packed with students who need her help with school-related issues, such as scheduling, classroom conflict, and college preparation. Furthermore, students have turned to her for counsel in out-of-school matters, such as disagreements with parents, relationship advice, and career aspirations. When Linda told her supervisor about her overwhelming caseload, her boss responded sympathetically but reminded her that everyone is in the same boat. Linda loves her students, but she is not sure whether she is physically and mentally able to keep up with all her duties.

Learning Exercise

1. How is Linda an example of an effective case advocate? Is there anything she should be doing differently to manage her caseload?
2. What are some ways that Linda can engage in policy advocacy to improve the situation at her school?

- **Red Flag Alert 12.34.** A teacher referred a student at risk of failing the sixth grade but was told repeatedly "to give it time." The school year ended, and the teacher was told that middle school would deal with it.
- **Red Flag Alert 12.35.** Two counselors at a middle school tried to develop an at-risk student group but had problems getting it going because they kept losing counseling staff. Eventually, *they* had to take on the caseloads of the departed staff.
- **Red Flag Alert 12.36.** A mother of a third-grade student requested her son to participate in an after-school reading enhancement program. She was given the application, and upon submission she was told that there was a four-month waiting period. The mother complained to the school, but the school was unable to do anything due to limited funding.

- **Red Flag Alert 12.37.** An elementary principal proposed a change in instruction as per the district. She decided to add a social skills curriculum that will be done in the morning for 45 minutes twice a week. The teachers were trained and given the materials. Upon unannounced class visits, the principal found out that half of the teachers were not following this directive.
- **Red Flag Alert 12.38.** Several parents from a local middle school went to the school with their concern regarding an increase in crime in the neighborhood. Parents expressed their concerns regarding the safety of the children as they walk to and from school.

Background

An estimated one-third of all school-age children, or around 13 million students, are bullied each year (The White House Office of the Press Secretary, 2011). According to a national survey in 2010, daily marijuana use is up among eighth, 10th, and 12th graders and, by some measures, has even surpassed cigarette smoking (National Institute on Drug Abuse, 2012). Suicide remains the third leading cause of death among 15- to 24-year-olds. Even more startling, the Centers for Disease Control and Prevention reported in 2007 that the rate in American teenagers jumped by 8% in 2003 to 2004, the largest increase observed in 15 years. For youths between 15 and 24, suicide is more likely than any other reason to be the cause of death. Risks factors include biomedical predisposition, depression, substance abuse, sexual orientation-related factors, poor coping and interpersonal skills, stressful life events, and suicide in family history (Peebles-Wilkins, 2006). These social problems exist in schools across all races, ethnic groups, and social classes. They have not significantly declined in number. School shootings have escalated, with one major incident every week in 2018, including at Marjorie Stoneman Douglas High School in Parkland, Florida.

Schools across the country frequently implement prevention programs designed to mitigate the many dangers and risks that youngsters face in the classroom, on the playground, and beyond. Services can include mental health hotlines, drug and substance abuse awareness programs, anti-bullying initiatives, remedial tutoring and sex education, and strategies to improve school culture, among other topics (Kamenetz, 2018). Schools can also adopt in-service training programs on behavior management, conflict resolution, and anger control. According to Whitted and Dupper (2005), schools need to develop cultural norms that prohibit bullying and that promote respectful and nonviolent behavior.

Professor Robert Balfanz at Johns Hopkins University discovered three variables that predict with accuracy which students in the sixth grade will not graduate from high school in schools mostly populated by low-income students. He called them the A, B, and C problems with A for attendance, B for behavior, and C for course performance. He discovered that students who missed more than 10% of classes (Problem A), had two more behavioral problems such as disrupting classes or attacking other students (Problem B), and received failing grades in math or English classes were far less likely to graduate from high school than their peers. He discovered that early intervention (a core strategy in prevention) markedly increased high school graduation rates of students with these problems in the sixth grade. The early intervention included a multidisciplinary team that worked intensively with these students until their problems abated—interventions that might last for relatively brief or long periods as needed (Balfanz & Fox, 2011). DePaoli et al. (2018) contend that this and other innovations have already reduced the number of "dropout factories," a term he uses to describe high schools where as many as 50% of students failed to graduate. Other researchers contend that interventions should start as early as the third grade. Social workers are uniquely equipped to participate in prevention programs, but the profession needs to market itself so that more social workers are hired by secondary schools.

Resources for Advocates

Initiatives/Recommendations. Many early intervention and prevention initiatives are being developed at local, state, and federal levels. All aim to address disparity at an early stage. Education advocates should keep abreast of these initiatives and utilize them to decrease and prevent gaps in services.

Under the NCLB initiative, school safety was identified as a top priority in terms of educational goals. As part of the Unsafe School Choice option, each state must establish a definition of what constitutes a "persistently dangerous" school, and students who attend that school can transfer to a safer school in the same district.

In March 2011, President Obama and First Lady Michelle held a White House Conference on Bullying Prevention to raise awareness of bullying in schools and recommend ways to make schoolchildren feel safer in their learning environments. To combat the harmful practice, six federal agencies, consisting of the Departments of Education, Health and Human Services, Justice, Defense, Agriculture, and the Interior, collaborated to form the Federal Partners in Bullying Prevention Steering Committee, which launched StopBullying.gov. Many states have their own anti-bullying laws, such as New Jersey.

Schools should adopt evidence-based practices, such as the Screening for Mental Health program called Signs of Suicide. This particular program has demonstrated effectiveness in reducing suicidal behavior and improving student outcomes, attitudes, and awareness (Aseltine & DeMartino, 2004).

It is important to place prevention strategies in schools in their broader community and national context. With respect to gun violence, for example, considerable research suggests that the sheer number of guns in a nation—not to mention military-style weapons—correlates with the number of mass killings in schools, businesses, and communities (CNN, 2018). As the students from Parkland suggest, gun control in the nation and in specific jurisdictions is necessary to make deep cuts in school killings. Educators can establish programs to improve school performance, but their efforts need to be combined with reductions in poverty among student populations and their families. Students who are chronically hungry or homeless will often have class performance and attendance issues no matter how well prevention programs are designed and implemented within schools.

POLICY ADVOCACY LEARNING CHALLENGE 12.4

CONNECTING MICRO, MEZZO, AND MACRO POLICY ADVOCACY TO PROMOTE PREVENTION

Averting Truancy

A school social worker (Mr. Jones) is working with a ninth-grade student (David) who is truant and at risk of failing. He's had multiple absences since the seventh grade but has not been flagged until now. He is failing multiple classes and now on the list to be processed for the Student Attendance Review Board (SARB). As defined by the California Board of Education (2012), SARB comprises members from different organizations providing services to families and youths. They are focused on helping truant students, determining the cause of the truancy and addressing the needs through available school and community resources.

A pre-SARB meeting was held to discuss the student, his history, and possible supports. Mr. Jones attended the meeting to support David and his father. The meeting revealed that his problems started in the fifth grade, when his mother died. His grades and attendance started deteriorating. He started complaining of being sick and asked to stay home. The paternal grandmother watched David and his two younger siblings while the father worked. The father had been working two jobs to make ends meet. He had attended a few parent-teacher meetings but had missed most of them. The grandmother had not been able to attend as well because of her age and some mobility issues. David had to stay home from school to help when his grandmother was sick. Both the father and the grandmother had shared some of the information with a teacher in the past. The grandmother's health had deteriorated the past two years, and David had to stay home more to help out. The family has very limited resources and was trying to cope with the situation. The SARB team was not aware of this information. Mr. Jones had been asked to meet with the family for a thorough assessment and to coordinate with the team for possible resources.

Social workers need to engage in macro policy advocacy to increase their inclusion on SARB teams because of their expertise in behavioral problems, family dynamics, and community resources. They need to increase funding for these teams in schools with high rates of truancy. Recall our earlier discussion in this chapter about early prevention strategies trial tested by Balfanz and Fox (2011) and DePaoli et al. (2018) that address not only attendance problems but also behavioral and course competency problems.

Learning Exercise

1. Are teachers trained in how to respond to community and family issues students bring into the classroom? What support could the school offer the family? What can the administration do to respond early to such issues?
2. Who could the school partner with to support David's family?
3. What impact does the student/family's ethnicity, race, socioeconomic status, and so on, have in identifying a student in need of services? How about access to resources?
4. What policy changes can the local school or district engage in for early identification of student support services needs?
5. Discuss some national policy reforms with respect to poverty, federal gun policies, homelessness, and safety net programs like the Supplemental Nutrition Assistance Program (SNAP) that could improve the educational performance of children and youth.

Core Problem 5: Engaging in Advocacy to Promote Affordable and Accessible Education—With Some Red Flag Alerts

- Red Flag Alert 12.39. A parent asked for a special education assessment but was told it is too late in the year to do so.
- Red Flag Alert 12.40. A parent was not able to attend an Individualized Education Program (IEP) meeting due to work constraints.
- Red Flag Alert 12.41. A school's IEP team decided on a program for the student without parent participation.
- Red Flag Alert 12.42. Letters and pamphlets were sent to students' homes in English only.
- Red Flag Alert 12.43. A foster child tries to enroll in a school without or with fragmented school records.
- Red Flag Alert 12.44. A foster child in the process of moving for the third time in a year wants to stay in the same school.
- Red Flag Alert 12.45. A student is suspended for the third time, resulting in a possible transfer or expulsion without support linkages or due process.
- Red Flag Alert 12.46. A parent does not respond to any school communication or attempts to meet.
- Red Flag Alert 12.47. Some teachers do not attend staff meetings or understand programs and policy changes at their school.
- Red Flag Alert 12.48. Some schools do not implement district mandates.

Background

The Condition of Education (2010) reports in 2007 and 2008 indicated that 16,122 schools were ranked as high-poverty schools. The majority of the students in these schools qualify for free or reduced-price meals. The report also states that there has been a 17% increase in high-poverty schools, identifying 20% of elementary schools and 9% of secondary schools to be high poverty (Aud et al., 2010). Most of these schools are located in urban settings and have, on average, 68% graduation rates compared with 91% graduation rates for low-poverty schools. They are Title I ESEA schools that need more services for their at-risk students: 12% to 15% of their students need IEPs (Aud et al., 2010). These high-poverty schools are faced with the heavy burden of responding to multiple needs. They are disadvantaged in accessing available resources to respond to those needs. They are faced with the tough choice of picking only those programs that they can finance to help decrease these achievement gaps, forcing them to decide which support services will best reduce the gaps and which ones to terminate. Or they may be able to provide only truncated services that end before students make major gains in achievement and well-being.

Families with limited resources also encounter multiple challenges when trying to engage and support the education of their children. They face financial constraints hindering them from taking advantage of some of the options and programs available. Many cannot use school voucher and charter school programs due to transportation and childcare issues. Many are intimidated from going to schools outside their communities due to perceived lifestyle issues. Extracurricular activities and school supplies are not included in voucher programs, placing additional financial pressures on these low-income families. Data from the Children's Defense Fund 2004 survey shows the following key facts (Allen-Meares, 2007, p. 223):

- Two in five eligible children do not participate in Head Start.
- One in three is behind a year or more in school.
- One in three is born to unmarried parents.
- One in four lives with only one parent.
- One in five is poor now.
- One in eight lives in a family receiving food stamps.
- One in eight has no health insurance.
- One in twelve has a disability.
- One in fourteen lives at less than half the poverty level.
- One in five is born to a mother who did not graduate from high school.
- One in seven never graduates from high school.

Parents' lack of information or knowledge about school programs, policies, and processes and procedures is a significant impediment in receiving equal educational opportunities. Parents are often unaware of what support programs to seek, the timeline involved, and to what services students are entitled, including special education, truancy prevention, enrollment, and school transfer programs. Some schools may not advertise programs they cannot afford to provide. There may also be language barriers that prevent parents from knowing about specific services. Many parents of low-income children are too intimidated to inquire about specific options.

Resources for Advocates

There are multiple statutes and constitutional rights, state and federal regulations, court rulings, and private and public initiatives supporting at-risk students who are from low-income families.

Head Start Program (1965). Head Start is a comprehensive child development program designed to support the cognitive and socio-emotional development of low-income children. Initially established in 1965, it has been reauthorized and expanded, servicing close to 30 million children (U.S. Department of Health & Human Services, Office of Head

Start, n.d.). It seeks to prepare at-risk children to enter school. Allen-Meares points out that two in five Head Start-qualified children do not participate (2007, p. 223). President Obama proposed a Preschool for All Initiative in 2013 through a cost-sharing agreement with states, but Congress has failed to take necessary action.

McKinney-Vento Act (2001). It requires all states to address the educational needs of homeless children by creating equal access and opportunities (Allen-Meares, 2007).

Title I, Part A of the Elementary and Secondary Education Act (ESEA 1965). This provides financial assistance to states and school districts to meet the needs of educationally at-risk students. It aims to provide extra instructional services and support services to students at risk of failing or already failing.

California Assembly Bill 490 (2003). This is a California law establishing the rights and regulations of foster care children regarding enrollment, school transfer, access to records, and many more. Its goal is to provide educational access, opportunities, and stability for foster care children (Youth Law Center/Children's Law Center of Los Angeles, 2003).

Charter School Laws/Voucher Program. Charter schools are publicly funded institutions but not fully regulated by state or local laws. According to Sipple (2007), about 40 states had passed a law allowing the formation of charter schools by 2005. On the other hand, the voucher program provides a fixed amount that could be used to pay the tuition to the parent's school of choice. This has been found constitutional by the U.S. Supreme Court in 2002 in *Zelman v. Simmons-Harris* (Sipple, 2007), opening the doors for states and local districts to do so if they choose. Betsy DeVos, the secretary of the U.S. Department of Education in 2018 in the Trump administration, has drawn praise from advocates of charter schools but criticism from their foes, who question whether they deplete funds for public schools.

Core Problem 6: Engaging in Advocacy to Promote Care for Students' Mental Distress—With Some Red Flag Alerts

- **Red Flag Alert 12.49.** A student needs mental health services, but her parent refuses to sign the consent for services.
- **Red Flag Alert 12.50.** A student is disengaged and isolated but does not exhibit behavior problems.
- **Red Flag Alert 12.51.** A group of middle school students are rumored to engage in cutting behavior.
- **Red Flag Alert 12.52.** An elementary student cries every day in school.
- **Red Flag Alert 12.53.** A teacher complains that a ninth grader who is being pulled out from her class once a week for counseling will fall behind other students.

POLICY ADVOCACY LEARNING CHALLENGE 12.5

CONNECTING MICRO, MEZZO, AND MACRO POLICY ADVOCACY TO PROMOTE AFFORDABLE AND ACCESSIBLE EDUCATION

Foster Children and Schools

Joe is a 14-year-old foster child who has been moved into foster homes four times in the last year. The last two were emergency moves, and the new foster parent does not have his school records. The foster mother tried to enroll him in a Southern California school but was told he cannot be enrolled because the school records are missing. She was told to come back when she has all his records. The foster mom went home and tried to ask Joe about his school history. Joe remembers the last two but was unclear about the other ones. Joe lived in different cities and even went to a school in another state at one point. The foster mother finally contacted children services and asked for his records. The social worker told her that she was still gathering his records but he should be allowed to enroll because of the AB 490 law. The foster mother did not know what this was, and so the social worker met both at the school to assist with enrollment. The school registrar was not aware of this law as well and had to check with the principal and the district office to proceed. Joe was finally enrolled after a week even though the records were not available.

Learning Exercise

1. Should Joe have been enrolled in school the first time he went?
2. What should the administration do to improve knowledge regarding existing and new policies?
3. What can child protective services agencies and schools do to respond better to foster care children's needs?
4. What responsibility did the foster care mother have in advocating for Joe?
5. What is the responsibility of the school staff regarding knowing and understanding school policy? How are they made accountable?

- **Red Flag Alert 12.54.** A student who often exhibits serious behavior issues is at risk of expulsion—or is handed over to criminal justice personnel only to join the school-to-prison pipeline.
- **Red Flag Alert 12.55.** A teacher walks away despite hearing a student inform another student that he is suicidal.
- **Red Flag Alert 12.56.** Students in junior colleges, colleges, and graduate schools who are depressed or anxious or have other mental problems are often not given mental health and other supports.
- **Red Flag Alert 12.57.** A school just lost its assigned designated instruction service (DIS) counselor and has difficulty identifying a new one. The school has two students with IEPs requiring weekly behavior support and DIS counseling.

- **Red Flag Alert 12.58.** A parent asks the school to let a community social worker/mental health worker access the student at school with parental consent. The school has no clear policy regarding this.
- **Red Flag Alert 12.59.** The district leaves it up to the school to decide what support services to purchase each year.

Background

"Education Takes a Beating Nationwide: More Layoffs, Bigger Classes, Fewer Programs" was the title of an article in the *Los Angeles Times* on July 31, 2011 (Ceasar & Watanabe, 2011). The article talks about recent budget cuts impacting schools nationwide. The cuts have led to reduced school days; non-instructional employee layoffs (school social workers, counselors, psychologist, and nurses); and the reduction or elimination of support services programs including but not limited to tutoring, summer school, after-school programs, prevention programs, and other health and human services programs (Ceasar & Watanabe, 2011). The impact of the budget cuts is felt all around, but the brunt of it was felt by those who are the most vulnerable populations in schools: lower-income and minority students. Funding augmentation from the federal government was achieved by the Obama administration during the Great Recession but disappeared when the Great Recession ended in 2009. It was also severely cut in the Budget Sequestration of 2013, which made across-the-board cuts in domestic spending. It was further reduced by budget cuts proposed by President Trump in 2017 and 2018.

Budget cuts in schools are a nationwide trend; the effects are not felt equally. One of the vulnerable populations that is directly affected are school-age children and adolescents needing mental health services. About 10 million or more youths are in need of mental services (Franklin, Harris, & Allen-Meares, 2007). Franklin, Harris, and Allen-Meares (2007) also highlight an existing discrepancy between the 70% of youths who expect to access mental health services in schools and the actual number who receive them. According to the U.S. Department of Health and Human Services, "a reality on school campuses is that less than one in five children in need of mental health services are receiving treatment" (Franklin, Harris, & Allen-Meares, 2007).

Unaddressed mental health problems of students negatively impact schools, parents, teachers, and of course, the students themselves. Schools have to deal with disruptions of their classes and poor student performance when they are under increasing pressure to produce academically proficient students. Parents often have to leave their jobs to care for children who do not receive care. Teachers are burdened with students with unaddressed mental health conditions. They are being asked to devote extra time for these students and at the same time teach their classes.

The mental health needs of children in specific schools vary significantly and are shaped by their demographics, culture, and available resources. Developing and delivering the appropriate mental health services within the school system is an enormous undertaking. Schools recognize the need to respond to this growing issue by having a wide variety of support services available to students. Current mental health services within schools take many forms from preventive to intensive wraparound care programs. The American Academy of Pediatrics (AAP, 2004) released a policy statement stating that one way to categorize a school's mental health program is by using the three-tiered model.

The three-tiered system categorizes the severity of mental health needs and the intensity of the services provided. The first tier is preventive, increasing resilience and school connectedness. It offers many programs targeting all types of students in an attempt to decrease risk factors and increase achievement (AAP, 2004). The second tier provides more specific services to students who have more identifiable mental health issues but are still able to function in school (AAP, 2004). Students in this tier might receive multiple services to help them reach their education goals. The third tier is the most intensive of the three. It is geared toward providing services to a small student population who are the most serious and at risk. These students will receive multiple intensive services from multiple professionals (AAP, 2004). This categorization helps create a framework for a continuum of care that is in line with the current model of response to intervention (RTI) used by educators for early identification and support of students with learning difficulties. RTI is a framework used in providing educational services to students (Fuchs & Fuchs, 2006).

Recent school shootings such as the mass killings in Parkland, Florida in the winter of 2018 have precipitated a national battle over the future of the nation's schools. Professors Ron Astor and Rami Benbenishty (2018) support three supportive ways to reduce school violence:

- They favor improving school climate by using ideas and measurements advocated by the National School Climate Center. It has developed an empirically validated tool (the CSCI school climate survey) that has been used by thousands of schools to measure the extent to which they have climates that are "safe, supportive, engaging, helpfully challenging and joyful K-12 schools." They have data that demonstrate that schools with climates with these characteristics have higher student achievement, lower dropout rates, decreased incidence of violence, and increased teacher retention.
- They favor engaging in social emotional learning (SEL) by complementing "the focus on academics with the development of social and emotional skills and competencies that are equally essential for students to thrive."
- They favor fostering a "compassionate community" that includes active participation in schools by community members, organizations, and agencies on many levels.

Other people favor views of law enforcement officials, as well as President Donald Trump, who want to "harden" schools by creating prison-like settings that include armed teachers and security guards. They are convinced that these strategies will not only kill hostile shooters but deter them from entering schools with these conditions. Many critics, including most teachers, favor the former approach. These incidents happen so rapidly that armed teachers and security guards will be unlikely to stop violence in time to save lives. They may even accidentally kill teachers and students in the tumult that occurs during mass shootings. Moreover, militarization of schools conflicts with efforts to develop a school climate that has been demonstrated to reduce violence.

Students from Parkland overwhelmingly reached yet another conclusion that is supported by empirical data. School violence is linked to the prevalence of military-style weapons in nations. If they can be easily obtained, some students will obtain them—and shooters will be able to discharge hundreds of bullets in five or 10 minutes. Students, teachers, and administrators, as well as millions of supportive citizens, need to keep unrelenting pressure on public officials to take military-style weapons off the streets.

The school-to-prison and school-to-deportation pipelines are manifestations of a hardening of schools that has led to tens of thousands of students, disproportionately students of color, ending up in juvenile justice facilities for minor misbehaviors and for unwarranted deportations. Despite years of criticism of these pipelines and some policy reforms, Black and Latino youth account for 92.4% of arrested youth, 88.6% of students receiving summonses, 88.8% of New York police juvenile reports, and 96% of students handcuffed during a mental health crisis (Center for Popular Democracy, 2017). Bill de Blasio, mayor of New York City, thought that policy reforms that he had initiated in 2010 had markedly reduced suspensions of students of color, only to realize in 2016 that they remained commonplace, partly due to opposition from the principal's union, the teachers' union, and charter school advocates. Police-student interventions remained most concentrated in low-income black populations by the 5,200 full-time police officers who work in schools for the New York Police Department. This is a lesson in not only racism but also the persistence of entrenched police-and-school practices (Joseph & CityLab, 2016).

Resources for Advocates

An AAP (2004) policy statement says that families do not address their mental health needs for multiple reasons. Some barriers include inadequate insurance coverage, lack of transportation, financial constraints, shortage in adequately trained mental health professionals, and the stigma attached to mental health problems (AAP, 2004). Schools encounter multiple barriers as they attempt to respond to the mental health needs. Limited resources force many schools to limit or cut plans for supportive services. Even when schools do set aside resources, administrators are unsure if those funds will be available the following year. Resource shortages make it even more difficult to determine priorities, such as deciding whether to focus on preventive or curative mental health programs and whether to contract these services to external agencies or provide them themselves.

Statutes and the Constitution. The Individuals with Disabilities and Education Improvement Act (IDEA) was reauthorized by Congress in 2004. Originally known as Public Law 92-142 (the Education for All Handicapped Children Act), it was passed by Congress in 1975 and has been amended and expanded several times to become what is now known as IDEA 2004. This legislation established a "bill of rights" for children with disabilities within the education system (Atkins-Burnett, 2007, p. 187). IDEA, as it stands today, authorizes funds and sets the guidelines as to how to meet the needs of children with disabilities within the education system. It specifies the different categories that qualify as disability, including autism, mental retardation, and emotional disturbance. It requires identification, assessment, and service provision to these children and adolescents. A more recent expansion to mental health service provision is the Affordable Care Act (ACA) of 2010. This legislation calls for an expansion for insurance coverage of mental health issues as well as an increase in the scope of coverage. Schools are in a position to explore ways to capitalize and promote improved mental health in light of this recent policy change, especially in a time where there is a strong push for the creation of school-based health services and wellness center programs within school campuses.

Policy Brief/Recommendations. The 2012 Condition of Education reports that about 6.5 million students received special education services from 2009 to 2010 (Aud et al., 2012). A 1999 report from the surgeon general states that one in five children and adolescents have emotional or behavioral problems sufficient to warrant a mental health diagnosis (U.S. Surgeon General, 1999). The Robert Wood Johnson Foundation (RWJF) also emphasizes the need for early childhood intervention programs to promote lifelong success and well-being (RWJF Commission to Build a Healthier America, 2014).

Not having clear policy mandates (except for IDEA) results in great variation in mental health service provision among school districts and states. Varying policy briefs, recommendations, and initiatives definitively state a need to address the mental health needs of school-age children. Education advocates will have to keep abreast of pertinent information and changes in available funding streams to engage in the discourse and access available resources as they advocate for funding and services. These circumstances and challenges make it even more important to know and be aware of the changing local, state, and federal policies and initiatives.

POLICY ADVOCACY LEARNING CHALLENGE 12.6

CONNECTING MICRO, MEZZO, AND MACRO POLICY ADVOCACY TO PROMOTE CARE FOR STUDENTS' MENTAL DISTRESS

Daryl is a ninth grader referred to the dean of discipline multiple times. He was sent to the dean due to classroom disruption, verbal altercation with the teacher and peers, refusing to do his work, not turning in his homework, not being in uniform, sleeping in class, shoving classroom furniture, and many more issues. The teacher stated he seems angry all the time and has a short fuse. The dean suspended him from class as well as from school several times, but the behavior continues. The teacher was frustrated and started complaining about the lack of support and how the other students are affected by his constant disruptions. The teacher talked to the dean almost every day to complain about the student and the situation. The teacher started talking to the school counselor about the student and mentioned wanting him out of her class and transferred to another teacher. The counselor did a little investigation and alerted the school social worker, Ms. Silva, about the student.

The school social worker asked the teacher to send the student to her the next time he is in class or before she sends him back to the dean. It took several days, but the school social worker finally met with the student. The student refused to talk to the school social worker and asked if he is in trouble. He was told he was not in trouble, and the school social worker continued to explain her role in the school and the services she offers. Daryl continued to be suspicious of the social worker and asked if he could go back to class. The school social worker agreed and asked if she could talk to him briefly once a week. She also gave him a business card and asked him to make sure to ask for her next time he gets in trouble. Daryl responded by saying, "Whatever, it makes no difference who I see. I am on my way out anyways."

The school social worker decided to find out more about Daryl. So she sought out the dean and the counselor. Ms. Silva found out that Daryl has had a history of barely passing his classes and has had several fails. He also had multiple absences. His records showed that his elementary school grades were significantly better and the decline started in sixth grade. When Ms. Silva asked if the counselor knew anything about his personal life, the counselor stated, "Not really. This is his first year at this school, and I do not really know him. I have a lot of students assigned to me, and I am still trying to get to know all of them, especially the ninth graders."

Ms. Silva then decided to meet with the dean to discuss Daryl. The dean recognized Ms. Silva but had not had many interactions with her. This was Ms. Silva's second year at the school, and she had tried to familiarize all the staff with her role and responsibilities, but it had been very challenging. Ms. Silva was assigned to coordinate and provide mental health services at the school. She saw students individually and in groups as well as linking them to additional services. She attended several staff meeting and sent memos explaining the services, but she had been inundated with a heavy caseload, so it had been difficult connecting with all of the school staff.

Ms. Silva proceeded to ask about Daryl after reminding the dean of her role. The dean explained the nature of her contacts with Daryl as well as the mother. She told Ms. Silva of the behavior problems and the multiple suspensions. She also stated that Daryl's records showed problems from the previous middle school and that she knew from the first referral that Daryl might become a frequent referral. The dean shared that it was difficult to schedule a meeting with the mother; she seemed to want to help Daryl but was frustrated with all the problems he had been causing. The dean never met the father—and Daryl refused to talk about him. All the dean knew was that Daryl had two young siblings and he spent a lot of time with them. She also shared that she was looking into an opportunity to transfer Daryl if the behavior continued.

Ms. Silva asked if the dean would be willing to try out a few interventions before going that route. She asked the dean to let her know the next time Daryl was referred to her for a behavior problem. Ms. Silva offered to collaborate in exploring ways to work with Daryl and turn it around. She also informed the dean that she had met Daryl and offered the services, but he was wary to accept them. The dean agreed to talk to the mother and get her consent for services.

Ms. Silva left the dean's office with a plan of saying hi to Daryl in the halls and meeting with him weekly to just check in. She also wanted to make sure the teacher was aware that she was available for support and to let her know when Daryl got referred to the dean again. Ms. Silva wanted to make sure to be present at the next meeting with the dean to help advocate for services rather than having Daryl transferred to another school.

Learning Exercise

1. How could Ms. Silva advocate for Daryl, or other students with discipline issues, or ninth graders transitioning from middle school?
2. Does the student/family's race, ethnicity, socioeconomic status, and so on impact the tolerance level and response to discipline issues?
3. How could the school social worker improve the identification of students with mental health needs and make sure they receive the services they need? What policy advocacy work is needed to make sure that this becomes an automatic process?
4. Are any federal, state, or local policies violated by any school staff?
5. What policy or program advocacy could Ms. Silva do to increase the mental health services at the school?

Core Problem 7: Engaging in Advocacy to Promote Education Linked to Students' Communities—With Some Red Flag Alerts

- Red Flag Alert 12.60. A social worker meets with a student at school but lacks a private space to meet.
- Red Flag Alert 12.61. A teacher inquires about the progress of a student he or she referred for counseling services.
- Red Flag Alert 12.62. A teacher refuses to release a student from class to meet with a service provider.
- Red Flag Alert 12.63. A community agency provider's requests to meet with a student during second period two times weekly is denied.
- Red Flag Alert 12.64. A community agency leaves health- and mental health-related messages about students with the front office worker for the school social worker or coordinator.
- Red Flag Alert 12.65. An administrator keeps tabling the school social worker or coordinator's presentation from the faculty meeting.
- Red Flag Alert 12.66. Teachers refer all problems in the classroom to the school social worker or coordinator.
- Red Flag Alert 12.67. An administrator fails to attend key collaborative meetings.
- Red Flag Alert 12.68. A student receives services from three different agencies at the school, presenting the need for collaboration and coordination.
- Red Flag Alert 12.69. A school has no after-school or summertime programs.
- Red Flag Alert 12.70. A school fails to include parents' views when making school policies.

Background

The U.S. Department of Justice (DOJ, 2009) reports that in 2008, about 2.11 million individuals under the age of 18 were arrested for various reasons. The same bulletin states, "Juveniles accounted for 16% of violent crime arrests and 26% of all property crime arrests in 2008" (DOJ, 2009, p. 1). The Anderson & Division of Vital Statistics reports (as cited in Furlong, Paige, & Osher, 2003), "In 2000, suicide and homicide were the third and fourth leading causes of death for children ages 10-14, and among individuals ages 15-19, homicide and suicide were the second and third leading causes of death." These data clearly demonstrate the issues students bring with them to school campuses and ultimately the classroom. Many schools realize they have limited resources to deal with all the problems with origins in students' families and communities, so they are increasingly partnering with community-based agencies and governmental agencies to create school-based service, wellness centers, community resource partnerships, and many more community-school partnerships.

Resources for Advocates

Although collaboration and community partnerships are advantageous to schools, they can also put a strain on an already thinly stretched system. They require countless hours of work, and for the most part these constitute additional duties in light of other duties (Dryfoos, 2005). Collaborators must be vested in the work, have specialized training, and fully understand all systems and stakeholders. It is often difficult to establish partnerships because of differing policies, operating procedures, ethical and legal mandates, and goals. Schools follow the regulations set forth by the Family Educational Rights and Privacy Act (FERPA) of 1974 when it comes to sharing school-related information, whereas community agencies follow the policies set forth by the Health Insurance Portability and Accountability Act (HIPAA) of 1996. All partners must learn to navigate these intricacies to protect students' rights as they engage in the collaboration to increase support services to students.

Federal agencies are also taking initiatives to link schools with community agencies. The Safe Schools Healthy Students Grants program (SSHS) is, for example, a collaboration between the federal Departments of Education, Justice, and Health and Human Services to promote and provide some resources for school-community preventive services, healthy child development, health services, safe school services, intensive mental health services, and use of evidence-based practices (Furlong et al., 2003). SSHS is only one of the many initiatives currently in existence. As another example, the Children's Health Program Act (CHIP) was created in 1997 and reauthorized in 2009 to help states insure low-income children who are ineligible for Medicaid but cannot afford private insurance. Schools collaborate and team with the federal CHIP to identify students who qualify for CHIP and educate parents about how to access it and use its benefits. Many schools are also partnering with community organizations for after-school programming.

POLICY ADVOCACY LEARNING CHALLENGE 12.7

CONNECTING MICRO, MEZZO, AND MACRO POLICY INTERVENTIONS TO LINK SCHOOLS TO COMMUNITIES

Opening Schools to Surrounding Communities

It is the first day of a new academic school year. Ms. Valdez, a school social worker, was assigned as the wellness facilitator in a high school. She has been identified as the new coordinator to implement a grant received by a local school cluster to simultaneously provide services to students and create a wellness center in a local high school. Administrators and a few community agencies that already provide services in the school planned the program. A new principal has now replaced the principal who helped write the grant.

The summer prior to the school year, a space at the high school was identified, and it was to be cleared so that it will be ready for the new program for the fall. Summer passed, and school started. The space was not ready, and it was full of supplies and furniture that the previous program left behind. Ms. Valdez was introduced to the new principal and the six assistant principals who jointly manage the school. Ms. Valdez was given the task to create a program following the mandates of the initiative but was not given any instructions on how to go about doing so. Ms. Valdez was not clear on where to start. She also

(Continued)

(Continued)

had to deal with suspicions from the school's staff who did know who she was and what her role was going to be. Ms. Valdez was assigned from the district office and was not chosen by the school administrator. The teachers were not fully involved in the planning and did not fully understand the grant.

Ms. Valdez was at a loss as to where to start. She decided to work on the physical space and at the same time get to know the school. Ms. Valdez and her interns worked on clearing the room of unwanted items and salvaging furniture to furnish the office. She did this with the help of the MSW interns assigned to her. At the same time, Ms. Valdez started meeting with as many school staff as she could on an individual basis. Her goal was to get to know them and their understanding of the program. She also wanted the staff to become familiar with her and learn her role in the school. Ms. Valdez did the same with all the different community agencies that provided services on campus, whether they were part of the initiative or not. She eventually set up a structure that encouraged outside agencies to check in at the wellness center and work with her when providing services to the students. Ms. Valdez learned about the school and the community culture when she decided to do this. She learned different facts and issues unique to the school as she got to know the school and the staff: The school and the community is a closed system.

- Faith-based organizations play a key role in this community.
- There had been several administrator turnovers over the years.
- There had been many programs that have come and gone over the years.
- The school had very limited participation in the actual program development and implementation while they took part in the planning.
- The school had their own vision how to develop and implement the program, which differs from the district's.

Ms. Valdez is now faced with three major tasks:

Deal with the above issues to pave the way for the program.

Align herself with multiple stakeholders to get buy-in and support for the program.

Create an infrastructure to foster a strong collaboration, start the delivery of services, and streamline access to the services.

The three tasks were the focus of the work for the first academic year. Ms. Valdez and the interns worked on building the relationship with the students, the school staff, and the community agencies. By the end of the school year, students were receiving services at school or linked to services in the community. The students were either referred by the teacher, by other staff, or self-referred. Unfortunately, parent participation was still at a minimum when school ended. It was by no means where it should be, but at this point most of the school staff and the majority of the community agencies were now familiar with Ms. Valdez. The wellness center had become a functioning resource center.

Learning Exercise

1. Who should have been a member of the planning, program development, and implementation?
2. How can Ms. Valdez work on aligning the school staff's vision and the district's vision for the initiative?
3. What could Ms. Valdez do to increase parent participation?
4. How can Ms. Valdez generate stakeholder participation and buy-in at this point in the implementation?
5. Is there any policy-related work that needs to be done or incorporated in the implementation to promote sustainability?

THINKING BIG AS POLICY ADVOCATES IN THE EDUCATION SECTOR

President Obama was determined to improve the nation's schools: Despite the enactment in 2001 of NCLB, students had lower scores on standardized math and English tests than many of their counterparts in Europe, Canada, and Asia. Obama devoted $80 billion of the stimulus bill to education; this was geared mostly to avoid laying off teachers during the recession. NCLB had set 2014 as the year when every child could read and do math at grade level, including persons of color, disabled students, and low-income students. Many states, too, had embraced the establishment of charter schools to give parents a choice of private schools, usually not-for-profit ones, that would establish their own curriculums and policies and whose teachers weren't usually members of teacher unions. But results of all of these efforts proved elusive.

President Obama and Arne Duncan, his secretary of education, replaced NCLB with an initiative known as Race to the Top, which asked states to produce plans that met goals established by the U.S. Department of Education grants from a pool of $4.3 billion. It was a clever strategy to avoid the political problems that had dogged educational reforms since the 1950s whenever school reformers proposed initiatives that appeared to allow federal officials to "interfere" with secondary education, which was widely viewed as "belonging" to states. By asking states to compete for funds based on innovations they had developed, Obama and Duncan avoided this problem.

A basic problem remained, however. Secondary education often remained behind the times. Rather than helping students grapple with ideas and math at a conceptual level, it often had a cookbook quality. Between 2008 and 2014, however, remarkable developments took place. Most states had accepted the concept of a Common Core curriculum that began in 2008, when Bill Gates was persuaded by two educational reformers to endorse a Common Core that emphasized problem-solving in reading and math. In the ensuing five years, the Gates Foundation dispersed tens of millions of dollars to school districts and public officials across the United States to support efforts to write and use curriculum based on Common Core precepts. In turn, the Obama administration made it known to school districts and states that they stood a better chance of obtaining grants from Race to the Top if they made the adoption of Common Core curriculum in their proposals.

Adoption of Common Core was widely accepted by liberals and conservatives—until 2014 when critics appeared. Some conservatives even named the Common Core "Obamacore." Some liberals doubted that evidence existed to show that changes in curriculum standards lead to better educational achievement. Advocates of the Common Core placed too little emphasis on issues other than curriculum improvement, such as the failure to fund schools sufficiently. In any event, the Trump administration rejected Common Core by leaving it up to states to develop curriculums.

If Gates and Obama emphasized curriculum reforms, a national movement to provide supportive assistance to students, particularly ones with learning and behavior problems, was absent even though evidence-based reforms demonstrated that teams of social workers, teachers, and other personnel had remarkable success in decreasing truancy, school dropouts, and low achievement. Once they saw warning signs in specific youth that they were falling behind or were becoming truant, they immersed these youth in supportive services. They visited their homes. They kept track of their truancy, finding them in their communities and shepherding them back to school. They involved their parents.

Stories fill the Internet about the cost to individuals, families, and American society of high dropout rates by low-income students, minority students, and others. Many evidence-based projects to cut truancy have been effective. Although high school graduation rates had increased to roughly 80% by 2018, graduation rates of low- and moderate-income students in junior colleges and colleges remained low. Junior college graduation rates were often less than 20% due to a constellation of factors, including students' inability to afford tuition and their need to support family members and relatives.

So why don't social workers take the lead in developing a national program that works to decrease dropouts, truancy, and poor performance by addressing students' mental health, economic, and social needs? Why shouldn't social workers take a leading role? Why not propose legislation at the national level to be funded by billions of dollars each year to detect students in trouble in the sixth grade or even earlier—and provide them with help from multidisciplinary teams that include social workers? Recall that we discussed earlier in this chapter the innovation developed by Professor Robert Balfanz at Johns Hopkins University, under Core Problem 4, that trial tested and implemented this innovation in scores of schools with evidence-based outcomes (Balfanz & Fox, 2011; DePaoli et al., 2018). Why not couple this program with massive increases in funding of the ESEA to increase resources of schools with high percentages of low-income students?

Discussion Questions

Develop a national program for cutting truancy and underachievement in which social workers have a leading role.

Estimate the likely cost of this program if it is applied to thousands of school districts with a large percentage of low-income children.

Give the program a name.

Send copies of the program to representatives and senators from your state.

Send copies of the program to the national director of NASW, and ask that person to assume a leadership role on Capitol Hill in pushing the program.

Couple this program with major increased funding of the ESEA.

Learning Outcomes

You are now equipped to:

- Develop an empowering perspective with respect to education
- Analyze the evolution of the American education sector
- Recognize problems in education that are created by an unequal nation
- Analyze the political economy of the education sector
- Analyze seven core issues in the American education sector
- Think big as policy advocates in the education sector

References

Allen-Meares, P. (2007). *Social work services in schools* (5th ed.). New York: Pearson.

American Academy of Pediatrics. (2004). School-based mental health services. *Pediatrics, 113*(6), 1839–1845. Retrieved from http://pediatrics.aappublications.org/content/113/6/1839.full

Aronowitz, M. (1984). The social and emotional adjustment of immigrant children: A review of the literature. *International Migration Review, 18*(2), 237–257.

Aseltine, R. H. Jr., & DeMartino, R. (2004). An outcome evaluation of the SOS suicide prevention program. *American Journal of Public Health, 94*(3), 446.

Astor, R. & Benbenishty, R. (2018). Blog tour: Three supportive ways school districts can create healthy schools and reduce threats of weapon violence. AASA: The School Superintendents Association. Retrieved from http://aasa.org/totalchild.aspx?id=42378&blogid=83505

Atkins-Burnett, S. (2007). Children with disabilities. In P. Allen-Meares (Ed.), *Social work services in schools* (pp.182–221). New York: Pearson.

Aud, S., Hussar, W., Johnson, F., Kena, G., Roth, E., Manning, E., . . . Yohn, C. (2012). The conditions of education 2012: Children and youth with disabilities (Indicator 9). Retrieved from http://nces.ed.gov/pubs2012/2012045.pdf

Aud, S., Hussar, W., Planty, M., Snyder, T., Bianco, K., Fox, M. A., . . . Hannes, G. (2010). The condition of education: Closer look 2010 (High-poverty schools). Retrieved from http://nces.ed.gov/programs/coe/analysis/2010-index.asp

Balfanz, R. & Fox, J. (2011). Early warning systems—foundational research and lessons from the field. Presented at the National Governors' Association, Philadelphia. Retrieved from https://docplayer.net/15154851-Early-warning-systems-foundational-research-and-lessons-from-the-field.html

California Department of Education. (2011, November 3). Charter schools. CalEdFacts. Retrieved from http://www.cde.ca.gov/sp/cs/re/cefcharterschools.asp

California Department of Education. (2012, September 17). School attendance review boards. Retrieved from http://www.cde.ca.gov/ls/ai/sb/

Catalan, J. (2014, August 1). Study: Are black students in integrated schools getting better education than segregated students? Retrieved from https://www.diversityinc.com/news/study-black-students-integrated-schools-getting-better-education-segregated-students

Ceasar, S. & Watanabe, T. (2011, July 31). Education takes a beating nationwide: More layoffs, bigger classes, fewer programs and higher tuition are nothing new to U.S. educators, but analysts say this year stands out. *Los Angeles Times*. Retrieved from http://articles.latimes.com/2011/jul/31/nation/la-na-education-budget-cuts-20110731

Center for Popular Democracy. (2017, September 27). Young people lead rally and march to reduce school-to-prison and school-to-deportation pipeline in New York City. Retrieved from https://populardemocracy.org/.../young-people-lead-rally-and-march-reduce-school-p

Chait, J. (2016, May 27). Teacher unions still haven't forgiven Michelle Rhee, don't care how well her policies work. *New York Magazine*.

Retrieved from http://nymag.com/daily/intelligencer/2016/05/teachers-still-havent-forgiven-michelle-rhee.html

Chan, S. & Leong, C. W. (1994). Chinese families in transition: Cultural conflicts and adjustment problems. *Journal of Social Distress and the Homeless, 3*(3), 263–281.

CNN (2018, March 9). America's gun culture vs. the world in 5 charts. Retrieved from https://www.cnn.com/2017/10/03/americas/us-gun-statistics/index.html

Cole, B. P. (1986). The black educator: An endangered species. *The Journal of Negro Education, 55*(3), 326–334.

Dee, T. S. (2005). A teacher like me: Does race, ethnicity, or gender matter? *The American Economic Review, 95*(2), 158–165.

Dee, T. S. (2016). *The state of racial diversity in the educator workforce*. Washington, DC: U.S. Department of Education, Office of Planning, Evaluation and Policy Development, Policy and Program Studies Service.

DePaoli, J., Balfanz, R., Atwell, M., & Bridgeland, J. (2018). *Building a grad nation: Progress and challenge in ending the high school dropout epidemic*. GradNation. Retrieved from http://gradnation.americaspromise.org/2018-building-grad-nation-report

Donlevy, V., Meierkord, A., & Rajania, A. (2016). *Study on the diversity within the teaching profession with particular focus on migrant and/or minority background: Annexes to the final report to DG education and culture of the European Commission, Luxembourg*. Retrieved from https://ec.europa.eu/education/news/20160309-study-teacher-diversity_en

Dryfoos, J. (2005). Full-service community schools: A strategy—not a program. *New Directions for Youth Development, 107*, 7–14. doi: 10.1002/yd.124

Dynarski, M. & Kainz, K. (2015, November 20). Why federal spending on low income students (Title 1) doesn't work. Brookings Institution. Retrieved from https://www.brookings.edu/research/why-federal-spending-on-disadvantaged-students

Elementary and Secondary Education Act of 1965, 20 U.S.C. 6301 et seq. (1965). Retrieved from http://www2.ed.gov/policy/elsec/leg/esea02/pg1.html

Fox, M. A., Connolly, B. A., & Snyder, T. D. (2005). *Youth indicators, 2005: Trends in the well-being of American youth* [NCES 2005-050]. U.S. Department of Education.

Franklin, C., Harris, M., & Allen-Meares, P. (2007). *The school services sourcebook: A guide for school-based professionals.* New York, NY: Oxford University Press.

Fuchs, D. & Fuchs, L. (2006). Introduction to response to intervention: What, why, and how valid is it? *Reading Research Quarterly, 41*(1), 93–99. doi:10.1598/RRQ.41.1.4

Furlong, M., Paige, L. Z., & Osher, D. (2003). The Safe Schools/Healthy Students (SS/HS) initiative: Lessons learned from implementing comprehensive youth development programs. *Psychology in the Schools, 40*(5), 447–456. doi: 10.1002/pits.10102

Ingersoll, R. & May, H. (2011). Recruitment, retention and the minority teacher shortage. *Educational Research*, 1–61. Retrieved from http://doi.org/10.1037/e546592012-001

Jansson, B. (2019). *Reducing inequality: Addressing the wicked problems across professions and disciplines.* San Diego: Cognella Academic Publishing.

Joseph, G. & CityLab. (2016, August 12). Why kindergarteners might still be suspended in

New York City. *The Atlantic*. Retrieved from https://www.theatlantic.com/education/archive/2016/16/08/why...newyork.../newyork.../495614/)

Kahlenberg, R. D. (2016, February 10). School integration's comeback. *The Atlantic*. Retrieved from https://www.theatlantic.com/education/archive/2016/02/breaking-up-school-poverty/462066/

Kamenetz, A. (2018, March 7). Here's how to prevent the next school shooting, experts say. NPR. Retrieved from https://www.npr.org/sections/ed/2018/03/07/590877717/experts-say-here-s-how-to-prevent-the-next-school-shooting

Karp, S. & Sanchez, A. (2018, Summer). The 2018 wave of teacher strikes: A turning point for our schools? *Rethinking Schools, 32*(4). Retrieved from https://www.rethinkingschools.org/articles/the-2018-wave-of-teacher-strikes

Kirby, L. D. & Fraser, M. W. (2004). *Risk and resilience in childhood: An ecolocal perspective*. NASW Press.

Mazama, A. & Lundy, G. (2012). African American homeschooling as racial protectionism. *Journal of Black Studies*, *43*(7), 723–748. Retrieved from http://doi.org/10.1177/0021934712457042

McDaniels, A. (2017, December 19). A new path for school integration. Center for American Progress. Retrieved from https://www.americanprogress.org/issues/education-k.../new-path-school-integration/

Moon, S. S. & Lee, J. (2009). Multiple predictors of Asian American children's school achievement. *Early Education and Development*, *20*(1), 129–147.

National College Access Network. (2018, May). Shutting low-income students out of public four-year higher education. Retrieved from http://www.collegeaccess.org/.../ShuttingLow-IncomeStudentsOutOfPublicFour-YearHigher...

National Institute on Drug Abuse. (2012). DrugFacts: High school and youth trends. *National Institutes of Health*. Washington, DC. Retrieved from http://www.nida.nih.gov/infofacts/hsyouthtrends.html

Ngo, B. (2006). Learning from the margins: The education of Southeast and South Asian Americans in context. *Race Ethnicity and Education, 9*(1), 51–65.

Orfield, G. & Eaton, S. (1996). *Dismantling desegregation: The quiet reversal of* Brown v. Board of Education. New York, NY: New Press.

Payan, R. M. & Nettles, M. T. (n.d.). Current state of English-language learners in the US K–12 student population. Retrieved from http://www.ets.org/Media/Conferences_and_Events/pdf/ELLsympsium/ELL_factsheet.pdf

Peebles-Wilkins, W. (2006). Evidence-based suicide prevention. *Children & Schools*, *28*(4), 195–196.

Plessy v. Ferguson, 163 U.S. 537, 539 (1896).

Rumberger, R. W. (2013, May).Poverty and high school dropouts: The impact of family and community poverty on high school dropouts. *The SES Indicator.* Retrieved from http://www.apa.org/pi/ses/resources/indicator/2013/05/poverty-dropouts.aspx

RWJF Commission to Build a Healthier America. (2014). Commission to build a healthier America recommends seismic shift in funding priorities to improve health, with emphasis on early childhood education, community revitalization and broader health care scope. Retrieved from http://www.rwjf.org/en/about-rwjf/newsroom/newsroom-content/2014/01/commission-to-build-a-healthier-america-recommends-seismic-shift.html

Sipple, J. W. (2007). Major issues in American schools. In P. Allen-Meares (Ed.), *Social work services in schools* (pp. 1–25). New York: Pearson.

Smrekar, C. & Goldring, E. (1999). *School choice in urban America: Magnet schools and the pursuit of equity. Critical issues in educational leadership series*. Williston, VT: Teachers College Press.

Sum, A., Khatiwada, I., & McLaughlin, J. (2009, October 1). The consequences of dropping out of high school: Joblessness and jailing for high school dropouts and the high cost for taxpayers. *Center for Labor Market Studies Publications*, 1–17. Retrieved from http://hdl.handle.net/2047/d20000596%5Cnhttp://iris.lib.neu.edu/cgi/viewcontent.cgi? article=1022&context=clms_pub

Sweet, L. (2009, July 16). Obama's NAACP speech. *Chicago Sun-Times*. Retrieved from http://blogs.suntimes.com/sweet/2009/07/obamas_naacp_speech.html

Tatum, B. (2017, September 14). Segregation worse in schools 60 years after *Brown v. Board of Education*. *Seattle Times*, Opinion Section.

Terkel, A. (2017, October 24). How Betsy DeVos became the most hated cabinet secretary. *Huffington Post*. Retrieved from https://www.huffingtonpost.com/entry/betsy-devos-most-hated-secretary_us_59ee3d3be4b003385ac13c9b

Thernstrom, A. & Thernstrom, S. (2003). *No excuses: Closing the racial gap in learning*. New York, NY: Simon & Schuster.

U.S. Department of Health & Human Services, Office of Head Start. (n.d.). History of Head Start. Retrieved from http://www.acf.hhs.gov/programs/ohs/about/history-of-head-start

U.S. Department of Justice. (2009). Juvenile justice bulletin: Juvenile arrests 2008. Retrieved from https://www.ncjrs.gov/pdffiles1/ojjdp/228479.pdf

U.S. Surgeon General. (1999). Report on mental health: Children and mental health, Chapter 3. Retrieved from https://profiles.nlm.nih.gov/ps/retrieve/ResourceMetadata/NNBBHS

Vance, J. D. (2016). *Hillbilly Elegy*. New York: Harper.

Villegas, A. M. & Irvine, J. J. (2010). Diversifying the teaching force: An examination of major arguments. *Urban Review*, *42*(3), 175–192. Retrieved from http://doi.org/10.1007/s11256-010-0150-1

Wells, A. S., Fox, L., & Cordova-Cobo, D. (2016). *How racially diverse schools and classrooms can benefit all students*. Retrieved from https://tcf.org/content/report/how-racially-diverse-schools-and-classrooms-can-benefit-all-students/

White House Office of the Press Secretary. (2011). President and First Lady call for a united effort to address bullying. Retrieved from http://www.whitehouse.gov/the-press-office/2011/03/10/president-and-first-lady-call-united-effort-address-bullying

White House Statements & Releases. (2013). Vice President Biden announces $100 million to increase access to mental health services. Retrieved from http://www.whitehouse.gov/the-press-office/2013/12/10/vice-president-biden-announces-100-million-increase-access-mental-health

Whitted, K. S. & Dupper, D. R. (2005). Best practices for preventing or reducing bullying in schools. *Children & Schools*, *27*(3), 167–175.

Wright, W. E. (2005). The political spectacle of Arizona's Proposition 203. *Educational Policy*, *19*(5), 662–700.

Yeh, C. J. & Inose, M. (2003). International students' reported English fluency, social support satisfaction, and social connectedness as predictors of acculturative stress. *Counselling Psychology Quarterly*, *16*(1), 15–28.

Youth Law Center/Children's Law Center of Los Angeles. (2003, December). Ensuring educational rights and stability for foster youth: AB 490 summary. Retrieved from http://www.youthlaw.org/fileadmin/ncyl/youthlaw/events_trainings/ab490/AB490_Summary.pdf

Zehr, M. A. (2011, March 3). Advocacy groups push for better translation services. *Education Week*. Retrieved from http://blogs.edweek.org/edweek/learning-the language/2011/03/advocacy_groups_push_for_bette.html

13

BECOMING POLICY ADVOCATES IN THE IMMIGRATION SECTOR

LEARNING OBJECTIVES

In this chapter you will learn to:

1. Develop an empowering perspective with respect to immigrants
2. Analyze the evolution of American immigration policies
3. Analyze the movement of persons across national boundaries
4. Recognize problems created for immigrants by extreme income inequality
5. Understand the political economy of American immigration
6. Get back to values and data
7. Analyze seven core problems in the immigration sector
8. Think big as advocates in the immigration sector

The United States is an immigrant nation. Immigrants are a marginalized population that encounters many social, economic, and legal issues, not to mention discrimination. Social workers often need to be advocates for immigrants no matter the policy sector in which they work.

DEVELOPING AN EMPOWERING PERSPECTIVE WITH RESPECT TO IMMIGRANTS

Imagine that you are a member of Congress tasked with the job of helping write immigration legislation in a nation with more than 11 million undocumented persons. You would have to decide whether to deport all of them, some of them, or none of them. You would have to decide how many of them can become citizens and in what time frame. You would have to decide what kinds of immigrants to allow into the nation, such as highly educated persons, persons with specific job-related skills, people of specific ages, people from specific nations, and people with health and mental health problems. You would have to decide whether to separate children from parents when they are detained at the border. You would make these decisions in a deeply polarized nation that includes anti-immigrant and pro-immigrant factions.

You would have to decide to what extent the nation should admit persons in distress, such as refugees from civil wars or other military conflict, dictatorships, family violence, epidemics, poverty, and starvation. What priority would you give to persons fleeing from natural disasters like drought, hurricanes, and earthquakes? You would also need to decide how many immigrants to admit with temporary work visas, such as to harvest crops or to clean up debris of three hurricanes that struck the United States and Puerto Rico in 2017. You would have to realize that immigrants from many nations risk their lives to enter the United States. Often driven by extreme poverty, political persecution, and family violence, persons coming from Central America and Mexico of all ages walk many miles just to reach the American border across deserts with temperatures below freezing at night only to have to walk long distances across deserts in the United States to reach cities like Tucson, Arizona. They encounter snakes, coyotes, fire ants, and Africanized bees. They often run out of water. They hire persons who guide them for a hefty fee. They run into drug dealers and others who rape women along the route. They encounter vans sent by the U.S. border patrol or by others to pick up corpses. If the United States increased the number of work visas, far more immigrants could gain legal entry for time-limited work—thus decreasing the number of deaths and injuries of undocumented immigrants.

Do note that entire segments of the American economy would come to a halt were immigrants excluded from the United States, including agriculture, tourism, restaurants, and construction. Many of these jobs have no takers from American citizens. Do realize, as well, that so-called DREAMers were brought to the United States as children by undocumented parents. They are a marginalized group whose rights are violated by police, neighbors, employers, the criminal justice system, immigration officials, and others. They are often described as criminals, rapists, and deadbeats when they have lower rates of crime and lower use of safety net programs than American citizens.

The demonization of immigrants took place during the congressional elections of 2018. Seeking to mobilize his base of support, President Donald Trump decided in the several weeks before the election to contend that the nation faced an imminent emergency: the arrival of a "caravan" of thousands of people from Central America to the Mexican/American border. He alleged that the caravan contained many criminals. He approved

Learning Exercise

To better understand the risk experienced by migrants, go to *Borderland*, a documentary series developed by Al Jazeera, at http://america.aljazeera.com/watch/shows/al-jazeera-america-presents-borderland.html

Read the brief article by Alessandra Stanley. Click on the Borderland link. Then click on "Learn about 3 migrants who died along the border" and "Episode 4." What did you learn from reading about how six Americans volunteered to cross the Mexican/American border en route to Tucson but with the assurance they would be rescued by stand-by volunteers if they could no longer walk.

the use of a video in campaign ads that showed hordes of people forcibly crossing the border and highlighted a Latino who had been placed in jail for numerous murders. The video was viewed as sufficiently inaccurate that even Fox News along with NBC and CNN decided to take it off the air. In truth, the caravan was roughly 1,000 miles from the U.S. border and mostly composed of women and children fleeing violence and poverty in El Salvador, Guatemala, and Nicaragua.

You would have to decide whether to give them, and other undocumented persons, a path to citizenship, as took place under Republican presidents (Ronald Reagan in 1986, who gave almost 4 million undocumented persons this path, and George W. Bush, who, in 2007, almost obtained passage of legislation that would have given 11 million undocumented persons this path).

The United States has a mixed immigration tradition. The United States is an immigrant nation where everyone has immigrant forebears or is an immigrant aside from indigenous Native Americans. Yet American immigration policies have discriminated against persons of color, Native Americans, people from Asia, Latinos, LGBTQ persons, Muslims, and refugees fleeing wars, starvation, religious persecution, and political persecution.

Realize, as well, that the current polarization between defenders and opponents of immigrants, as exists in the Trump presidency, also existed in our prior history such as in the mid-1920s, when the Congress enacted legislation that almost excluded Italians, Mexicans, and Asians.

ANALYZING THE EVOLUTION OF AMERICA'S IMMIGRATION POLICIES

Colonial authorities, specific states, and the federal government enacted immigration policies from the inception of the republic as this brief overview demonstrates.

- Congress allowed immigrants to enter the United States in the early republic only with a minimal health test, no literacy test, no annual limits, and no limits on specific nations.
- The American Constitution vested power over immigration with the federal government, not individual states, when it was enacted in 1789.

- Congress enacted legislation in 1790 that restricted citizenship to white Caucasians that led to non-citizenship for persons of African descent, Native Americans, and (later) persons with Asian descent.
- Tens of millions of immigrants flooded into the United States as it industrialized from 1860 to 1920.
- The unfair treatment of Spanish-speaking persons in the American Southwest in the wake of the Treaty of Guadalupe Hidalgo after the United States conquered Mexico in 1848 led to the loss of their land titles. Many of them worked subsequently as laborers on ranches, farms, and mines in the United States.
- Many immigrants experienced considerable discrimination in the 19th and early 20th centuries, including persons from Ireland, China, Japan, Italy, Mexico, and Eastern Europe, along with emancipated slaves, African Americans, Jews, and Catholics. Not technically immigrants because they were indigenous, Native Americans were cruelly attacked and exploited as they, as well as persons of Mexican and Spanish descent, were pushed off land as white Americans moved westward.
- Harsh legislation was enacted against immigrants from China and Japan in 1907.
- The Immigration Act of 1924 cut immigration from Eastern and Southern Europe, as well as Asia, while increasing immigration from Northern Europe in an effort to decrease non-white immigrants as well as Catholic and Jewish immigrants.
- The Bracero program was initiated in 1942. It allowed a specific number of Mexicans to enter the United States to meet labor shortages on farms and ranches. It was accompanied by supposed protections for them, such as guaranteeing them a suitable wage and preserving their civil rights, but these were mostly ignored by American employers such as ranchers. The program ended in 1964.
- The Immigration Act of 1965, the Indochina Migration and Refugee Assistance Act of 1975, and the 1980 Immigration Act abolished quotas for specific nations and regions—and led to a huge upswing in immigration from Asia, as well as Central America, the Caribbean, the Middle East, Russia, and Eastern Europe.
- The Immigration Act of 1986, enacted with bipartisan support, provided amnesty to 3 million undocumented immigrants, mostly from Mexico and Central America if they could prove they had lived in the United States for four years.
- Proposition 187 was enacted in California in 1994. It declared immigrants to be ineligible for medical, social, and educational services—but was not implemented due to court challenges. The Personal Responsibility and Work Opportunity Reconciliation Act of 1996 banned access by undocumented immigrants to many medical, safety net, and social services—and remains in effect in 2019.

- Distrust of Muslim immigrants from the Middle East flared upward in the wake of the attack on the Twin Towers in New York City on September 11, 2001.
- The Great Recession of 2007–2009 increased resentment against immigrants who were widely seen as taking jobs from American citizens when unemployment exceeded 10%. The rise of the Tea Party, which was formed in 2009, led to the election of many Tea Party members in succeeding years who favored restricting the flow of immigrants into the United States.
- Due to bipartisan agreement between leaders of both parties, Congress appeared ready to enact sweeping immigration reforms in 2007 that would have legalized as many as 12 million undocumented persons but could not surmount partisan gridlock.
- President Barack Obama promised to secure sweeping immigration reforms during both his first and second terms but could not succeed due to intense opposition from conservatives including most Republicans. Conservatives spearheaded anti-immigrant policies in many states including Arizona, North Carolina, Alabama, and Georgia such as ones that allowed police to check the citizenship of anyone who "looked like" an undocumented person. The U.S. Supreme Court overturned many of these policies when it ruled that the U.S. Constitution vested power over immigration with the federal government rather than with the states.
- Thousands of children and youth came over the border between Mexico and the United States—many from Central American nations like Guatemala and Nicaragua—seeking to reunite with their parents and relatives in 2014. They were housed in makeshift settings in the United States as Congress refused to pass a request for more than $3 billion by President Obama to meet their housing and medical needs. The president promised to decide by the end of the summer of 2014 whether to allow them entry or to deport them to their nations of origin while accusing Republicans of blocking major immigration reforms. These makeshift settings still existed in 2019.
- President Donald Trump promised to "Build the Wall" between the United States and Mexico in the presidential campaign of 2016 but had made little progress in securing funding for it from Mexico and the U.S. Congress by mid-2018.
- Heated controversies erupted during Trump's first two presidential years including over the fate of Deferred Action for Childhood Arrivals (DACA), the rights of "sanctuary cities," deportation polices of Immigration and Customs Enforcement (ICE), border enforcement policies of U.S. Customs and Border Protection (CBP), and immigration bans imposed on six to eight Middle Eastern nations with majority Muslim populations. In each of these cases, President Trump sought to decrease immigration into the United States only to be countered by immigrant advocates and federal courts.
- The U.S. Supreme Court finally approved President Trump's ban of Muslim immigrants in 2018 but only by a 5–4 margin.

- President Trump's immigration policies led to the separation of many immigrant mothers from their children at the border, leading to deportation of many mothers absent their children in 2018 as well as the detention of hundreds of children separated from their parents even after a federal judge ordered the Trump administration to reunite these children with their parents.

ANALYZING THE MOVEMENT OF PERSONS ACROSS NATIONAL BOUNDARIES

We cannot understand American immigrants without placing them in a global context. Persons move between many nations as documented by the fact that roughly 214 million people lived outside their nations of birth worldwide—or 3.1% of the world's population in 2011 (International Organization for Migration [IOM], 2011). These migrants are evenly split between males and females. They send payments (often called remittances) to family members that total $440 billion per year.

The United States has attracted the greatest number of immigrants relative to other countries, with 38.5 million foreign-born persons or 12.5% of its total population, including many from Mexico (29.9%), the Philippines (4.5%), India (4.3%), China (3.7%), Vietnam (3%), El Salvador (3%), Korea (2.6%), Cuba (2.6%), Canada (2.1%), and the Dominican Republic (2.1%) (Migration Policy Institute, 2011). About half of these immigrants are limited English proficient persons (LEPs) who report on surveys that they speak English "not at all" or "not well" as compared with "well." The immigrants speak many languages, although 62% speak Spanish. These immigrants are evenly split by gender. About one-fourth of the immigrants have a bachelor's degree or higher, whereas about one-third lack a high school diploma. Roughly two in five of them are naturalized U.S. citizens, with remaining ones split among legal permanent residents, unauthorized (or undocumented) immigrants, and legal residents on temporary visas as students and workers (Migration Policy Institute, 2011). Immigrants from Mexico are concentrated in California (37.5%), Texas (20.9%), Illinois (6%), Arizona (5.2%), and Georgia (2.4%). Immigrants constitute roughly 16% of the American workforce (Migration Policy Institute, 2011).

Households that have one or more of their heads born in other nations vary widely in their economic status. Roughly one-fourth of the 26 million households are middle class with total annual incomes between $47,000 and $79,000. These immigrant adults tend to have high school educations and some postsecondary credential short of a bachelor's degree (IOM, 2011). Many immigrants fall beneath federal poverty standards. Immigrants cross the age spectrum, including about 16.9 million children at or under age 17 (Migration Policy Institute, 2011).

More than 1 million persons become new, lawful, permanent residents each year, with roughly one-half immediate relatives of U.S. citizens, one-fifth through a family-sponsored preference, one-tenth through employment-based preference, 16% from a refugee or asylum status, and 4% as diversity lottery winners who win annual lotteries established by the federal government to increase entry from nations with low rates of immigration to the United States (Migration Policy Institute, 2011). However, immigration of undocumented persons markedly decreased during the Great Recession of 2007 to 2009—and remained low even in 2018.

Roughly 11.2 million unauthorized immigrants lived in the United States in 2010—or 28% of the nation's foreign-born population and 5% of the nation's workers (Pew Hispanic Center, 2011). Immigration of unauthorized persons ebbs and flows with economic conditions, falling from 850,000 per year from March 2000 to March 2005, for example, to only 300,000 per year during the economic downturn in March 2007 to March 2009 (Pew Hispanic Center, 2011). The economic well-being of unauthorized immigrants in the United States markedly declined during the Great Recession with downturns in agriculture, tourism, and industry, leading some of them to return to Mexico and other nations. Their economic distress was experienced, as well, by broader Hispanic, African American, and white households, whose wealth respectively declined by 66%, 53%, and 16% from 2005 through 2009 (Pew Hispanic Center, 2011). Immigration of Latinos to the United States from Mexico had reversed by 2012 as more of them exited than entered the United States (Pew Hispanic Center, 2012). Despite claims by candidate and President Donald Trump, immigration remained low by historical standards from 2012 through 2019.

Immigrants are both "pulled" and "pushed" across national boundaries. They are pulled by economic factors when they seek improved economic conditions in another nation. Immigrants constitute about 6% of the populations in rich nations as opposed to just 1% in poor countries. A considerable proportion of immigrants send remittances to family members who remain in their native lands to provide them with needed capital. India received the largest volume of remittances ($21.7 billion in 2004), followed by China ($21.3 billion) and Mexico ($18.1 billion).

Many immigrants are pushed by intolerable economic and social conditions in their native lands including violations of human rights, wars between nations, civil wars, genocide, tribal or sectarian conflict, domestic violence, and famine. We can distinguish among refugees, asylum seekers, and internally displaced persons (IDPs). Refugees are persons who are pushed away from their homes and can document that they have been persecuted for reasons of race, religion, nationality, membership in a particular social group, or political opinion. Some LGBTQ persons seek refugee status due to discrimination and physical attacks in their native lands. Refugees include persons displaced by wars or disturbances of public order: Many persons emigrated to the United States during civil wars in Nicaragua, El Salvador, Cuba, Afghanistan, and Iraq, whereas many others came to the United States in the wake of the Vietnam War. Refugees include women subjected to family violence or to other forms of violence, although the Trump administration tried to deny refugee status to women subjected to family violence. Poverty is not grounds for asylum. Asylum seekers are persons whose claim to be refugees has not yet been documented. IDPs are persons who must migrate due to natural disasters, tribal warfare, or internal discord. The United Nations High Commissioner for Refugees (UNHCR), which coordinates refugee protection, estimated that 10.5 million refugees existed in 2010. Most refugees flee to neighboring nations, but many seek refuge in distant lands, such as Vietnamese persons who came to the United States in the wake of the Vietnam War in the 1970s and later. Internally displaced persons include persons who lose their homes during hurricanes. Droughts, floods, tsunamis, and earthquakes have often led to the destruction of food and housing in many nations—leading many persons to flee to other nations or to refugee camps funded by nongovernmental organizations, the UN, or specific governments.

Many immigrant women come to the United States due to human trafficking that includes sex trafficking and forced use of persons for labor or services through use of force, fraud, or coercion (ACLU, n.d.). Some parents insist that their oldest child migrate to provide them with needed resources.

ANALYZING THE LEGAL STATUS OF IMMIGRANTS IN THE UNITED STATES

Migrants to other nations are a heterogeneous group with respect to their legal status. Many of them are undocumented who cross borders without the approval of host nations. They often exist in a state of limbo. Employers frequently exploit them. They often do not report burglaries and physical attacks because they fear deportation. They often work and even pay taxes but find it difficult to access social programs, health, and education—whether because host nations deny them these benefits or they fear deportation if they claim benefits. Undocumented persons have some rights, however, as we discuss subsequently.

Persons need visas to legalize temporary stays through petitions filed with the U.S. Citizenship and Immigration Services (USCIS) (http://FAQ.VisaPro.com/). Family visas include ones for persons with close relations with a U.S. citizen, such as fiancées, spouses, and children—as well as visas for persons with more distant family relationships with a U.S. citizen. Green cards grant lawful permanent residency to persons including permission to live and work in the United States. Holders must maintain permanent resident status and can be removed from the United States if certain conditions of this status are not met. Roughly 140,000 work visas are issued each year, such as ones for business visitors; registered nurses; persons of extraordinary ability in the arts, athletics, business, education, or science; and agriculture. Some work visas are only given to temporary immigrants who are sponsored for a specified period by an American employer. A system of priorities exists. First priority is given to persons with extraordinary ability, researchers, professors, and multinational managers or executives. Second priority goes to professionals holding advanced degrees. Third priority is given to skilled workers, professionals, and unskilled workers. Fourth preference goes to certain special immigrants, including Iraqi and Afghan interpreters and translators and persons recruited for U.S. armed forces outside the nation. Fifth priority goes to immigrant investors and entrepreneurs.

Student visas are given to academic or language students, exchange visitors, or vocational or non-academic students. The Trump administration threatened to decrease student visas given to Chinese students in 2018. Other visas include ones for persons seeking asylum from persecution by governments in other nations as well as ones for persons who come to the United States from human trafficking. Some persons have uncertain immigration status, such as persons who wait for extended periods in the United States as they seek citizenship—waits that can last months or years. American citizens who wish to adopt children from other nations petition the USCIS.

Family members and some employers file petitions to help persons emigrate to the United States, often becoming their financial sponsors by filing an affidavit of support in which they promise to support the immigrant and even to repay certain benefits that the immigrant uses as required by the 1996 welfare law (National Immigration

Law Center, 2005). Sponsors' income is sometimes added to immigrants' income to determine their eligibility for specific public benefit programs, even for 10 or more years after immigrants enter the nation under the 1996 legislation. (Exceptions include survivors of domestic violence and immigrants who would become hungry or homeless without assistance.)

Persons who criticize undocumented immigrants, or even ones with work visas, often fail to understand benefits that they bring to the United States. Immigrants often take jobs that naturalized Americans do not want, particularly low-paying and undesirable jobs in agriculture, tourism, health care, and industry—or as caregivers for disabled and elderly people in their homes or in institutions. These jobs often expose immigrants to health and other risks. These immigrants pay sales taxes and social security taxes. Many researchers conclude that immigrants contribute more resources to the American economy than they take from it because they make relatively little use of American safety net programs, whether because they are ineligible or because they fear deportation if they do use them or because they are not familiar with them. Americans will increasingly need labor from immigrants due to the aging of the American population in coming decades.

The United States pursues conflicting policies. On the one hand, it invites the inflow of undocumented persons through its immigration policies. By not providing sufficient work visas to fill existing jobs that are not claimed by American citizens, however, the United States creates an employment vacuum that is filled by undocumented persons who want and need employment. Needing their labor, employers often do not check workers' documents to ascertain if they have entered the United States legally—and pressure immigration authorities not to conduct raids on their workforces. Having enticed them to enter the United States, Americans often harass undocumented workers. ICE raids places of work and even homes of undocumented persons. It deports undocumented persons. Although the federal government is given the power to regulate immigration by the U.S. Constitution, local police officials often report immigrants to ICE. Employers often defraud immigrants by not paying them the minimum wage—or not paying them at all—because they believe immigrants will not complain to federal agencies for fear they will be deported. Federal legislation forbids immigrants to use many of the nation's health and social service programs, limiting them, for example, only to emergency medical care, obstetrics, and treatment for diseases that can be passed to others such as tuberculosis or HIV/AIDS. Many immigrants do not use these programs because they fear they will be reported to ICE by them even when these fears are groundless.

A crisis took place at the Mexican border in the spring and summer of 2018. Border personnel on the Southern border had often given mere misdemeanors to persons and families that illegally crossed the border in prior years but had allowed them to proceed because they knew that most of them would take jobs in agriculture, tourism, caregiving agencies for seniors, restaurants, and other positions that American citizens were unwilling to take. With Trump's approval, Attorney General Jeff Sessions decided to detain them, to hold adults in temporary centers until they could be deported, and to send thousands of children to holding centers across the nation. When a federal judge ruled that the children had to be reunited with their parents, it became clear that the Trump administration did not know where some of the children were housed, much less how to locate their parents, who had often been deported. (The Trump administration had secretly sent the children to many detention centers around the nation.) Polls indicated that that vast majority of

Americans disliked separating children from parents for ethical reasons and because it traumatized parents and children. President Trump reluctantly backed down but did not meet deadlines imposed by a federal court in July 2018 even by mid-November 2018.

Immigration became the central campaign issue in the congressional elections of 2018. If Democrats mostly opposed detention of immigrants seeking asylum as well as separation of children from their parents, President Trump insisted that Democrats had organized a caravan of immigrants from Central America to the United States. Although many immigration experts insisted that this caravan was at least 1,000 miles from the border and that it mostly consisted of children and their mothers who would seek asylum to escape violence and poverty, Trump portrayed it as an "invasion" that would break through barricades at the border and wreak havoc in the United States. In fact, numerous caravans of immigrants had sought asylum in preceding years even if relatively few of them received asylum.

Immigrants often believe, or are told, that they are ineligible for many services. Policy advocates need to inform them that they should not assume they are ineligible for many services and that they are not obligated to convey their immigration status when they apply for many of them. Nor are they required to state their immigration status when accosted by ICE officials in their homes and neighborhoods.

Sweeping immigration reform is needed that legalizes immigrants sufficient to meet the nation's economic needs and preserve immigrants' dignity and human rights. It should provide amnesty to more than 10 million immigrants who have "paid their dues" by working in the United States for extended periods. It should give temporary work visas to immigrants needed for the nation's economy. It should protect their rights to the minimum wage of local, state, and federal governments. It should prosecute people who harass or physically harm immigrants.

Many groups have provided micro, mezzo, and macro policy advocacy for immigrants throughout American history by seeking to improve their legal status. (See Table 13.1 for some contemporary advocacy groups.)

TABLE 13.1 ■ Some Advocacy Groups for Immigrants

American Civil Liberties Union (ACLU). The ACLU has fought for rights of immigrants in courts, legislatures, and regulatory bodies: http://www.aclu.org

American Friends Service Committee (AFSC). The AFSC operates advocacy programs for immigrants along the Mexican border: http://www.afsc.org

Asian American Justice Center (AAJC). Founded in 1991, AAJC (formerly the National Asian Pacific American Legal Consortium) works to "advance the human and civil rights of Asian Americans through advocacy, public policy, public education, and litigation": http://www.advancingequality.org

Center for Immigration Studies (CIS). Founded in 1985, the CIS is an independent, nonpartisan, nonprofit research organization devoted to research and policy analysis dealing with the economic, social, demographic, fiscal, and other implications of immigration for the United States: http://www.cis.org/

Centro Humanitario para los Trabajadores. This organization provides work opportunities and safe working conditions for immigrants: http://www.centrohumanitario.org

International Organization for Migration (IOM). IOM is the world's primary intergovernmental organization focused on the issue of migration with 116 member states: http://www.iom.int

Mexican American Legal Defense and Educational Fund (MALDEF). A leading advocate for Latinos: http://www.maldef.org

Migrants Rights International (MRI). MRI is a membership organization of migration and human rights experts and practitioners: http://www.migrantsrightsinternational.org/

Migration Policy Institute (MPI). Founded in 2001, MPI conducts research on the movement of people worldwide and provides analysis for local, national, and international migration and refugee policies: http://www.migrationpolicy.org and http://www.migrationinformation.org

National Council of La Raza (NCLR). NCLR is a national Hispanic civil rights advocacy organization in the United States: http://www.nclr.org

National Immigration Forum (NIF). NIF advocates for public policies for immigrants and refugees: http://www.immigrationforum.org

National Immigration Law Center (NILC). NILC is a leading expert on immigration, public benefits, and employment laws affecting immigrants and refugees: http://www.nilc.org

National Network for Immigrant & Refugee Rights (NNIRR). The NNIRR develops and coordinates plans of action on important immigrant and refugee issues: http://www.nnirr.org/

Office of the United Nations High Commissioner for Refugees (UNHCR). The agency has helped approximately 50 million refugees looking to restart their lives: http://www.unhcr.org

Pew Hispanic Center. The center conducts research on migration flows from Mexico, Central America, and South America: http://www.pewhispanic.org

Population Reference Bureau (PRB). PRB works to inform people around the world about population, health, and the environment: http://www.prb.org

RECOGNIZING PROBLEMS FOR IMMIGRANTS CREATED BY EXTREME INEQUALITY

Immigrants work disproportionately in low-income, unskilled, or semi-skilled positions in American society, including janitorial work, the tourist industry, agriculture, construction, restaurants, and the health system. They receive even lower wages than their counterparts in Canada and Europe, where economic inequality is not as marked as in the United States. Their economic marginalization is often accentuated by widespread prejudice against undocumented persons—even as most Americans agree that immigrants' labor is indispensable to the American economy. Undocumented immigrants are often stereotyped as using social programs excessively, committing crimes, and having babies to obtain citizenship for them even when considerable research indicates these stereotypes are false (ACLU, 2008).

The sheer level of poverty and marginal economic status for many American citizens stimulates hostility toward immigrants by American citizens. Many persons "on the edge" fear that immigrants take away their employment because they often cluster in low-wage jobs in tourism, non-unionized industrial jobs, agriculture, construction, and other areas—even when many indigenous citizens are unwilling to work in many of these positions. Leaders of American trade unions have often been hostile to immigrants because they observe how employers hire immigrants to bust unions.

Extensive prejudice exists against many immigrants. Most of them are persons of color. Some of them are Muslims. False rumors often circulate that they have high rates of crime when they have lower rates of crime than American citizens.

UNDERSTANDING THE POLITICAL ECONOMY OF THE IMMIGRATION SECTOR

Controversy swirls around immigrants in the United States. Suspicion of immigrants was triggered by the bombings of the Twin Towers on September 11, 2001, that heightened fear of terrorists. Recall that many statutes have helped millions of immigrants become U.S. citizens from 1789 to the present. Given the sole power to regulate immigration in the U.S. Constitution, the federal government often implemented a relatively permissive immigration policy from 1789 to the present—allowing tens of millions of immigrants from scores of nations to enter the nation, including ancestors of everyone reading this book aside from indigenous Native Americans who pre-dated 1789. (We discuss subsequently laws that restricted immigration.) A surge of immigration laws occurred in the 1960s during and after the Vietnam War. President Ronald Reagan, many Republicans, and Democrats enacted the Immigration Reform and Control Act of 1986 that granted a path to citizenship for more than 3 million immigrants. President George W. Bush teamed with Republican Senator John McCain and Democratic Senator Ted Kennedy to seek enactment of comprehensive immigration reform in 2007 and 2008 that would have given more than 10 million undocumented immigrants a path to citizenship—legislation that was almost enacted.

President Obama decided that the United States should expand the entry of refugees into the United States in light of foreign events that included a civil war in Syria and a catastrophic drought in Africa. These catastrophes led more than a million refugees to attempt to reach European nations, often by migrating to Turkey, Italy, and Spain en route to other European nations—not including ones that encamped in refugee camps in Syria, Somalia, and other nations. These were desperate people whose entire neighborhoods were destroyed from mortar shells and bombing in Syria and who neared starvation in Somalia and other African nations. Many drowned in their fragile and overcrowded boats. Alarmed by the sheer magnitude of destruction and death, Obama's State Department admitted 84,995 refugees in 2015 and 2016, including nearly 39,000 Muslim refugees. But a large partisan gap developed as 87% of Trump supporters believed the United States does not have the responsibility to accept Syrian Muslims versus only 27% of supporters of Clinton. President Trump drastically cut admissions of refugees from the Middle East. He campaigned against immigrants during the presidential campaign of 2016. He urged building a wall on the Mexican/American border. He urged complete cessation of admittance of Muslim immigrants. He used lurid language to describe undocumented immigrants coming to the United States from Mexico and Central America, citing gangs, murders, and rape. He never discussed the contributions of immigrants to the American economy or the facts that they committed fewer crimes and made less use of safety-net programs than American citizens.

Trump opposed admitting most or all of these refugees to the United States on grounds they would become terrorists or criminals. Obama and Clinton countered by arguing that they and presidential predecessors had implemented careful and years-long vetting of refugees before they were admitted to the United States. They demonstrated that no vetted refugee had become a terrorist because all terrorists, aside from ones who bombed the Twin Towers in 2001, were homegrown children of immigrants who were U.S. citizens. Nor did a refugee or immigrant from the Middle East engage in terrorism during the first year and one-half of the Trump presidency.

Trump's assault on immigrants continued after his election in 2016. On three separate occasions, federal courts overturned his quest to ban Muslim refugees from specified Middle Eastern nations, but finally the U.S. Supreme Court approved a modified ban in June 2018.

President Obama confronted a major immigration crisis in the United States with respect to youth brought to the United States when they were children by undocumented parents. They had become "Americans" in the fullest sense but not citizens. They had attended American schools. Many of them had entered or graduated from colleges. Many were in the workforce. Many conservatives demanded that they be deported as undocumented immigrants, even though polls showed that most Americans favored giving them a path to citizenship. Obama proposed a federal Development, Relief, and Education for Alien Minors (DREAM) Act that would give them this path but could not get it enacted by the gridlocked Congress. The nonpartisan Congressional Budget Office (CBO) predicted that that the DREAM Act, if enacted, would increase government tax revenues by $2.3 billion from the federal taxes that employed DREAMers would pay over the next 10 years (CBO, 2010). Using his presidential powers, Obama issued a regulation in 2012 called DACA. It enabled DREAMers to receive renewable two-year work permits from the Department of Homeland Security to stay in the United States if they gave their fingerprints, places of employment or study, and their addresses. Roughly 844,000 of them registered. Many DACA youth feared that ICE might use their personal information to find and deport them.

President Trump vacillated about whether to grant DACA youth a path to citizenship. He eventually relented to conservative Republicans who opposed it. No path to citizenship occurred before the 2018 midterm elections or even during the Trump presidency due to Trump's ambivalence and continuing opposition from conservative Republicans.

Many economic enterprises have supported immigration because they see it as essential to their viability. These include ranchers in the Southwest; farmers in California, the West, the Midwest, and the South; owners of hotels and motels throughout the nation; agricultural processing industries; nursing and convalescent homes; and day care centers. They include technology companies who vigorously recruit workers in high-tech firms and from universities in India, East Asia, Russia, and Europe.

Trump and his attorney general, Jeff Sessions, confronted "sanctuary cities" that had decided not to refer undocumented persons to ICE. Their police forces and prisons would not inform ICE of names and addresses of undocumented immigrants unless they had committed violent crimes. Why should we provide this information, they argued, when the U.S. Constitution gives the federal government—not states and local communities—the power to regulate immigration? Trump issued a regulation five days after taking office that local governments that refused to cooperate with federal immigration authorities would be ineligible for federal grants. A federal judge ruled in April 2017 that Trump's regulation was unconstitutional and could not be enforced because it violated the 10th Amendment to the U.S. Constitution that protects states from federal interference. Additional court hearings ensued in which lawyers representing cities argued that the injunction is "a weapon to defund sanctuary cities that don't comply with the policy the president prefers." These lawyers did not trust Attorney General Jeff Sessions, who argued the federal government would make only minor cuts in federal grants to cities. Further legal proceedings were under way in May 2018 (Dolan, 2018).

Trump went on an anti-immigrant crusade in the spring of 2018 to energize his base of support. He pointed to a "migrant caravan" that consisted mostly of women, children, and youth from Central America that moved northward through Mexico to the border. Although Trump portrayed it as a massive incursion on the American border, such caravans had existed for at least the five prior years with persons in danger of death from political persecution and murder in the northern triangle of Honduras, Guatemala, and El Salvador, one of the most violent regions in the world. Like prior caravans, it was relatively small in size, traveled as a group for safety, and sought asylum at the border in accordance with international and American law. Relatively few of them obtained asylum in prior years. The commissioner of U.S. Customs and Border Protection claimed he lacked staff to process even this small number of persons (Semple & Jordan, 2018).

Trump ordered states to deploy National Guard troops to the border to curb illegal immigration even though no crisis existed (Lima, 2018). He renewed his attacks on sanctuary cities even though a panel of Republican-appointed judges upheld an injunction from another court that blocked the Justice Department from enforcing new grant conditions (Gerstein, 2018). Trump attacked "dumb laws" that lead to "big flows of people" trying to cross the U.S.-Mexico border (Griffiths, 2018). The Trump administration sought to block the ability of immigrant pregnant teens held in federal custody to have abortions only to be told by a federal court that this policy is unconstitutional (Associated Press, 2018).

Trump declared "zero tolerance" of undocumented migration across the Southern border of the United States in May 2018. Without divulging his plan to the mass media or even some of his top aides, Trump and Stephen Miller, Trump's senior policy adviser and designer of his hardline approach to immigration, hatched a scheme to detain thousands of families at the Southern border, separate parents from their children, and deport the parents as soon as possible. Prior to this plan, large numbers of families entered the United States and were slapped with a misdemeanor and allowed to proceed into the United States, where many of them took jobs in agriculture and other industries. They put their plan into place in June 2018, secretly flying several thousand children to organizations in New York and other cities while detaining their parents near the border. Once the plan was publicized, a massive protest took place across the nation including Republicans and Democrats, evangelical voters, and religious groups that questioned the morality of separating parents from children. Trump backtracked in light of opposition from members of his own party but then issued statements and a regulation that were unclear. A federal judge ruled that separation of immigrant children from their parents was unconstitutional and issued a deadline to reunite families that Trump had failed to meet by mid-July 2018.

BACK TO VALUES AND DATA

All of us have to decide where we stand on immigration. Do we stand with persons who contend that the United States is an immigrant nation? Do we believe that Americans have an ethical duty to help refugees who escape civil wars, droughts, family violence, and dictatorships; persons fleeing from poverty; and innocent children brought into the United States by undocumented parents? Do we support immigrants, as well, because they make important contributions to the nation's economy? Or do we take the side of opponents of immigration by viewing them as threats to the nation's safety as well as the jobs of many citizens? In making these decisions, we need to examine data, which indicates low levels of crime among immigrants, low use of the nation's safety net and health programs as compared with American citizens, and often obtaining employment in jobs that American citizens do not want.

In the longer term, political forces such as the growing power of Latino voters will impact immigration policies. Latinos comprise roughly 20% of the American population. An upsurge of voting by Latinos in local, state, and federal elections has markedly changed the political dynamics of immigration. More than 70% of Latinos voted for President Barack Obama in 2012 as well as Democratic candidates at all levels of government. Were immigrants suddenly to disappear, large sections of the American economy would suffer, including agriculture, technology, tourism, senior care, and construction. Advocates of immigrants organized massive marches in the first two years of Trump's presidency. He was unable, however, to persuade Mexican authorities to pay for the wall. Advocates of immigrants contended, as well, that immigrants had not committed terrorist acts except for the attack on the World Trade Center on September 11, 2001.

More than 80% of Americans supported leniency for DACA youth because they are innocent victims—and many were in high school and college. Federal courts repeatedly overruled Trump's decisions, including a federal judge who ruled that the government cannot revoke DACA recipients' work permits or other protections without giving them notice and a chance to defend themselves—and another judge who ruled that the government had to continue renewing the work permits while legal challenges went through the courts (Shoichet & Kopan, 2018).

ANALYZING SEVEN CORE PROBLEMS IN THE IMMIGRATION SECTOR

Core Problem 1: Engaging in Advocacy to Promote Ethical Rights, Human Rights, and Economic Justice—With Some Red Flag Alerts

Immigrants often confront situations where their rights are violated as the following Red Flag Alerts indicate. They will often need legal assistance as revealed by examining cases handled by volunteer attorneys and law students at the Stanford University law clinic (see http://www.law.stanford.edu/immigrants-rights/-clinic/).

- **Red Flag Alert 13.1.** Specific immigrants are victimized by employers, realtors, moneylenders, or the police and need assistance in finding remedies from immigration attorneys or public officials.
- **Red Flag Alert 13.2.** Specific immigrants are detained in centers that do not meet standards of decency and need assistance in finding remedies from immigration attorneys or public officials.
- **Red Flag Alert 13.3.** Specific immigrants experience threats to their privacy through illegal searches and seizures or surveillance by Homeland Security and need assistance in finding remedies from immigration attorneys or public officials.
- **Red Flag Alert 13.4.** Specific immigrants are wrongly denied specific benefits or services and need assistance in finding remedies from immigration attorneys, legal aid, or public officials.
- **Red Flag Alert 13.5.** Staff in agencies that disperse public benefits wrongly act as immigration enforcers by demanding documents, asking unnecessary questions, and issuing unnecessary warnings to specific immigrants—and these immigrants need help in finding remedies from immigration attorneys, legal aid, or public officials.
- **Red Flag Alert 13.6.** Specific refugees fail to receive visas or eventual naturalization, so they need help from immigration attorneys, legal aid, or public officials.
- **Red Flag Alert 13.7.** Specific immigrants are wrongly placed in the "unqualified" category established by the welfare legislation of 1996 and wrongly deemed to be ineligible for specific federal public benefit programs, so they need help from immigration attorneys, legal aid, public officials, or advocacy groups.
- **Red Flag Alert 13.8.** Specific immigrants placed in the "unqualified" category established by the 1996 welfare legislation are wrongly declared to be ineligible for protection of their lives and safety from work-safety regulations, so they need help from immigration attorneys, legal aid, or public officials.
- **Red Flag Alert 13.9.** Specific immigrants placed in the "unqualified" category established by the 1996 welfare legislation are wrongly deemed to be ineligible for Meals on Wheels, shelters for homeless persons, summer food programs, medical care from public systems of care, public education, and services from specific not-for-profit agencies.
- **Red Flag Alert 13.10.** Specific immigrants fail to ask for a reliable immigration attorney if ICE arrests them.
- **Red Flag Alert 13.11.** Specific immigrants are deported without a hearing before a judge so they need legal counsel (roughly 160,000 immigrants are annually deported under these conditions).
- **Red Flag Alert 13.12.** Specific immigrants are subjected to racial profiling including searches and seizures without sufficient cause or denial of benefits without sufficient cause.

- **Red Flag Alert 13.13.** Specific immigrants or immigrant communities are subjected to intrusive police actions such as searches without probable cause, including knocking on doors in the middle of the night to catch "fugitive aliens" (Weiland & Sylvester, 2008).
- **Red Flag Alert 13.14.** Specific immigrants are not granted asylum when they face genuine threats of detention, injury, and death if they return home.
- **Red Flag Alert 13.15.** Victims of human trafficking are often lured to the United States through false promises of better income and lives but are prevented from returning to their homelands or leaving prostitution by coercion from their captors (ACLU, 2011).
- **Red Flag Alert 13.16.** Immigrants lack legal representation when they are threatened with deportation due to minor or old drug offenses.
- **Red Flag Alert 13.17.** Immigrants do not realize that they may qualify for a U visa, which protects victims of crime even if they are undocumented.
- **Red Flag Alert 13.18.** Roughly 5.5 million children have undocumented immigrant parents—and about 75% of these children are U.S. citizens because they were born in the United States. More than 100,000 parents with citizen children have been deported from the United States since 1999 (USDHS, 2009). These citizen children are subject to hardships and trauma that often negatively impact them, such as fear of deportation and possible separation from deported parents (Chaudry et al., 2009). A gap in immigration law means that the interests of these citizen children are not considered in many deportation proceedings despite that fact that nations such as Spain, as well as the United Nations, have or favor regulations not to separate family members through deportation (Finno, 2010). Social workers should become advocates for children.

Background

In the wake of the attack on the World Trade Center on September 11, 2001, surveillance at the nation's borders, particularly the border with Mexico, was tightened with National Guard troops, sensors, cameras, and aircraft. Immigrants often died as they crossed more remote deserts. They paid higher fees to so-called coyotes that escorted them across the border. These innovations condemned increasing numbers of immigrants to deaths and injuries (ACLU, 2009). The Department of Homeland Security (DHS) instituted the National Security Entry-Exit Registration System (NSEERS) in the wake of 9/11, requiring men and boys from Arab- and Muslim-majority nations to register with DHS. NSEERS led to deportations of thousands of Muslims for civil immigration violations and "brought to abrupt end to their productive jobs, property ownership and community ties including to U.S. citizen family members" (ACLU, 2011). Some legislators demanded a national ID to identify undocumented persons, that is, a national biometric system of required Social Security cards that would serve as an employment verification system but could also serve in travel, voting, financial transactions, student identification, and other areas (ACLU, 2010). Would this innovation, some persons wondered, intrude on the privacy of American citizens while not preventing hiring of undocumented immigrants through false documents and corrupt employers? Other threats to privacy arose, including unauthorized use of credit card, social networking, travel, and mobile phone surveillance by Homeland Security.

It is sometimes difficult to determine what rights immigrants possess due to ambiguities and flux in existing laws (Leicher, 2004). When immigrants seek a change in their immigration status, for example, such as obtaining permanent residency through a green card, immigration officials sometimes deny their requests because they might become "public charges." It is often unclear, however, how they decide that someone will become a charge—and someone may be denied a green card due even to relatively brief use of food stamps. In fact, few government agencies have sought reimbursement for immigrants' use of public benefits (National Immigration Law Center, 2005).

Immigrants often discover that specific protections available to American citizens do not exist for them in federal legislation (such as minimum wage protections) or exist in state laws that provide lesser protections. Federal and state authorities do not enforce many protections, however, against sexual harassment, employer negligence that leads to injuries, and workers' right to workmen's compensation (ACLU, 2006). Living in a state of legal and social limbo, it is not surprising that undocumented immigrants are often victimized because the perpetrators realize that law enforcement and government officials often will not come to their defense or enforce their rights because they are a marginalized population—and because immigrants often do not assert their rights for fear of deportation. Immigrants are abused by many employers who pay them even less than the minimum wage and expose them to dangerous work conditions, poor housing, and lack of basic services in this legal limbo. They are subject to hate crimes. They are victims of harsh legislation as reflected by welfare and immigration legislation in 1996 by the Congress and harsh policies recently enacted by legislatures in Arizona, Georgia, Alabama, and South Carolina.

Resources for Advocates

Principles in the U.S. Constitution, state constitutions, and legal precedents are often used to block implementation of harsh provisions against immigrants, including unlawful search and seizure, invasion of privacy, and racial profiling—as well as the constitutional provision that gives the federal government rather than states the sole authority over immigration matters. Many immigration advocates have argued that use of local law enforcement officials to act as immigration agents invites discrimination against people who look foreign but who are American citizens and legal residents.

Numerous ethical issues arise with respect to immigrants. Employers, realtors, moneylenders, and the police often victimize them. They often have their privacy invaded through searches and seizures at variance with current laws. Their families are often divided in deportation proceedings. They are often detained for extended periods in centers that do not meet standards of decency. They fear deportation even from using services and programs to which they are entitled.

An informal contract existed in the United States prior to 2000. Although border patrols made it difficult for undocumented immigrants to enter the United States, those who made it could usually stay if they worked hard—and millions of them received amnesty in 1986 if they could prove residency of four years or more. In the 21st century, however, the U.S. government made immigration far more difficult when it installed or used fences, patrols, aircraft, and sensors at borders that led to higher rates of injury and death as more immigrants used hazardous routes across deserts—and had to pay large sums to so-called coyotes to escort them. Raids on employers have increased, as well, in recent years—leading to increasing numbers of deportations. Many experts contend that immigrants' ethical rights can be protected only by enacting sweeping immigration reforms that allow many persons to migrate to the United States for time-limited employment as legal immigrants under a regulated system that protects their rights and gives them access to programs needed for their safety and well-being.

The term *legal permanent residents* (LPRs) describes persons who have been granted the right to reside in the United States, such as persons with green cards.

Both qualified and unqualified immigrants are eligible for many programs that do not require income eligibility but are required to protect persons' lives and safety. They are eligible, for example, for federal and state safeguards of workers' safety on their jobs, Meals on Wheels, and shelters for homeless persons.

Considerable variation exists among states regarding immigrants' eligibility for specific programs. Whereas the 1996 welfare legislation excludes many immigrants from "federal public benefits," for example, many federal agencies have not specified which specific programs are covered—so state and local agencies are not required to verify immigration status for some programs. The welfare legislation also exempts not-for-profit charitable organizations from obtaining proof of eligibility for such benefits (Los Angeles Coalition to End Hunger and Homelessness, 2010).

States often have considerable latitude in their treatment of immigrants. They can decide whether to grant benefits to immigrants excluded from federal benefits by funding them with state funds. They can decide not to require verification of immigration status in those programs not specifically identified as needing it by federal agencies.

Immigrants' advocates must remember that despite harsh American laws, all immigrants, including undocumented ones, may qualify for prenatal care, emergency and minor consent Medicaid, immunizations for children, Women, Infants, and Children (WIC), school breakfast and lunch, summer food, medical care from the public system of care, public education, help from shelters, and services from many not-for-profit agencies (Los Angeles Coalition to End Hunger and Homelessness, 2010).

Immigrants who are arrested by ICE should remain silent, ask to speak to a reliable immigration attorney, know their alien registration number if they have one and place it where family members can find it, prepare a form or document that authorizes another adult to care for their minor children, and tell family members who do not want to be questioned by ICE to stay away from the place where they are detained (Los Angeles Coalition to End Hunger and Homelessness, 2010).

The Illegal Immigration Reform and Immigrant Responsibility Act of 1996 allowed immediate deportation of LPRs even for minor offenses such as shoplifting instead of prior regulations that required offenses that could lead to five or more years in jail. It restricted the use of waivers to allow many convicted immigrants to remain in the United States on grounds of their close family relations or the length of time they had been in the United States (Morawetz, 2000). It removed the ability of immigration judges to not deport parents because they have children who are U.S. citizens because they were born in the United States—and immigration judges have usually ruled that citizen children should accompany their deported parents because they lack the ability to make decisions about where to live (Demleitner, 2003). It allowed deportees to be kept in jails for months and even up to two years. It allowed the secretary of Homeland Security to permit designated state and local law enforcement officers to perform immigration law enforcement functions rather than relying only on federal officials. The deportation of undocumented persons and other persons who do not have green cards is even more likely than LPRs whether because they have committed crimes or lack visas. They must show that they or their citizen children or spouses would suffer "exceptional and extremely unusual hardship" if they were deported (Demleitner, 2003).

The plight of citizen children of undocumented persons received considerable publicity during raids of worksites by ICE from 1996 to 2008 because many of their parents had been in the United States for long periods and had not committed crimes. Their families were immediately disrupted. Children often did not even have childcare after their parents' arrest. They lacked income with the arrest of breadwinners. They often left the United States abruptly because many arrested parents chose to immediately leave the United States to avoid court hearings and sentences that would make them ineligible to reenter the United States at a later point in time (Capps, Casteneda, Chaudry, & Santos, 2007). Many immigrants feared even to request legal assistance for fear of retaliation by ICE. Families that chose not to leave the United States at once suffered considerable turmoil. Although local agencies and schools often offered them assistance, parents were often detained for months as family and community members took care of their children.

Partly due to public outcries, worksite raids were greatly decreased in 2008 as more emphasis was given to fining employers who did not check applicants' documents. But they increased again during the Obama administration until it decided in 2012 to focus the efforts of ICE on finding and deporting immigrants who had committed crimes.

Core Problem 2: Advocacy for Quality Services for Immigrants—With Some Red Flag Alerts

Research that establishes evidence-based practices and policies for immigrants is evolving, so we often do not know what services work well for them, such as best practices with newcomer youth (Delgado, Jones, & Rahni, 2005). Service providers are sometimes prejudiced against them, such as assigning them to low-achieving classrooms in schools, not offering them translation services, or refusing to serve them. Advocates should insist that immigrants receive comparable services to other persons whenever possible.

- **Red Flag Alert 13.19.** Immigrants are assigned to lower-quality services than others due to prejudice by service providers against them (Fores, 2005).
- **Red Flag Alert 13.20.** Immigrants are not given evidence-based services.

Core Problem 3: Advocacy for Culturally Competent Services for Immigrants—With Some Red Flag Alerts

- **Red Flag Alert 13.21.** LEP immigrants fail to receive translation services in specific social-service settings.
- **Red Flag Alert 13.22.** Persons do not obtain services that are sensitive to their specific culture in specific settings.
- **Red Flag Alert 13.23.** Staff are not diversified in specific service settings.

POLICY ADVOCACY LEARNING CHALLENGE 13.1

CONNECTING MICRO, MEZZO, AND MACRO POLICY ADVOCACY TO ADVANCE IMMIGRANTS' RIGHTS

Advocacy for Immigrant Children

Children and parents are often separated during deportation proceedings. Although ICE officials are mandated to identify and locate parents' children, they sometimes do not. Advocates should insist that children not be separated from their parents such as by allowing parents to remain in the community with electronic monitoring and asking that children be allowed to stay with parents in deportation centers. To the extent children are separated from parents, advocates should seek to have children live with family members where the children can reside pending deportation outcomes. Child welfare social workers should be informed of deportation cases that involve children to be advocates for them.

Micro policy advocacy can be linked with macro policy advocacy to change state or federal law to make clearer that ICE and child welfare officials should work together to keep families intact, including allowing undocumented parents with citizen children to remain in the United States legally. If citizen children often choose to live in the United States when they become adults, why not enhance their well-being by minimizing dislocation and trauma caused by deportation?

Advocacy for immigrant children became a national issue in 2014 as President Obama and the Congress faced thousands of children from Central America who crossed the U.S. border at Mexico seeking asylum as refugees—often without their parents (personal interview with Ms. Maria Garcia, National Public Radio, August 14, 2012). The children often secured attorneys who sought asylum for them under U.S. law that allows refugee status for persons who are victims of natural disasters or wars. Yet the immigration system was stacked against these children because they often did not meet these two conditions, and U.S. law established a ceiling of 5,000 asylum grants. The system also presumes an asylum seeker does not qualify for asylum—a presumption of guilt, unlike U.S. criminal law in which persons have a presumption of innocence. President Obama said he would come up with a solution by the end of the summer of 2014. If he had the political will, he could invoke the Immigration Act of 1990 that allows temporary protected status for persons who seek refugee status.

Blocked by Congress from enacting a DREAM Act to grant children who had been brought to the United States by their parents illegally the right to remain in the United States as they finished their educations and held employment for several years, President Obama established his own program using his executive authority. It granted these persons two years of work before any deportation proceeding could be held if they paid a fee, had no criminal offenses, and were self-supporting. Only about half of the eligible population applied by 2014 because they could not afford the fee or feared that the application process would give information to ICE that would deport them.

Background

The United States is possibly the most culturally diverse society in human history when we include first-generation immigrants and their second-generation descendants—or roughly, for example, 16.9 million children with at least one immigrant parent and 14.6 million second-generation children. Two challenges confront social workers and other professionals: They need to be able to converse with immigrants, and they need to be culturally competent. Conversing with immigrants and their children is difficult when about 41% (25.1 million) are LEP by self-reporting that they speak English "not at all" or "not well." At least 224 languages have been identified, for example, in Los Angeles County, not including differing dialects, whereas only 92 of them have been identified among students at the Los Angeles Unified School District (Los Angeles Almanac, 2011).

Conversing with persons in foreign languages presents daunting challenges. Professionals need to recognize specific words and understand immigrants' questions and assertions at a deeper level, including shades of meaning, subtle expressions, and hidden emotions. Conversations that lack deeper levels of understanding are often superficial ones, particularly when persons discuss important topics like whether to have a life-threatening surgery, to relinquish custody of a child, or to resolve marital conflict. Most of us can sustain conversations at these deeper levels only in a single language—or possibly another language at most. A shortage of professionals who can converse at high levels of proficiency often exists in specific settings such as in mental health clinics. Agencies and programs often lack resources to hire staff who are proficient in specific languages—or to fund translation services over telephones such as ones provided by AT&T.

It is difficult to recruit persons from different ethnic backgrounds even into the profession of social work to decrease shortages of culturally competent staff significantly. Persons from some ethnic groups do not know about social work. The need is particularly acute with respect to recruitment of Spanish-speaking staff in light of the sheer number of Spanish-speaking first- and second-generation immigrants in the United States.

In the case of mental health services, for example, "undocumented persons have the least access and many times the highest need for mental health services" (Derr, 2016). Many immigrant cultures lack understanding of mental health problems even when many of their members possess them.

Resources for Advocates

Title VI of the 1964 Civil Rights Act requires that no person be excluded from federally funded programs because they cannot understand English—and this act has been reinforced by presidential executive orders, court rulings, and policy guidance (Kao & Jansson, 2011). A mélange of state and local laws also exist. California, for example, has more than 150 laws germane to language access as compared with some states that have fewer than 10 of them. The Department of Health and Human Services issued guidance in 2003 that required health providers to give language assistance services by using four factors, including the number of LEP persons served, the frequency that specific LEP persons interact with their programs, the nature or importance of those programs, and available resources. Federal laws and ones in some states generally state that agencies should provide translation services to those populations that constitute 5% or more of their client base. It is widely accepted that minor children should never be used as translators—and that relatives should rarely assume this role and only with the concurrence of immigrant clients.

Hospitals and some health clinics make extensive use of phone-based translation services partly because of statutes and court rulings that require them to obtain patients' informed consent before providing them with medical care. These clinics and hospitals can lose their accreditation and their Medicare and Medicaid funding—and face litigation from patients—if they do not obtain informed consent of their patients. Phone-based translation services are often helpful but have limitations. They are cumbersome because patients and health personnel converse with a bilingual speaker who is not in their presence and who cannot see the patient's body language or other cues.

Core Problem 4: Advocacy for Preventive Services for Immigrants—With Some Red Flag Alerts

- Red Flag Alert 13.24. Immigrants are wrongly informed that they are not eligible for specific programs that may prevent poverty.
- Red Flag Alert 13.25. Immigrants are wrongly informed that they will suffer possible deportation if their children use educational, preschool, and other programs for which they qualify.
- Red Flag Alert 13.26. Immigrants needlessly jeopardize their eligibility for specific programs by inadvertently providing unnecessary information.

Resources for Advocates

Many Americans have been so preoccupied with alleged lawbreaking of adult immigrants that they fail to realize that they often are parents with young children who are citizens because they were born in the United States. (Four million children of immigrants attended schools in 2010—or nearly one student in every public classroom in the United States.) When Americans restrict access of immigrant adults to safety net programs, they are also restricting access of children to these programs—consigning many of them to live in grinding poverty (Yoshikawa, 2011).

Researchers have shown that immigrant children have lower cognitive skills and poorer development when their parents are undocumented partly because their parents are preoccupied with surviving, often holding multiple and onerous jobs; often suffer from anxiety and depression stemming from their poverty and undocumented status; and fear deportation that might even lead to separation from their citizen children. It is likely that children would benefit from two immigration reforms: giving parents temporary and legal work visas and providing them with a path to eventual citizenship (Yoshikawa, 2011). With legal status, parents would be more likely to use safety net programs and search for preschool and center-based childcare programs. They would be more likely to search for better and higher-paying jobs for themselves. They would be less likely to be victimized by employers if they had legal status, including receiving at least the minimum wage. Exclusion of adult undocumented immigrants from GED examinations, as well as from publicly funded education and job training, precludes them from improving their work skills and finding better jobs.

Immigrants and their children often do not realize that they qualify for specific programs in the United States—or fear they may be deported if they step forward to use them. Citizen children of undocumented parents are eligible for more programs than undocumented children of undocumented parents. Eligibility for specific programs sometimes hinges on immigrant-specific visas. Eligibility sometimes hinges, too, on whether specific programs, such as Head Start, have available slots due to funding shortages. Undocumented adults are eligible for specific health benefits but sometimes don't claim them because they are not eligible for other health benefits. Confusion often stems, too, from variations among states. Some states (but not most states) have enacted programs that enable undocumented youth to attend colleges modeled on the DREAM Act introduced in the U.S. Senate in 2001 and reintroduced on May 11, 2011, but defeated by congressional conservatives—and then considered and enacted by such states as California in 2011 to be effective in 2013 (McGreevy & York, 2011). The California legislation gives undocumented students who graduate from high school access to state scholarships and loans for low-income students who attend junior colleges and four-year colleges and universities.

Steadfast opposition of many conservatives to policies to enhance preventive programs for immigrants has blocked many reforms. Liberals counter that children, often brought to the United States at an early age by undocumented parents, should not be punished for acts beyond their control. They argue that preventive programs could help millions of citizen children of undocumented parents who will remain in the United States to be more productive persons.

Some federal regulations deter prevention of immigrants' health problems, such as the five-year ban on immigrants' eligibility for Medicaid and the Children's Health Insurance Program (CHIP) from the point they receive green cards. Policies that deter immigrants who are victims of domestic violence from obtaining safety net benefits, such as the five-year ban or liability of sponsors, make it more difficult for them to leave abusive relationships (National Immigration Law Center, 2005).

All immigrants are eligible for an array of preventive programs linked to their immigration status. Many programs do not have immigration requirements so undocumented persons often qualify. These include prenatal care, emergency Medicaid and Medicaid for minors, immunizations for children, Special Supplemental Nutrition Program for WIC, school breakfast and lunch, summer food, county health care, public education, food pantries, shelters, and services from many nonprofit agencies.

The Los Angeles Coalition to End Hunger and Homelessness (2010) advises undocumented persons as follows:

> These programs don't have immigration requirements. . . . If anyone asks you about your immigration status, be careful. You do not need to tell anyone that you or anyone else who lives with you are undocumented. Your workers do not need to ask about your immigration status if you are not getting benefits for yourself. If they do ask you, simply tell them that you are a "not qualified immigrant" ("not qualified" is not the same as undocumented). That is all they need to know.

The Los Angeles Coalition to End Hunger and Homelessness (2010) also advises immigrants to write "none" on forms, or leave them blank where schools or child centers request social security numbers—and states they may not give the form or information to a government agency. It also advises immigrants, who often do not get correct information or become discouraged, to "be strong . . . insist on talking to a supervisor, and seek out the help of someone who will advocate for you. Insist on speaking to someone who is fluent in your language or call Legal Aid."

Persons who have LEP are entitled to interpreters free of charge, whether from Departments of Public Social Services, Health Services, or the Social Security Administration. These include bilingual workers or telephone interpreter services as well as possible translation or explanations of documents written in English.

Children born in the United States automatically become American citizens—or roughly 4 million children who constitute about one-third of all immigrants' children and roughly one student in each classroom of every American elementary school (Yoshikawa, 2011). These children are entitled to participate in programs available to children of American citizens subject to availability of funding for them in specific jurisdictions. These programs include Head Start; publicly subsidized childcare; kindergarten, primary, and secondary education; and the Supplemental Nutrition Assistance Program (SNAP). These children are eligible for health benefits from Medicaid, CHIP, and family health benefits provided by employers of their parents.

The cognitive development of these children is strongly linked to the extent they and their families take advantage of early-childhood programs as indicated by national studies after adjusting for social class, parental education, and family structure (Yoshikawa, 2011). Conflicting data exist about behavioral development of children of undocumented parents, but a major California study suggests that these children are at risk of delayed development because their parent or parents confront formidable obstacles as undocumented persons. They must often work long hours in two jobs. They experience psychological distress from their marginal legal positions. They are sometimes isolated with weak support systems. They often experience extreme poverty and food insecurity due to low wages and poor working conditions that are exacerbated by lack of enforcement of wage and workplace regulations as well as accumulated debts. They often live in crowded apartments shared by several families. Their parents are often unaware that they are eligible for programs like Head Start, government-subsidized infant childcare, and income-enhancing programs (Yoshikawa, 2011). Undocumented parents and their children are less likely to have a usual source of health care or to use health care at levels of children of documented parents. Their parents are not eligible for public housing or Section 8 subsidized housing.

The children of undocumented immigrants who were born in their nation of origin, but who migrate to the United States, are not entitled to citizenship. They have been allowed to attend public schools—and schools have not divulged their undocumented status to immigration authorities. Recent legislation by some states, such as Alabama, requires school administrators to identify students and parents who are undocumented and to report them to federal immigration authorities—although these laws are subject to legal challenge in the courts.

Clinics and hospitals are required to offer medical treatment to anyone who possesses life-threatening health conditions or who is in child labor under the Emergency Medical Treatment and Active Labor Act (EMTALA) of 1986 as well as laws in many states. These patients cannot be transferred to other health institutions or their communities until they are medically stabilized. They cannot make treatment conditional upon income or citizenship. EMTALA guidelines are available from the Centers for Medicare and Medicaid Services (CMS) at http://www.cms.hhs.gov/manuals/Downloads/som107ap_v_emerg.pdf. Undocumented adults are also eligible for some public health programs offered by specific states or local units of government, such as prevention programs for tuberculosis and HIV/AIDS.

The Fair Labor Standards Act (FLSA) of 1938, which established the federal minimum wage, maximum hours, and overtime pay standards, covers all American workers, including undocumented ones, except volunteers and independent contractors. Yet more than one-third of undocumented workers in New York City work below the federal minimum wage, work more than 40 hours per week, and do not receive overtime pay required by the FLSA to be at least 1.5 times normal pay (Yoshikawa, 2011). (Work hours of these undocumented parents routinely exceed 54 hours a week.) Undocumented workers are usually employed, as well, in jobs with low job autonomy—and both low wages and low job autonomy are linked to low child cognitive ability at 36 months (Yoshikawa, 2011). These poor work conditions and pay are caused by poor enforcement of the FLSA because the number of federal inspectors decreased by 31% between 1980 and 2007 (Yoshikawa, 2011).

Work conditions of undocumented persons would likely improve if more of them become unionized, as occurred with undocumented janitors in Los Angeles since the late 1990s in the wake of registration of many Latino voters and an increase in pro-Latino political leaders. But this will occur only if immigrants are organized politically and if community organizations are developed that advocate their needs.

Immigrants often prefer non-welfare programs like WIC that give them non-cash resources such as food and nutritional counseling that help their children—as well as obstetrics and child-birth medical services and primary care health clinics. They often prefer to receive these services from immigrant-friendly organizations in their communities, such as ones in Chinatowns of New York and Los Angeles. They prefer programs that do not require extensive paperwork and complex applications. These programs need to be expanded to include parenting and child development programs (Yoshikawa, 2011).

POLICY ADVOCACY LEARNING CHALLENGE 13.2

CONNECTING MICRO, MEZZO, AND MACRO POLICY INTERVENTIONS TO PREVENT SOCIAL PROBLEMS

Improving Cognitive Skills of Immigrants' Children

Research demonstrates that immigrant children benefit cognitively from participating in preschool and educational programs. Advocates can use micro policy advocacy to obtain their enrollments in these programs while also working against enactment of policies that require administrators of these programs to ask parents where they are documented.

We have discussed how participation in preschool and center-based childcare enhances cognitive levels of immigrant children of undocumented parents. Advocates can engage in micro policy advocacy to link specific immigrant families to these services. They can also work to form community organizations in communities with high concentrations of immigrant families that can provide micro policy advocacy and develop outreach to immigrants to inform them of benefits of participating in these programs and helping increase their access to them (Yoshikawa, 2011).

Core Problem 5: Advocacy for Affordable Services for Immigrants—With Some Red Flag Alerts

- **Red Flag Alert 13.27.** Immigrants are wrongly informed that they do not qualify for safety net programs.
- **Red Flag Alert 13.28.** Undocumented parents, who may not qualify for specific safety net programs, are not informed that their citizen children do qualify for benefits.
- **Red Flag Alert 13.29.** Many immigrants do not know they are eligible for specific programs. In California, for example, macro policy advocates developed and distributed a brochure informing them of their right in California to resources from the state's California Cash Assistance Program for Immigrants (CAPI), which provides monthly financial assistance to certain elderly, blind, or disabled non-citizens not eligible for Supplemental Security Income (SSI) due to their immigration status (http://law.stanford.edu/immigrants-rights-clinic/).
- **Red Flag Alert 13.30.** Micro policy advocates help immigrants surmount adverse financial impacts that they encounter due to specific federal policies, like the five-year bans on use of CHIP and Medicaid by green card holders and time limits placed on use of SSI by some refugees. Although these time limits were temporarily extended in 2008 by the Bush administration, they have now expired, so almost 50,000 time-limited non-citizen SSI recipients lost their SSI benefits after August 2011—with half on SSI due to disabilities and half to extreme age, including "Kurdish victims of Saddam Hussein, Jews who were persecuted in Russia, Hmong tribesmen who fought for the U.S. in Vietnam, and victims of sex trafficking . . . (http://www.nilc.org, "SSI for Refugees, Asylees, and Other Humanitarian Immigrants").

Resources for Advocates

Widespread confusion often exists about eligibility of immigrants for specific programs due to "the complex interaction of immigration and welfare laws, differences in eligibility criteria for various state and federal programs, and lack of adequate training on the rules by agency personnel" (National Immigration Law Center, 2005). Many eligible immigrants are mistakenly denied services or benefits. Immigrants often need expert advocacy from immigration attorneys or from staff of organizations that specialize in assisting immigrants in specific jurisdictions.

A distinction was established in 1996 between "qualified" and "unqualified" immigrants. Qualified immigrants include persons with green cards, refugees, asylees, persons granted withholding of deportation or removal, Cuban and Haitian entrants, persons paroled into the United States for at least one year, conditional entrants, and certain spouses and children who are victims of domestic violence (Los Angeles Coalition to End Hunger and Homelessness, 2010). The term *LPR* describes persons who have been granted the right to reside permanently in the United States, such as persons with green cards. The federal welfare law does not define federal public benefits precisely, but the

U.S. Department of Health and Human Services included Medicaid, CHIP, Medicare, Temporary Assistance for Needy Families (TANF), Foster Care, Adoption Assistance, the Child Care and Development Fund, and the Low-Income Home Energy Assistance Program in 1998. It must be noted, however, that Congress imposed restrictions on use of TANF, Medicaid, and CHIP by distinguishing between those who entered the United States before, on, or after August 22, 1996, when the welfare legislation was enacted. The welfare law barred most qualified immigrants from receiving SSI, SNAP, non-emergency Medicaid, TANF, and CHIP during the five years after they secured qualified status—while exempting refugees, victims of trafficking, veterans, Cuban/Haitian entrants, and Amerasian immigrants from this requirement (National Immigration Law Center, 2005). (About 20 states use state funds to give some of these benefits to immigrants subject to the five-year ban.)

Unqualified immigrants include undocumented immigrants, immigrants with temporary protected status (TPS), immigrants who are permanently residing under color of law (PRUCOL) who are known by immigration authorities to be in the United States but whom the authorities do not plan to deport, persons in the United States on a temporary non-immigrant visa, applicants for U visa/interim relief, and victims of trafficking. Unqualified immigrants are not eligible for most federal public benefit programs. When a federal agency does designate a program as a federal public benefit for which unqualified immigrants are ineligible, federal law requires that state or local agencies verify all immigrants' immigration and citizenship status.

The Affordable Care Act was not kind to immigrants. Roughly one-third of immigrants lacked private or public insurance in 2009 before its enactment in 2010. (Roughly half of these 13.4 million uninsured immigrants were undocumented, whereas roughly one-third were LPRs, and another fifth were naturalized citizens.) Rather than expanding coverage to include them, the ACA did not grant undocumented immigrants health coverage, so they are largely restricted to care for emergency conditions, obstetrics, and some public health preventive programs such as for tuberculosis. Their lack of health insurance often disqualifies these immigrants, moreover, from mental health services unless they develop emergency mental health conditions that qualify them for care in emergency rooms.

Even when immigrants are eligible for specific services and resources, they often avoid them for fear that providers will report them to ICE. Immigrants often need micro policy advocacy to help them navigate the complex rules that the American welfare state has established—and to dispel false fears of deportation when they are eligible for benefits and services.

Many agencies that provide public benefits mistakenly believe that their personnel are supposed to act as immigration enforcers, such as by demanding immigration documents and social security numbers, asking unnecessary questions on application forms, and issuing unnecessary warnings on walls of waiting rooms (National Immigration Law Center, 2005). A series of federal guidance to federal benefit providers have narrowed the questions that agency personnel can ask immigrants and their family members. When agency personnel exceed these limits, they not only violate immigrants' privacy but frighten them away from using services and benefits to which they are entitled.

POLICY ADVOCACY LEARNING CHALLENGE 13.3

CONNECTING MICRO, MEZZO, AND MACRO POLICY INTERVENTIONS TO HELP IMMIGRANTS MEET SURVIVAL NEEDS

Expanding Benefits to Immigrants

Micro policy advocates help green card holders meet their survival needs when they face five-year bans on use of Medicaid and CHIP, such as by informing them that they can use public systems of health care as well as emergency medical services. Macro policy advocates work to eliminate these five-year bans from federal law, following the lead of the National Immigration Law Center. Macro policy advocates can fight for other policies, such as the Immigrant Children's Health Improvement Act (ICHIA) to allow states to provide Medicaid and CHIP to lawfully present children and pregnant women regardless of entry into the United States. They can seek to overturn specific provisions in restrictive immigration legislation enacted in 1996 and 1997.

Core Problem 6: Advocacy to Address Immigrants' Mental Health Problems—With Some Red Flag Alerts

- Red Flag Alert 13.31. Immigrants and refugees who have been subjected to traumatic events in their homelands have not been diagnosed or treated for posttraumatic stress disorder (Marshall, 2005).
- Red Flag Alert 13.32. Some hospital emergency rooms may not medically stabilize immigrants who come to them with mental health problems that endanger their lives or the lives of other persons.
- Red Flag Alert 13.33. Specific immigrants lack support from immigrants from their nation of origin, such as membership in churches, social groups, or community groups. This isolation may cause mental distress (Yoshikawa, 2011).
- Red Flag Alert 13.34. Specific immigrants are wrongly denied mental health services from specific mental health agencies, whether not-for-profit or public ones.
- Red Flag Alert 13.35. Some hospital emergency rooms may not provide quality services to immigrant women who have been sexually or physically abused (http://law.stanford.edu/immigrants-rights-clinic/).
- Red Flag Alert 13.36. Specific immigrants are given inferior mental health services due to prejudice of service providers.
- Red Flag Alert 13.37. Some immigrants do not obtain mental health services because they cannot afford fees charged by not-for-profit and public agencies at a time when many of them have budget shortfalls. Or they may encounter long waits due to staff cuts caused by these budget shortfalls.

- **Red Flag Alert 13.38.** Immigrants do not receive services for substance abuse from public and not-for-profit counseling agencies due to long waits caused by funding cuts or failure of their staff to prioritize them.
- **Red Flag Alert 13.39.** Immigrants lack legal representation when they are threatened with deportation for a minor drug offense that requires no jail time (http://law.stanford.edu/immigrants-rights-clinic/).

Resources for Advocates

The marginalized status of immigrants, as well as their poverty, causes or exacerbates mental problems for a significant number of them. Unauthorized immigrants have the poorest access and the greatest need for mental health services, even though many of them experience anxiety, depression, and other mental health disorders due to their poverty, marginal status, and fears of deportation (Salinas, 2009). Many immigrants and refugees have been subjected to trauma in their native lands, including rape, murder, and civil wars—often causing posttraumatic stress disorder and other mental conditions (Marshall, 2005).

Translation services are unavailable in many not-for-profit agencies because they are not subject to government regulations when they receive no federal or state funds and because budget shortfalls make it difficult to purchase telephone translation services.

Some barriers to services derive from immigrants' culture. Many Asian and Asian American persons, as well as immigrants from many other nations, do not like to admit or talk about mental health problems with family members or with mental health professionals. They often do not even visit mental health clinics. They often rely upon herbs and other natural substances. Mental health conditions may be most likely among immigrants who lack strong supports from other immigrants from their nations of origin (Yoshikawa, 2011).

All immigrants qualify for in-kind programs that protect their lives and safety including child and adult protective services. All immigrants qualify for mental health services for life-threatening mental health conditions, such as psychoses, depression, and anxiety, that make it likely they have attempted suicide or are at risk of attempting it. Some immigrants qualify for mental health services funded by Medicaid. Immigrants qualify for mental health services provided by many not-for-profit agencies. Some immigrants qualify for mental health services funded by states and local units of government. Some immigrants cannot afford fees that are charged by mental health clinics that have become more onerous with cuts in government funding as well as donations to not-for-profit agencies.

Core Problem 7: Advocacy to Link Services for Immigrants to Their Communities—With Some Red Flag Alerts

- **Red Flag Alert 13.40.** Immigrants receive services and benefits from professionals who do not know where they live, their living conditions, or the nature of their work. Lack of knowledge of these realities makes it difficult for professionals to be micro, mezzo, or macro policy advocates.

POLICY ADVOCACY LEARNING CHALLENGE 13.4

CONNECTING MICRO, MEZZO, AND MACRO POLICY INTERVENTIONS TO HELP IMMIGRANTS OBTAIN MENTAL HEALTH SERVICES

Improving Translation Services in Mental Health Agencies

Assume that you work for a not-for-profit mental health counseling agency and have engaged in micro policy advocacy to secure translation services for immigrants by referring them to other agencies due to the paucity of in-house translation services. You decide that you want to develop a plan for increasing translation services in your agency. How might you engage in mezzo policy advocacy to achieve approval of this plan by the executive director? How might you advance to macro policy advocacy to change city- or state-wide regulations related to translation services for immigrants in mental health settings?

Barriers

Many professionals possess limited knowledge of the geographic location of immigrants even in their immediate areas. A list of the 20 American cities with the most immigrants as a percentage of their populations illustrates the sheer number of immigrants (https://www.thedailybeast.com/20-us-cities-with-the-most-immigrants), as seen in Table 13.2. Visit this website and identify which immigrant groups are most numerous in these 20 cities. An advocate working in these cities would want to identify specific ethnic enclaves within them to better understand their needs and resources.

Urban enclaves of immigrants exist in every metropolitan area in every state as well as smaller towns and rural areas in light of the spread of Latinos and other immigrants throughout the United States. Immigrant farmworkers and ranch workers, disproportionately from Mexico and Central America, have proven indispensable to American growers and ranchers for more than a century. Some immigrants are migrants who relocate frequently, such as farmworkers who move to new locations as crops are planted, cultivated, and harvested.

Staff in agencies and programs who help these populations need to use information from throughout this chapter to provide advocacy services for immigrants who are linked to communities and enclaves where they reside. Too often, immigrants are invisible persons who come to and from their jobs but are otherwise anonymous persons who make relatively little use of organized services.

TABLE 13.2 ■ American Metro Areas With the Highest Proportion of Immigrants

City Percentage Number

1. Miami-Fort Lauderdale-Pompano Beach, FL: 36.9%, 1.995 mil
2. San Jose-Sunnyvale-Santa Clara, CA: 36.31%, 0.650 mil
3. Los Angeles, Long Beach-Santa Ana, CA: 34.28%, 4.394 mil
4. San Francisco-Oakland-Fremont, CA: 29.50%, 1.246 mil
5. New York-Northern New Jersey-Long Island, NY-NJ-PA: 28.1%, 5.3 mil

(Continued)

TABLE 13.2 ■ American Metro Areas With the Highest Proportion of Immigrants (Continued)

6. Chicago, Naperville-Joliet, IL: 17.64%, 1.676 mil
7. Dallas-Fort Worth-Arlington, TX: 17.73%, 1.09 mil
8. Washington-Arlington-Alexandria, DC-VA: 20.23%, 1.074 mil
9. Houston-Sugar Land-Baytown, TX: 21.39%, 1.120 mil
10. Las Vegas-Paradise, NV: 21.81%, 1.471 mil
11. Riverside-San Bernardino-Ontario, CA: 22.04%, 0.894 mil
12. San Diego-Carlsbad-San Marcos, CA: 22.64%, 0.672 mil
13. Sacramento-Arden-Arcade-Roseville, CA: 17.21%, 0.358 mil
14. Phoenix-Mesa-Scottsdale, AZ: 16.63%, 0.692 mil
15. Boston-Cambridge-Quincy, MA: 15.94%, 0.716 mil
16. Orlando-Kissmee, FL: 15.85%, 0.321 mil
17. Seattle-Tacoma-Bellevue, WA: 15.57%, 0.514 mil
18. Austin-Round Rock, TX: 14.63%, 0.228 mil
19. Atlanta-Sandy Springs-Marietta, GA: 12.84%, 0.674 mil
20. Denver-Aurora, CO: 12.6%, 0.309 mil

Source: Adapted from Florida, Richard (29 July 2010). U.S. Cities with the Most Immigrants. Daily Beast. Retrieved from: https://www.thedailybeast.com/us-cities-with-the-most-immigrants

POLICY ADVOCACY LEARNING CHALLENGE 13.5

CONNECTING MICRO, MEZZO, AND MACRO ADVOCACY INVERVENTIONS

Invisible Persons

Ramiro Gomez, a Latino folk artist, makes life-sized, cardboard cutouts of immigrant laborers who serve as nannies, gardeners, valet workers, and housekeepers in wealthy areas of Los Angeles. He uses acrylic paint to depict these persons, giving them names. He places them in these areas, such as near George Clooney's home just prior to a fundraiser attended by President Barack Obama. Gomez contends, "We see the beautiful homes. The hedges are trimmed; the gardens are perfect; the children are cared for. We've come to expect it to be this way. But who maintains all this? Who looks after it? And do we treat the workers with the dignity they deserve? Do we stop and notice them?" Sometimes the police, hotel staff, or property owners remove this folk art, which is attached to trees or propped against hedges. The Secret Service asked they be removed from Clooney's neighborhood. Most pieces only make it for a day or two.

Learning Exercise

1. Are immigrants sometimes "invisible," as well, among professionals who serve them or see them?
2. Can professionals effectively link their services to the communities of immigrants if they do not know where immigrant enclaves exist?

THINKING BIG AS POLICY ADVOCATES IN THE IMMIGRATION SECTOR

Immigration policy issues are frequently associated with political conflict. When President Obama, along with a bipartisan group of senators, crafted immigration legislation in early 2013, they knew they would be tested in coming months as Republicans, Democrats, trade unions, agricultural interests, tourism companies, construction companies, persons living in states and communities bordering Mexico, Latino leaders, African American leaders, the Mexican government, and other groups entered the fray. Unfortunately, immigration reform fell victim to political gridlock, so no progress had been made by the congressional elections of 2014. Nor has progress been made through 2018. Yet immigration reform almost was enacted in 2007 as liberal Democrat Ted Kennedy teamed with President George Bush and Republican Senator John McCain. The Comprehensive Immigration Reform Act of 2007 (also known as the Secure Borders, Economic Opportunity and Immigration Reform Act of 2007) proposed to provide amnesty to more than 10 million immigrant residents by creating a Z visa to allow everyone in the United States without a valid visa on January 1, 2008, to have the legal right to remain in the United States for the rest of their lives as well as receive a Social Security number. Holders of this visa would be eligible for a green card once they had paid a $2,000 fine and back taxes for some of the period in which they worked. As with other green card holders, they could begin the process of becoming a U.S. citizen five years later. The bill strengthened border enforcement; funding 20,000 border patrol agents, 105 camera and radar towers, and 300 miles of vehicle barriers. It replaced the employer-sponsored facet of the immigration system with a point-based merit system based on a combination of education, job skills, family connections, and English proficiency.

Several points of contention proved fatal to its passage. Some conservatives argued it needed even more focus on border enforcement and sanctions against employers hiring "illegals" and more stringent standards for obtaining a Z visa. Some liberals argued that it gave excessive concessions to highly skilled immigrants as compared with immigrants with fewer skills, failed to sufficiently support family reunification of immigrants to allow relatives from outside the United States to obtain green cards, rolled back protections for immigrant victims of domestic violence and human trafficking, and created an immigration system based on the lives of men by not giving points for caring for children and elderly or disabled family members.

President Barack Obama promised that he would enact sweeping immigration legislation when he ran for the presidency in 2008 but failed to undertake this task to the chagrin of many Latinos and immigration advocates, partly because he was preoccupied with policies to address the Great Recession of 2007 to 2009 as well as the ACA and banking regulations. He renewed this pledge in his Inaugural Address and his State of the Union Address in early 2013—but it soon became obvious that Republicans would not cooperate with him and Democratic leaders.

Why shouldn't social workers be leaders in seeking comprehensive immigration reforms during the coming years?

Discussion Questions

1. What kind of sweeping immigration legislation should social workers support in light of their code of ethics with special reference to vulnerable populations and social justice?
2. What provisions should protect immigrants' civil rights so that immigrants do not continue to be subject to poor working conditions and low wages?
3. Can the social work profession be a leader in developing support for comprehensive immigration reform?
4. Drawing on materials in this chapter, list groups likely to oppose major immigration reforms like granting amnesty to roughly 10 million persons and establishing a visa status for many working immigrants that allow them to remain in the United States for extended periods.
5. Identify some power resources that advocates of immigration reform will need to use to obtain immigration reforms, including the growing electoral power of Latino voters and female voters.
6. What services and educational opportunities might accompany immigration reform to speed the ability of immigrants to enter the workforce, obtain an education, obtain medical care, and obtain civil rights?

Learning Outcomes

You are now equipped to:

- Develop an empowering perspective with respect to immigrants
- Analyze the evolution of American immigration policies
- Analyze the movement of persons across national boundaries
- Recognize problems created for immigrants by extreme income inequality
- Understand the political economy of the immigration sector
- Get back to values and data
- Analyze seven core problems in the immigration sector
- Think big as advocates in the immigration sector

References

ACLU (2006). *Undocumented workers bring plea for non-discrimination to human rights body.* Retrieved from http://www.aclu.org/immigrants-rights/undocumented-workers-bring-plea-non-discrimination-human-rights-body

ACLU (2008). *Immigration myths and facts.* Retrieved from http://www.aclu.org/immigrants-rights/immigration-myths-and-facts

ACLU (2009). *U.S.-Mexico border crossing deaths are a humanitarian crisis, according to report from the ACLU and CNDH.* Retrieved from http://www.aclu.org/immigrants-rights/us-mexico-border-crossing-deaths-are-humanitarian-crisis-according-report-aclu-and

ACLU (2010). *Immigration reform must respect civil liberties, says ACLU.* Retrieved from http://www.aclu.org/immigrants-rights/immiogration-reform-must-respect-civil-liberties-says-aclu

ACLU (2011). *DHS announces indefinite suspension of controversial and ineffective immigrant registration and tracking system.* Retrieved from http://www.aclu.org/immigrants-rights/dhs-announces-indefinite-suspension-controversial-and-ineffective-immigrant-regist

ACLU (n.d.). *Human trafficking: Modern enslavement of immigrant women in the United States.* Retrieved from http://www.aclu.org/immigrants-rights

Associated Press (2018, March 30). Court: Trump administration can't block immigrant teens from obtaining abortions. *Politico.* Retrieved from https://www.politico.com/story/2018/03/30/trump-administration-immigrant-teen-abortions-court-492827

Capps, R., Casteneda, R. M., Chaudry, A., & Santos, R. (2007). *Paying the price: The impact of immigration raids on America's children.* Washington, DC: Urban Institute.

Chaudry, A., Capps, R., Pedroza, J. M., Castaneda, R. M., Santos, R., & Scott, M. (2009). *Facing our future: Children in the aftermath of immigration enforcement.* Washington, DC: Urban Institute.

Congressional Budget Office. (2010). Cost Estimate S. 3992 Development, Relief, and Education for Alien Minors Act of 2010. Retrieved from http://www.cbo.gov/ftdocs/119xx?doc11991/s3992.pdf

Delgado, M., Jones, K. & Rohani, M. (2005). *Social work practice with refugee and immigrant youth.* Boston: Pearson.

Demleitner, N. V. (2003). How much do Western democracies value family and marriage: Immigration law's conflicted answers. *Hofstra Law Review, 31,* 270–280.

Derr, A. (2016, March). Mental health service use among immigrants in the United States: A systematic review. *Psychiatric Services, 67*(3), 265–274.

Dolan, M. (2018, April 11). Federal appeals court considers "sanctuary city" case. *Los Angeles Times.* Retrieved from http://www.latimes.com/local/lanow/la-me-ln-sanctuary-9th-circuit-20180410-story.html

Finno, M. (2010). *Immigration enforcement in the U.S. and its impact on family separation and child trauma: The Spanish Model as a solution* [Paper written for doctoral policy class]. School of Social Work, University of Southern California.

Fores, G. (2005). She walked from El Salvador. *Health Affairs, 24,* 506–510.

Gerstein, J. (2018, April 19). Appeals court rules against Trump policy punishing sanctuary cities. *Politico.* Retrieved from https://www.politico.com/story/2018/04/19/appeals-court-ruling-trump-sanctuary-cities-537823

Griffiths, B. (2018, April 1). Trump complains about "dumb" immigration laws. *Politico.* Retrieved from https://www.politico.com/story/2018/04/01/trump-slams-immigration-mexico-493075

International Organization for Migration (IOM). (2011). *Facts & figures.* Retrieved from http://www.iomint/jahia/jahia/about-migration/facts-and-figures/lang/en

Kao, D. & Jansson, B. S. (2011). Advocacy to promote culturally competent health services. In B. Jansson (Ed.), *Improving healthcare through advocacy* (pp. 179–210). Hoboken, NJ: John Wiley & Sons.

Leicher, H. (2004). Ethnic politics, policy fragmentation, and dependent health care access in California. *Journal of Health Politics, Policy, and Law, 29*, 177–201.

Lima, C. (2018, April 19). Trump threatens to cut funding from California troop deployment. *Politico*. Retrieved from https://www.politico.com/story/2018/04/19/trump-california-troop-funding-537369

Los Angeles Almanac. (2011). *Languages spoken at home by individual Los Angeles communities*. Retrieved from http://www.laalmanac.com/LA/la10b.htm

Los Angeles Coalition to End Hunger and Homelessness. (2010). *The people's guide to welfare, health, and other services* (33rd ed.). Retrieved from http://www.lacehh.org/tpg/documents/english10PeoplesGuide.pdf

Marshall, G. (2005). Mental health of Cambodian refugees two decades after resettlement in the United States. *Journal of the American Medical Society, 294*, 571–579.

McGreevy, P. & York, A. (2011, October). Brown signs California Dream Act. *Los Angeles Times*, p. 1.

Migration Policy Institute. (2011). *Frequently requested statistics on immigrants and immigration in the United States*. Migration Information Source. Retrieved from http://www.migrationinformation.org/feature/display.cfm? ID=818

Morawetz, N. (2000). Understanding the impact of the 1996 deportation laws and the limited scope of proposed reforms. *Harvard Law Review, 113*, 1950–1954.

National Immigration Law Center. (2005). *Overview of immigrant eligibility for federal programs*. Resource Manual: Low-Income Immigrant Rights Conference.

Pew Hispanic Center. (2011). *The toll of the Great Recession*. Retrieved from http://pewhispanic.org/reports/report.php?ReportID=145

Pew Hispanic Center. (2012, April 23). *Net migration from Mexico falls to zero—and perhaps less*. Retrieved from http://www.pewhispanic.org/2012/04/23/net-migration-from-mexico-falls-to-zero-and-perhaps-less/

Preston, J. (2013, January 13). Obama will seek citizenship path in one fast push. *New York Times*, pp. 2, 21.

Semple, K. & Jordan, M. (2018, April 29). Migrant caravan of asylum seekers reaches U.S. border. *New York Times*. Retrieved from https://www.nytimes.com/2018/04/29/world/americas/mexico-caravan-trump.html

Shoichet, C. & Kopan, T. (2018, February 27). Court hands DACA recipients another victory. CNN. Retrieved from https://www.CNN.com.

U.S. Department of Homeland Security (DHS). (2009). *Removals involving illegal alien parents of United States citizen children*. Washington, DC: Office of the Inspector General.

Weiland, J. & Sylvester, A. (2008, May 6). Unlawful immigration raids should trouble all Americans. *San Francisco Daily Journal*, p. 6.

Yoshikawa, H. (2011). *Immigrants raising citizens: Undocumented parents and their young children*. New York: Russell Sage.

BECOMING POLICY ADVOCATES IN THE CRIMINAL JUSTICE SECTOR

This chapter was coauthored by Gretchen Heidemann, Ph.D., and Elain Sanchez Wilson, MPP

LEARNING OBJECTIVES

In this chapter you will learn to:

1. Understand the evolution of the health care system in the United States
2. Understand how mass incarceration is powerfully linked to economic inequality in the United States
3. Describe the political economy of the corrections sector, including powerful players and interests as well as underrepresented ones
4. Identify seven problems encountered in the criminal justice sector as well as the policies, regulations, and organizational factors pertinent to them
5. Develop Red Flag Alerts for each of the seven problems at the micro advocacy level
6. Identify policy resources available to policy advocates with respect to each of the seven core problems
7. Identify strategies for moving from micro policy advocacy to mezzo and macro policy advocacy in the criminal justice sector
8. Engage the eight challenges in the multilevel policy empowerment framework in the criminal justice sector

Note: Gretchen Heidemann, Ph.D., and Elain Sanchez Wilson, MPP, made invaluable contributions to this chapter in this book's first edition. Bruce Jansson was the sole author of the second edition, editing and updating contributions from each original contributor.

Many policy makers and advocates are reconsidering the harshness of the American penal system that places far more people in jails than any other industrialized nation. American prisons hold 22% of all incarcerated persons in the world even though Americans make up only 4.6% of the world's population (Lussenhop, 2016). With many competing domestic and international priorities, many politicians on both sides of the aisle are looking for ways to cut the costs of running jails and other institutions that house roughly 2.3 million inmates in the United States. If a growing number of "prison abolitionists" want to terminate all prisons, other critics seek incremental reductions of inmates as well as prison reforms (Arrieta-Kenna, 2018). If the United States does reconfigure its penal system so that it focuses on rehabilitating its prisoners and downsizing prisons, social workers will likely become a larger part of its workforce. As Gumz (2004) noted, "The tenets of social work practice—the innate dignity of the individual, self-determination of the client, confidentiality, moral neutrality, and social justice—were, and are, challenged by a criminal justice system that values order, control, and punishment" (p. 451). As you read this chapter, reflect on ways that social workers can engage in addressing the seven core problems of the policy advocacy framework in the criminal justice sector.

DEVELOPING AN EMPOWERING PERSPECTIVE ABOUT CRIMINAL JUSTICE

Many Americans subscribe to a punitive mission for their criminal justice system. This is understandable: No one likes to have property stolen, monies robbed, and lives ended through criminal action. Some people are violent. Some people do commit fraud. People and pharmaceutical companies do market illegal and addictive drugs. Bank officials caused the onset of the Great Depression in 1929 and the Great Recession in 2007 through rampant speculation and bank fraud. Some people commit crimes so heinous that many onlookers favor a death penalty.

We all agree that the United States needs a network of laws, as well as police forces and courts, to detect criminal behavior. It needs courts to enforce these laws. It needs to incarcerate some criminals for varying lengths of time. Yet the United States can ill afford a criminal system that houses 2.3 million persons. Nor should the nation rely almost exclusively on a system that fails to rehabilitate hundreds of thousands of inmates who leave correctional institutions with almost no job preparation, housing, mental health services, and medical care or that houses hundreds of thousands of person awaiting court hearings simply because they cannot afford bail (Wagner, 2015).

A good case can be made that the United States incarcerates many people who commit relatively minor property crimes, smoke marijuana, or have driving infractions. The Brennan Center for Justice estimates that as many Americans have criminal records during their lifetimes as college diplomas, such as the 70 million persons arrested and fingerprinted by local, state, or federal agencies (Jansson, 2019a; Friedman, 2015). Most of the 70 million persons are not incarcerated, but many do end up in correctional institutions. This "arrest epidemic" morphed into mass incarceration from the mid-1970s to the present, increasing numbers of persons in jails and prisons from roughly 600,000 inmates in 1975 to roughly 2.3 million inmates in 2016.

An excessively strict code of justice has other shortcomings on several counts. Take the decision decades ago to declare use and distribution of marijuana to be a criminal offense under federal law—a decision that led hundreds of thousands of persons to be incarcerated, often for many years in federal prisons. Or take the movement that enacted "three strikes and you are out" in many states, leading to incarceration of many persons who had committed three crimes even if all of them were nonviolent, like petty theft. Realize, too, that bias and prejudice often enter into the arresting process. Law enforcement arrested and incarcerated large numbers of African Americans for using or distributing crack over many decades while not arresting or incarcerating many white Americans for using cocaine powder. They incarcerated large numbers of Latinos as well. Corporations that financed for-profit prisons became a huge industry and hired many lobbyists to pressure local, state, and federal officials to fund more prisons. Unions of jail and prison workers pressured local, state, and federal officials to build jails and prisons. Wanting more jobs in their jurisdictions, many politicians from semi-rural and rural areas pressured public officials to build prisons to create jobs for their constituents.

Americans have often resorted to incarceration rather than other forms of punishment even when data shows that incarceration is less effective than community-based penalties, such as specified hours of community service and keeping persons in their homes with electronic ankle bracelets. Data shows that persons who are incarcerated are more likely than other persons to re-commit crimes that bring them back to jails and prisons (Kann, 2018). Employees of prisons and jails often expect inmates to return. Other inmates socialize newcomers to crime. Incarceration often destroys family relationships. Inmates often find it impossible to find jobs when they exit prisons and jails because employers do not want to hire ex-prisoners or people who lack employment records. Jails and prisons, moreover, cost extraordinary amounts of money to build and maintain.

It is difficult to decide what penalties to impose on people for specific criminal behavior. How many years, for example, should an alcoholic driver receive who causes severe injuries to a pedestrian versus a person who robs a bank versus someone who uses an illegal drug? When is leniency acceptable versus strict enforcement to the letter of the law? All too often, judges resort to excessive sentences for a variety of reasons. They are not versed in psychology or social work, so they often wrongly assume that persons cannot be rehabilitated. Courts often lack sufficient numbers of psychologists, social workers, and psychiatrists to determine if specific persons are good candidates for immediate release, community service, or speedy release. They are often harsher in their sentences for persons of color than for whites. Released inmates often cannot find health and mental health services, much less help in finding jobs and housing. A huge number of inmates are housed in local jails awaiting a court hearing but unable to afford bail.

Considerable controversy has long existed about the most extreme sentence—the death penalty. Critics contend that many murders are crimes of passion rather than calculation—and many of them take place when the murderer is high on drugs or alcohol. Some murders are defensive in nature, such as when women murder spouses or partners who are physically abusing them or threatening to kill them. Some murders are unintentional such as when someone shoots a rifle into the air only to have the bullet kill someone when it descends. Some wardens of prisons trust some convicted murderers sufficiently that they have them babysit their children—a decision that suggests that some, even many, murderers on death row have been rehabilitated. Yet other murderers, such as Charles Manson, a cult figure who planned and executed the murders of movie actress Sharon Tate and others in the 1960s, appeared unrepentant even after spending 48 years in prison (Gilbert, 2017).

Many critics of the death penalty contend that life imprisonment is preferable to killing human beings, no matter how "humane" the drugs or other means of killing may be. Pope Francis came to this conclusion in late 2018 when he vowed to oppose death penalties around the world. Many Americans also came to this conclusion when they learned that many prisoners on death row and many executed prisoners were exonerated by DNA analysis or by belated confessions of other persons.

A study estimated that DNA and belated confessions exonerated one in 25 persons who were sentenced to death, resulting in an untold number of innocent persons to be executed and many others to "languish in prison and never be freed" (Levy, 2014). Many nations and 18 American states have abandoned the death penalty in favor of life imprisonment—and Colorado, Oregon, and Washington have formed moratoriums on executions imposed by their governors. Yet others, such as Norway, do not favor life imprisonment except in extraordinary circumstances.

Probation officers are charged with working with inmates after they have been released to the community as well as with persons who receive sentences that require them to work for a specific number of hours in community service. Probation departments often lack sufficient staff to monitor ex-inmates, much less to help them obtain job training, social services, and resources from the safety net sector. Many probation officers restrict their work to discovering whether released prisoners commit crimes rather than helping them find jobs and housing.

Other factors skew the American criminal justice system toward punishment rather than (also) rehabilitation. Correctional personnel are often primarily charged with keeping law and order in correctional institutions rather than helping prisoners surmount mental health, economic, familial, and other problems that led them to commit a crime in the first instance. They often view themselves as law enforcement officers rather than persons who work with prisoners so that they thrive in society after their release. They have to deal with prisoners who live (in effect) in cages, have minimal exercise, receive little education and job training, and realize they will leave prison with almost no money and no imminent employment. Some end up on the streets.

From Mass Incarceration to Lower Numbers of Inmates

The momentum toward mass incarceration began in the United States in the mid-1970s when roughly 600,000 persons were incarcerated, but it grew markedly from 1980 to 2010 when it had roughly 2.4 million inmates. It was a mass incarceration because the United States came to have a far greater percentage of its population in jails and prisons than any other industrialized nation. If the United States has a prison population of 655 per 100,000 persons, other industrialized nations had far fewer. In descending order, they have these levels: New Zealand (219), Australia (167), Scotland (143), England and Wales (140), Portugal (127), Canada (114), France (104), Austria (98), Italy (96), Greece (94), Switzerland (81), Ireland (79), Germany (78), Norway (74), Denmark (59), the Netherlands (59), Sweden (57), Finland (52), and Japan (45) (Kann, 2018). The United States has 2.3 million prisoners or almost the size of Chicago's population. It has more than three times the number of inmates as a percentage of its population than the next closest industrial nation—New Zealand.

Many Americans contend that they cannot learn lessons from other industrialized nations because they are so different from the United States. They ascribe their low incarceration rates to their small populations and their socialist economies. None of these arguments are valid. France and Germany have substantial populations, respectively 67 million and 81 million persons. None of these nations are socialist: They all have capitalist economies and rely heavily on exports. Many nations had death penalties and large populations of inmates but chose to move in the direction of rehabilitation, such as Norway in the 1990s.

Take the case of Norway, which has the official goal of getting inmates out of its penal system (Benko, 2015). It has none of the sights that are characteristic of American prisons, such as coils of razor wire and towers with snipers. Halden, a Norwegian prison, seeks to prepare inmates for life in the community. It has no death penalty and a maximum sentence of 21 years, even for Anders Breivik who killed 77 people and injured hundreds more in an attack in Oslo and a nearby summer camp—although Norway allows additional years of incarceration if an inmate is judged to be dangerous to others. Halden develops post-release plans for each inmate that include help in securing homes, jobs, and access to a supportive social network. It spends $93,000 per inmate as compared with $31,000 per inmate in the United States. Yet the United States could save roughly $45 billion per year if it incarcerated prisoners at the same rate as Norway. But it wasn't always this way; the Norwegians had a criminal justice system akin to the American one as recently as 1998. It was then that they set the goal of using education, job training, and therapy to rehabilitate prisoners. They chose in 2007 to help inmates find housing and jobs with a steady income *before* they are released in a process they called "reintegration." By contrast, American inmates are often placed on buses with petty cash, no jobs, and no housing when they are discharged, not to mention inadequate job and educational preparation in prison.

Every feature of Halden is geared to reduce stress, mitigate conflict, and minimize interpersonal friction (Benko, 2015). It is placed in a park with trees and blueberry bushes. It seeks interpersonal relationships between staff and inmates to maintain safety within the prison. It has no surveillance cameras. Inmates move around unaccompanied by guards. Its isolation cell has not been used in five years. Of its 251 prisoners, half are imprisoned for violent crimes including murder, rape, and assault—and one-third for smuggling and selling drugs. Although two-thirds are Norwegian, many prisoners come from different nations such as from the Middle East; some prisoners are transferred to psychiatric facilities if they cannot live with the other prisoners. A special unit focuses on addiction recovery. Prisoners cook many of the meals. Each inmate's cell has a small refrigerator to store food obtained from communal suppers and weekly visits to the prison grocery shop. Furniture is similar to that in college dorms in the United States rather than the lack of furniture found in cells of most American jails and prisons. If inmates violate rules, they experience swift, consistent, and evenly applied consequences such as cell confinement and loss of TV.

A brief reform movement to make the American criminal justice system more humane was initiated by President Lyndon Johnson as part of his War on Crime when he established the President's Commission on Law Enforcement and Administration of Justice in 1967. The commission's members concluded that the American correctional institutions were "at best barren and futile . . . and are the poorest possible preparation for [inmates'] successful re-entry into society" (Benko, 2015). It advised the nation to develop model correctional institutions that would resemble "a normal residential setting [. . . with] doors rather than bars [where] inmates would eat at small tables in an informal atmosphere [with] classrooms, recreation facilities, day rooms, and perhaps a shop and library." The federal Bureau of Prisons (BOP) constructed model correctional centers in the mid-1970s that placed groups of 44 prisoners in self-contained units with single-inmate cells, a day room, only a single, unarmed correctional officer, wooden and upholstered furniture, porcelain toilets, and other amenities (Benko, 2015).

The tide soon shifted, however, in a coercive direction. Drawing on correctional data, a researcher simplistically concluded that rehabilitation did not work—a conclusion that was rapidly accepted by many public officials. By the time the researcher concluded that his first analysis was erroneous and that data confirmed that "some treatment programs" do rehabilitate inmates, the damage was done (Benko, 2015). The nation quickly moved toward mass incarceration. President Richard Nixon's War on Drugs led to incarceration of many drug dealers and addicts, including for use of marijuana, which was classified as a Schedule 1 substance with high potential for abuse and with no accepted medical use even under the direction of a physician. It joined cocaine and heroin as drugs whose users and dealers were criminalized (Jansson, 2019b). A war-on-crime movement arose among public officials at all levels of government that imposed minimum sentences on persons as well as longer sentences, three-strikes laws, and prosecuting juveniles as adults.

These coercive policies were disproportionately imposed on African Americans, Latinos, and Native Americans who came to be a large majority of inmates. President Ronald Reagan's anti-drug abuse acts established mandatory minimum sentences for drug possession. President Bill Clinton enacted the Violent Crime Control and Law Enforcement Act that included a three-strikes mandatory life sentence for repeat offenders, hiring of 100,000 new police officers, providing $9.7 billion for funding of prisons, and expanding death penalty-eligible offenses. Whereas Clinton's legislation also provided $6.1 billion for prevention programs designed by "experienced police officers," it overwhelmingly reinforced mass incarceration by causing a large increase of prisoners in federal prisons. His legislation also stripped Pell Grant college scholarships from prisoners as well as required prisoners to be evicted from public housing if they were involved in any criminal activity (Lussenhop, 2016).

It is not surprising that many leaders of African American, Latino, and Native American advocacy groups view the American correctional system negatively. The vast majority of inmates came from these three groups for decades to the present time. Many white persons who are convicted of crimes can pay their bail and hire attorneys that allow them to exit the correctional system pending a court hearing, whereas persons of color depended on public defenders who have huge caseloads. Judges' and juries' biases against persons of color often lead them to give them longer sentences than white people who committed the same crime.

A swing toward a less coercive correctional system is currently under way in some states and at the national level. California sought to cut the number of inmates in its correctional institutions when it enacted "realignment legislation" in 2011 by enacting Assembly Bill 109. Large numbers of prisoners in state jails who had not committed acts of violence or serious crimes or sexual crimes were turned back to California's 58 counties—or 30,000 inmates by 2016 (California Radically Revamping, 2016). Researchers sought to determine whether placing more inmates on the streets increased or decreased crime—or had no effect on levels of crime. Researcher David Roodman, after he examined existing data, decided that "decarceration has zero net impact on crime." But he acknowledged that this "estimate is uncertain, but at least as much evidence suggests that decarceration *reduces* crime as increases it." He argued that "while imprisoning people temporarily stops them from committing crime outside prison walls, it also tends to increase criminality *after* release. As a result, 'tough-on-crime' initiatives can reduce crime in the short run but cause offsetting harm in the long run" (Lopez, 2018b).

Yet another factor has to be considered. Some states are cutting the number of inmates in state prisons but not in local jails. Or some states reduce prison and jail populations in their large cities, such as New York City and Rochester in New York State, but not in jails of relatively small towns and counties. Moreover, huge increases in incarceration are occurring in some states and some counties. Moreover, some states "simply reclassified felony crimes as misdemeanors, emptying prisons while filling up jails." Or some states send more people to prison than jail because the state foots the bill for the former (Exstrum, 2018). It is premature in late 2018, then, to declare a significant erosion in mass incarceration in the United States. However, the U.S. prison population fell from a peak of 1.6 million in 2009 to 1.5 million people in 2016 (Lopez, 2018c).

Bipartisan support for reduction in the federal prison system existed in 2018 after President Obama had initiated the release of some federal prisoners who had not committed violent offenses and only had minor drug offenses. The U.S. Senate Judiciary Committee passed the Sentencing Reform and Corrections Act by a 16–5 vote in early 2018.

The bill would decrease mandatory minimum sentences so that judges could use their discretion in setting sentences below a certain threshold. It would make retroactive reductions in the disparity in sentences between crack and powder cocaine to reduce the harshness of sentences for some African Americans for use of drugs. It would increase educational and other preventive programs in federal prisons. But even this modest legislation ran into opposition from law-and-order persons and groups including Jeff Sessions, the U.S. attorney general, as well as law enforcement groups (Watkins, 2018). The House enacted legislation that allowed eligible inmates in federal prisons to earn time out of prison, expanding access of prisoners to educational and other programs and strengthening protections for female prisoners. Many Democrats found the Senate and House bills to be too cautious: They believed major reductions in sentences were needed to stop the flow of persons into prisons in the first place, including Senators Kamala Harris and Corey Booker, who are recently seated on the Senate Judiciary Committee as well the American Civil Liberties Union (ACLU), the Leadership Conference on Civil and Human Rights, and the NAACP Legal Defense Fund. President Donald Trump opposed sentencing reforms, not to mention Attorney General Jeff Sessions. Eric Young, president of the Council of Prison Locals, insisted that measures to rehabilitate prisoners are unsuccessful unless legislation includes sufficient new revenues to allow these measures to be successful (Lopez, 2018a). The First Step Act was finally signed into law in late 2018, relaxing "three strikes laws," easing crack sentences, and allowing inmates in federal prisons with good behavior to leave prisons earlier.

Major prison reforms are unlikely to be enacted unless the flow of inmates into prisons is decreased through major sentencing reforms, the prevention of future crimes is achieved by massive increases of education and counseling within prisons, and major supportive employment, housing, counseling, and medical services are provided to released prisoners. It is highly unlikely that major prison reforms will be enacted unless congressional elections of 2018 and 2020 and the presidential election of 2020 provide strong majorities of public officials who want to end mass incarceration. It would be helpful, too, to have some public officials who want to abolish prisons to put pressure on their moderate colleagues to enact reforms that are not just incremental adjustments. They include "young, mostly black lawyers, academics, artists, authors, and community organizers" that often have family members who were incarcerated or who themselves were incarcerated. Georgetown law professor Allegra McLeod views their work as a continuation of the earlier movement to abolish slavery. As Alexandria Ocasio-Cortez noted, "There are more African-Americans under correctional control than were enslaved in 1850—that is, before the Civil War" (Arrieta-Kenna, 2018).

ANALYZING THE EVOLUTION OF THE AMERICAN CRIMINAL JUSTICE SECTOR

Figure 14.1 depicts the evolution of the correctional systems in America, beginning with antiquated notions of "an eye for an eye" as the appropriate punishment for crime up through the present era of mass incarceration in the United States.

FIGURE 14.1 ■ Evolution of the Criminal Justice Sector

1750 BC
King Hammurabi of Babylon's Code imposes "an eye for an eye, and a tooth for a tooth" as punishment for crime.

Early 1800s
Elam Lynds, warden of Auburn State Prison in New York, subjects prisoners to harsh and humiliating punishments in the name of discipline, including floggings, prison stripes, and lockstep formation.

1890s
During the reformatory era, criminals are largely seen as disadvantaged persons who require better education and training, especially in vocational and occupational skills.

1920–1930s
The medical model becomes popular; offenders are viewed as being reformable with proper diagnosis and treatment. Treatment ideology encourages offenders to realize the benefits and rewards of positive behavior.

1933
Alcatraz becomes the nation's first "supermax" prison, housing the country's most notorious convicted felons.

1935–1960
Public sentiment shifts against prisoners, due in part to the excessive debauchery of the Prohibition Era and reduced opportunities for inmate labor. Rioting takes place in many prisons due to overcrowding, neglect, harsh conditions, and lack of opportunities for rehabilitation.

1960–1980
New correctional models emerge, such as prevention ideology (which acknowledges the role of the environment in cultivating criminal behavior) and the reintegration model (which posits that an offender's active participation in the community would result in better future outcomes upon his or her release). These models are not widely applied.

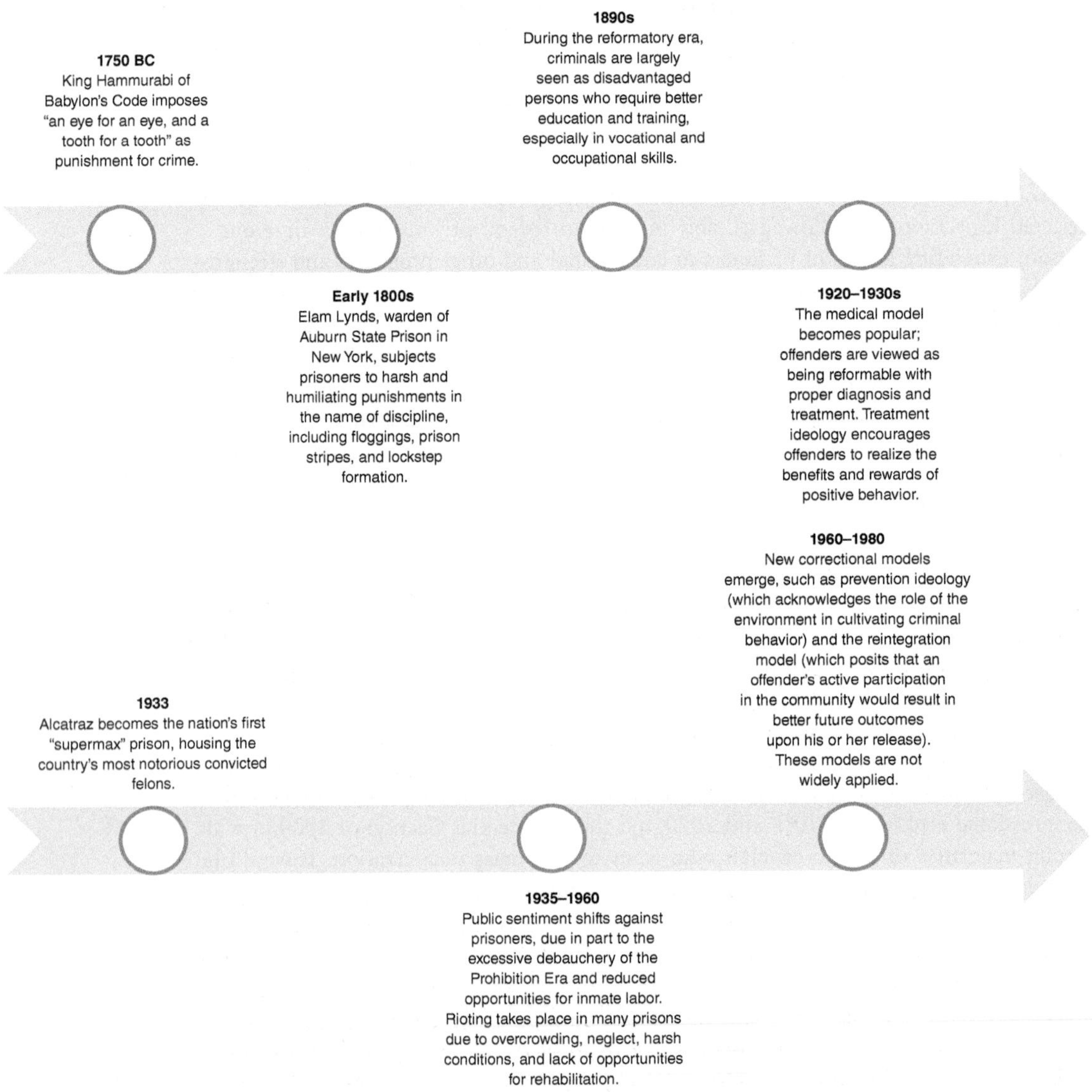

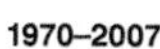

Of note in the history of corrections in America are prevailing notions of the reformability of those who violate the law; shifts in what behaviors constitute criminal behavior; and the appropriate amount of public spending on enforcement, punishment, and surveillance versus services to address social ills and public health problems that contribute to crime.

IDENTIFYING CONNECTIONS BETWEEN MASS INCARCERATION AND ECONOMIC INEQUALITY

An understanding of the criminal justice sector would be incomplete absent a discussion of mass incarceration and its connection to two phenomena: the deindustrialization that occurred in the 1970s and 1980s and the so-called War on Drugs.

Unemployment rates skyrocketed in the mid-1970s to early 1980s when many manufacturing facilities pulled out of urban centers and moved their operations overseas where labor was cheaper (Western & Wildeman, 2009). Unemployment benefits and the shrinking welfare state were insufficient to handle the social problems that arose from mass unemployment among men, who at that time were often the family breadwinner. The drug trade became a source of economic opportunity where a vacuum of legitimate employment existed.

When President Richard Nixon characterized the abuse of illicit substances as "America's public enemy number one" in 1971, a wave of anti-drug legislation ensued (Nixon, 1971). The Comprehensive Crime Control Act of 1984 and the Anti-Drug Abuse Act of 1986 reclassified drug users as "criminals" rather than persons in need of treatment, imposed tougher and longer mandatory sentences for drug-related crime (even for first-time offenders), provided the opportunity for higher-level traffickers and dealers to receive lesser sentences by turning in their accomplices, and targeted low-income communities through the crack and powder cocaine disparity (Sudbury, 2002). Individual states also enacted mandatory sentencing laws (i.e., mandatory minimums) that removed judicial discretion and required automatic prison terms for drug offenses.

This tough-on-crime mentality thus criminalized certain segments of the U.S. population—namely poor persons and persons of color who used and sold illicit substances—and contributed to skyrocketing levels of incarceration. According to the Bureau of Justice Statistics (BJS), in 1980, U.S. prisons held roughly 330,000 inmates. By 1990 that number had doubled to 774,000, and by 2015 the number of people incarcerated in the United States topped 2.2 million (Beck & Gilliard, 1995; Carson & Golinelli, 2013). Today, the United States incarcerates more of its citizens than any other country in the world, and its prison population makes up nearly one-quarter of the world's 8.5 million total prisoners (Nation Master, n.d.). These dramatic statistics have led some to call this the "era of mass incarceration" (Alexander, 2010; Western & Wildeman, 2009; Wacquant, 2002).

Racial and ethnic minorities were disproportionately swept up in the broad net cast by the "war on drugs." While white Americans represent about 73% of the total U.S. population, they represent only 32% of the prison population. African Americans, on the other hand, compose only about 13% of the total U.S. population but 38% of all prisoners, and while Latinos/Hispanics compose 16% of the total U.S. population, they make up 22% of the prison population (Guerino, Harrison, & Sabol, 2011; Humes, Jones, & Ramirez, 2011). Indeed, one in three African American males will go to prison at some point in his lifetime. Some have argued that the disproportional representation of racial and ethnic minorities in our nation's criminal justice system is akin to, and even an outgrowth of, the slavery and Jim Crow eras (Alexander, 2010; Blackmon, 2009; Davis, 2003; Oshinsky, 1996).

Many advocacy groups are working to reform the criminal justice system, advocate for more just correctional policies, and address the effects of mass incarceration on individuals, families, and communities. We provide just a snapshot below of them.

Some Criminal Justice Advocacy Groups

American Friends Service Committee (AFSC). The AFSC runs numerous projects that advocate for prisoners' rights. Its STOPMAX program works to eliminate the use of isolation and segregation in U.S. prisons through grassroots organizing, public education, and policy advocacy.

All of Us or None. This is a national organizing initiative of prisoners, former prisoners, and felons that combats many forms of discrimination that persons confront as the result of felony convictions.

American Civil Liberties Union (ACLU). The ACLU aims to ensure that American prisons, jails, and juvenile facilities comply with the Constitution, federal law, and international human rights principles.

Amnesty International. This organization seeks to protect human rights, stop torture, defend women's rights, and abolish the death penalty.

NAACP Legal Defense Fund. This organization seeks to advance social justice and fairness in the criminal justice system with a focus upon African Americans.

National Coalition to Abolish the Death Penalty. This organization serves as a clearinghouse for advocacy groups seeking to end the death penalty, including its network of 100 state and national affiliates.

National Legal Aid and Defender Association. This organization provides resources for advocates seeking equity in the criminal justice system and as a resource to persons seeking more information about equal justice in the United States.

Pew Research Center. This organization conducts research on the criminal justice system.

The Sentencing Project. This project seeks a fair and effective criminal justice system by seeking alternatives to incarceration.

The Vera Institute of Justice. This is a leading nonprofit research and advocacy organization.

ANALYZING THE POLITICAL ECONOMY OF THE CRIMINAL JUSTICE SECTOR

Powerful political forces contributed to mass incarceration from the mid-1970s to the present that included numerous powerful actors that were far more powerful than groups that wanted to downsize and humanize correctional facilities. These groups supported the "prison-industrial complex" and included trade unions representing correctional staff who worked in jails and prisons, private corporations that built and maintained for-profit jails and prisons, corporations that provided clothes and meals to inmates, and lobbyists that obtained funding for new correctional facilities. Politicians also lobbied to have correctional facilities placed in the small towns and cities that they represented to provide jobs in these jurisdictions (Henderson, 2015).

For example, GEO and Corrections Corporation of American had given more than $10 million to political candidates, spent nearly $25 million on lobbying, and secured $3.3 billion in annual revenue for their for-profit correctional institutions from 1989 to 2015. The prison population of for-profit correctional institutions doubled between 2000 and 2010 because of their lobbying (Cohen, 2018).

The criminal justice system is made up of a complex web of entities, including law enforcement agencies, courts, attorneys, local jails, state and federal prisons, and probation and parole entities. We briefly describe here the main players and how an individual typically is processed through the criminal justice system.

Law enforcement agencies include local police and sheriff, highway patrol, and federal agencies that are tasked with enforcing laws at local, state, and federal levels. Law enforcement is typically the first point of contact for individuals involved in the criminal justice system, and usually this contact comes in the form of an arrest. According to the Federal Bureau of Investigation (2010), law enforcement officials made an estimated 13,687,241 arrests nationwide (not including traffic violations) in 2009. The offense categories representing the largest proportion of cases were larceny theft at 10% (fewer than 1.3 million arrests), driving under the influence at 11% (fewer than 1.4 million arrests), drug abuse violations at 12% (fewer than 1.6 million arrests), and property crimes at 13% (fewer than 1.7 million arrests) (U.S. Department of Justice, 2010).

Individuals are processed through the criminal justice system via courts. All persons suspected of committing a crime are officially charged with that crime (typically by a prosecutor) and offered an opportunity to either admit guilt, accept a plea deal, or take their case to trial to let a jury decide. Whereas the vast majority (94%) of persons charged with a felony plead guilty (Rosenmerkel, Durose, & Farole, 2009), those who deny guilt must retain either a defense attorney or a public defender to represent them. Many advocates argue that the quality of defense is greatly limited for individuals of limited means, thus increasing the likelihood that poor persons will be convicted compared with persons who can afford private representation.

A felony or misdemeanor conviction is typically accompanied by a sentence, which is issued by a judge. Sentences typically match the severity of the crime, and can range from restitution and fines to community service or probation to serving time in a local jail or state or federal prison. Many advocates argue that the imposition of mandatory minimum sentences for drug-related crime (discussed previously), "three-strikes-and-you're-out" laws, and other similar sentences that remove judicial discretion are unfitting of the crime.

Jails and prisons are major figures in the political economy of the criminal justice system. More than 2 million people are currently confined in U.S. jails and prisons combined. Future sections of this chapter will discuss a vast array of issues confronting these entities—which are designed to house and punish convicted persons—including overcrowding, lack of proper health and mental health care, sexual assault of inmates, and solitary confinement. Jails and prisons stand in stark contrast to the social welfare entities found in other sectors and discussed in Chapters 7 through 13 of this text that are designed to serve, treat, and assist individuals.

Finally, the political economy of the criminal justice sector includes the entities of probation and parole. Probation offices (operated by counties) and parole offices (operated by states) supervise persons who are sentenced to serve a portion of time in the community, either instead of or in addition to jail or prison time. The BJS reports that nearly 5 million adults were under supervision of probation or parole at year-end 2010; the equivalent of about one out of every 48 adults in the United States (Glaze & Bonczar, 2011).

We cannot understand the political economy of the criminal justice system without describing its institutions at the local, state, and federal levels. The largest numbers are housed in local jails, the next largest numbers in state prisons, and the smallest number in federal prisons (Wagner & Rabuy, 2017).

Local Level. Six hundred thousand prisoners are housed in 3,163 local jails. Fifty-three thousand youth reside in juvenile correctional facilities that include 18,079 in detention centers, 11,025 in residential treatment facilities, 12,013 in long-term facilities, and 4,656 in adult prisons and jails (Sawyer, 2018). Other local facilities include 76 Indian Country jails, immigration detention centers, and civil commitment centers and prisons in the U.S. territories. Local jails often hold people who haven't yet been convicted of a crime as they await court hearings. Ninety-nine percent of the total jail growth in the United States was in pretrial detention—and only about half of these inmates were actually convicted of a crime (Wagner & Rabuy, 2017). They reside in jail because they cannot afford to pay bail, a fee they must pay to be released until the court takes action. Critics contend this system is unfair: Affluent people pay the bail, but low-income persons disproportionately languish in jail, even for six months or longer until the court hearing, even though two-thirds of them will be declared innocent or guilty of a minor crime such as petty theft—so-called misdemeanors with sentences under one year. Most persons in jail are there for pretrial detention. This long absence from families and jobs exacts a horrible toll on those in jail and their families, who often become impoverished, which leads to some of them becoming incarcerated for theft or other crimes. The proprietors of the bail bond companies obtain profits of $2 billion each year (Kann, 2018).

Jail populations can be reduced markedly in the following ways. Bail can be eliminated for many, even most, persons incarcerated in jails pending hearings. The amount of bail can be markedly reduced so that more prisoners can afford it. Sentences can be reduced markedly, not only with respect to the time of incarceration but replacing incarceration with community service. More judges can be hired to reduce the interlude between incarceration and court action. Data can be used to reduce or eliminate bail for persons highly unlikely to flee who also are charged with minor infractions.

State Level. About 1.330 million prisoners are housed in 1,719 state prisons (Wagner & Rabuy, 2017). They are incarcerated for violent offenses (704,000), drug offenses (208,000), theft and other property offenses (253,000), and public order offenses (154,000) including driving while intoxicated and illegal use or possession of weapons.

Federal Level. About 197,000 inmates are housed in 102 federal prisons. Roughly one-half of them (97,000) are incarcerated for drug offenses (Wagner & Rabuy, 2017). There are more than 1 million drug possession arrests each year, more for drug possession than for drug sales. So nonviolent drug offenses are a defining characteristic of federal prisons. About 57,000 persons are kept in prison due to immigration offenses, including 16,000 persons held by the Bureau of Prisons and 41,000 held by Immigration and Customs Enforcement (ICE). (Some states have relatively few drug possession arrests, such as New York, but other states have not ended the War on Drugs, such as Oklahoma.) About 71,000 are incarcerated for public order, including illegal use or possession of weapons, immigration violations, and other public order cases. Few persons are in federal prisons for violent offenses (14,000 for homicides, robberies, and other violent crimes). Only 12,000 have committed property crimes such as burglaries and fraud (Wagner & Rabuy, 2017).

Our discussion suggests that a combination of immigration reforms, decriminalizing drugs, bail reforms, sentencing reforms, release of nonviolent offenders, and other reforms could, taken together, markedly reduce inmates at these three levels of government. But they have to be coupled with reforms that decrease levels of re-incarceration of released inmates (on the back end of the incarceration system), decrease levels of youth pouring into incarceration (on the front end of the incarceration system), and reforms of correctional institutions so that they rehabilitate inmates rather than socializing them to a life of crime.

The roughly 600,000 men and women who leave incarceration each year face formidable challenges. Many cannot afford housing. Many cannot find employment partly because of the stigma of incarceration, partly because many have low levels of education, partly because they have been incarcerated under harsh conditions, and partly because they experienced childhood and other trauma before they were incarcerated. Many of them were exposed to drugs and neglect as children and youth. Many need counseling but do not receive it. Many need health care that addresses addiction and physical and mental problems. Inmates who stay out of prison for three years are considered successes—and the longer they are out, the more likely that they will stay out (Rodriguez & Bernstein, 2018).

ANALYZING SEVEN CORE PROBLEMS IN THE CRIMINAL JUSTICE SECTOR

Core Problem 1: Advocacy to Improve Prisoners' and Former Prisoners' Ethical Rights—With Some Red Flag Alerts

The concepts of ethical rights (Problem 1), quality care (Problem 2), and culturally appropriate services (Problem 3) are difficult to apply to the criminal justice sector, given that persons involved in corrections are viewed not as consumers but rather as deviants or persons who have violated the social contract and are thus subject to prevailing notions of punishment and retribution. This is not to say that the criminal justice system itself is not bound by important quality assurance mechanisms. Indeed, the Bill of Rights, which comprises the first 10 amendments of the U.S. Constitution, serves to protect the natural rights of U.S. citizens. It guarantees, among other things, protection against cruel and unusual punishment (Eighth Amendment); due process of law, including the right to a speedy trial and right to legal counsel (Fifth and Sixth Amendments); protection from unreasonable searches and seizures, and the right of people "to be secure in their persons, houses, papers, and effects" (Fourth Amendment); and a public trial by an impartial jury, both in criminal cases (Sixth Amendment) and in common law civil suits (Seventh Amendment).

Yet, the rights enshrined in the Constitution initially only applied to landowning white men. Women and racial minorities were excluded. It took additional Constitutional amendments and numerous Supreme Court cases to extend the same rights to all U.S. citizens. Even since the passage of the Thirteenth, Fourteenth, and Fifteenth Amendments, which abolished slavery and granted rights and citizenship to former slaves, and the Nineteenth Amendment, which extended rights to women, countless advocates have challenged the criminal justice system on grounds that the rights ensured in the U.S. Constitution and the Bill of Rights have been violated. One historical example is Dorothea Dix, who advocated for better conditions for indigent, mentally ill detainees in the mid-1800s (see Problem 6). This section will discuss current efforts to ensure the rights of those involved in the criminal justice system.

Some advocates have developed strategies for enhancing human rights in prison management such as the third edition of a handbook for prison staff (Coyle & Fair, 2018). It draws upon the principle that all persons have basic rights. It draws upon principles enunciated by Nelson Mandela and numerous policy positions of the United Nations. It provides invaluable advice to social workers who work in prisons and who engage in micro and mezzo policy advocacy in them, not to mention macro policy advocacy to establish legislation that prohibits solitary confinement and other assaults on the dignity of prisoners.

- **Red Flag Alert 14.1.** A prisoner is labeled as a gang member and is threatened with solitary confinement unless they inform officials about gang leadership and planned activities.
- **Red Flag Alert 14.2.** A prisoner held in solitary confinement for five years becomes increasingly disoriented and lethargic, yet they have not been seen by a certified mental health professional.
- **Red Flag Alert 14.3.** An illiterate prisoner with limited English speaking ability is given a form that authorizes their inclusion in a pharmaceutical study.
- **Red Flag Alert 14.4.** Food is routinely withheld from a prisoner who refuses to work.
- **Red Flag Alert 14.5.** A female prisoner is groped by a male guard during a routine strip search.
- **Red Flag Alert 14.6.** Prison guards look the other way while a male inmate is sexually assaulted by other inmates.

Background

Like many of the populations discussed in this text, prisoners housed in this country's correctional facilities are vulnerable to experiencing violations of their ethical rights. Yet, unlike students, seniors, and foster youth, prisoners carry with them a unique stigma; it is not politically incorrect for society to ignore or even discriminate against incarcerated individuals. There are numerous areas within the criminal justice sector—from the point of arrest through incarceration to reentry—in which ethical violations are rife. We touch on just a few of them here.

One of those areas includes the use of controversial stop-and-frisk policies as a law enforcement mechanism for maintaining social order. At the present time in New York City, lawsuits have been filed against the New York Police Department for their use of the practice, which primarily targets young men of color. These tactics have been ruled by district courts to violate residents' constitutional rights.

A second area of great concern to many advocates is the sexual assault that takes place within the walls of many prisons and is perpetrated by both guards and fellow inmates (often while guards look the other way). The advocacy group Just Detention reports that 200,000 adults and children are sexually abused behind bars every year (Just Detention, n.d.).

Long-term segregation (i.e., solitary confinement) is another area that is considered a violation of prisoners' rights by some. Depending on the institution, the experience of solitary confinement can range from limited contact with other persons, including the general population of prisoners and family visitors, to complete sensory deprivation for

23 hours each day. Prisoners can be held in solitary confinement for months and years. Although its devastating psychological impacts have been widely documented (Arrigo & Bullock, 2008), only recently have policies regulated this practice in some states. Human Rights Watch (2012) is currently working to end the practice of solitary confinement for incarcerated juveniles under the age of 18, whereas the ACLU seeks reforms to solitary confinement practices generally across the United States (ACLU, n.d.b).

Other violations of ethical rights include (but are not limited to) the disproportionate imposition of capital punishment sentences on members of ethnic minority groups; prison overcrowding that leads to inadequate health and mental health services; the elimination of educational and vocational training programs for inmates; the denial of employment, housing, and educational opportunities to persons with drug convictions; and lifetime disenfranchisement of formerly incarcerated people in many states (i.e., eliminating their right to vote because of a prior conviction).

Resources for Advocates

Court Rulings. Supreme Court decisions have played key roles in the formulation of correctional policies regarding prisoners' ethical rights. In its 1964 *Cooper v. Pate* decision, the Court held that prisoners are persons whose rights are protected by the Constitution, and they can challenge the conditions of their confinement. In *Hudson v. Palmer* (1984), prison officials are permitted to search cells and confiscate items without violating the Fourth Amendment's prohibition of unreasonable search and seizures. Also, in the 1974 *Wolff v. McDonnell* case, the Court ruled that inmates have due process rights under the 14th Amendment when being disciplined.

The constitutionality of the death penalty has also been discussed and debated in the courts. In *Furman v. Georgia* (1972), the Supreme Court found that capital punishment was imposed unpredictably and infrequently. As such, the Court ruled that the death penalty amounted to cruel and unusual punishment and should be ceased. However, four years later, in *Gregg v. Georgia*, the Court allowed executions to continue after a few states revised their death penalty statutes. Specifically, the Court upheld death penalty laws that require the sentencing judge or jury to take into account specific aggravating and mitigating factors in deciding which convicted murderers should be sentenced to death. In recent cases, the Court has ruled that the execution of mentally disabled individuals was unconstitutional (*Atkins v. Virginia*, 2002) as well as the execution of offenders for crimes committed under the age of 18 (*Roper v. Simmons*, 2005).

Legislation. In 1996, Congress enacted the Prison Litigation Reform Act in an effort to curb the increase in prisoner litigation that was finding its way into the federal courts, including claims of physical and sexual abuse, mistreatment of confined individuals, and prison officials' indifference to inmate-to-inmate assault. Under the law, prisoners must exhaust all available administrative avenues before challenging a condition of their confinement. Prisoners are required by the act to also pay court filing fees in full, either up front or through monthly installments, and this process requires the coordination and cooperation of the prison in handling the transaction (ACLU, n.d.a). Lawsuits or appeals that a judge rules are frivolous or without merit count as strikes; after three strikes, prisoners are unable to file additional lawsuits without up-front payment of filing fees. Also, prisoners seeking monetary damages for mental or emotional injury must provide proof of physical injury.

POLICY ADVOCACY LEARNING CHALLENGE 14.1

CONNECTING MIRCO, MEZZO, AND MACRO POLICY INTERVENTIONS TO HONOR RIGHTS OF PRISONERS AND EX-OFFENDERS

Many ex-prisoners discover on their release from prison that they are still punished by local officials, public programs, employers, and ordinary citizens. They are sometimes denied access to public housing, rejected by employers, denied housing by landlords, and allowed only to have general assistance with monthly allotments lower than other welfare programs such as SSI and TANF. Unlike Norwegian prisons that help them obtain housing and jobs before they are released, American prisons give them only a small stipend upon their release—almost guaranteeing they will become penniless and often homeless.

Discuss some options that might be considered by American correctional facilities including those in the following exercise.

Learning Exercise

1. Providing guaranteed jobs in public agencies on their release so that they can establish an employment record to enable a much higher likelihood of gaining employment in the private sector
2. Creating rent subsidies to allow them to fund monthly rents in the private sector for a year as they establish themselves in the community
3. Reform local probation services by adding social workers and nurses to their staff so that released prisoners have mentors rather than only probation officers who often focus only on law enforcement

Core Problem 2: Engaging in Advocacy to Promote Quality Programs for Prisoners and Ex-Prisoners—With Some Red Flag Alerts

Although certain rights are granted to citizens who come into contact with the criminal justice system via the U.S. Constitution, quality care is a notion largely foreign to corrections. A meta-analysis of the effects of custodial versus non-custodial sentences found that the rate of re-offending after a non-custodial sanction is actually lower than after a custodial sanction (Villettaz, Killias, & Zoder, 2006). In other words, prison does not decrease future criminal activity. Rather, it may actually perpetuate it through the psychological damage it exacts and by denying former prisoners opportunities for legitimate employment and self-sufficiency. Researchers and advocates from a broad spectrum of ideologies have concluded that the system, as currently constituted, does not serve the best interests of the public but rather extracts from it valuable resources that could better be spent on social programs to address problems of drug abuse, social disorder, and poverty.

This state of affairs—a criminal justice system that routinely violates civil and human rights, does not effectively respond to social problems but may actually exacerbate them, and costs taxpayers enormous valuable resources—has moved some advocates to want to fundamentally reform the criminal justice system.

- **Red Flag Alert 14.7.** A poor person suffering from drug addiction is sentenced to life in prison for petty theft—a nonviolent, non-serious third felony offense.
- **Red Flag Alert 14.8.** A crime victim wants to confront the perpetrator, and the offender wants to make amends, but avenues through which to do so are not available in their jurisdiction.
- **Red Flag Alert 14.9.** A city council member wishes to implement criminal justice reforms in his or her jurisdictions based on evidence and best practices but is lambasted by her opponents for being soft on crime.
- **Red Flag Alert 14.10.** Resources in one's jurisdiction are spent on corrections at a rate many times the amount spent on social programs, such as housing, Women, Infants, and Children (WIC), and drug treatment.

Background

Those who seek to transform the basic structure of the criminal justice system are met with formidable foes. These include powerful special interests as well as conservative scare tactics and the previously mentioned law-and-order mentality that hinders elected officials from accomplishing significant reforms for fear of being labeled soft on crime.

Special Interests. Prisons are expensive to operate, yet they are quite profitable to many parties. First, rising incarceration rates require greater numbers of correctional staff. According to the BJS, a total of 2.5 million persons were employed in the nation's justice system in 2007, and nearly a half million of those jobs were in state corrections (Kyckelhahn, 2011). Additionally, vendors who supply food, uniforms, commissary items, and collect phone calls from inmates to their families benefit from lucrative contracts with prison authorities (Dannenberg, 2011). Moreover, major manufacturers and retailers benefit from the free or cheap labor of the nation's 1.6 million inmates (Winter, 2008).

Examination of lobbying efforts on behalf of these special interests suggests that their level of investment is high. The California Correctional and Peace Officer's Association (CCPOA), a 31,000-member strong prison guards union, spends approximately $8 million of the annual $23 million it raises through membership dues on lobbying (Kowal, 2011). In a December 2009 report, the Pennsylvania State Corrections Officers Association touted the union's efforts to support a $500 million Pennsylvania prison expansion effort that built five new prisons with 12,000 total beds (The Winter Group, 2009). Advocates seeking to reform the criminal justice system in their jurisdictions can expect to be met by these and other powerful special interests.

Tough-on-Crime Ideologies. At the time of the initiation of the War on Drugs and the massive prison buildup, considerable media attention was focused on the so-called dangerous class (Golden, 2005). Nightly news programs portrayed images of inner-city residents, usually black men, as violent outlaws, drug dealers, rapists, and murderers. Black women, on the other hand, were portrayed as "welfare queens" and crack addicts who exposed their fetuses (so-called crack babies) to drugs. These spectacles of the gangster and the unfit mother were given tremendous national media attention while similar attention was not given to white-collar crime (Golden, 2005). This prompted a dramatic rise in the fear of crime, and more and more politicians sought to align their platforms with popular views. Tough-on-crime rhetoric was subsequently translated into policies such as mandatory minimum sentencing for drug-related crime, and "three strikes, you're out" legislation, which mandated a life sentence for any person convicted of a third felony (Sudbury, 2002).

Resources for Advocates

There has been a shift in recent years away from the tough-on-crime stance, and toward a smart-on-crime approach. In a 2009 address, Attorney General Eric Holder argued,

> Getting smart on crime requires talking honestly about which policies have worked and which have not, without fear of being labeled as too hard or, more likely, as too soft on crime. Getting smart on crime means moving beyond useless labels and instead embracing science and data, and relying on them to shape policy. And it means thinking about crime in context—not just reacting to the criminal act, but developing the government's ability to enhance public safety before the crime is committed and after the former offender is returned to society. (Vera Institute, 2009)

Efforts have been initiated at local, state, and federal levels to reverse some of the trends of the harsh sentencing era and to challenge prevailing systems of punishment.

Restorative Justice. A growing movement favors the restorative justice approach as an alternative to incarceration. Restorative justice emphasizes repairing the harm caused by crime. Howard Zehr, a pioneer of the American restorative justice movement, defines restorative justice as processes "to involve, to the extent possible, those who have a stake in a specific offense to collectively identify and address harms, needs, and obligations, in order to heal and put things as right as possible" (Zehr, 2002). In the United States, numerous restorative justice programs exist (see, e.g., Bishop, 2012, and Mills, Maley, & Shy, 2009). Advocates tout the benefits of the approach as humanizing in comparison with the harsh and stifling conditions of prison. A meta-analysis of restorative justice programs found that they were more effective than traditional sanctions in reducing recidivism and ensuring restitution compliance (Latimer, Dowden, & Muise, 2005).

Prison Abolition. Some advocates argue that prisons should be done away with entirely. The preeminent organization working to abolish the prison system and replace it with a system of community accountability is Critical Resistance (n.d.). Formed in 1997 by activists challenging the idea that imprisonment and policing are the best solutions for social, political, and economic problems, Critical Resistance seeks to build a movement to abolish the prison industrial complex. Critical Resistance defines the prison industrial complex as "the system of surveillance, policing, and imprisonment that government, industry and their interests use as solutions to economic, social, and political problems" (http://www.criticalresistance.org). Through local chapters, Critical advocates for systems of harm prevention that build community and provide for basic needs and systems of accountability that address the root causes of harm.

POLICY ADVOCACY LEARNING CHALLENGE 14.2

CONNECTING MICRO, MEZZO, AND MACRO POLICY INTERVENTIONS DISCUSSING THE ETHICS OF THREE STRIKES

Visit the website of Families to Amend California's Three Strikes at http://www.november.org/razorwire/rzold/13/1323.html.

Select the rationale for amending three strikes and well as the opposition. Select three of the arguments in favor of amending three strikes and three of the arguments opposing amending of three strikes. In small groups, complete the following exercise.

Learning Exercise

1. Do you favor the arguments for amending it?
2. Do you favor the arguments for not amending it?
3. How does your discussion bear on larger ethical issues about punishment and deterrence?
4. In what ways can social workers intervene at micro and mezzo levels to assist individuals from engaging in repeated crimes?
5. In what ways can social workers intervene at the macro level to change three strikes laws in their state?
6. Twenty-nine states had three-strikes laws in the spring of 2018. Use the web to discover if your state has one. Go to https://www.legalmatch.com/law-library/article/three-strikes-laws-in-different-states.html.
7. Consider writing a letter to the editor of a leading newspaper or write an op-ed.

Core Problem 3: Engaging in Advocacy to Promote Culturally Competent Services and Policies for Prisioners and Ex-Prisoners—With Some Red Flag Alerts

The U.S. criminal justice system is not set up to "serve" those who come into contact with it but rather to punish them. However, there is growing recognition that certain vulnerable groups, such as women, the elderly, the indigent, and mentally ill persons, have unique needs and circumstances and require specialized attention at stages of the criminal justice process. This section will address advocates' attempts to reform the criminal justice system by changing specific laws and policies pertaining to marginalized or vulnerable groups. Specific focus will be paid to women, the elderly, and the ill as issues concerning mentally ill prisoners will be discussed later.

- **Red Flag Alert 14.11.** A woman is given a mandatory minimum sentence for her role as an accomplice to her abusive boyfriend who sells drugs out of the couple's home. She was aware of the drug activity but not directly involved. He was given a reduced sentence of probation for turning in his accomplices (including her).
- **Red Flag Alert 14.12.** A woman inmate is handcuffed and shackled to her bed while giving birth, and the child is immediately removed and placed in foster care.

- **Red Flag Alert 14.13.** A woman inmate serving a two-year sentence for a first-time drug offense is not informed that the state has moved to permanently terminate her parental rights.
- **Red Flag Alert 14.14.** An elderly man serving a life sentence for murder is diagnosed with brain cancer and will probably die within six months. His in-prison medical care costs $100,000 per year.
- **Red Flag Alert 14.15.** Persons of color comprise the vast bulk of prison populations in the United States including African Americans, Latinos, and Native Americans. They receive disproportionate convictions and sentences as compared with the white population.

Background

Three kinds of prisoners and ex-prisoners have drawn considerable attention in policy debates about the criminal justice sector: women, the elderly and persons with medical problems, and persons of color.

Women. Women are the fastest-growing prisoner population in the United States, and their rate of incarceration has increased by 840% since 1980 (Gilliard & Beck, 1994; West & Sabol, 2010). Women are more likely to be sentenced for drug-related crime and less likely to be sentenced for violent crime than men (Guerino, Harrison, & Sabol, 2011). Women prisoners report significantly higher rates of prior physical and sexual abuse as well as higher rates of mental illness and substance abuse compared with their male counterparts (Harlow, 1999; James & Glaze, 2006; Mumola & Karberg, 2006). In addition, women prisoners are more likely than their male counterparts to have lived at or below the poverty level, less likely to have been employed prior to their incarceration, more likely to have been receiving welfare assistance prior to incarceration, and more likely than men to engage in criminal behavior for economic reasons (Lewis, 2006; O'Brien & Harm, 2002; Covington & Bloom, 2006). Yet current sentencing laws and standard management strategies, which include surveillance, searches, restraints, infractions, and isolation, are based on male characteristics and crime and fail to take into account women's characteristics, responsibilities, and roles in crimes (Covington & Bloom, 2003).

Female prisoners are also more likely to be parents of dependent children than are male prisoners (Glaze & Maruschak, 2008). Correctional practices and policies governing contact between prisoners and their children often impede, rather than support, the maintenance of family ties (Hairston, 2004). Most children are unable to visit their mothers due to the remote locations of prisons. Those who are able to visit often experience long waits, rude treatment by staff, and physical environments that restrain mother-child interaction (Arditti & Few, 2006). Phone calls placed to correctional facilities or received as collect calls from prisons are up to six times more expensive than the typical long-distance call, making this form of communication difficult for poor families (Dannenberg, 2011). Moreover, pregnant inmates in many states face the dehumanizing prospect of being shackled to their beds during childbirth and of having their infant taken from them within hours of delivery (Women's Prison Association, n.d.).

The Elderly and the Ill. Prisoners sentenced to death, to life without the possibility of parole, or to consecutive or lengthy sentences (such as those mandated by three strikes laws) often grow old and/or ill in prison. In recent years, the number and proportion of older inmates has grown. There were more than 25,000 inmates age 65 and older in state and federal prisons at year-end 2010 (Guerino et al., 2011), and 89% of all inmate deaths are attributable to medical conditions (Mumola, 2007). For inmates who serve at least 10 years in state prison, the mortality rate due to illness is triple that of inmates who serve less than five years (Mumola, 2007).

The ever-expanding prison population places a strain on prison health care systems to provide basic medical services as well as specialty services for these populations. This strain is not only felt by prison physicians and nurses; elected officials struggle to justify rising correctional budgets to weary taxpayers. The ultimate impact, however, is experienced by a population that is too often invisible to society—the inmates themselves. Among them, terminally ill prisoners are perhaps the most vulnerable of all. Typically, terminally ill prisoners are isolated, without access to visitors, and often fear dying alone (Snyder, van Wormer, Chadha, & Jaggers, 2009). Social workers who are able to visit terminally ill patients report that these prisoners are subjected to body cavity searches after visits (Snyder et al., 2009). Many terminally ill prisoners report feeling shame that they will die as a prisoner (Wahidin, 2004). The National Institute of Justice in 2004 reported that only half of the 50 U.S. states operate hospice programs in their prisons (Anno, Graham, Lawrence, & Shansky, 2004).

Persons of Color. We have discussed how male persons of color constitute the overwhelming majority of inmates in the United States at greatly disproportionate levels as compared with the white population. Considerable evidence suggests that they are subject to the death penalty, as well as long sentences, far more than whites who commit similar crimes.

Resources for Advocates

Policy Reforms Targeting Females. In 1997 the U.S. BOP issued a formal policy on the management of female offenders that mandated all BOP programs and services to "consider and address" the unique treatment needs of female offenders (General Accounting Office, 1999). It also requires each applicable BOP facility to develop and document programs and services that meet women's needs, prepare them to function in an institutional environment, and return them to the community. The BOP's gender-responsive mandate, although difficult to enforce, nonetheless recognizes that a one-size-fits-all correctional system is insensitive to the differential needs and characteristics of women who often leave behind children for whom they were the primary caregiver and who often become involved in crime for reasons related to economic marginality and drug addiction. The gender-responsive philosophy promotes "creating an environment through site selection, staff selection, program development, content, and material that reflects an understanding of the realities of the lives of women in criminal justice settings and addresses their specific challenges and strengths" (Covington & Bloom, 2006, p. 19).

Unfortunately, this ideal model is far from the norm. In 2010, of the 93,000 total women incarcerated in state prisons, nearly one in 12 was confined in California's super-max women's prisons. The largest women's prison in the world—Central

TABLE 14.1 ■ Advocacy Groups Focusing on Women and Children

Women's Prison Association: http://www.wpaonline.org/
The Family and Corrections Network: http://fcnetwork.org/
Legal Services for Prisoners with Children: http://www.prisonerswithchildren.org/
The Center for Community Alternatives: http://www.communityalternatives.org/
The Justice Policy Institute: http://www.justicepolicy.org/index.html
The Sentencing Project: http://www.sentencingproject.org
The Vera Institute of Justice: http://www.vera.org/

California Women's Facility (CCWF)—had a design capacity of 2,004 and a population of 3,736 on September 30, 2011. Its neighbor across the street—Valley State Prison for Women (VSPW)—is the second-largest women's prison in the world. It had a design capacity of 1,980 and a population of 3,496 on September 30, 2011 (California Department of Corrections and Rehabilitation, 2011). Far from being nurturing environments designed to address women's trauma and promote healthy relationships, these facilities have instead been described as "suicide cities" (Olson, 2007). In writing about her experience as an inmate at CCWF, Olson recounts how the overcrowded conditions inside the facility led to frequent lockdowns and inadequate mental health care, all of which precipitated a number of attempted and successful suicides (Olson, 2007). Clearly, there is much work to be done to properly address women prisoners' unique needs. Advocates including the Women's Prison Association and others are working to ensure that the needs of women prisoners and their families are addressed (see Table 14.1).

Policy Reforms Targeting the Elderly and Ill. Advocates concerned with issues confronting elderly and terminally ill prisoners often argue for compassionate release of these vulnerable populations. Compassionate release—a program that allows some eligible, seriously ill prisoners to die outside of prison before sentence completion—is permitted under the Sentencing Reform Act of 1984 (Williams, Sudore, Greifinger, & Morrison, 2011), and all but five states have some mechanism through which dying prisoners can seek release (Anno et al., 2004). Advocates of compassionate release argue both that releasing prisoners with life-limiting illnesses who no longer pose a threat to society is the ethical and moral thing to do and that the financial costs to society of continuing to incarcerate such persons outweigh the benefits. Medical eligibility guidelines vary by jurisdiction, but most states require that the prisoner have a diagnosed terminal or severely debilitating medical condition that cannot be appropriately cared for within the prison and that the prisoner pose no further threat to society (Williams et al., 2011).

However, a recent study of compassionate release in practice found that only 36 requests for compassionate release in the federal BOP in 2008 made it to the final review stage, and 27 were approved. The Vera Institute of Justice cites narrow eligibility criteria, complicated and lengthy referral and review processes, political considerations, and public opinion as barriers to compassionate release (Chiu, 2010). Those in the medical field concur that eligibility guidelines are too stringent, requiring a short prognosis (e.g., six months) that excludes prisoners with severe dementia, those in a persistent vegetative state, or those with end-stage organ disease who may actually live longer, although they pose no threat to society (Williams et al., 2011). However, the power of the court of public opinion and political motivations for the denial of compassionate releases should not be discounted. Advocates, including family members of dying prisoners, as well as those in the medical profession, continue to fight for the rights of terminally ill and severely incapacitated prisoners by challenging the necessary qualifications for compassionate release.

Prison Reforms Targeting Persons of Color. Considerable attention has been given to greatly decreasing incarceration rates of African American, Latino, and Native American males by decreasing them for nonviolent crimes—and by greatly shortening sentences for nonviolent offenders. Some states have abolished the death penalty in the wake of evidence that many innocent males of color receive this sentence.

POLICY ADVOCACY LEARNING CHALLENGE 14.3

CONNECTING MICRO, MEZZO, AND MACRO POLICY INTERVENTIONS—SHOULD THEY BE RELEASED?

Picture 92-year-old Nick leaning on his cane, out of breath, in the quad at the largest men's prison in Massachusetts. Nick is making one of his three trips a day across the prison complex to get his life-sustaining medications and is forced to ask another prisoner to dig nitroglycerin out of his pocket so he can address the heart episode he is experiencing. Nick was sentenced to life in prison more than 40 years ago for murdering his wife when he caught her with another man. He was drunk at the time. He had no prior criminal history, and has expressed remorse for his crime.

Also picture Frank, a 70-year-old diabetic and Vietnam War veteran, in his wheelchair. Frank has had both legs amputated and is unable to see well enough to write his son to tell him about a guard in the assisted care facility at the prison who would not allow him to be wheeled over to a church service. Frank was sentenced to life in prison 27 years ago on a third strike for armed robbery. His previous convictions include unarmed robbery and petty theft. As a young father, Frank had been laid off and resorted to stealing to support his wife and three kids. His wife is now deceased. Like Nick, Frank has also expressed remorse for his crimes.

Both men are model prisoners with no infractions in the past 25 years. The family members of both men are fighting for their release, but Massachusetts is one of only a few states that does not have some type of compassionate or medical release law that allows seriously ill prisoners to be released to more appropriate care. The cost to care for aging Massachusetts prisoners ranges from $75,000 to $115,000 per prisoner per year as opposed to about $44,000 for a healthy man or woman. The managed care these men would receive outside of prison walls is a fraction of the cost of in-prison care.

Material drawn from The Real Cost of Prisons Project website: http://www.realcostofprisons.org/writing/Muise_Compassionate_Release.pdf

Learning Exercise

Consider the following questions from the points of view of (1) a terminally ill prisoner's doctor, (2) a terminally ill prisoner's son or daughter, (3) a member of the victim's family, and (4) a concerned taxpayer:

- Should prisoners like Frank and Nick be granted compassionate release?
- What guidelines do you think should be used to determine the release of elderly or severely or terminally ill prisoners?
- Who should ultimately get to decide the fate of terminally ill prisoners?
- Do the risks of releasing prisoners like Nick and Frank outweigh the benefits in terms of the health care savings?
- How might you engage in micro, mezzo, and/or macro level policy advocacy to improve the situation of terminally ill prisoners?

Core Problem 4: Engaging in Advocacy to Promote Prevention for Prisoners and Ex-Prisoners—With Some Red Flag Alerts

Prison not only affects an individual during the time he or she spends there; it also carries lasting financial, emotional, and social effects. For example, black men without high school diplomas have a 60% chance of being sent to jail, which is associated with a reduction in their annual employment by nine weeks and a 40% drop in their yearly income (Romano, 2011). The stigma of a criminal record can haunt an individual for life, even despite his or her best efforts to change. Therefore, it is crucial that efforts be made to strengthen preventive services targeted at those at risk of entering the system. Additionally, preventive services can do much to reduce recidivism and promote successful reentry into society.

- **Red Flag Alert 14.16.** A prisoner who has been in jail for years is up for parole and will return to his or her impoverished neighborhood with few skills or resources.
- **Red Flag Alert 14.17.** A soon-to-be-released prisoner has never had any visits from family or made calls to home. His or her only known support system was the gang of which he or she was previously a member.
- **Red Flag Alert 14.18.** A prisoner is unable to attain his or her certification while incarcerated because the vocational training program was cut due to budgetary constraints.

Background

The implementation of evidence-based practices has been shown to reduce recidivism rates by 50% (Andrews et al., 1990). Early intervention programs have also kept vulnerable youth from entering into gangs and their spiraling life of crime. Replacing a once-size-fits-all approach with a flexible model for responding to ex-offenders' individual situations can yield more positive outcomes. For example, the National Institute of Corrections and Crime and Justice Institute presented eight evidence-based practices that aim to improve reentry outcomes (Bogue et al., 2004), including assessing risks and needs, enhancing intrinsic motivation, targeting interventions, using cognitive behavioral treatment methods for skill-building purposes, increasing positive reinforcement, engaging with natural communities, measuring practices, and providing feedback.

Gang Involvement. According to figures from the Office of Juvenile Justice and Delinquency Prevention, there were 800,000 gang members active in more than 3,000 gangs in the year 2000. A number of risk factors are correlated with gang involvement, including but not limited to low household incomes, single-parent households, low academic achievement, identification as learning disabled, and accessibility to marijuana (Hill, Howell, Hawkins, & Battin-Pearson, 1999). Early childhood, school-based, and after-school programs have been used to prevent adolescents from entering into gangs as well as to disrupt and dismantle existing gangs. However, the most effective programs involve the community at large and require multiagency coordination and integration among a variety of stakeholders, such as youth services, police, parole, and grassroots organizations (Howell, 2006). Listen to a Ted talk by Father Gregory Boyle, who pioneered creative programs to help members of gangs obtain jobs and reduce violence. Go to https://www.youtube.com/watch?v=ZGYKu70qi48.

Recidivism. Upon their release from prison, ex-offenders face the challenge of figuring out where to live, where to work, and where to seek or continue their medical care. The lack of resources and employment opportunities, limited supervision, and inadequate collaboration between parole agencies and community partners lead many to commit crimes and find themselves back in jail. The Bureau of Justice Statistics published a 2002 report on recidivism rates of prisoners who had been released in 1994. The study showed that in the first six months, 30% had been rearrested; within the first year, 44%; and within three years, 67.5% (Langan & Levin, 2002).

Resources for Advocates

The Racketeer Influenced and Corrupt Organizations (RICO) Act has been used against youth and adult gang members by 17% of local prosecutors in large counties and less than 10% of prosecutors in small counties (Howell, 2006). A number of states, including California, Florida, and Illinois, have enacted policies that impose stiffer penalties for crimes affiliated with identified gangs. For example, under California's Street Terrorism, Enforcement, and Prevention (STEP) Act of 1988, gang members receive written documentation that they have been identified as part of a particular gang and can face harsher penalties for any crimes committed. In Hawaii's Youth Gang Response System, prosecutors focus on high-level gang leadership, while youth members have access to prevention and education services. Furthermore, many cities have turned to curfew laws to curb

POLICY ADVOCACY LEARNING CHALLENGE 14.4

CONNECTING MICRO, MEZZO, AND MARCRO POLICY INTERVENTIONS—CREATING JOBS FOR GANG MEMBERS

Father Gregory Boyle in East Los Angeles has pioneered the creation of jobs for former gang members and prisoners, called "Homeboys," in scores of restaurants in communities as well as airplane terminals. Not content with helping released males with restaurants, he created jobs for released women called "Homegirls," including at Los Angeles International Airport, where they make sandwiches for outgoing passengers. Discuss other employment options that the United States might develop, such as developing jobs in community agencies in after-school programs, creating projects that beautify low-income neighborhoods, creating pocket parks in inner cities, creating community gardens, and installing solar panels. These projects have to be supervised and funded by foundations and local governments, so social workers would need to engage in mezzo and macro policy advocacy to establish them.

gang activity. Yet, this policy does little to address the fundamental problem of gang recruitment and membership.

A number of states have passed measures seeking to reduce ineffective and inappropriate policies that favor incarceration rather than treatment in the case of those convicted of simple drug possession. For example, Arizona passed the Drug Medicalization, Prevention, and Control Act in 1996, allowing nonviolent drug offenders to undergo drug treatment and education services rather than incarceration. In November 2000, California passed the Substance Abuse and Crime Prevention Act, also known as Proposition 36. Citing that substance abuse treatment "is a proven public safety and health measure," the act enabled first- and second-time nonviolent drug possession offenders to undergo a treatment program instead of jail time. It continued to state that "non-violent, drug dependent criminal offenders who receive drug treatment are much less likely to abuse drugs and commit future crimes, and are likelier to live healthier, more stable and more productive lives" (National Families in Action, 2000). Furthermore, the act pointed out that replacing incarceration with appropriate community-based treatment not only improves community safety and health but also saves taxpayer dollars.

Core Problem 5: Engaging in Advocacy to Promote Reductions in Resources Spent on Prisons and Increases in Spending on Affordable and Accessible Services for Prisoners and Ex-Prisoners—With Some Red Flag Alerts

As we have seen, the seven problems identified in other social welfare sectors are difficult to apply to corrections because its purpose—punishment of offenders—is antithetical to service provision. This section will instead discuss the exorbitant cost of the U.S. correctional system to taxpayers and what advocacy groups are doing to shift resources away from corrections and into prevention and community-based services that address the root causes of crime.

- **Red Flag Alert 14.19.** A 14-year-old male from the inner city is exposed daily to violence and gangs. The city in which he lives has done little to invest in gang prevention or reduction programs, parks, recreational facilities, or other opportunities for youth. Instead, the city has invested in maintaining a strong police presence in inner-city neighborhoods.
- **Red Flag Alert 14.20.** A formerly incarcerated woman who was recently released from state prison returns to the community to find there are no services to assist her with housing, job training, or child reunification. Her parole officer seems to be more interested in the results of her mandatory drug test than in assisting her to get back on her feet. This dearth of programs to help prisoners and ex-prisoners extends as well to males, persons of color, youthful persons, and other kinds of prisoners and ex-prisoners.

FIGURE 14.2 ■ Spending on Prisons in the United States From 1992 to 2012

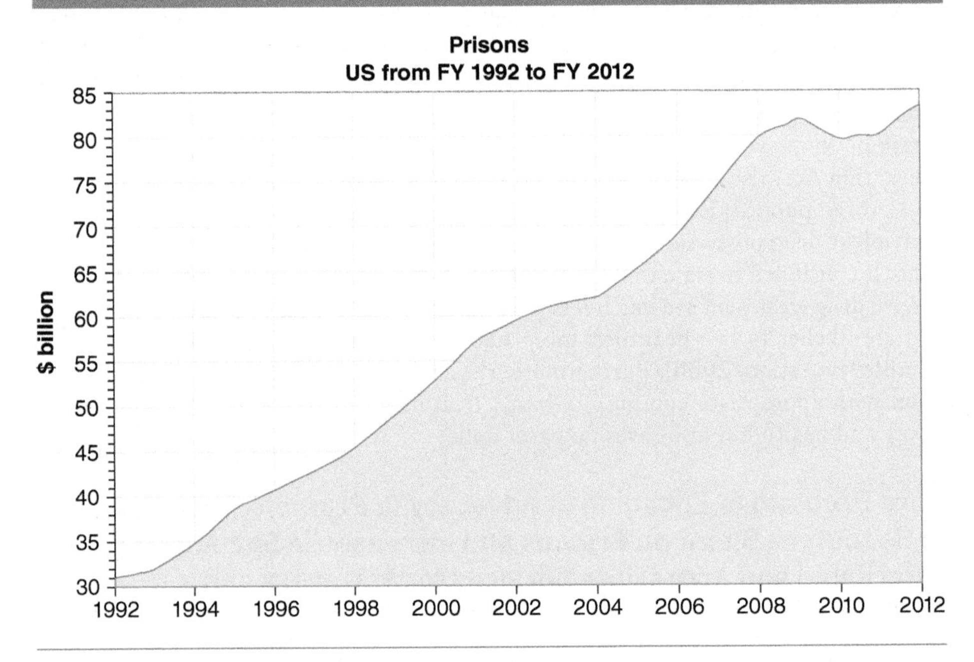

Source: usgovernmentspending.com

POLICY ADVOCACY LEARNING CHALLENGE 14.5

CONNECTING MICRO, MEZZO, AND MACRO POLICY INTERVENTIONS

Impact of Prisons on Communities That Surround Them

View the four-minute trailer for the film Prison Town, USA (http://www.pbs.org/pov/prisontown/). Download the discussion guide to learn more about the town portrayed in the film and the impacts of the Susanville prison on residents of the town. Pay particular attention to the economic impact that building the prison had on the community (p. 9 of the Discussion Guide). Think about the possible micro, mezzo, and macro implications of building a prison in your town.

Resources for Advocates

Local, state, and federal budgets are considered by many to be social contracts. Influencing how governments invest public dollars is no simple task. Yet many advocates and groups are attempting to do just that in hopes that resources will be channeled away from prisons and into social programs and education.

One of the most active of such spending reform advocacy groups is Californians United for a Responsible Budget (CURB). CURB is a coalition of more than 40 organizations that work to reduce spending on prisons. The organization seeks a "budget for humanity" that stops all prison and jail construction, reduces prison overcrowding through sentencing reform and reentry support, and instead invests in education, affordable housing, jobs, and mental and medical health care. According to its website, CURB has helped defeat more than 140,000 new prison and jail beds proposed since 2004 (Californians United for a Responsible Budget, n.d.).

Core Problem 6: Engaging in Advocacy to Promote Care for Mental Distress of Prisoners and Ex-Prisoners—With Some Red Flag Alerts

Dorothea Dix was a famous advocate for mentally ill persons in the United States in the middle portion of the 19th century. She launched a legendary battle to rescue them from poorhouses, where they were often shackled to their beds and forced to deal with vermin and feces (Jansson, 2019b). Fast-forward 150 years from the mid-1800s to the early 21st century when contemporary prisons have become "the new poorhouses."

What would Dix find if she were to reappear and visit the jails of today? According to the Department of Justice, the number of U.S. residents being held in federal and state correctional facilities and municipal jails soared to 2.2 million by the end of 2005 even if they have declined in succeeding years, as we have discussed earlier in this chapter (Lopez, 2018c). Although this statistic in itself is alarming, of greater concern is the issue highlighted by Human Rights Watch, the largest human rights organization in the United

States, that of those incarcerated in prisons today, about half the jail and prison population in the United States, roughly 1,254,800 men and women, have mental health problems in local, state, and federal jails—and many of them are subjected to physical abuse by correctional staff and fellow prisoners (Fellner, 2007; Human Rights Watch, 2015).

- **Red Flag Alert 14.21.** A new inmate with a history of substance abuse and depression was not told about the institution's mental health services.
- **Red Flag Alert 14.22.** A prisoner is shown to be developmentally disabled.
- **Red Flag Alert 14.23.** An inmate has been experiencing panic attacks for the past week. She or he requested an evaluation by a qualified mental health professional four days ago.
- **Red Flag Alert 14.24.** A mental health professional asks a prisoner about his or her past suicide attempts in front of other prisoners.
- **Red Flag Alert 14.25.** A prison guard routinely places inmates in solitary confinement as a means of punishment.
- **Red Flag Alert 14.26.** A prisoner receives no continuing or follow-up treatment services upon parole.

Background

As a result of inadequate and inaccessible mental health treatment in the community, people with mental illness often engage in behavior deemed illegal, thus thrusting them into the criminal justice system—a system not designed to deal with mental health issues. Availability of adequate mental health services in the community prior to the criminal offense could have acted as a deterrent to behavior that resulted in incarceration. Further, even though prisons are housing men and women suffering from serious mental health disorders, including schizophrenia, bipolar disorder, and major depression, Bureau of Justice records indicate that the federal prisons have provided treatment to only about 24% of the inmates identified with mental health problems (Fellner, 2007).

Prisons have become the last stop for many Americans with severe mental illness (Gilligan, 2001). In fact, the country's largest psychiatric inpatient facility is the Los Angeles County Jail, which has 3,400 inmates with mental illness (Torrey, 1999). It comes as no surprise that the correctional system is not conducive to treating persons with psychiatric symptoms. Not only are correctional employees oftentimes ill equipped to recognize or help prisoners with mental illness, but the prison environment itself can intensify emotional and physical distress. For example, the American Correctional Association has recognized that holding mentally challenged individuals in isolation can exacerbate their problems and bring about additional mental problems.

Within the population of incarcerated persons with mental illness, a vast majority also battle substance use disorders; one study found that 90% of mentally ill inmates struggled with substance abuse at some point in their lifetimes. Only 11% of inmates with substance use disorders receive any type of treatment during incarceration; few of those receive evidence-based care (The National Center on Addiction and Substance Abuse at Columbia University, 2010). Not to be forgotten is the increased number of offenders in community settings, such as those in parole, probation, or other forms of supervised release.

Moreover, other major players in the criminal justice system—such as police officers who are dispatched to respond to mentally ill persons as well as judges and attorneys who prosecute and sentence mentally ill defendants—are oftentimes not sufficiently trained to deal with those with psychiatric issues. The recent beating of mentally ill homeless person Kelly Thomas by two Fullerton, California, law enforcement officers is an egregious example of these inadequacies in training. Persons who exhibit signs of a mental disorder are associated with a 67% increased likelihood of being arrested compared with persons without an apparent disorder (Teplin, 2000). Explicitly stated guidelines, outlining protocols for situation assessment and safety management, should be developed to educate law enforcement on how to respond appropriately.

Last, another important issue involves releasing mentally ill individuals back into society without a mandate to participate in continuing treatment. Although some persons are offered discharge planning services, many reenter communities without having their post-release mental health needs addressed.

Resources for Advocates

In taking inmates into custody, the United States government has a special legal obligation to protect them from harm (Cohen & Gerbasi, 2005). According to the American Psychiatric Association (2000),

> The fundamental policy goal for correctional mental health is to provide the same level of mental health services to each patient in the criminal justice process that should be available in the community. This policy goal is deliberately higher than the "community standard" that is called for in various contexts.

In other words, incarcerated persons should experience no discrimination in their mental health services.

No court-mandated guidelines exist in terms of a single mental health service delivery model. Instead, national organizations have created their own sets of policy recommendations. The American Psychiatric Association and the National Commission on Correctional Health Care, for example, published a comprehensive set of guidelines to expand and improve mental health services for prisoners (Hills, Siegfried, & Ickowitz, 2004).

Court Rulings. Although numerous court rulings have held that inmates should receive equal mental health care to that available in the community, the vast majority of prisons are unable to offer a comprehensive array of services. Because of constrained resources and time, corrections personnel must prioritize individuals with the most serious, dangerous, or disruptive conditions. Persons with milder problems or adjustment disorders are likely to receive delayed treatment or no treatment at all.

After the passage of the Ku Klux Klan Act in 1871, inmates were able to sue correctional officials for neglecting to provide constitutionally adequate care. In particular, the federal statute enabled incarcerated persons to sue providers for violating their Eighth Amendment rights if their care or lack of care amounted to cruel and unusual punishment. Under the 1976 *Estelle v. Gamble* case, the U.S. Supreme Court ruled that corrections personnel must exhibit "deliberate indifference" to an incarcerated person's "serious illness or injury." Deliberate indifference can be shown by prison doctors' responses to individuals' needs, correctional personnel who intentionally deny or delay access to care, or corrections officers who intentionally interfere with a prisoner's treatment.

Mental health needs fall under medical needs, according to lower courts. Incarcerated persons are entitled to psychological or psychiatric treatment if a physician observes that the prisoner's symptoms are indicative of a serious disease or injury that could be significantly remedied and the denial of care could substantially cause harm to the individual.

Federal Bureau of Prisons. The BOP (n.d.) has noted that the suicide rate at its facilities is lower than that of the United States as a whole. In a 2010 national study commissioned by the Justice Department's National Institute of Corrections, research showed that the suicide rate in county jails has had a dramatic decrease during the past 20 years. According to the BOP, its staff members receive annual training—and some semi-annual training—on suicide prevention. Inmates are provided with information on mental health services upon admission. Since 1982, the bureau has used the following five-step suicide prevention program (Hills et al., 2004):

1. The initial screening of all inmates for suicidal potential
2. Criteria for the treatment and housing of suicidal inmates
3. Standardized record keeping, follow-up procedures, and collection of data relevant to suicides
4. Staff training
5. Periodic reviews and audits

In the first decade after the five-step program's launch, inmate suicides within the BOP decreased by 43% (White & Schimmel, 1995). Still, suicide remains as the third-leading cause of death for prison inmates. No mandated standards exist for suicide prevention policies, but organizations like the American Correctional Association and National Commission on Correctional Health Care have issued their own suicide prevention plans. Included in these plans are more frequent observation and monitoring of inmates with violent tendencies or mental illness, intake screening assessments for at-risk inmates to ensure that suicidal inmates are not placed in isolation, staff training, and procedures for notifying family, prison administrators, and other authorities.

Core Problem 7: Engaging in Advocacy to Promote Linkages With Communities of Prisoners and Ex-Prisoners—With Some Red Flag Alerts

All prisoners except those sentenced to death or life without the possibility of parole are eventually released (Hughes & Wilson, 2002). Only in the past two to three decades has attention been focused on the topic of prisoner reentry. Few formalized interventions or evidence-based practices exist to address the needs of this growing population. Instead, formerly incarcerated people return to their communities to face a host of barriers to successful reentry (discussed as follows). These barriers wreak devastating consequences for individuals, families, communities, and society as a whole and contribute to soaring rates of recidivism.

POLICY ADVOCACY LEARNING CHALLENGE 14.6

CONNECTING MICRO, MEZZO, AND MACRO POLICY ADVOCACY INTERVENTIONS TO ADDRESS MENTAL ILLNESS AMONG PRISONERS

Go to PBS archives of *Frontline* and access its documentary "The New Asylums" (2006) at https://www.thirteen.org/programs/frontline/frontline-new-asylums/. After watching the one-hour documentary, discuss these questions:

1. Why do such a large proportion of inmates suffer from substance abuse and mental health problems?
2. Why is it difficult to engage many inmates in mental health interventions in prison settings?
3. Why don't prisons provide better mental health and substance abuse services?
4. Do you think released prisoners receive follow-up mental health and substance abuse treatment in the community?
5. Does evidence exist regarding whether mental health and substance abuse services decrease the likelihood that prisoners will be repeat offenders?
6. What micro-, mezzo-, and/or macro-level advocacy interventions might help reduce the number of people suffering from substance abuse and mental health issues from becoming involved in the criminal justice system?

- **Red Flag Alert 14.27.** A formerly incarcerated woman wishes to begin the process of reunification with her children and files papers with the court to obtain visitation rights.
- **Red Flag Alert 14.28.** A formerly incarcerated person is frustrated because he or she is unable to find work and believes he or she is being discriminated against by potential employers.
- **Red Flag Alert 14.29.** A formerly incarcerated person is homeless on the streets and does not qualify for public housing because of a drug conviction.
- **Red Flag Alert 14.30.** A formerly incarcerated woman is unable to obtain food stamps or welfare to support herself and her two children because of a drug conviction.
- **Red Flag Alert 14.31.** A formerly incarcerated person is unsure about his or her right to vote in his or her state.

Background

No federal guidelines exist with regard to the release of prisoners. Many released inmates are sent back to their communities without any form of identification, which would have been confiscated by authorities upon arrest. Without ID, former prisoners must first locate a birth certificate and social security card, which can take weeks if they do not have copies stored in safe locations. Those with diagnosed mental health conditions are typically released with only two weeks' worth of medication (O'Shea, 2012), and obtaining an appointment to see a psychiatrist or mental health professional can take a month or more.

The majority of released prisoners are released on parole and are typically required to report to a parole officer within 24 hours. Requirements of parole vary from state to state, but in virtually all cases, persons on parole do not enjoy the rights of free citizens. Parolees are entitled to only limited Fifth Amendment due process rights before having their parole revoked; and in most jurisdictions, parolees are subject to search and seizure at any time without a warrant and without cause. Such searches, however, must be made on "reasonable" suspicion of criminal activity (Koshy, 1987). In addition, parolees typically cannot leave their county of residence or go beyond a proscribed area (such as a 50-mile radius) without the prior approval of a parole officer. They must inform the parole officer of any change of address or of new employment, and they are restricted from owning or carrying weapons. They must also comply with individualized requirements, such as to complete substance abuse treatment programs.

Parole can be revoked for any violation of parole conditions, and parole officers enjoy wide discretion in determining parole revocations (California Department of Corrections and Rehabilitation, 2012). Nationwide, approximately 23% of all adults on parole return to incarceration due to parole revocation (Glaze & Bonczar, 2011).

Social workers are likely to encounter formerly incarcerated people in various settings, including in child welfare, health care settings, WIC programs, public assistance offices, drug rehabilitation and treatment programs, homeless shelters, and mental health facilities. Social workers should be aware of the formidable reentry barriers that formerly incarcerated people face. Supplemental material for this chapter documents that many policy barriers exist to successful reunification, including barriers to housing, public assistance, employment, education, civic participation, and reunification.

Resources for Advocates

In recognition of the reentry crisis posed by half a million inmates returning to communities unprepared and without support every year, Congress enacted The Second Chance Act in April 2008. It authorized $25 million in grants to state, local, and tribal agencies and community organizations in 2009 to provide vital services for recently released prisoners and an average of $82 million in each subsequent year (Reentry Policy Council, 2012). Since 2009, more than 300 government agencies and nonprofit organizations from 48 states have received grant awards for reentry programs serving adults and juveniles.

Formerly incarcerated people and their allies feel, however, that these efforts do not go far enough. Instead, they are working to change policies that hinder successful reentry, as well as public perceptions of formerly incarcerated people, in hopes of diminishing barriers and reclaiming their civil rights. One of those groups is All of Us or None (AOUON), a national organizing initiative of former prisoners that works "to combat the many forms of discrimination that we face as the result of felony convictions" (All of Us or None, n.d.). Through local chapters, AOUON is working to implement strategies that ensure former prisoners are able to participate in the democratic process through public education and voter registration drives. They are also working to eliminate barriers to

POLICY ADVOCACY LEARNING CHALLENGE 14.7

CONNECTING MICRO, MEZZO, AND MACRO POLICY INTERVENTIONS

View the video *Enough Is Enough* about formerly incarcerated people (http://www.facebook.com/video/video.php?v=1371013354970). In small groups, discuss the following:

1. At what point do you think a person has paid his or her debt to society?
2. Should restrictions be placed on convicted felons, even after they have completed their sentences (including parole/probation)? If so, what types, and why?
3. Should people with previous convictions be allowed to reacquire rights they possessed prior to their conviction?
4. What types of micro-, mezzo-, and/or macro-level advocacy might you engage in to assist formerly incarcerated people in becoming successful after release?

employment through their Ban the Box campaign, which seeks to remove the question about prior conviction from applications for employment. Moreover, AOUON's clean slate work provides training on the legal remedies available to people with convictions, such as expungements and certificates of rehabilitation.

THINKING BIG AS POLICY ADVOCATES IN THE CRIMINAL JUSTICE SECTOR

Comprehensive reforms in the criminal justice system are finally "in the air." A discussion finally began during the last year of the Obama presidency and the first two years of the Trump presidency. Conservatives, such as representatives of the Koch brothers, and liberals, such as Democrats who favored drastic cuts in rates of imprisonment, began talking about ways to cut the level of incarceration.

Develop a broad program to end mass incarceration. It can only happen through (1) placing fewer persons in correctional facilities through changes in sentencing, ending the prison pipeline in schools, and helping more youth complete high school and enter higher education; (2) making reforms in the jail and prison system, such as reforming bail so that low-income people are not incarcerated for months before their cases are heard by courts and reforming prisons so they move in the direction of Norway's jails, making jails and prisons instruments of rehabilitation; and (3) reforming services that ex-prisoners encounter to help them find work, find and afford housing, and return to high school or higher education. Ascertain whether your jurisdiction has "banned the box" as discussed at the National Employment Law Project website: https://nelp.org/wp-content/uploads/2015/03/Bantheboxcurrent.pdf?nocdn=1. Take any one of these three options, and develop strategies for accomplishing real reform.

Learning Outcomes

You are now equipped to:

- Describe how members of specific populations receive harsh and often inequitable treatment from the criminal justice sector
- Identify key eras in the evolution of the criminal justice sector
- Identify and analyze the seven problems in the criminal justice sector
- Apply the eight challenges of the multilevel policy empowerment framework to the criminal justice sector with respect to the thinking big exercise

References

Alexander, M. (2010). *The new Jim Crow: Mass incarceration in the age of colorblindness.* New York, NY: New Press.

All of Us or None. (n.d.). Retrieved from http://www.allofusornone.org

American Civil Liberties Union (ACLU). (n.d.a). Know your rights: The Prison Litigation Reform Act. Retrieved from: http://www.aclu.org/images/asset_upload_file79_25805.pdf

American Civil Liberties Union (ACLU). (n.d.b). *We can stop solitary.* Retrieved from https://www.aclu.org/we-can-stop-solitary

American Psychiatric Association. (2000). *Psychiatric services in jails and prisons: A task force report of the American Psychiatric Association* (2nd ed.). Washington, DC: Author.

Andrews, D. A., Zinger, I., Hoge, R. D., Bonta, J., Gendreau, P., & Cullen, F. T. (1990). Does correctional treatment work: A clinically relevant and psychologically informed meta-analysis. *Criminology, 28,* 369–404.

Anno, J., Graham, C., Lawrence, J., & Shansky, R. (2004). *Correctional health care: Addressing the needs of elderly, chronically ill and terminally ill inmates.* Washington, DC: National Institute of Corrections.

Arditti, J. & Few, A. (2006). Mothers' reentry into family life following incarceration. *Criminal Justice Policy Review, 17*(1), 103–123.

Arrieta-Kenna, R. (2018, August 15). "Abolish prisons" is the new "Abolish ICE." Politico. Retrieved from https://www.politico.com/.../2018/08/15/abolish-prisons-is-the-new-abolish-ice-21936...

Arrigo, B. A. & Bullock, J. L. (2008). The psychological effects of solitary confinement on prisoners in supermax units: Reviewing what we know and recommending what should change. *International Journal of Offender Therapy and Comparative Criminology, 52*(6), 622–640.

Beck, A. & Gilliard, D. (1995). *Prisoners in 1994.* Washington, DC: Bureau of Justice Statistics.

Benko, J. (2015, March 26). The radical humaneness of Norway's Halden Prison. *New York Times.* Retrieved from https://nyti.ms/1HMjNqD

Bishop, J. (2012, February 2). Restorative justice provides new path for prisoners. *Vox Magazine.* Retrieved from http://www.voxmagazine.com/stories/2012/02/02/restorative-justice-provides-new-path-prisoners/

Blackmon, D. A. (2009). *Slavery by another name: The re-enslavement of black Americans from the Civil War to World War II*. Harpswell, ME: Anchor Publishing.

Bogue, B., Campbell, N., Carey, M., Clawson, E., Faust, D., Florio, K., . . . Woodward, W. (2004). Implementing evidence-based practice in community corrections: The principles of effective intervention. Washington, DC: National Institute of Corrections. Retrieved from http://www.nicic.org/pubs/2004/019342.pdf

Bureau of Justice Statistics. (2006). Bureau of justice statistics [Press release]. Retrieved from http://www.ojp.usdoj.gov/bjs/pub/press/pripropr.htm

Bureau of Prisons (BOP). (n.d.). Inmate mental health. Retrieved from https://www.bop.gov/inmates/custody_and_care/mental_health.jsp

California Department of Corrections and Rehabilitation. (2011). *Prison census data as of June 30, 2011*. Sacramento, CA: Offender Information Services Branch.

California Department of Corrections and Rehabilitation. (2012). *Division of Adult Parole Operations: Parole requirements*. Retrieved from http://www.cdcr.ca.gov/parole/Parole_Requirements/index.html

California radically revamping prison system. (2016, March 15). *San Francisco Chronicle*. Retrieved from https://www.sfchronicle.com/opinion/editorials/article/California-radically-revamping-prison-system-6885920.php

Californians United for a Responsible Budget. (n.d.). Californians United for a Responsible budget. Retrieved from http://www.curbprisonspending.org

Carson, A. & Golinelli, D. (2013). *Prisoners in 2012: Trends in admissions and releases, 1991–2012*. Washington, DC: Bureau of Justice Statistics.

Chiu, T. (2010). *It's about time: Aging prisoners, increasing costs, and geriatric release*. New York: Vera Institute of Justice.

Cohen, F. & Gerbasi, J. (2005). Legal issues. In C. L. Scott & J. B. Gerbasi (Eds.), *Handbook of correctional mental health* (pp. 259–283). Washington, DC: American Psychiatric Publishing.

Cohen, M. (2018, April 28). How for-profit prisons have become the biggest lobby no one is talking about. *Washington Post*. Retrieved from https://www.washingtonpost.com/.../how-for-profit-prisons-have-become-the-biggest-...

Covington, S. & Bloom, B. (2003). Gendered justice: Women in the criminal justice system. In B. Bloom (Ed.), *Gendered justice: Addressing female offenders*. Durham, NC: Carolina Academic Press.

Covington, S. & Bloom, B. (2006). Gender-responsive treatment and services in correctional settings. *Women & Therapy, 29*(3–4), 9–33.

Coyle, A. & Fair, H. (2018). *A human rights approach to prison management* (3rd ed.). London: University of London, Institute for Criminal Policy Research. Retrieved from https://www.ncjrs.gov/App/Publications/abstract.aspx?ID=262855

Critical Resistance. (n.d.). Retrieved from http://www.criticalresistance.org

Dannenberg, J. (2011, April). Nationwide PLN survey examines prison phone contracts, kickbacks. *Prison Legal News, 22*(4). Retrieved from https://www.prisonlegalnews.org/23083_displayArticle.aspx

Davis, A. (2003). *Are prisons obsolete?* New York: Seven Stories Press.

Exstrum, O. (2018, June 14). The era of mass incarceration isn't over. This new report shows why. *Mother Jones.* Retrieved from https://www.motherjones.com/crime-justice/2018/06/the-era-of-mass-incarceration-isnt-over-this-new-report-shows-why/

Federal Bureau of Investigation. (2010, September). *Arrests by race, 2009.* Retrieved from http://www.fbi.gov/ucr/cius2009/data/table_43.html

Fellner, J. (2007). Prevalence and policy: New data on the prevalence of mental illness in US prisons. Retrieved from http://hrw.org/English/docs/200701/10/usdom15040.htm

Friedman, M. (2015). Just facts: As many Americans have criminal records as college diplomas. Retrieved from https://www.brennancenter.org

General Accounting Office. (1999). *Women in prison: Issues and challenges confronting U.S. correctional systems.* Washington, DC: Author.

Gilbert, S. (2017, November 20). The real cult of Charles Manson. *The Atlantic.* Retrieved from https://www.theatlantic.com/entertainment/archive/2017/11/the-real-cult-of-charles-manson/546206/

Gilliard, D. & Beck, A. (1994). *Prisoners in 1993.* Washington, DC: Bureau of Justice Statistics.

Gilligan, J. (2001). The last mental hospital. *Psychiatric Quarterly, 72*(1), 45–61.

Glaze, L. & Bonczar, T. (2011). *Probation and parole in the United States, 2010.* Washington, DC: Bureau of Justice Statistics.

Glaze, L. & Maruschak, L. (2008). *Parents in prison and their minor children.* Washington, DC: Bureau of Justice Statistics.

Golden, R. (2005). *War on the family: Mothers in prison and the families they leave behind.* New York, NY: Routledge.

Guerino, P., Harrison, P., & Sabol, W. (2011). *Prisoners in 2010.* Washington, DC: Bureau of Justice Statistics.

Gumz, E. J. (2004). American social work, corrections and restorative justice: An appraisal. *International Journal of Offender Therapy and Comparative Criminology, 48*(4), 449–460.

Hairston, C. (2004). Prisoners and their families: Parenting issues during incarceration. In J. Travis & M. Waul (Eds.), *Prisoners once removed: The impact of incarceration and reentry on children, families, and communities.* Washington, DC: The Urban Institute Press.

Harlow, C. (1999). *Prior abuse reported by inmates and probationers.* Washington, DC: Bureau of Justice Statistics.

Henderson, A. (2015, February 22). 9 Surprising industries getting filthy rich from mass incarceration. *Salon.* Retrieved from https://www.salon.com/2015/02/22/9_surprising_industries_getting_filthy_rich_from_mass_incarceration_partner/

Hill, K. G., Howell, J. C., Hawkins, J. D., & Battin-Pearson, S. R. (1999). Childhood risk factors for adolescent gang membership: Results from the Seattle Social Development Project. *Journal of Research in Crime and Delinquency, 36*(3), 300–322.

Hills, H., Siegfried, C., & Ickowitz, A. (2004). *Effective prison mental health services: Guidelines to expand and improve treatment* [NIC Accession Number 018604]. Washington, DC: United States Department of Corrections National Institute of Corrections.

Howell, J. C. (2006). *Youth gang programs and strategies.* Washington, DC: U.S. Department of Justice Office of Juvenile Justice and Delinquency Prevention. Retrieved from https://www.ncjrs.gov/pdffiles1/ojjdp/171154.pdf

Hughes, H. & Wilson, D. J. (2002). *Reentry trends in the United States*. Washington, DC: Bureau of Justice Statistics.

Human Rights Watch. (2012). *Teens in solitary confinement*. Retrieved from http://www.hrw.org/news/2012/10/10/us-teens-solitary-confinement

Human Rights Watch. (2015, May 15). Callous and cruel. Retrieved from https://www.hrw.org/...and.../use-force-against-inmates-mental-disabilities-us-jails-and

Humes, K., Jones, N., & Ramirez, R. (2011). *Overview of race and Hispanic origin: 2010*. Washington, DC: U.S. Census Bureau.

James, D. & Glaze, L. (2006). *Mental health problems of jail and prison inmates*. Washington, DC: Bureau of Justice Statistics.

Jansson, B. (2019a). *Reducing inequality: Addressing the wicked problems across professions and disciplines*. San Diego: Cognella Academic Press.

Jansson, B. (2019b). *The reluctant welfare state*. Boston: Cengage.

Just Detention. (n.d.). Learn the basics. Retrieved from http://www.justdetention.org/en/learn_the_basics.aspx

Kann, D. (2018, July 10). America still locks more people than anywhere else. CNN. Retrieved from https://www.cnn.com/2018/06/28/us/mass-incarceration-five-key-facts/index.html

Koshy, S. (1987). The right of [all] the people to be secure: Extending fundamental Fourth Amendment rights to probationers and parolees. *Hastings Law Journal, 39*, 449.

Kowal, T. (2011, June 5). The role of the Prison Guards Union in California's troubled prison system. *The League of Ordinary Gentlemen*. Retrieved from http://ordinary-gentlemen.com/blog/2011/06/the-role-of-the-prison-guards-union-in-californias-troubled-prison-system/

Kyckelhahn, T. (2011). *Justice expenditures and employment, FY 1982–2007: Statistical tables*. Washington, DC: Bureau of Justice Statistics.

Langan, P. A. & Levin, D. J. (2002). Recidivism of prisoners released in 1994. *Federal Sentencing Reporter, 15*(1), 58–65.

Latimer, J., Dowden, C., & Muise, D. (2005). The effectiveness of restorative justice practices: A meta-analysis. *The Prison Journal, 85*(2), 127–144.

Levy, P. (2014, April 28). One in 25 sentenced to death in the U.S. is innocent, study claims. *Newsweek*. Retrieved from https://www.newsweek.com/one-25-executed-us-innocent-study-claims-248889

Lewis, C. (2006). Treating incarcerated women: Gender matters. *Psychiatric Clinics of North America, 29*(3), 773–789.

Lopez, G. (2018a, May 22). The first step act: Congress's prison reform bill, explained. *Vox*. Retrieved from https://www.vox.com/policy-and-politics/2018/.../first-step-act-prison-reform-congres...

Lopez, G. (2018b, June 14). A massive review of the evidence shows letting people out of prison doesn't increase crime. *Vox*. Retrieved from https://www.vox.com/policy-and-politics/2017/9/25/.../study-mass-incarceration

Lopez, G. (2018c, January 11). The U.S. prison population fell in 2016—for the 3rd year in a row. *Vox*. Retrieved from https://www.vox.com/policy-and-politics/2018/1/11/16880166/prison-rate-mass-incarceration-2016

Lussenhop, J. (2016, April 18). Clinton crime bill: Why is it so controversial? *BBC News Magazine*. Retrieved from https://www.bbc.com/news/world-us-canada-36020717

Mills, L., Maley, M. H., & Shy, Y. (2009). Circulos de Paz and the promise of peace: Restorative justice meets intimate violence. *NYU Review of Law & Social Change, 33*, 127–152.

Mumola, C. (2007). *Medical causes of death in state prisons, 2001–2004*. Washington, DC: Bureau of Justice Statistics.

Mumola, C. & Karberg, J. (2006). *Drug use and dependence, state and federal prisoners, 2004*. Washington, DC: Bureau of Justice Statistics.

Nation Master. (n.d.). Crime statistics: Prisoners by country. Retrieved from http://www.nationmaster.com/graph/cri_pri-crime-prisoners

The National Center on Addiction and Substance Abuse at Columbia University. (2010). *Behind bars II: Substance abuse and America's prison population*. New York, NY. Retrieved from http://www.casacolumbia.org/articlefiles/575-report2010behindbars2.pdf

National Families in Action. (2000). A guide to drug-related state ballot initiatives. Retrieved from http://www.nationalfamilies.org/guide/california36-full.html

Nixon, R. (1971, June 17). Remarks about an intensified program for drug abuse prevention and control. The American Presidency Project. Retrieved from http://www.presidency.ucsb.edu/ws/?pid=3047

O'Brien, P. & Harm, N. (2002). Women's recidivism and reintegration: Two sides of the same coin. In J. Figueira-McDonough & R. Sarri (Eds.), *Women at the margins: Neglect, punishment, and resistance.* New York: Haworth.

Olson, S. (2007, November 17). Suicide city. *Indy bay.* Retrieved from http://www.indybay.org/newsitems/2007/11/17/18461835.php

O'Shea, B. (2012, February 18). Psychiatric patients with no place to go but jail. *New York Times*, p. A25A.

Oshinsky, D. (1996). *Worse than slavery.* New York: Free Press.

Reentry Policy Council. (2012). *The Second Chance Act.* Retrieved from http://www.reentrypolicy.org/government_affairs/second_chance_act

Rodriguez, J. & Bernstein, N. (2018, April 7). Life after "17 to life." *New York Times.* Retrieved from https://www.nytimes.com/2018/04/07/opinion/sunday/life-after-17-to-life.html

Romano, A. (2011, September 19). Jim Webb's last crusade. *Newsweek, 158*(12). Retrieved from https://www.newsweek.com/jim-webbs-criminal-justice-crusade-67347

Rosenmarkle, S., Durose, M., & Farole, D. (2009, December). Statistical tables: Sentences in state courts, 2006—statistical tables. Retrieved from https://www.bjs.gov/content/pub/pdf/fssc06st.pdf

Sawyer, W. (2018, February 27). Youth confinement: The whole pie. Retrieved from https://www.prisonpolicy.org/reports/youth2018.html

Snyder, C., van Wormer, K., Chadha, J., & Jaggers, J. (2009). Older adult inmates: The challenge for social work. *Social Work, 54*(2), 117–124.

Sudbury, J. (2002). Celling black bodies: Black women in the global prison industrial complex. *Feminist Review, 70*, 57–74.

Teplin, L. A. (2000). Keeping the peace: Police discretion and mentally ill persons. *National Institute of Justice Journal.* Retrieved from http://www.ncjrs.org/pdffiles1/jr000244c.pdf

Torrey, E. F. (1999, Autumn). Reinventing mental health care. *City Journal.* Retrieved from http://www.city-journal.org/html/9_4_a5.html

U.S. Department of Justice. (2010). *Uniform crime report: Arrests, by race, 2009.* Washington, DC: Author.

Vera Institute of Justice. (2009, July 9). Remarks as prepared for delivery by Attorney General

Eric Holder at the Vera Institute of Justice's third annual justice address. Retrieved from http://www.vera.org/? q=events/justice-address-2009

Villettaz, P., Killias, M., & Zoder, I. (2006). The effects of custodial vs. non-custodial sentences on re-offending: A systematic review of the state of knowledge. *Campbell Systematic Reviews, 13.* Retrieved from https://www.campbellcollaboration.org/library/custodial-vs-non-custodial-sanctions-re-offending-effects.html

Wagner, P. (2015, August 14). Jails matter, but who is listening? Retrieved from https://www.prisonpolicy.org/blog/2015/08/14/jailsmatter/

Wagner, P. & Rabuy, B. (2017, March 14). Mass incarceration: The whole pie 2017. Retrieved from https://www.prisonpolicy.org/reports/pie2017.html

Wahidin, A. (2004). *Older women in the criminal justice system: Running out of time*. London: Jessica Kingsley.

Watkins, E. (2018, February 15). Advance criminal justice reform bill. *CNN*. Retrieved from https://www.cnn.com/2018/02/15/politics/sentencing-prison-reform-senate-grassley-sessions/index.html

West, H. & Sabol, W. (2010). *Prisoners in 2009.* Washington, DC: Bureau of Justice Statistics.

Western, B., & Wildeman, C. (2009). Punishment, inequality, and the future of mass incarceration. *Kansas Law Review, 57,* 851–877.

White, T. & Schimmel, D. (1995). *Suicide prevention in federal prisons: A successful five step program.* In L. M. Hayes (Ed.), *Prison suicide: An overview and guide to prevention.* Mansfield, MA: National Center for Institution and Alternatives.

Williams, B., Sudore, R., Greifinger, R., & Morrison, R. S. (2011). Balancing punishment and compassion for seriously ill prisoners. *Annals of Internal Medicine, 155,* 122–126.

Winter, C. (2008, July/August). What do prisoners make for Victoria's Secret? *Mother Jones*. Retrieved from http://www.motherjones.com/politics/2008/07/what-do-prisoners-make-victorias-secret

The Winter Group. (2009). *PSCOA: Past accomplishments & current projects*. Retrieved from http://www.pscoa.org/wp-content/uploads/WG_Dec_2009.pdf

Women's Prison Association. (n.d.). Laws banning shackling during childbirth gaining momentum nationwide. Retrieved from http://66.29.139.159/pdf/Shackling Brief_final.pdf

Zehr, H. (2002). *The little book of restorative justice*. Intercourse, PA: Good Books.

INDEX